AF904632

Nano-Optics: Principles Enabling Basic Research and Applications

NATO Science for Peace and Security Series

This Series presents the results of scientific meetings supported under the NATO Programme: Science for Peace and Security (SPS).

The NATO SPS Programme supports meetings in the following Key Priority areas: (1) Defence Against Terrorism; (2) Countering other Threats to Security and (3) NATO, Partner and Mediterranean Dialogue Country Priorities. The types of meetings supported are generally "Advanced Study Institutes" and "Advanced Research Workshops". The NATO SPS Series collects together the results of these meetings. The meetings are co-organized by scientists from NATO countries and scientists from NATO's "Partner" or "Mediterranean Dialogue" countries. The observations and recommendations made at the meetings, as well as the contents of the volumes in the Series, reflect those of participants and contributors only; they should not necessarily be regarded as reflecting NATO views or policy.

Advanced Study Institutes (ASI) are high-level tutorial courses to convey the latest developments in a subject to an advanced-level audience.

Advanced Research Workshops (ARW) are expert meetings where an intense but informal exchange of views at the frontiers of a subject aims at identifying directions for future action.

Following a transformation of the programme in 2006, the Series has been re-named and re-organised. Recent volumes on topics not related to security, which result from meetings supported under the programme earlier, may be found in the NATO Science Series.

The Series is published by IOS Press, Amsterdam, and Springer, Dordrecht, in conjunction with the NATO Emerging Security Challenges Division.

Sub-Series

A.	Chemistry and Biology	Springer
B.	Physics and Biophysics	Springer
C.	Environmental Security	Springer
D.	Information and Communication Security	IOS Press
E.	Human and Societal Dynamics	IOS Press

http://www.nato.int/science
http://www.springer.com
http://www.iospress.nl

Series B: Physics and Biophysics

Nano-Optics: Principles Enabling Basic Research and Applications

edited by

Baldassare Di Bartolo
Department of Physics, Boston College, Chestnut Hill, Massachusetts, USA

John Collins
Department of Physics and Astronomy, Wheaton College, Norton, Massachusetts, USA

and

Luciano Silvestri
Department of Physics, Boston College, Chestnut Hill, Massachusetts, USA

Springer

Published in Cooperation with NATO Emerging Security Challenges Division

Proceedings of the NATO Advanced Study Institute on Nano-Optics:
Principles Enabling Basic Research and Applications
Erice, Sicily, Italy
4–19 July, 2015

Library of Congress Control Number: 2017931968

ISBN 978-94-024-0869-0 (PB)
ISBN 978-94-024-0848-5 (HB)
ISBN 978-94-024-0850-8 (e-book)
DOI 10.1007/978-94-024-0850-8

Published by Springer,
P.O. Box 17, 3300 AA Dordrecht, The Netherlands.

www.springer.com

Printed on acid-free paper

All Rights Reserved
© Springer Science+Business Media Dordrecht 2017
This work is subject to copyright. All rights are reserved by the Publisher, whether the whole or part of the material is concerned, specifically the rights of translation, reprinting, reuse of illustrations, recitation, broadcasting, reproduction on microfilms or in any other physical way, and transmission or information storage and retrieval, electronic adaptation, computer software, or by similar or dissimilar methodology now known or hereafter developed.
The use of general descriptive names, registered names, trademarks, service marks, etc. in this publication does not imply, even in the absence of a specific statement, that such names are exempt from the relevant protective laws and regulations and therefore free for general use.
The publisher, the authors and the editors are safe to assume that the advice and information in this book are believed to be true and accurate at the date of publication. Neither the publisher nor the authors or the editors give a warranty, express or implied, with respect to the material contained herein or for any errors or omissions that may have been made.

Optimus est magister, qui docendo animos discipulorum delectare potest.

Andrea Pozzo – Painter and architect (1642–1709)

Preface

This book presents an account of the NATO Advanced Study Institute on "*Nano-Structures for Optics and Photonics – Nano-optics: Principles Enabling Basic Research and Applications*" held in Erice, Sicily, Italy, from July 4 to 19, 2015. This meeting was organized by the International School of Atomic and Molecular Spectroscopy of the Ettore Majorana Foundation and Centre for Scientific Culture. Sponsors of the meeting have been the Italian Ministry of University and Scientific-Technological Research, Sicilian Regional Government, and Boston College. The extraordinary opportunities of nanotechnology were predicted by Richard Feynman (Nobel Prize in Physics) in a landmark talk called "There Is Plenty of Room at the Bottom" on December 12, 1959, before the American Physical Society meeting held at Caltech. Throughout the last decade, nanoscience and nanofabrication have prompted a remarkable progress in the field of optics and photonics. These in turn have become important driver and toolbox of nanotechnologies, e.g., resulting in novel super-resolution approaches in optical microscopy and lithography. Such a synergic interplay has favored the development of important optics and photonics subfields, such as three-dimensional optical lithography and microscopy beyond the Abbe diffraction limit, optical diagnostics and biosensing, bio- photonics, optical data and ultracompact and ultrafast optical telecommunication devices, energy-efficient lighting, efficient sustainable solar energy conversion, a completely new level of molding the flow of light and controlling light-matter interaction nearly on the atomic scale, nano-antennas, artificial optical materials ("metamaterials") assembled from these antennas, and nano-plasmonic structures. Transformation optics and "flat optics" expand the concepts of metamaterials and metasurfaces, respectively, toward intentionally spatially inhomogeneous structures, e.g., for invisibility cloaking and flat lenses or flat polarization optics. The Institute introduced the students to the fascinating field of nanoscience and nanotechnology and provided a comprehensive overview of experiments and theory and basic physics and applications as well as nanofabrication and optical characterization. It has brought together physicists, chemists, biologists, and engineers. According to the best tradition of the past institutes, the consideration of fundamentals was the underlying principle of the 2015 school, furnishing the background to reach the frontiers of research in a systematic and didactic fashion. The participants were protagonists of the meeting: they had the opportunity to present and discuss their research work with people from around the world thanks to slots devoted to short seminars or posters. Moreover, a few roundtable discussions were conducted during the course enabling the organizers and lecturers to get immediate feedback from the participants about the organizational aspects of the course, the scientific contents, and the ability of the lecturers to teach complex scientific topics in a didactic, stimulating, and inspiring way. On the last day of the course, all participants had the opportunity to fill out a questionnaire comprising an evaluation form for the course, organization, and lectures as well as to express their views, criticisms, and advices. The results of both roundtable discussions and questionnaire are summarized in the following pages. In general, high scientific level of the course, friendly atmosphere, and social moments (dinners, excursions, poster sessions, roundtables) promoting interaction between participants (students and professors) as well as lecturers available for discussion during the whole time of the meeting were pointed out.

Director of the International School of Atomic and Molecular Spectroscopy of the "Ettore Majorana" Center
Boston, USA
Dec, 2015

Baldassare (Rino) Di Bartolo

Fig. 1 The directors

Fig. 2 The team

Fig. 3 NATO ASI Summer School, Erice 2015, July 4–19, 2015

Contents

Part I Lectures

Part III Short Seminars

Part I
Lectures

Chapter 1
Light-Matter Interactions: A Coupled Oscillator Description

Martin Frimmer and Lukas Novotny

Abstract The semiclassical theory of light-matter interactions describes the interaction between a classical electromagnetic field with a quantum mechanical two-level system. We show that the quantum mechanical two-level system can be modeled by a system of two coupled classical harmonic oscillators whose eigenstates are split in frequency according to the coupling strength and play the roles of the two levels of the quantum mechanical two-level system. The effect of the light field on the mechanical system is modeled as a modulation of the spring constants of the individual oscillators. Using this fully classical model, we derive the Bloch equations for a two-level system and discuss the mechanical analogues of Rabi oscillations and coherent control experiments.

1.1 Introduction

One of the main thrusts of contemporary physics is quantum engineering, aiming to exploit the properties of quantum systems for information storage, processing and transmission. The fundamental building block of any quantum device is the quantum mechanical two-level system (TLS). In practice, atoms, ions, molecules and solid-state defect centers have been identified as near ideal representations of such a TLS. With the energy-level splittings in these systems corresponding to optical frequencies, light fields provide a handle to control the internal dynamics of such a TLS. To achieve maximum fidelity of the operations on the quantum system, the interaction strength between the electromagnetic field and the TLS has to be maximized by maximizing the field strength at the position of the TLS. Nanophotonics has developed a powerful toolbox to control light at the subwavelength scale, allowing the confinement of electromagnetic radiation to volumes smaller than the limit imposed by diffraction. With the quality factors of nanophotonic resonators increasing, accompanied by shrinking mode volumes, the interaction strength between a single quantum emitter and a nanophotonic resonator is reaching

M. Frimmer • L. Novotny (✉)
Photonics Laboratory, ETH Zürich, 8093 Zürich, Switzerland
e-mail: lnovotny@ethz.ch

© Springer Science+Business Media Dordrecht 2017
B. Di Bartolo et al. (eds.), *Nano-Optics: Principles Enabling Basic Research and Applications*, NATO Science for Peace and Security Series B: Physics and Biophysics, DOI 10.1007/978-94-024-0850-8_1

a level where coherent quantum mechanical effects are observable. Various coherent control schemes, such as Rabi oscillations [1], Hahn echoes [2], and Ramsey fringes [3], have been developed in the past, based on a semiclassical description of the interaction between a TLS and an electromagnetic field. In this model, the TLS is described quantum mechanically, whereas the light field is of purely classical nature. While this semiclassical theory has been extremely successful in describing the physical reality, it does not provide an intuitive handle to understand the evolution of the quantum mechanical TLS. Interestingly, over the past decades, classical analogues have been constructed for several quantum phenomena [4], such as strongly driven two-level systems [5], electromagnetically induced transparency [6], rapid adiabatic passage [7], and Landau-Zener transitions [8]. However, a classical Newtonian model describing the internal dynamics of a quantum system driven by an external field has been missing to date.

Here, we present a classical model for the interaction of a quantum mechanical TLS with a classical optical field. We construct a *mechanical atom*, consisting of a pair of coupled classical harmonic oscillators. The coupling gives rise to two eigenmodes, split in frequency according to the strength of the coupling between the bare oscillators. These eigenmodes play the role the two states of a quantum mechanical TLS. The interaction between the mechanical atom and a driving field is reflected in the modulation of the spring constants of the bare oscillators. Under this parametric driving, the Newtonian equations of motion describing the evolution of our mechanical atom take the exact same form as the optical Bloch equations derived from a semiclassical model based on the Schrödinger equation. Accordingly, our model provides as intuitive classical approach to understanding the coherent dynamics of a quantum mechanical TLS.

1.2 Semiclassical Treatment

Figure 1.1 conceptually illustrates the interaction between light and matter in a semiclassical framework. The matter part is described by a TLS, from now on termed *atom* for simplicity, with an electronic ground state $|g\rangle$ and an excited state $|e\rangle$. The two atomic states are separated by the energy $\hbar\omega_0$, with ω_0 the transition frequency. The spontaneous decay rate of the atom is γ_A. The interaction of the two systems is characterized by the coupling rate g, which derives from the interaction Hamiltonian as $g = H_{\mathrm{int}}/\hbar$. In the dipole approximation the interaction Hamiltonian can be written as $H_{\mathrm{int}} = -\mathbf{p} \cdot \mathbf{E}$, with $\mathbf{p}$ denoting the transition dipole between $|g\rangle$ and $|e\rangle$. Furthermore, $\mathbf{E}(t) = -\mathbf{E}_0 \cos \omega t$ is a classical time-harmonic electric field. Under the influence of the optical field, the wave function of the atom can be written as a superposition of its ground and excited state

$$|\psi\rangle = a(t)|g\rangle + b(t)|e\rangle \;, \tag{1.1}$$

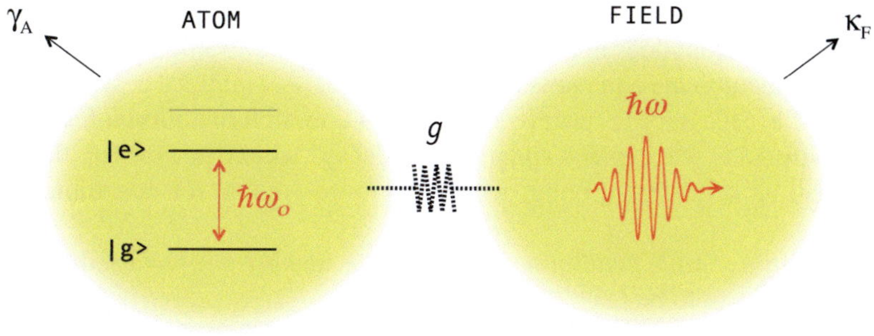

Fig. 1.1 Schematic of light-matter interactions. The optical field is characterized by the frequency ω and the atom (matter) is represented by two electronic states $|g\rangle$ and $|e\rangle$ separated by the energy $\hbar\omega_0$. The interaction of the two systems is characterized by the coupling rate g. The excited-state spontaneous decay rate of the atom is γ_A

where $a(t)$ and $b(t)$ are complex time dependent coefficients. They are found by inserting Eq. (1.1) into the time dependent Schrödinger equation $\mathrm{i}\hbar\,\partial_t|\psi\rangle = \hat{H}|\psi\rangle$. The detailed procedure is described in textbooks on quantum optics (see, for example, Ref. [9]) and we only outline the main aspects here. It is convenient to offset the energy scale, such that the energies of ground state and excited state are $E_g = -\hbar\omega_0/2$ and $E_e = +\hbar\omega_0/2$, respectively, and then move to the rotating frame, that is, performing the transformation

$$\begin{aligned} a(t) &= \bar{a}(t)\exp\left[-\mathrm{i}\omega\, t/2\right], \\ b(t) &= \bar{b}(t)\exp\left[+\mathrm{i}\omega\, t/2\right]. \end{aligned} \tag{1.2}$$

Inserting Eqs. (1.1) and (1.2) into the Schrödinger equation and performing the rotating wave approximation (i.e. assuming $\omega \sim \omega_0$), we obtain

$$\mathrm{i}\hbar \begin{bmatrix} \dot{\bar{a}} \\ \dot{\bar{b}} \end{bmatrix} = \frac{\hbar}{2} \begin{bmatrix} \delta & g \\ g & -\delta \end{bmatrix} \begin{bmatrix} \bar{a} \\ \bar{b} \end{bmatrix}, \tag{1.3}$$

where we have defined the detuning δ between the driving frequency and the transition frequency

$$\delta = \omega_0 - \omega\,, \tag{1.4}$$

and the coupling rate

$$g = \mathbf{p}\cdot\mathbf{E}_0\,/\,\hbar\,, \tag{1.5}$$

which is also denoted as the classical Rabi frequency. Note that the spontaneous decay rate of the atom does not appear in the semiclassical framework and has to be inserted by hand into the equations of motion in (1.3). A quantized description of the electric field is necessary to retrieve spontaneous emission in a formal fashion [10]. For our purposes, we neglect spontaneous decay, which places our discussion into the regime of strong driving, where $\Omega_R \gg \gamma_A$ holds for any finite driving field.

Using arbitrary initial conditions $\bar{a}(t=0) = \bar{a}_0$ and $\bar{b}(t=0) = \bar{b}_0$, the solutions of Eq. (1.3) turn out to be [11]

$$\bar{a}(t) = \left[-\frac{\mathrm{i}g}{\Omega_R} \sin\left(\frac{\Omega_R t}{2}\right) \bar{b}_0 + \left\{ \cos\left(\frac{\Omega_R t}{2}\right) - \mathrm{i}\frac{\delta}{\Omega_R} \sin\left(\frac{\Omega_R t}{2}\right) \right\} \bar{a}_0 \right],$$
$$\bar{b}(t) = \left[-\frac{\mathrm{i}g}{\Omega_R} \sin\left(\frac{\Omega_R t}{2}\right) \bar{a}_0 + \left\{ \cos\left(\frac{\Omega_R t}{2}\right) + \mathrm{i}\frac{\delta}{\Omega_R} \sin\left(\frac{\Omega_R t}{2}\right) \right\} \bar{b}_0 \right], \quad (1.6)$$

where we have introduced the generalized Rabi-frequency

$$\Omega_R = \sqrt{g^2 + \delta^2}. \quad (1.7)$$

Equations (1.6) together with (1.2) and (1.1) are the general solutions to the problem of a two-level atom interacting with a time-harmonic optical field $\mathbf{E}(t) = -\mathbf{E}_0 \cos \omega t$. Before discussing the properties of the solutions we first turn to our purely classical model of the mechanical atom.

1.3 Classical Treatment

We now replace the quantum mechanical atom with states $|g\rangle$ and $|e\rangle$ by a classical mechanical atom made of two coupled classical harmonic oscillators. As shown in Fig. 1.2, each oscillator consists of a mass m suspended by a spring with spring constant k and the oscillators are coupled by a spring with stiffness κ. The interaction with the optical field is described by a parametric process, that is, the optical field modulates the spring constant k of one of the oscillators by an amount $\Delta k(t)$ and that of the other by $-\Delta k(t)$. The solution of this coupled system has been recently derived in Ref. [12]. Here we outline the main steps and show that the solutions of this system are identical to those found for the semiclassical treatment in the previous section.

In terms of the coordinates x_A and x_B of the two oscillators, the equations of motion are

$$m\ddot{x}_A + [k + \kappa - \Delta k(t)]x_A - \kappa x_B = 0\,,$$
$$m\ddot{x}_B + [k + \kappa + \Delta k(t)]x_B - \kappa x_A = 0\,, \quad (1.8)$$

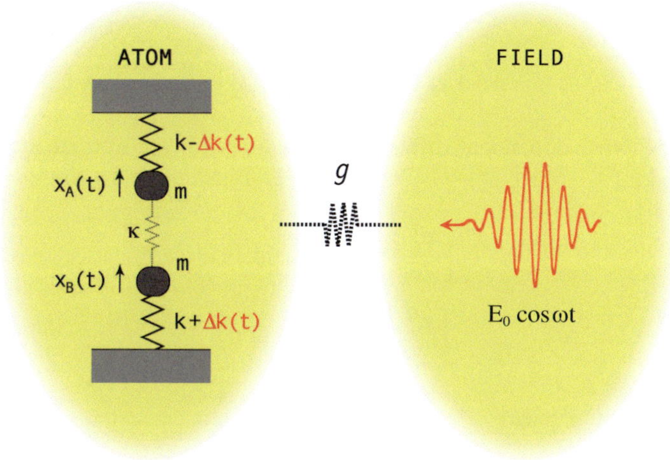

Fig. 1.2 Light interacting with a mechanical atom. The mechanical atom consists of two coupled mechanical oscillators with equal masses m and spring constants k. The two masses are coupled by a spring with stiffness κ. The interaction with the optical field gives rise to a parametric modulation of the spring constants Δk

where m are the masses, k the spring constants, $\Delta k(t)$ a time-dependent modulation of the spring constants, and κ the coupling spring constant. We introduce the carrier frequency Ω_0, the detuning frequency Ω_d and the coupling frequency Ω_c as

$$\begin{aligned} \Omega_0^2 &= [k+\kappa]/m, \\ \Omega_d^2 &= \Delta k/m, \\ \Omega_c^2 &= \kappa/m, \end{aligned} \tag{1.9}$$

and represent the coupled differential equations in (1.8) in matrix form as

$$\left[\frac{d^2}{dt^2} + \Omega_0^2\right]\begin{bmatrix} x_A \\ x_B \end{bmatrix} + \begin{bmatrix} -\Omega_d^2 & -\Omega_c^2 \\ -\Omega_c^2 & \Omega_d^2 \end{bmatrix}\begin{bmatrix} x_A \\ x_B \end{bmatrix} = 0\,. \tag{1.10}$$

This system of equations describes the full dynamics of the undriven coupled oscillator problem.

1.3.1 *Eigenmodes of the Coupled Oscillator*

We now solve for the eigenmodes of the system and their respective eigenfrequencies in absence of modulation ($\Omega_d = 0$). To diagonalize the matrix in Eq. (1.10) we write

$$\begin{bmatrix} x_A \\ x_B \end{bmatrix} = \begin{bmatrix} 1 & 1 \\ 1 & -1 \end{bmatrix} \begin{bmatrix} x_+ \\ x_- \end{bmatrix}, \tag{1.11}$$

which yields two independent differential equations for the normal mode coordinates x_+ and x_-

$$\begin{aligned} \left[\frac{d^2}{dt^2} + \Omega_+^2\right] x_+ &= 0, \\ \left[\frac{d^2}{dt^2} + \Omega_-^2\right] x_- &= 0, \end{aligned} \tag{1.12}$$

with the eigenfrequencies

$$\Omega_\pm = \left[\Omega_0^2 \mp \Omega_c^2\right]^{1/2}. \tag{1.13}$$

Here, Ω_+ denotes the frequency of the symmetric eigenmode ($x_+ = x_A + x_B$), which is lower than the frequency Ω_- of the antisymmetric eigenmode ($x_- = x_A - x_B$). The frequency splitting is

$$\omega_0 = \Omega_- - \Omega_+ \approx \frac{\Omega_c^2}{\Omega_0}, \tag{1.14}$$

where we made use of $\Omega_c \ll \Omega_0$. Thus, the splitting is proportional to the coupling strength κ.

1.3.2 Interaction with the Optical Field

We now transform the equations of motion (1.10) to the basis x_+, x_- and obtain

$$\left[\frac{d^2}{dt^2} + \Omega_0^2\right] \begin{bmatrix} x_+ \\ x_- \end{bmatrix} + \begin{bmatrix} -\Omega_c^2 & -\Omega_d^2 \\ -\Omega_d^2 & \Omega_c^2 \end{bmatrix} \begin{bmatrix} x_+ \\ x_+ \end{bmatrix} = 0\,. \tag{1.15}$$

We demand that the interaction of the mechanical atom with the optical field oscillating at frequency ω gives rise to a modulation of the oscillators' spring constants

$$\Delta k(t) = 2m\Omega_0 g \cos(\omega t), \tag{1.16}$$

where g is the coupling rate. To understand the evolution of the eigenmodes we write

$$\begin{aligned} x_+ &= \mathrm{Re}\,\{a(t)\exp[\mathrm{i}\Omega_0 t]\}, \\ x_- &= \mathrm{Re}\,\{b(t)\exp[\mathrm{i}\Omega_0 t]\}, \end{aligned} \tag{1.17}$$

where each mode is rapidly oscillating at the carrier frequency Ω_0 and modulated by the slowly varying complex amplitudes $a(t)$ and $b(t)$, respectively. Upon inserting (1.17) into the coupled equations of motion (1.15) we assume that the amplitude functions $a(t)$ and $b(t)$ do not change appreciably during an oscillation period $2\pi/\Omega_0$, which allows us to neglect terms containing second time derivatives (slowly varying envelope approximation). With this approximation we arrive at the following equations of motion for the eigenmode amplitudes

$$\mathrm{i}\begin{bmatrix}\dot{a}\\ \dot{b}\end{bmatrix} = \frac{1}{2}\begin{bmatrix}\omega_0 & -\Omega_d^2/\Omega_0\\ -\Omega_d^2/\Omega_0 & -\omega_0\end{bmatrix}\begin{bmatrix}a\\ b\end{bmatrix}. \tag{1.18}$$

In a next step we apply the transformation

$$\begin{aligned} a(t) &= \bar{a}(t)\exp[-\mathrm{i}\omega\, t/2], \\ b(t) &= \bar{b}(t)\exp[+\mathrm{i}\omega\, t/2]. \end{aligned} \tag{1.19}$$

Here, $\bar{a}$ and $\bar{b}$ are the slowly varying amplitudes of the symmetric and antisymmetric eigenmodes in a coordinate frame rotating at the driving frequency. This transformation generates terms $\exp[\pm 3\mathrm{i}\omega t/2]$ in (1.18), which are rapidly oscillating and which we neglect since they average out on the time scales of interest (rotating wave approximation). In terms of the detuning $\delta = \Delta\Omega - \omega$ we then obtain

$$\mathrm{i}\begin{bmatrix}\dot{\bar{a}}\\ \dot{\bar{b}}\end{bmatrix} = \frac{1}{2}\begin{bmatrix}\delta & g\\ g & -\delta\end{bmatrix}\begin{bmatrix}\bar{a}\\ \bar{b}\end{bmatrix}, \tag{1.20}$$

which is identical with (1.3) derived in the previous section using semiclassical theory. Hence, the solutions of (1.20) are given by (1.6). We conclude that the semiclassical theory of light-matter interaction can be reproduced by a pair of parametrically modulated classical oscillators. Note that both theories make use of the rotating wave approximation, which is reflected by the fact that (1.6) only retains dynamics on the time scale given by the generalized Rabi-frequency Ω_R and neglects any fast dynamics on time scales set by the optical frequency ω. Accordingly, our solutions are only valid for coupling rates g and detunings δ small enough to ensure $\Omega_R \ll \omega$.

1.4 Arbitrary Interaction Phases

In Sect. 1.2 we have considered an optical field of the form $\mathbf{E}(t) = -\mathbf{E}_0\cos(\omega t)$. We now allow for an arbitrary phase offset ϕ_0, such that $\mathbf{E}(t) = -\mathbf{E}_0\cos(\omega t - \phi_0) = -\mathbf{E}_0\cos\phi_0\cos(\omega t) - \mathbf{E}_0\sin\phi_0\sin(\omega t)$. This turns the interaction Hamiltonian into

$$H_{\mathrm{int}} = -\mathbf{p}\cdot\mathbf{E} = \hbar\left[g_x\cos(\omega t) + g_y\sin(\omega t)\right], \tag{1.21}$$

where we defined the coupling rates as $g_x = g\cos\phi_0$ and $g_y = g\sin\phi_0$, with $g = \mathbf{p}\cdot\mathbf{E}_0/\hbar$. Correspondingly, in the classical oscillator model considered in Sect. 1.3, the modulation of the spring constant becomes

$$\Delta k = 2m\Omega_0\left[g_x\cos(\omega t) + g_y\sin(\omega t)\right]. \tag{1.22}$$

As a result of the phase ϕ_0, the systems of equations (1.3) and (1.20) turn into

$$\mathrm{i}\hbar\begin{bmatrix}\dot{\bar{a}}\\ \dot{\bar{b}}\end{bmatrix} = \frac{\hbar}{2}\begin{bmatrix}\delta & g_x - \mathrm{i}g_y\\ g_x + \mathrm{i}g_y & -\delta\end{bmatrix}\begin{bmatrix}\bar{a}\\ \bar{b}\end{bmatrix}. \tag{1.23}$$

1.4.1 Representation in Terms of Pauli Matrices

Let us write the eigenstates of the atom as vectors

$$|g\rangle = \begin{bmatrix}1\\0\end{bmatrix}, \quad |e\rangle = \begin{bmatrix}0\\1\end{bmatrix}, \tag{1.24}$$

which turn the state of the atom in the rotating frame into

$$|\bar{\psi}\rangle = \bar{a}(t)|g\rangle + \bar{b}(t)|e\rangle = \begin{bmatrix}\bar{a}\\ \bar{b}\end{bmatrix}. \tag{1.25}$$

The evolution of $|\bar{\psi}\rangle$ is governed by the Schrödinger equation $\mathrm{i}\hbar\partial_t|\bar{\psi}\rangle = \bar{H}|\bar{\psi}\rangle$ with $\bar{H}$ being the Hamiltonian in the rotating frame. Inserting (1.25) yields

$$\mathrm{i}\hbar\begin{bmatrix}\dot{\bar{a}}\\ \dot{\bar{b}}\end{bmatrix} = \bar{H}\begin{bmatrix}\bar{a}\\ \bar{b}\end{bmatrix} \tag{1.26}$$

and a comparison with (1.23) gives

$$\bar{H} = \frac{\hbar}{2}\begin{bmatrix}\delta & g_x - \mathrm{i}g_y\\ g_x + \mathrm{i}g_y & -\delta\end{bmatrix}. \tag{1.27}$$

In terms of the Pauli matrices

$$\sigma_x = \begin{bmatrix}0 & 1\\ 1 & 0\end{bmatrix}, \quad \sigma_y = \begin{bmatrix}0 & -\mathrm{i}\\ \mathrm{i} & 0\end{bmatrix}, \quad \sigma_z = \begin{bmatrix}1 & 0\\ 0 & -1\end{bmatrix}, \tag{1.28}$$

we can cast $\bar{H}$ in the compact form

$$\bar{H} = \frac{\hbar g_x}{2}\sigma_x + \frac{\hbar g_y}{2}\sigma_y + \frac{\hbar \delta}{2}\sigma_z = \hbar \mathbf{R} \cdot \boldsymbol{\sigma}/2 , \tag{1.29}$$

where we introduced the rotation vector $\mathbf{R} = [g_x, g_y, \delta]^T$ and defined $\boldsymbol{\sigma} = [\sigma_x, \sigma_y, \sigma_z]^T$.

1.5 The Bloch Equations

Let us represent the solutions $\bar{a}$ and $\bar{b}$ in a vectorial form that was originally introduced by Felix Bloch in the context of nuclear magnetic resonance [13]. To do so, we introduce the Bloch vector $\mathbf{s} = (s_x, s_y, s_z)^T$ with the components

$$\begin{aligned} s_x &= \bar{a}\bar{b}^* + \bar{a}^*\bar{b} &&= 2\,\mathrm{Re}\{\bar{a}\bar{b}^*\} &&= 2|\bar{a}||\bar{b}|\cos\phi , \\ s_y &= \mathrm{i}(\bar{a}\bar{b}^* - \bar{a}^*\bar{b}) &&= -2\,\mathrm{Im}\{\bar{a}\bar{b}^*\} &&= -2|\bar{a}||\bar{b}|\sin\phi , \\ s_z &= \bar{a}\bar{a}^* - \bar{b}\bar{b}^* &&= |\bar{a}|^2 - |\bar{b}|^2 . \end{aligned} \tag{1.30}$$

The Bloch vector $\mathbf{s}$ encodes in its three real-valued components the state of the atom, which is represented by the amplitudes $|\bar{a}|$, $|\bar{b}|$ and the relative phase ϕ. Importantly, every state $(\bar{a}, \bar{b})$ of the oscillator system can be multiplied by an arbitrary phase factor $\exp[\mathrm{i}\varphi]$ without changing the corresponding Bloch vector $\mathbf{s}$. Discarding this absolute phase of the complex amplitudes $\bar{a}, \bar{b}$ reduces the degrees of freedom from four (two real amplitudes and two phases for $\bar{a}$ and $\bar{b}$) to three, such that the state of the oscillator system can be represented in the three dimensional Bloch vector space.

For a normalized system ($|\bar{a}|^2 + |\bar{b}|^2 = 1$) the tip of the Bloch vector always lies on a unit sphere, called the Bloch sphere, illustrated in Fig. 1.3. The north pole of the Bloch sphere $\mathbf{s} = (0, 0, 1)^T$ corresponds to the state vector $(\bar{a}, \bar{b}) = (1, 0)$, the ground state $|g\rangle$ of the atom, according to Eq. (1.1). Accordingly, for the excited state $|e\rangle$, corresponding to $(\bar{a}, \bar{b}) = (0, 1)$, the tip of the Bloch vector is located at the south pole of the Bloch sphere $\mathbf{s} = (0, 0, -1)^T$. All points on the equator of the Bloch sphere correspond to equal superpositions of $|g\rangle$ and $|e\rangle$, but with varying relative phase ϕ. For example, the state $(\bar{a}, \bar{b}) = (1, 1)/\sqrt{2}$ lies at the intersection of the x-axis and the Bloch sphere $\mathbf{s} = (1, 0, 0)^T$, whereas the state $(\bar{a}, \bar{b}) = (1, \mathrm{i})/\sqrt{2}$ lies at the intersection with the y-axis $\mathbf{s} = (0, 1, 0)^T$.

It is instructive to express the dynamics of the light-matter interaction in terms of the Bloch vector $\mathbf{s}$. Using Eqs. (1.27) and (1.30) we can easily show that the time evolution of the Bloch vector is given by

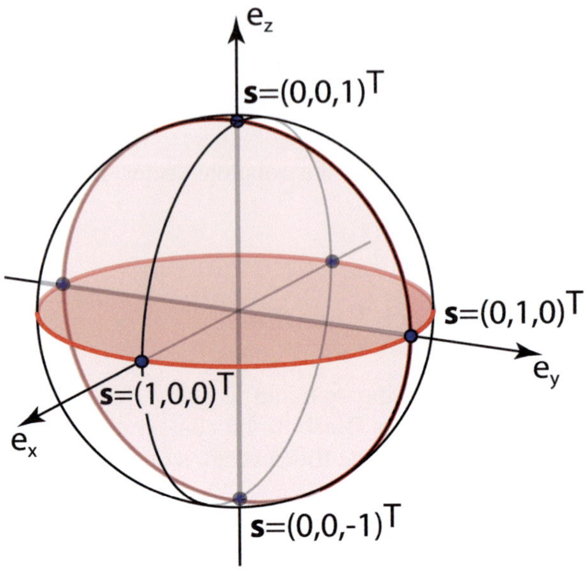

Fig. 1.3 The Bloch sphere. A pair of complex amplitudes $(\bar{a}, \bar{b})$ is represented by the Bloch vector $\mathbf{s} = (s_x, s_y, s_z)^T$. Normalized amplitude pairs ($|\bar{a}|^2 + |\bar{b}|^2 = 1$) lie on the surface of the Bloch sphere, which has unit radius. All amplitude pairs $(\bar{a}, \bar{b}) \exp[i\varphi]$ with arbitrary φ are mapped onto the same point $\mathbf{s}$

$$\frac{d}{dt}\begin{bmatrix} s_x \\ s_y \\ s_z \end{bmatrix} = \begin{bmatrix} 0 & -\delta & g_y \\ \delta & 0 & -g_x \\ -g_y & g_x & 0 \end{bmatrix} \begin{bmatrix} s_x \\ s_y \\ s_z \end{bmatrix}. \tag{1.31}$$

This system of equations can be represented in compact form as

$$\dot{\mathbf{s}} = \mathbf{R} \times \mathbf{s}, \tag{1.32}$$

where $\mathbf{R} = (g_x, g_y, \delta)^T$. The equation of motion (1.32) describes the precession of the Bloch vector $\mathbf{s}$ around the rotation vector $\mathbf{R}$ with the angular frequency Ω_R defined in Eq. (1.7), where Ω_R equals the length of $\mathbf{R}$.

A point on the Bloch sphere entirely defines the state of the atom. According to Eq. (1.32) we can bring the atom from any starting point to any other point on the Bloch sphere simply by choosing the right rotation vector $\mathbf{R}$ and waiting for the right time to achieve the desired amount of rotation. This idea is at the core of the concept of coherent control. We note that we have neglected spontaneous emission both in our semiclassical treatment and in the mechanical atom. Due to spontaneous emission, a quantum two-level system will always end up in its ground state after a long time. Spontaneous emission is a process that is genuinely quantum mechanical in nature and requires a fully quantized theory for the electromagnetic field. Even Bloch added the decay constants semi-phenomenologically in his treatment of nuclear spins [9].

In 1937 Rabi studied the dynamics of a spin in a static magnetic field that is modulated by a radio frequency field and he found that the spin vector is periodically oscillating between parallel and anti-parallel directions with respect to the static magnetic field [1]. These oscillations are referred to as Rabi oscillations, or Rabi flopping. To illustrate Rabi oscillations we assume a resonant ($\delta = 0$) driving with $g_y = 0$, such that the Bloch vector, starting at the north pole $\mathbf{s} = (0, 0, 1)^T$, rotates around the axis $\mathbf{R} = g\mathbf{e}_x$ at a frequency $\Omega_R = g_x$ according to Eq. (1.32). After a time $t_\pi = \pi/g$ the Bloch vector will have rotated to the south pole $\mathbf{s} = (0, 0, -1)^T$. Obviously, parametric driving for a time t_π (called π-pulse) inverts our system. Accordingly, after parametrically driving the system for a time $t_{2\pi} = 2\pi/g$ it has returned to its initial state at the north pole of the Bloch sphere. For a continuous parametric driving, starting at $\mathbf{s} = (0, 0, 1)^T$, the system is oscillating between its two eigenmodes at the resonant Rabi-frequency $\Omega_R = g_x$.

If we start out at the north pole $\mathbf{s} = (0, 0, 1)^T$ but use a resonant driving with $g_x = 0$, the Bloch vector will rotate around the axis $\mathbf{R} = g\mathbf{e}_y$ at a frequency $\Omega_R = g_y$. Thus, by selectively switching on and off the parameters g_x, g_y and δ we can rotate the Bloch vector around any axis and by an arbitrary amount.

1.6 Conclusions

We have reviewed the semiclassical theory of light-matter interaction and presented a classical oscillator model that yields identical equations of motion for the amplitudes of a two-level system. The oscillator model comprises two coupled harmonic oscillators with spring constants that are modulated by the external optical field. The coupling strength κ between the two oscillators sets the transition frequency ω_0 and defines the detuning δ. The correspondence between the classical and the quantum system is established by using the slowly varying envelope approximation, which casts the Newtonian equations of motion of the coupled oscillators into a form resembling the Schrödinger equation for a two-level atom. The mechanical analogue lends itself to visualize coherent-control experiments, such as Rabi oscillations, Ramsey fringes and Hahn echoes. In a continuation of the work presented here it would be interesting to consider a model of *strong* light-matter interactions, in which the driving optical field is itself modeled by a mechanical oscillator.

References

1. Rabi, I. I. (1937). Space Quantization in a Gyrating Magnetic Field. *Physics Review, 51*, 652.
2. Hahn, E. L. (1950). Spin Echoes. *Physics Review, 80*, 580.
3. Ramsey, N. F. (1950). A Molecular Beam Resonance Method with Separated Oscillating Fields. *Physics Review, 78*, 695.
4. Dragoman, D., & Dragoman, M. (2004). *Quantum-classical analogies* (The Frontiers Collection). Berlin/New York: Springer.
5. Spreeuw, R. J. C., van Druten, N. J., Beijersbergen, M. W., Eliel, E. R., & Woerdman, J. P. (1990). Classical realization of a strongly driven two-level system. *Physics Review Letter, 65*, 2642.

6. Alzar, C. L. G., Martinez, M. A. G., & Nussenzveig, P. (2002). Classical analog of electromagnetically induced transparency. *American Journal of Physics, 70*, 37.
7. Shore, B. W., Gromovyy, M. V., Yatsenko, L. P., & Romanenko, V. I. (2009). Simple mechanical analogs of rapid adiabatic passage in atomic physics. *American Journal of Physics, 77*(12), 1183.
8. Novotny, L. (2010). Strong coupling, energy splitting, and level crossings: A classical perspective. *American Journal of Physics, 78*(11), 1199
9. Allen, L., & Eberly, J. H. (1987). *Optical resonance and two-level atoms* (Dover books on physics series). New York: Dover
10. Haroche, S., & Raimond, J. (2006). *Exploring the quantum: Atoms, cavities, and photons* (Oxford Graduate Texts). Oxford/New York: OUP Oxford
11. Scully, M., & Zubairy, M. (1997). *Quantum optics*. Cambridge/New York: Cambridge University Press.
12. Frimmer, M., & Novotny, L. (2014). The classical Bloch equations. *American Journal of Physics, 82*(10), 947.
13. Bloch, F. (1946). Nuclear Induction. *Physics Review, 70*, 460.

Chapter 2
Luminescence Spectroscopy of Nanophosphors

Maura Cesaria and Baldassare Di Bartolo

Abstract The properties of phosphor materials in nanoscale form, usually termed "nanophosphors", present particular aspects that are worthy of extensive investigations at both fundamental and applicative level. Nanophosphors are mainly represented by compounds doped with rare earths, whose *4f* optically active electrons are protected from static and dynamic perturbations by the *5s* and *5p* subshells.

In this paper we first review the spectroscopy of rare earth ions and the properties of bulk phosphors (physical mechanisms underlying luminescence processes, thermal line broadening and line shift) and then consider the corresponding scenario at the nanoscale and discuss the role that spectroscopy may play in the investigation of nanophosphors. The paper aims at giving a background to a reader new to the spectroscopy of nanophosphors and useful points of discussion as well as guidelines for further studies.

2.1 Introduction

The term "luminescence", first introduced by Eilhard Wiedemann in 1888 [1], classifies phenomena involving emission of electromagnetic radiation (photons) from materials excited by energy absorption processes. Luminescent materials are commonly referred to as "phosphors" and can be classified as organic or inorganic depending on their composition. Inorganic phosphors, that will be the focus of this paper, usually consist of a host inorganic compound and a dopant element (activator) that acts as emitting source.

Luminescence at a given wavelength requires the absorbed energy to be "localized" on discrete states of emitting centers. The energy absorbed by a system without being emitted as luminescence is lost in non-radiative processes (typically

M. Cesaria
Department of Mathematics and Physics "Ennio De Giorgi", University of Salento, Lecce, Italy

B. Di Bartolo (✉)
Department of Physics, Boston College, Chestnut Hill, MA, USA
e-mail: dibartob@bc.edu

© Springer Science+Business Media Dordrecht 2017
B. Di Bartolo et al. (eds.), *Nano-Optics: Principles Enabling Basic Research and Applications*, NATO Science for Peace and Security Series B: Physics and Biophysics, DOI 10.1007/978-94-024-0850-8_2

heat transfer) which can compete with luminescence and affect the luminescence yield and efficiency (number of photons emitted divided by the number of photons absorbed).

In practice, inorganic phosphors are usually nanomaterials, microcrystalline powders, transition metal (TM) or rare earth (RE) compounds and TM- or RE-doped host crystalline structure (host lattice or bulk of the phosphor) [2, 3]. Selection of a phosphor host lattice and dopants depend on the applicative requirements: basic properties to be considered are excitation energy and spectrum, absorption strength, chemical environment, temperature, nature and concentration of the dopant as well as its solubility, mobility, oxidation state, stability and location of the charge-transfer state.

Phosphors have been extensively investigated during the last decade due to their increasingly important role in telecommunications, high-performance displays and lighting devices, solid-state laser materials, data storage, radiation detection, and medical applications. For example, RE-based bulk phosphors have been successfully employed as display devices and compact solid state lasers [4–7]. In the field of phosphor-based luminescence, RE ions (in particular trivalent lanthanides) by far are the most used class of localized luminescent centers and interest in their performances/applications has widely grown beyond the starting demand for optical sources and amplifiers operating at wavelengths compatible with optical fiber technology (coincidence between the Er^{3+} emission band around 1535 nm and the principal low-loss window in the absorption spectrum of alumino-silicate optical fiber).

Notably, since RE elements possess electronic configuration with *4f* electrons effectively shielded by the outer *5s* and *5p* subshells, the energy levels of RE ions are barely perturbed by the local chemical microenvironment. All of this implies unique emission properties of RE compounds such as richness of energy levels, emission wavelengths mainly dependent on the RE's electronic configuration, combination (at certain ratio) of different REs to carry out phosphors with tunable luminescence, excellent photo-stability, long luminescence lifetime, narrow emission lines, and high color purity in contrast to quantum dots (QDs) and organic dyes [8–11]. Moreover, under excitation of QDs by ultraviolet (UV) or visible light, auto-fluorescence and photo-damage to biological samples may be induced causing low signal-to-noise ratio and limited sensitivity. As an alternative to QDs, recently up-conversion (UC) nano-phopshors (i.e., nano-sized phosphors) are being developed (see Ref. [12] and references therein).

With the development of nanoscience and nanotechnology as well as the increasing trend to miniaturization, RE-based nano-phosphors have been deeply investigated and used in a wide range of applications, such as bio-probes for photoluminescent bio-imaging [13–15], sensing [16], FRET-based sensing [17], bio-detection [18, 19], bio-imaging [20, 21], drug delivery [22], photovoltaic devices [23], photochemical reactions and transducers for photoactivation [24], catalysis [25], solar cells [26], nano-thermometry [27] and data storage devices. [28, 29]. Unique properties of nanophosphors include high photostability, near infrared (NIR) excitation with low power density (nearly 100 mW cm^{-2}), large tissue penetration depth, minimum background noise for luminescent applications in vivo, high resistance to photo-bleaching and photo-blinking and micro/millisecond lifetimes [30–35] and size-dependent quantum efficiency with luminescence lifetime decreasing by a few orders of magnitude for decreasing size [36].

Wide literature is available dealing with physics of bulk phosphors, emission from centers of luminescence (optically active ions), basic radiative and non-radiative processes (energy- transfer (ET), UC, phonon-assisted relaxation, cross-relaxation, concentration quenching) [3, 37, 38]. Reduction of the dimension to the nanoscale involves changes in the spectroscopic properties of luminescence materials as compared to the bulk counterpart. Indeed, despite the shielding of the electronic configuration of the *f*-levels in RE ions, the luminescent properties of nanophosphors can be affected by size, shape, crystal structure, and chemical environment/surrounding as well as electron-lattice coupling. The phonon energy plays an important role in non-radiative transitions due to multiphonon relaxation between closely spaced energy levels. Moreover, increased surface to volume ratio with decreasing size introduces surface states responsible of luminescence quenching with loss channels.

In this review paper we introduce fundamentals of the spectroscopy of REs and an overview of the physical mechanisms currently invoked to interpret luminescence in phosphors as well as the spectral features such as thermal line broadening and line shift with particular attention to size-dependent spectroscopy. This topic is very fascinating because size-dependent quantum confinement can remarkably affect both radiative and nonradiative electronic transitions in nanoparticles. In the case of RE ions, spatial confinement to the nanoscale has effects on spectroscopy mainly due to electron-phonon interactions [39–42].

2.2 Luminescence and Physics of Phosphors

In this paragraph we introduce (i) a general background on luminescence and the commonly used nomenclature, (ii) an overview of the processes and decay pathways currently invoked to interpret luminescence in phosphors and (iii) loss channels causing reduction of the luminescence efficiency.

2.2.1 Luminescence: Basics and Nomenclature

The term "luminescence" refers to emission of photons from an excited material. Indeed, the primary process to occur is absorption of energy that can be induced by several sources (electromagnetic, electric current, charged energetic particles or chemical) and causes excitation (i.e., transition from a lower to a higher energy state) of the system followed by photon or heat emission through spontaneous decay to lower-energy states. Since heat emission resulting from non-radiative de-excitation

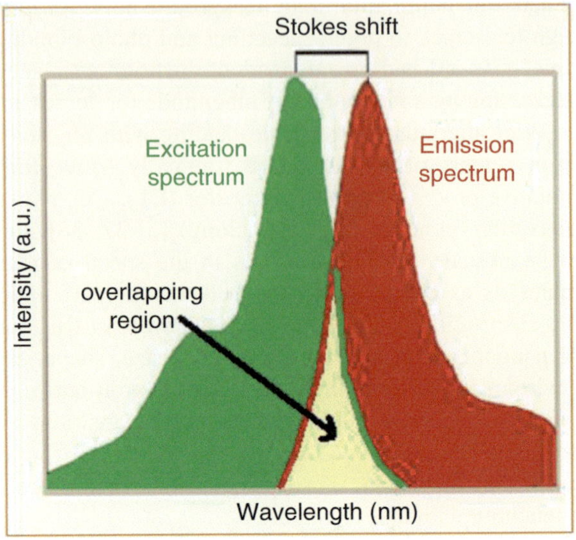

Fig. 2.1 General sketch of the absorption and emission spectra of a system with Stokes shift

mechanisms doesn't lead to luminescence, the luminescence yield depends on the energy losses. When the energy difference between the states involved in a decay is large, the probability associated with transfer of the transition energy to the lattice is low and a radiative transition occurs.

Luminescence can be classified by a prefix referring to the excitation source. For example, the terms photo-luminescence, electro-luminescence, cathode-luminescence, thermo-luminescence, chemi-luminescence, X-ray luminescence, bio-luminescence and mechano-luminescence are used in the case of excitation induced by photons, electric current, cathodes, heating, chemical reaction, X-rays, biochemical reaction and mechanical energy, respectively. The nomenclature "optical excitation of luminescence" is introduced when emission is induced by absorption of UV or visible light.

Generally, based on the emission time-scale after the removal of the excitation source, photo-luminescence phenomena can be divided into two main classes, termed "fluorescence" and "phosphorescence". While fluorescence occurs over a short time lapse (10^{-9} s < t < 10 ms), phosphorescence is characterized by much longer decay time (t > 0.1 s up to hours or even days) due to the role of spin-forbidden transitions (long lived excited state) and phonon-related processes [43].

In most cases, the well-known "Stokes principle" holds, i.e., the emitted photons have a longer wavelength (i.e., lower energy) than the absorbed radiation and the difference (in wavelength or frequency units) between the absorption peak and the emission peak of the same electronic transition is named Stokes shift. Figure 2.1 depicts the situation envisaged by the Stokes shift: excitation and emission spectra are shifted one with respect to other (Stokes shift) and have an overlapping region.

Even if the usual fluorescence follows the Stokes law, the photo-luminesce radiation may also have the same wavelength as the absorbed radiation (resonance fluorescence) or shorter wavelength (i.e., higher energy) than the absorbed radiation (anti-Stokes photoluminescence and photon-energy UC) [4, 44]. The energies emitted in anti-Stokes processes are usually in excess of excited energies by only a few k_BT (k_B Boltzmann's constant, T absolute temperature). Instead, in the case of UC (i.e., conversion of infrared (IR) radiation to visible and ultraviolet (UV) emission) the emitted photon energy exceeds the exciting photon energy by 10–100 k_BT. The RE-doped materials are extensively used as UC materials in that REs mainly have *4f* states insensitive to outer influences (crystalline field of the ligands or time variation of the crystal field), which results in transitions within the *4f* core hardly affected from the chemical environment.

2.2.2 Luminescence Features of Phosphors

Sharp-line emission spectra with spectral positions weakly or hardly influenced by chemical environment can be obtained at room temperature by doping semiconductors or oxides by TM or RE elements acting as optical emission centers. In the following we focus our discussion on this class of materials, that are usually microcrystalline doped or un-doped powders and, recently, nanomaterials consisting of TM or RE compounds and TM- or RE- doped host crystalline structures [2, 3]. Indeed, to excite the luminescence, the absorption of energy takes place by either the host lattice (band absorption) or dopants, called activators (emitting centers) or sensitizer (if energy is then transferred to activator ions) depending on the effectiveness of their absorption. The luminescence spectrum of an optical center (emitting dopant) can depend or not on the host lattice. Emission at spectral positions hardly influenced by the chemical environment can occur and not in the case of RE ions and TM ions, respectively. To interpret and tune luminescence from optical centers the degree of shielding of the optically active electrons of the TM or RE ions by the outermost electronic shells must be considered.

The class of TM elements consists of elements with unfilled outermost *d*-orbitals. Such valence *d* electrons form bonds with ligands affected by the host crystal field and are involved in the optical transitions of TMs. In general, absence of shielding from the crystal in the case of the valence *d* electrons involves a large Stokes shift in the energy and large splitting of the atomic (unperturbed) levels from the host crystal field.

RE metals are, by definition, the elements of the Group IIIB in the periodic table (elements Sc, Y, La and the 14 lanthanides Ce–Lu) (Fig. 2.2). The term 'rare earth' is often used in a more restricted sense as a synonym only for the lanthanides because their properties most depend on the occupancy of the *4f* electron shell (from 0 (La) to 14 (Lu) there are 14 lanthanides with atomic numbers $Z=58$–71). Trivalent lanthanides have an outer electronic configuration $5s^2 5p^6 4f^n$ where

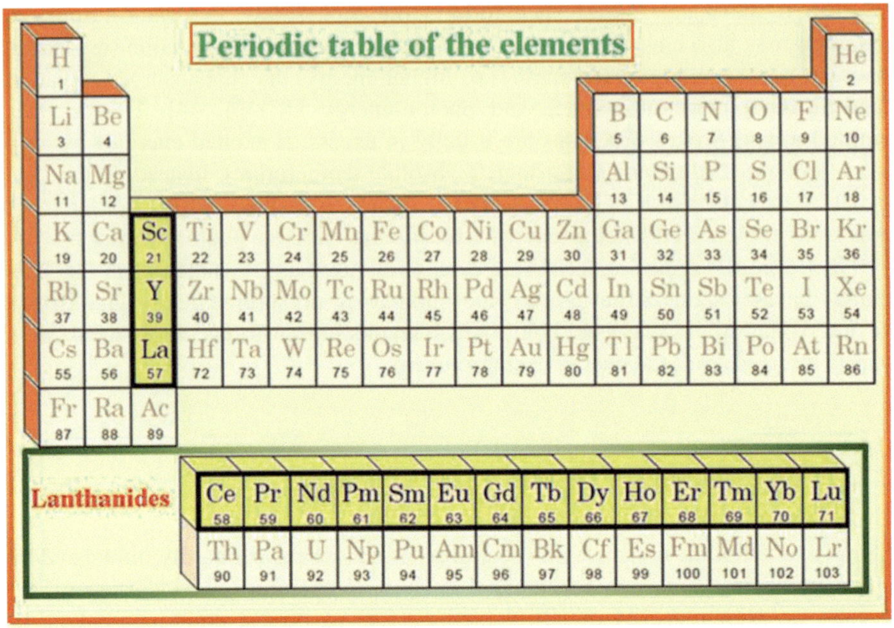

Fig. 2.2 Periodic chart of the elements where the location of the rare earth elements is evidenced

n ranges from 1 (Ce ion) to 13 (Yb ion) and refers to the number of electrons in the un-filled *4f* shell. The filling of such a shell obeys the Hund's rules: (i) the term with the highest spin number S has the lowest energy and if there are several terms with the same S, the one with the highest angular momentum quantum number L has the lowest energy; (ii) because of spin–orbit coupling, the terms ^{2S+1}L are split into levels $J = L + S$, $L + S - 1, \ldots, |L - S|$, where for less than half–filled shells, the term with the smallest J lies lowest in energy; (iii) accounting for the spin–orbit coupling, for more than less half–filled shells the term with the largest J has the lowest energy. Since the valence electrons $4f^n$ are shielded by the $5s^2 5p^6$ outer less bounded configuration, then the $^{2S+1}L_J$ states (S, L, J spin, orbital and total angular momentum, respectively, with $J = L–S,..L + S$) of the free RE ions are weakly affected from the host crystals, meaning that optical spectra of RE ions in crystals resemble the ones of the corresponding free ions.

Each host lattice, in addition to providing particular doping sites, provides different spatial distances and different local coordination structures of luminescent centers. For example, the character of the bonding between lanthanides and surrounding ligands is more ionic than in the case of the TMs. As a consequence, the number of possible coordination numbers of REs is larger than the ones of TMs, resulting in a richer variety of symmetries and spectroscopy [45]. Further, the phonon energy, which affects ET and UC energy transfer pathways, is also lattice-dependent [46–50].

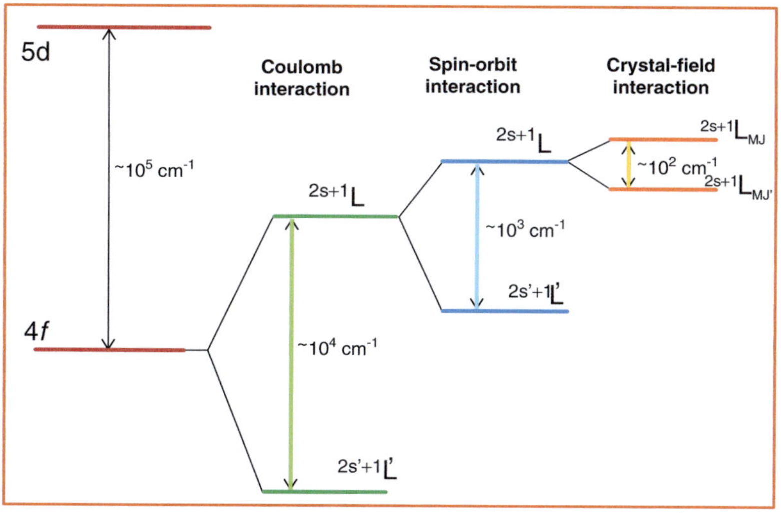

Fig. 2.3 Contributions to the splitting of energy levels in a 4f 5d configuration

Therefore, to understand the luminescence properties of phosphors, the relationship between the electronic structure of TM and RE ions and the transitions allowed by selection rules and crystal symmetry must be considered.

Basically, there are two types of RE-emission: *f-f* transitions (involving ground state $4f^n$ and excited state $4f^n$) and *d-f* transitions (involving ground state f^n and excited state $4f^{(n-1)}$–$5d$). The first emission type is characterized by emission lines with FWHM < 10 nm and is independent of the host lattice because the 4f-electrons are well shielded by the outer *s* and *p* orbitals. As the sketch reported in Fig. 2.3 shows, the splitting contributions of the energy levels $4f^n$ are, in order of decreasing influence, Coulomb interaction, spin-orbit interaction and crystal-field interaction. The listed interactions are the perturbations which more effectively let sketch the energy level scheme of the triply ionized lanthanides inserted in a crystal lattice.

In general, in the ground state intra-configuration *f-f*, the electric dipole transitions are forbidden for centrosymmetric sites by the parity conservation rule. Since the parity doesn't change within the *4f* shells, the only transitions that can occur are the magnetic dipole transitions, which are weaker than the electric dipole ones, called "forced dipole transitions". In the case of non-centrosymmetric sites, *f-f* transitions are observed because of the mixing of the *4f* states with states of different (opposite) parity belonging to a different electronic configuration. The mixing may be due to the interaction of REs with crystal field or to the interaction of RE with lattice vibrations [51–53]. Forced dipole transitions involving states described by mixed parity wave functions exhibit intensity directly depending on the degree of mixing and sharp emission peaks.

As a consequence, even if the energies of the emission lines stemming from *f-f* transitions are almost independent on the host lattice, the relative emission intensity change remarkably depending on site symmetry and selection rules:

(i) if the site of the RE ion has inversion symmetry, then magnetic dipole transitions are allowed (i.e., the selection rule is $\Delta J = 0{,}1$ and when $\Delta J = 0$, being forbidden any transition involving another state with $J = 0$, only $\Delta J = |1|$ is allowed);
(ii) in absence of inversion symmetry, electric dipole transitions are also allowed.

The second emission class (*d-f* transitions) is characterized by emission lines with FWHM ranging from 30 to 100 nm and is strongly dependent on the host lattice because the *5d*-orbitals (more extended and delocalized than *f*-orbitals) are influenced from the surrounding crystal field. As a consequence of the crystal field effect on the *5d* electron, $4f^{n-1}5d$ configurations of RE ions in solids are very different from those of free ions.

The *f-f* and *f-d* transitions of REs show weak and strong ion-lattice coupling strength, electric dipole oscillator strengths of 10^{-6} and 10^{-1}–10^{-2}, lifetimes of 10^{-2}–10^{-5}s and 10^{-8}–10^{-6}s, respectively.

Host-dependent luminescence can occur in the case of many TM ions, ions with a "lone pair" of *s* electrons (s^2 ions, such as Ga^+, In^+, Tl^+, Ge^{2+}, Sn^{2+}, Pb^{2+}, As^{3+}, Sb^{3+}, Bi^{3+}), and RE ions showing *d-f* emission. Broad emission bands involving the partly filled *d*-shell of TMs (*d-d* transitions) and the *5d* shell and the *4f* shell of REs (*d-f* transitions) can be observed as well as emission from s^2 ions. In general, in presence of characteristic luminescence, broad emission bands (widths exceeding 50 nm) result when the character of the dopant-ligand chemical bond is different between the ground state and the excited state due to a change in the equilibrium distance of the emitting ion with respect to its surrounding. In the case of charge transfer transitions (taking place between different kinds of orbitals or between different ions) very broad emission spectra can result because of changes in the charge distribution on the emitting center induced by excitation.

On the other hand sharp emission bands are characteristic of transitions between electronic states that don't participate in the chemical bonding, such as *f-f* transitions on RE ions which originate within the manifold of *f*-electronic shells and are almost independent on the chemical environment. The key condition to observe sharp emission bands is having ground state and excited state with the same chemical bonding character.

In regard to the influence of the crystal field, shift of the luminescence emission can result from a change of the covalence-degree of the dopant-ligand bond in the case of *4f-5d* transitions because the crystal-field interaction splits the *d*-level, depending on symmetry and crystal field strength. In the case of transitions between *f*-electronic states, even if the spectral position of the emission lines is almost independent on the choice of the host lattice, the relative emission intensity of the several possible optical transitions may change considerably due to the selection rules [51, 54]. Spin and parity forbidden transitions on RE ions are rather slow (i.e., in the ms range).

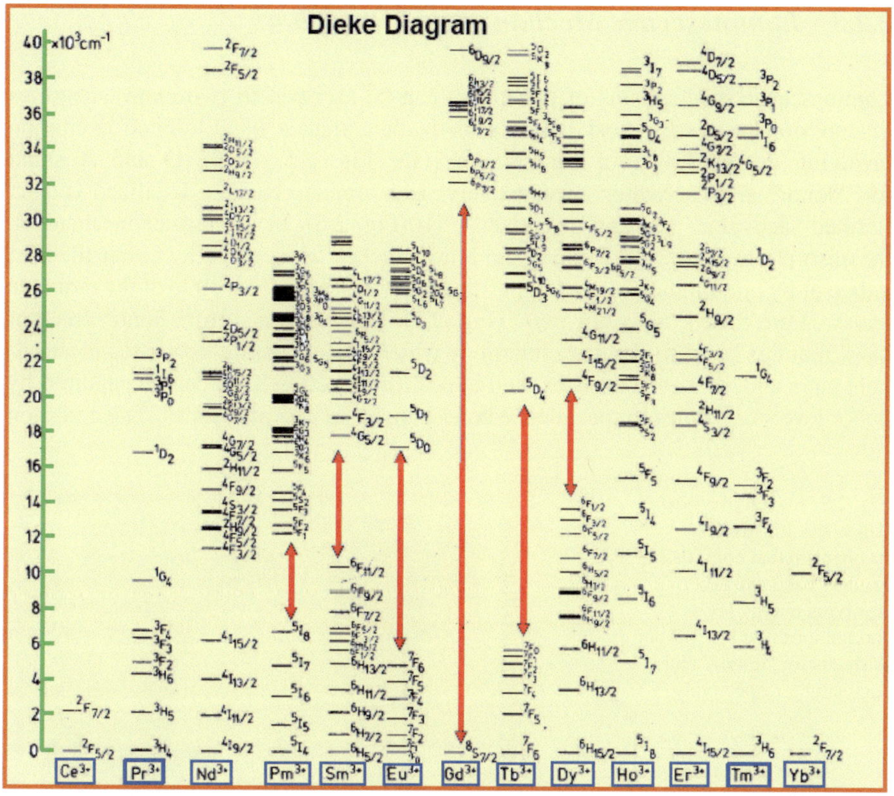

Fig. 2.4 Sketch of the Dieke's diagram listing the energy levels for all lanthanides in the energy range 0–40,000 cm^{-1}

If emission spectra with spectral positions hardly influenced by their chemical environment are required, then lanthanides are a proper choice. Indeed, measurements of absorption and luminescence spectra of the lanthanide series in a $LaCl_3$ crystal performed by Dieke and co-workers in the 1950s and 1960s resulted in a level diagram, known as "Dieke's diagram" (Fig. 2.4), that lists the energy levels for all lanthanides in the energy range 0–40,000 cm^{-1} [55]. Further measurements extending to the energy levels of the $4f^{\,n}$ electronic configurations of lanthanides in various host lattices demonstrated the general features and information given by the Dieke's diagram remain unchanged by changing host lattice. Dieke's energy-level diagram shows the energies of the states of the $(RE)^{3+}$ ions with the magnitude of the crystal-field splitting indicated by the width of the state and the energy location of the corresponding free ion energy level ($^{2S+1}L_J$, S and J quantum number of spin and total angular momentum) given by the center of gravity of each J-multiplet. Hence, prediction, identification or assignment of the emission lines of REs in listed crystals and new hosts can be easily made based on the Dieke's diagram.

2.2.3 Luminescence Mechanisms of Phosphors

Luminescence mechanisms of phosphors can be ascribed to two main categories of emission processes: host lattice sensitization, that is band-to-band excitation involving energy levels of foreign ions (see Fig. 2.5, where D and A stand for "donor" and "activator", respectively) and luminescence of localized centers (termed "activator" (A) and"sensitizer" (S)) (Fig. 2.6). In the former mechanism, the absorption of energy to excite the luminescence takes place by either the host lattice or by intentionally doped impurities that when are responsible of the radiative emission are called "activator ions" (Fig. 2.5). Luminescence from donor-acceptor pairs requires band-to-band excitation in which the excitation generates electron-hole pairs and, then, electrons excited into the conduction band can be captured by donor levels and holes in the valence band transfer to acceptor states. The emission

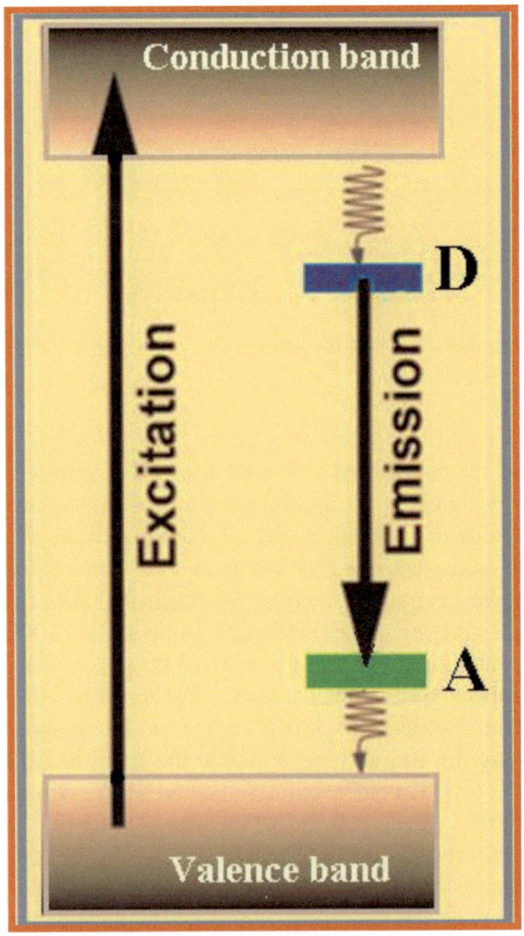

Fig. 2.5 Sketch of a host lattice sensitization mechanism that consists in band-to-band excitation involving energy levels of foreign ions (the label D and A stands for "donor" and "activator", respectively

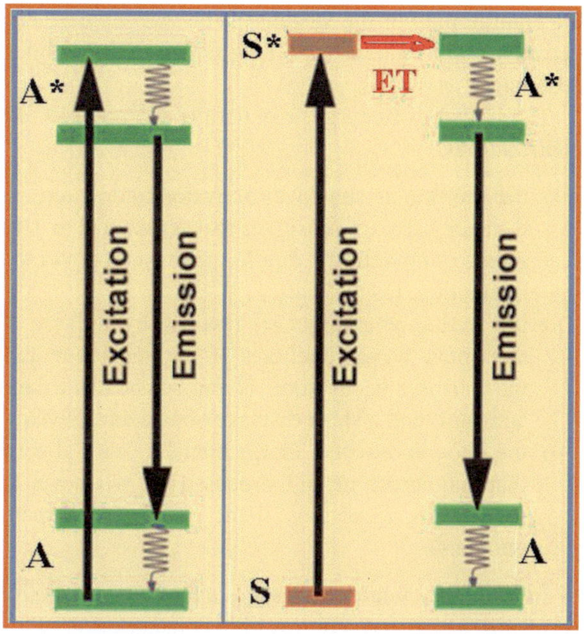

Fig. 2.6 Sketch of processes of luminescence by localized centers (termed "activators" (A) and"sensitizers" (S)) involving an activator excited state (indicated as A*), a sensitized excited state (indicated as S*) and an energy transfer (ET) process (a sensitizer S absorbs the exciting energy and transfers it to the activator A)

is induced by the donor-acceptor pair and can depend on the relative distance donor-acceptor (the smaller the distance, the higher the emission energy) due to electron transfer between donors and acceptors.

In the case of luminescence by localized centers, an activator element A is a center localized within the host lattice (dopant) that is able to generate the designed emission by a transition starting from an its excited state (indicated as A* in Fig. 2.6). A sensitizer element S is a localized center added to the host lattice that absorbs the exciting energy and transfers it to the activators by ET.

In what follows we shall concentrate on the effective luminescence mechanisms of phosphors as well as non-radiative processes.

2.2.3.1 Energy Transfer

One of the basic mechanisms of luminescence spectroscopy is energy transfer (ET), consisting of transfer of energy from an excited center (S) to an unexcited center (A). The obvious way in which this transfer can take place is radiatively, by the emission of a photon by an excited optical center and the absorption of the emitted photon by an unexcited center. In this case the excited center is unaffected by the transfer process, i.e. its lifetime is not affected by the transfer.

A more common ET process does not involve any photon and is due to the interaction between the charged particles of a sensitizer center and the charged particles of an activator center. This interaction represents a perturbation affecting the Hamiltonian of the system of the two separate centers and establishes a bridge between them. When the ET involves similar centers, the ET process is termed "resonant."

The transfer of energy is generally irreversible. It may consist of the following contributions:

(i) the multipolar sensitizer-activator interaction, with a rate depending on the distance between sensitizer and activator R as $1/R^n$ with $n = 6, 8, 10$ for dipole-dipole interaction, dipole-quadrupole interaction, quadrupole-quadrupole interaction, respectively,
(ii) the exchange interaction favored by a large overlap integral between the electronic wave functions of the sensitizer and activator, a small value of the intrinsic decay time of the sensitizer, a large absorption strength of the activator, and a short distance sensitizer-activator ($R < 0.5$–1 nm).
(iii) the phonon-assisted energy transfer, with absorption or emission of phonons that may assist any ET process (i.e., whenever the energy difference between the transition energies of the centers is compensated by a process involving phonons).

In practice, while exchange interaction is active over very short distances (<0.5 nm–1 nm) and requires high concentration of luminescent centers, multipolar interactions are affected by the strength of the optical transitions and can occur for center to center separations ranging from 2 to 10 nm (low doping concentration).

2.2.3.2 Non-radiative Processes

Excited states release their energy via either of two competitive ways, i.e., by light emission or phonon emission, depending on the energy difference with respect to the closest lower energy state.

Non-radiative processes in phosphors are very relevant to their functioning and can be useful or harmful according to circumstances. To illustrate the role played by phonons, a typical situation is represented as an illustration in Fig. 2.7, where a three-level system is depicted under the action of electromagnetic radiation. The frequency of the electromagnetic external perturbation is assumed to correspond to the energy-gap between levels labeled 1 and 3. In these conditions the probability associated with a $1 \rightarrow 3$ radiative transition is equal to that associated to the opposite transition (i.e., $3 \rightarrow 1$), assuming that the $1 \leftrightarrow 3$ energy gap is so large to make a $3 \rightarrow 1$ non-radiative transition impossible. Once a center is in level 3, it has two chances: it may go down to level 1 or to level 2. The latter level, being closer to level 3, favors a fast non-radiative process that brings the center preferentially to level 2 producing an accumulation of centers in this level. It is worth noticing at this point that small gaps and large gaps favor non-radiative and radiative processes, respectively, because (i) the bridging of small gaps requires the interaction of few phonons and (ii) the probability of a radiative transition is proportional to the cubic power of the gap-energy. In addition, it is also appropriate to note the pervasive

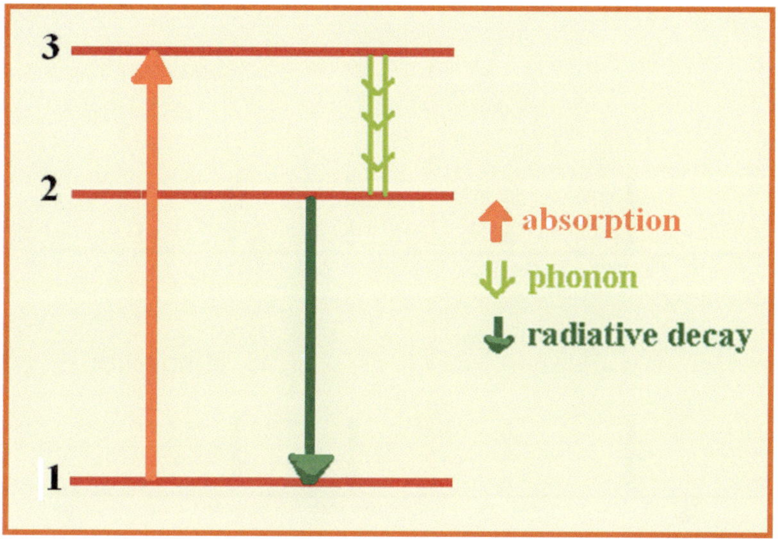

Fig. 2.7 Contribution from phonon states to the decay transitions of a three-level system under the action of electromagnetic radiation with frequency corresponding to the energy-gap between levels labeled 1 and 3

presence of phonons due to the fact that their density of states is proportional to the inverse of the cubic power of acoustic wave speed [38]. Conclusively, phonon intervention is desirable and obtainable in small gap $3 \rightarrow 2$ type transitions and not desirable as well as not generally present in large gap $2 \rightarrow 1$ transitions. Furthermore, systems with low Debye temperature, having phonons of small energy, make the non-radiative process less probable.

2.2.3.3 Vibronic Transitions

Another class of transitions driven by the optical center-lattice interaction are the "vibronic transitions" that in emission can involve a photon and simultaneously create or absorb one or more phonons [56–58]. The vibrations characteristic of a host material may perturb the wavefunction of an optical center. Such modifications are characteristic of the host solid material and may give rise to "vibronic transitions" or "phonon sideband"of the center. They are particularly active in the case of the (rare) sharp lines of TM centers.

With respect to the zero-phonon line, the vibronic features appear on the high (low) energy side in absorption (emission) at low temperatures and on both the high and low energy side at high temperatures [59]. The zero-phonon line refers to transition involving completely relaxed (no excited phonon states) excited and ground states. Figure 2.8 shows the spectrum of two phonon vibronic lines (k and k′ are the momentum phonon modes with frequency ω_k and $\omega_{k'}$, respectively)

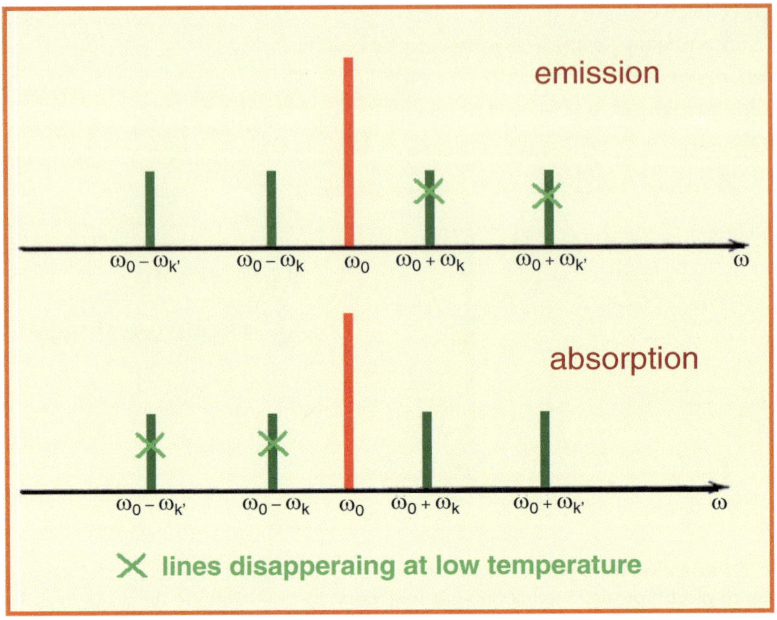

Fig. 2.8 General features of a vibronic spectrum in emission and absorption at low temperature

in emission (top panel) and absorption (bottom panel). The line associated with frequency ω_0 is termed "zero-phonon line" and is determined by the intrinsic difference in energy levels between the ground and excited states. Vibronic transitions in emission can take place at very low temperature, because phonons can always be emitted spontaneously even at zero temperature. At very low temperatures they have frequencies smaller than that of the no-phonon transition as demanded by the law of conservation of energy. Vibronic emission at frequencies greater than the no-phonon transition cannot take place in absence of phonons, i.e. at very low temperatures. At higher temperature vibronic transitions in emission can take place at frequencies greater than the frequency of the no-phonon transition. The opposite is true for vibronic transitions in absorption. At very low temperatures phonons can be emitted and the transitions which start from the ground state and end in an excited state can take place at frequencies higher than the no-phonon transition. At higher temperatures transitions can take place at frequencies lower than the frequency of the no-phonon transition.

2.2.3.4 Cross Relaxation

Another important ET process is cross relaxation, where a center loses a certain amount of energy ΔE following a transition to a lower energy state E (which may also be the ground state E_{GS}) and a center gets excited to a state higher in energy than ΔE. Cross relaxation may take place between the same centers in a given material or between two different centers with two pairs of energy levels separated by the same energy. Such a mechanism of ET may lead to energy migration, e.g. diffusion between sensitizers in the case of identical levels without energy loss, or to self-quenching with energy loss or change of the emitted photons. Indeed, cross-relaxation is responsible for luminescence quenching at concentrations of the activator above a critical value in a given material (i.e., cross relaxation between the same lanthanides) or between two differing elements with two pairs of energy levels equally spaced in energy [2, 60, 61]. In the case of self-generated quenching, the average lifetime of the luminescence decreases with increasing concentration of the emitting center. Usually, such a concentration quenching process occurs whenever the average distance between luminescence centers is short enough according to the following relationship (Auzel's model) between dopant concentration (c) and fluorescence lifetime ($\tau(c)$) and [62]:

$$\tau(c) = \frac{\tau_0}{1 - \left(\frac{c}{c_0}\right) \exp\left(-\frac{N}{3}\right)}$$

where τ_0 is radiative transition lifetime, c_0 is a constant and N is the generated phonon number for quenching via non radiative relaxation. The optimal dopant concentration for a RE ion is the one yielding the maximum luminescence intensity and can be different in various host lattices. It must be determined experimentally.

2.2.3.5 Up-Conversion Processes

The term up-conversion (UC) refers to nonlinear processes which convert infrared exciting radiation to visible and ultraviolet emission (i.e., the continuous absorption of two or more low-energy photons leads to the emission of photons more energetic than the exciting ones (anti-Stokes emission)). Materials exhibiting UC processes are known as up-converters In practice, UC lets avoid undesirable effects, such as creation of color centers and excitation of deep traps, that can occur under UV excitation of doped solids.

The main mechanisms of UC can be classified as follows: "excited-state absorption" (ESA), "up-conversion by energy transfer" (ETU) and "two-photon absorption" (TPA) (Figure 2.9 shows simplified representations of some UC processes).

In the case of ESA, which involves a single center, two photons are absorbed sequentially and the intermediate state is real, that is initial absorption leads to an intermediate excited state which lives long enough to let absorption of a second photon to reach a higher excited state. Following the absorption of the first photon, radiationless relaxations to a lower energy may or may not occur.

ETU processes result from successive energy transfers between centers at different sites starting with an activator in an excited state. Figure 2.9 shows one photo-excited center that transfers to a second center acting as up-converter. Since mutual interaction between centers is involved, the concentration of sensitizer and activator is required to be not too low to allow energy migration between centers.

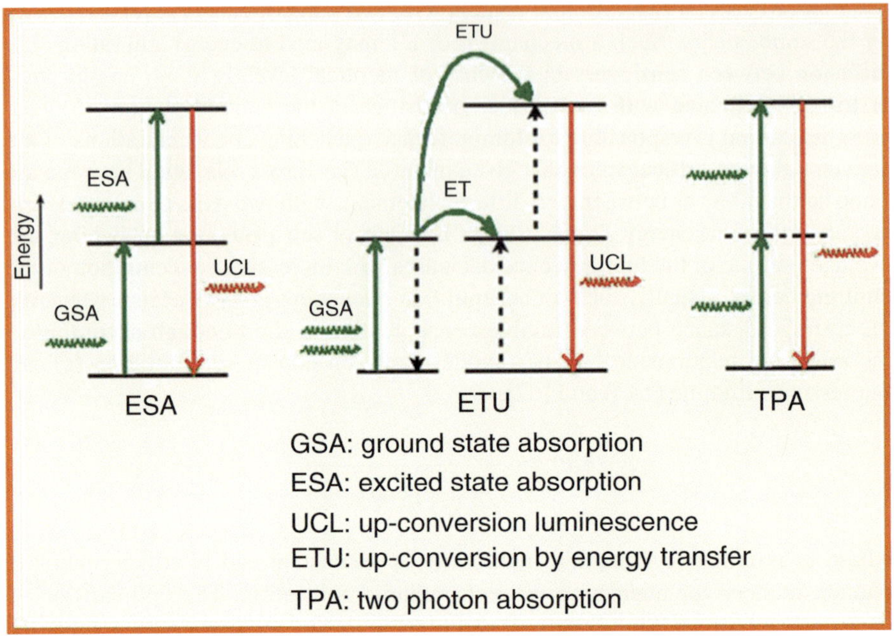

Fig. 2.9 Sketch of the mechanisms corresponding to the main UC processes

TPA processes involve the absorption of two photons with an intermediate virtual state, i. e. there is no real intermediate excited state. Simultaneous absorption of two photons results.

All the above processes, involving more then one photon, are non-linear.

In addition to the mentioned UC mechanisms, there are less efficient others such as "second harmonic generation" (SHG), where two photons do not interact by any excited energy state, and "hot-band absorption", where an electron in a vibrationally excited level of the ground state is preferentially excited and emission then proceeds back to a lower vibrational level again of the ground state (UC by only a few units of thermal energy). Since f levels of lanthanide ions with relatively long lifetimes are suitable intermediate excited states in the UC process, UC materials are mostly RE-based phosphors. It is worth noticing that the UC emission efficiency of RE ions requires strategy to be enhanced due to small absorption cross section of RE ions resulting from the parity-forbidden 4f-4f intra-configurational transitions of RE ions (see Ref. [11] and references therein).

2.2.3.6 Luminescence Quenching

Luminescence quenching, that is the loss of radiative efficiency, can be ascribed to several processes, such as thermal quenching, concentration quenching (non radiative ET pathways dominates luminescence if the concentration of luminescent centers is beyond a threshold value), ET involving unknown non-radiative centers (killing centers, disorder, vacancies, sensitizer or activator ions that have changed their valence), when both the luminescent center and the host lattice absorb at the excitation wavelength or the host lattice itself shows optical absorption in the spectral range where emission occurs.

Notably, a large Stokes shift (difference between absorption and emission wavelength) reduces concentration quenching by reduced ET. On the other hand, a very large Stokes shift increases the probability for thermal quenching because the crossing at a lower energy between the ground state parabola and excited state parabola lets the ground state be gained non-radiatively. Also, the same reason implies that thermal quenching becomes less probable when the emission energy increases.

According to the listed luminescence processes in phosphors, the most relevant mechanisms and decay pathways resulting in luminescence degradation are processes in which:

(i) the absorbed energy does not reach the activator ion (competitive absorption, ET to defects or non-luminescent impurity ions, excited state absorption (ESA) and Auger processes),
(ii) the absorbed energy reaches the activator center, but non-radiative channels dominate the radiative return to the ground state (crossing of excited and ground-state parabola, multi-phonon relaxation, cross-relaxation, photoionization, ET to quenching sites) and
(iii) the emitted radiation is re-absorbed by the luminescent material (self-absorption due to spectral overlap between excitation and emission band, additional absorption bands (formation of color-center) in the spectral range where the activators or sensitizers also absorb due to degradation of the material)

To enhance the luminescence efficiency, the following is demanded:

(i) strong absorption by active centers,
(ii) large band-gap host to avoid absorption of the emitted radiation,
(iii) dopant concentration below the threshold of concentration quenching,
(iv) highly crystalline systems and low defect density,
(v) high purity of the samples,
(vi) homogeneous distribution of activators and sensitizers, and
(vii) low phonon frequencies.

2.3 Spectroscopy of Nanophosphors

Luminescence can result from atomic transitions, organic materials, QDs, TM- and RE- doped semiconductors and insulators. The last two classes of materials (QDs and TM- and RE- doped host lattices) are particularly interesting both at fundamental and applicative levels.

QD nanocrystals, also termed zero-dimensional systems or quantum-box or artificial atoms, result by scaling down the size of a system to the nanometer-level along all three directions (L_x,L_y,L_z) in such a way that electrons "feel" the presence of the boundaries. Being all three degrees of freedom quantum confined, the electron wavefunctions are no longer plane-waves and the introduced "geometrical" constraints cause a size-dependent electronic dispersion which is characterized by a sequence of quantized energy levels (Fig. 2.10a) each labeled by three quantum numbers (typically referred to as n_x,n_y,n_z). This phenomenon, known as "quantum-size effect", remarkably changes the electronic dispersion. In this respect Fig. 2.10 sketches the quantum confinement effects in the case of a semiconductor material: discrete conduction and valence energy states (Fig. 2.10b) and a blue-shift of the fundamental band-gap tunable by size (increasing band gap with decreasing size) (Fig. 2.10c). Indeed, as compared to the bulk counterpart, discrete energy levels (termed sub-bands) arise at the band edges with band-gap energy and spacing between sub-bands increasing for decreasing size of the QDs (Fig. 2.10c). Since a QD exhibits distinct narrow-line spectra, i.e., an atom-like energy spectrum (δ-function-like density of states (DOS)) (Fig. 2.10a), QDs are often termed "artificial atoms".

Although significant progress has been achieved in using QDs as single-wavelength emitting materials and biomarkers (real-time detection and bio-imaging), their application still presents drawbacks. For example, since real samples consist of arrays/distributions of QDs instead of a single QD, a serious concern of the fabrication process is being able to control size- and shape- distribution of QDs in order to preserve the characteristic of an artificial atom. Tunable emission wavelength and sharp-line emission spectra with spectral positions weakly or strongly influenced by the chemical environment can be obtained at room temperature by phosphors based on semiconductors or oxides doped by TM or RE elements acting as optical emission centers. Therefore, unlike semiconductor QDs, the emission bands of RE-based phosphors exhibits fluorescence less affected from the particle size. On the other hand, since excitation of QD samples by UV or visible light may induce auto-fluorescence and photo-damage to biological samples, UC-nano-phosphors, such as inorganic host doped with trivalent RE ions, can be a valid alternative to QDs as biomarkers (see Introduction). In fact, RE ions are excellent supporters for photon UC emissions due to the several energy levels introducing energy transfer pathways and possibility to couple pairs of RE ions. Some main advantages of UC nanophoshors are their excellent penetration depth in biosystems and absence of autofluorescence from backgrounds.

On the other hand, nanotechnology and nanoscience exploits integration (high density of devices distributed over small surfaces) and bulk phosphor materials do not meet this requirement. As compared to bulk phosphors, nanophosphors have a high surface to volume ratio and surface atoms with fewer and less coordinated neighbors, which changes the crystal field environment of an optical center as well as its peak position, emission intensity and lifetime. Basically, localized centers inserted in nanophosphors may differ from their counterpart in bulk in what we could call the "dynamical" characteristics, i.e., probability of radiative transitions, probability of non-radiative transitions, lifetime of excited states, thermal line

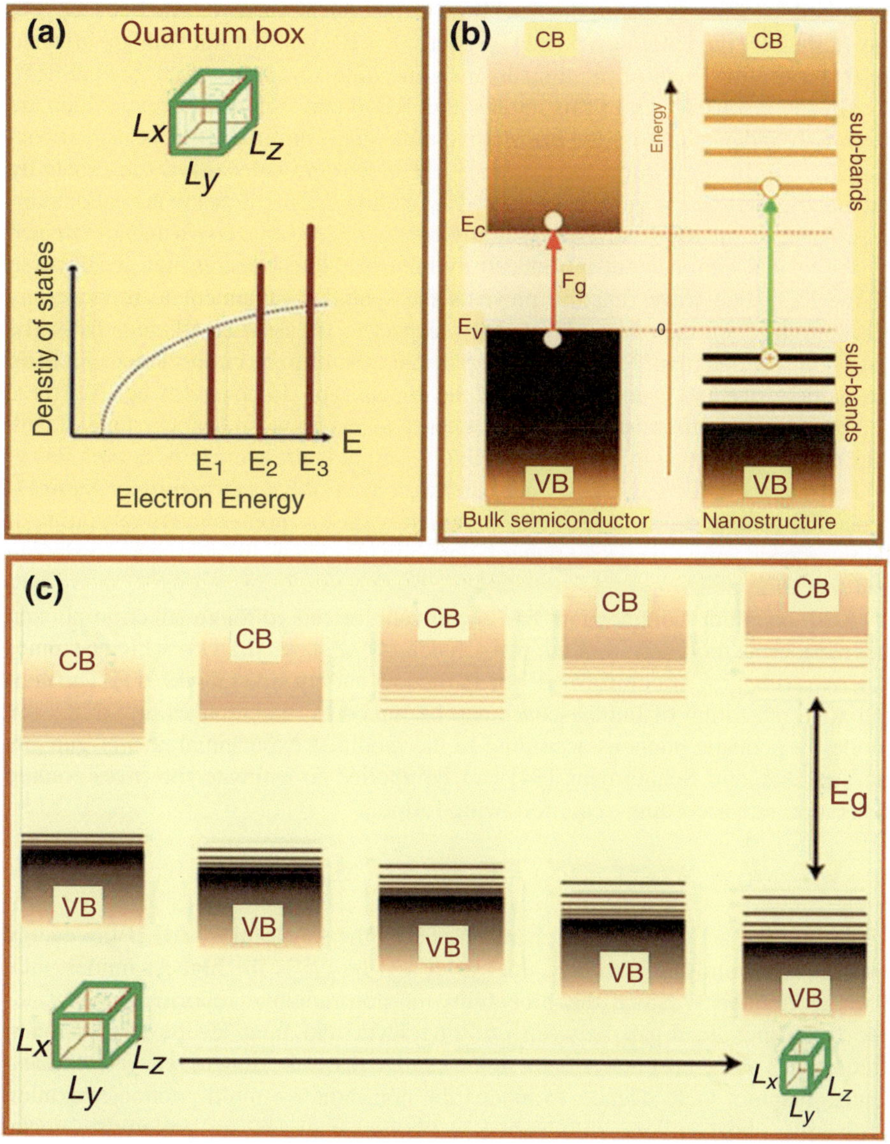

Fig. 2.10 Quantum confinement effects in the case of a semiconductor material: (**a**) atom-like energy spectrum (δ-function-like density of states), (**b**) discrete conduction and valence states with a blue-shift of the fundamental band-gap and (**c**) size-dependent electronic dispersion tunable for decreasing size

broadening and thermal line shifting. Therefore, fluorescence spectroscopy of the emission spectra of localized centers in nano-particles is particularly adept at uncovering the effects of spatial confinement on the spectral properties of TM- and RE- dopants in nano-systems.

In bulk phosphors, due to the shielding of the optically active centers occupying the *4f* shell by the outermost *5s* and *5p* shells, the RE ions are not strongly affected by the crystalline field of the ligands or by the time variations of the crystal field. As already discussed, all of this implies that RE dopants present spectra which are seemingly independent on the host lattice with sharp luminescence peaks at room temperature. In semiconductor materials, the electronic delocalized (i.e., spatially extended) states can be quantum confined by scaling the size down to the nanometer-scale. Instead, the RE *4f* strongly spatially confined electronic levels are not affected by the spatial confinement introduced by nanostructures, namely the localization of the RE f-like wave function prevents the spatial confinement to produce any relevant effect on the energy levels with respect to the isolated RE ion. Thus, the emission wavelength of RE-doped nanoparticles results to be almost independent on the particle size and is characteristic of the dopant type. In general, whenever there is no observable difference in the peak energy levels between nanocrystal and bulk material, it means that the local crystalline structure is the same in both samples.

Instead, the effects of the spatial confinement on the spectroscopy of luminescence RE-centers can be revealed by parameters such as non-radiative relaxation of luminescence, emission decay-pattern, quantum efficiency, thermal line broadening and line shift.

Also quantum confinement has significant effects through electron-phonon interactions as modified from the phonon density of states (PDOS) which becomes discrete with a lower cutoff frequency in the low energy side [38–40, 63]. The non-radiative relaxation of luminescence can be linked to the cutoff of low-frequency modes of acoustic phonons according to the modified exponential energy-gap law of Van Dijk and Schuurmans [64] can be applied to estimate the multi-phonon relaxation rate according to the following formula

$$W_{NR}\,(T = 0K) = \beta_{el} \exp\left[-\alpha\,(\Delta E - 2\hbar\omega_{max})\right]$$

where β_{el} and α are constants characteristic of the host lattice, ΔE is the energy gap between emitting level and final level and $\hbar\omega_{max}$ is the highest fundamental phonon frequency. Since the probability of non-radiative relaxations decreases for increasing band-gap between emitting level and final levels, large phonon frequencies of the host lattice favor non-radiative losses leading to reduced luminescence quantum yield. Hence, to reduce the non-radiative multi-phonon relaxation probability, host lattices with very low phonon energies are preferred. On the other hand, low-frequency phonons contribute significantly to radiationless decays between the closely spaced crystal-field states. Size confinement reduces the density of phonon states at low frequencies, but does not change their density at high frequencies. Therefore, if the non-radiative decay takes place via the emission of high-frequency phonons, size confinement should have no significant effect on it. The fact that the emission decay-pattern has been found in several nanosystems to be non-exponential and size-dependent [65] indicates the relevant role played by the surfaces while scaling dimensions down to the nanometer level.

Size-confinement has no significant effect on nonradiative decay and thermalization between electronic states with energy spacing near or higher than the Debye energy of the lattice.

Moreover, size-dependent quantum efficiency is another interesting property of doped nanophoshors. As first observation, the room-temperature photoluminescence quantum efficiency of ZnS nanocrystals doped with Mn was found to remarkably increase with decreasing size R (from $<1\,\%$ at $R = 70$ Å to nearly $18\,\%$ at $R = 3.5$ Å [36]. and the luminescence lifetime of the Mn^{2+} $^4T_1 \rightarrow {}^6A_1$ transition was observed to be faster than the corresponding Mn^{2+} radiative transition in the bulk crystals by more than five orders of magnitude (ns versus ms) while retaining a high radiative efficiency, approximately $20\,\%$. The efficiency increased with decreasing size was ascribed to improved electron-hole localization resulting from spatial confinement.

Quantities such as probability of radiative transitions, probability of non-radiative transitions, lifetime of excited states, thermal line broadening and thermal line shifting, are sensitive to small changes of the wave functions of the states involved in a transition that affect the coupling of the emitting ion to the optical oscillators of the vacuum and to the lattice vibrations. While the spectrum of the vacuum oscillators is the same in all cases, the spectrum of the lattice vibrations depends on the host material. A perturbation having negligible effect on the energy levels of a system, may have relevant effects on the probability of transition between the apparently unperturbed levels. This is best exemplified by the energy levels of RE ions in solids.

The basic features of the behavior of spectral lines of optically active ions in crystals can be described based on a few basic concepts such as line broadening and line shifting with temperature. Since the interaction with lattice is temperature dependent, spectral position, splitting, and lifetimes of the emitting levels can be a function of the temperature. As a general guideline, luminescence is favored at lower temperature and non-radiative processes become dominant at high temperatures. This behavior is termed "thermal quenching".

The width $\Delta\omega$ of a spectral line is influenced by several contributions: inhomogeneous broadening, direct one-phonon processes, spontaneous decay, multiphonon decay, Raman scattering and strain.

About the contribution to the resultant spectral line width by the listed mechanisms: the total contribution from direct one-phonon processes, multiphonon decay and radiative decay amounts, in general, to less 0.01 cm^{-1} and Raman scattering gives a contribution of a few cm^{-1}. In bulk materials the dominant contribution to the thermal broadening is from Raman scattering at temperature above 70 K too.

A transition line shape is homogeneously or inhomogeneously broadened depending on whether the lines are from the same center or from different but closely spaced centers. Usually, a homogeneous broadening is defined by the random perturbation of the optical centers, such as interactions with lattice phonons. In the case of homogeneous broadening the transition line shape is described by a Lorentzian function g_L (Fig. 2.11). Inhomogeneous broadening of the line

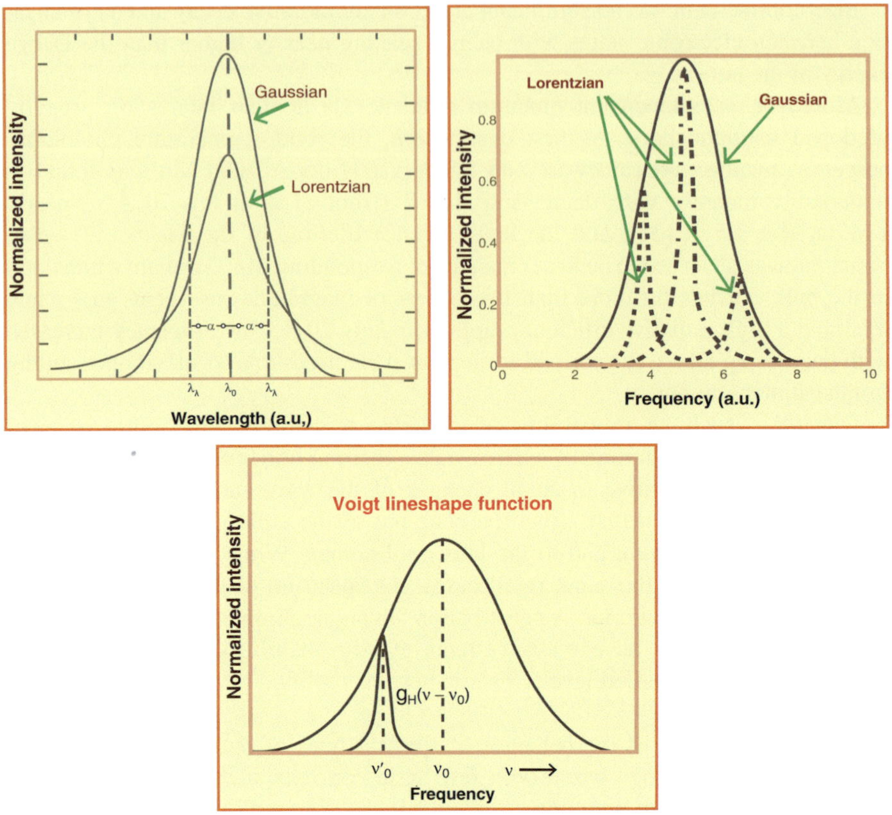

Fig. 2.11 Gaussian and Lorentzian function broadened emission line spectral profile describing an inhomogeneously and homogeneously broadened optical transition, respectively. Overall line shape resulting from a convolution of Lorentzian and Gaussian profiles (Voigt function)

shape during a transition originates from a local site-to-site change in the lattice environment surrounding the optical center. In this case, the spectral line is a superposition of all individual, homogenously broadened lines corresponding to different optical centers and the line shape of an inhomogeneously broadened optical transition is described by a Gaussian function (Fig. 2.11).

In practice, since at room temperature the individual contributions to the overall line shape due to homogeneous and inhomogeneous broadening may be different or comparable, the resultant overall line shape is best described by the so-called Voigt function, i.e,. a convolution of Lorentzian and Gaussian profiles (Fig. 2.11).

If it is assumed that the phonon density of states is described by a Debye distribution, integration over all allowed phonon frequencies leads to the following expression of the line width (spectral line broadening) as a function of temperature:

$$\Delta E\left(cm^{-1}\right) = \Delta E_o + \overline{\alpha}\left(\frac{T}{T_D}\right)^7 \int_0^{\frac{T}{T_D}} \frac{x^6 e^x}{(e^x - 1)^2} dx$$

where ΔE_o, $\overline{\alpha}$ and T_D represents the residual line width, a constant parameter related to the electron-phonon coupling and the Debye temperature, respectively. They are treated as adjustable parameters to be estimated by fitting the experimental data to the above formula [38, 66].

The thermal shift of a sharp emission line, i.e. a line whose width is much smaller than the Debye frequency, is due to the process of emission and reabsorption of virtual phonons. It can be expressed as follows

$$\delta E\left(cm^{-1}\right) = \alpha\left(\frac{T}{T_D}\right)^4 \int_0^{\frac{T}{T_D}} \frac{x^3}{(e^x - 1)} dx$$

where α and T_D (Debye temperature) are adjustable parameters to be estimated by fitting this formula to the experimental data [57, 66]. The parameter α is a parameter related to the electron-phonon coupling.

To summarize, in general the following can be observed:

(i) the width $\Delta\omega$ of a sharp spectral line (i.e., $\Delta\omega \ll \omega_D$, where ω_D is the Debye frequency) increases with temperature and generally the line shape presents a Lorentzian profile (Fig. 2.11);
(ii) the dominant broadening mechanism is the Raman scattering of phonons that produces a homogeneous broadening and a Lorentzian profile of the line shape;
(iii) while going down in temperature at about 77 K the shape of the spectral line is generally of the Voigt type (Fig. 2.11), i. e. a convolution of Lorentzian and Gaussian profiles, at very low temperature the line is Gaussian (Fig. 2.11)
(iv) the residual width of the line at very low temperatures is a measure of how widespread are the sites available to the dopant ions and the randomness of these sites produces an inhomogeneous broadening and a Gaussian profile
(v) the position of a spectral line varies with temperature (model based on the emission and absorption of virtual phonons) in such a way that it can in principle move towards higher or lower wavelengths as the temperature rises.
(vi) a fitting of the spectral line shift versus temperature can be performed by introducing two adjustable parameters: the Debye temperature and a coupling coefficient related to the electron-lattice coupling;
(vii) a fitting of the temperature dependence of the width the spectral line by using the Raman scattering model can be done by introducing three adjustable parameters: the Debye temperature, a coupling coefficient and a residual line width (ΔE_o) at low temperature;

Another quantity to be considered in fluorescence transitions is the lifetime of the optical center, that is the inverse of the probability of the transition under examination, which is the sum of several probabilities related to radiative and non-radiative independent processes. In general, the luminescence decay-time at low temperatures (liquid nitrogen temperatures (~77 K) or helium temperatures (\10 K)) is longer because phonon interaction is ruled out and, therefore, the measure estimates a purely radiative lifetime.

System's size scaled down to the nanometer level involves a changed phonon DOS, i.e., the Debye approximation of continuous PDOS may no longer be valid at the nanoscale, where in the low frequency region phonon modes become discrete and present a cutoff frequency. A restricted number of phonon modes implies a restriction of non radiative relaxation channels which inhibits decays between states spaced by a small energy gap.

Since coupling electron-phonon is involved in the expression of the line width and the phonon density of states depends on particle size, then the line width can be size-dependent through size-dependent electron-phonon coupling. In this respect, controversial experimental results are reported in the literature [38]. Theoretical analysis of the dependence of the phonon density of states and electron-phonon coupling terms on particle size suggests that the electron-phonon coupling is not sensitive to the particle size and the effects of the particle size on broadening and shifting of the spectral lines can be likely observed only in the limit of low temperatures and very small particles (much smaller than 50 nm) [38]. Moreover, as a result of the spatial confinement, the discrete phonon density of states (i.e., reduced density of phonon states) can inhibit phonon-assisted ET and, hence, concentration quenching of luminescence. Moreover, other effects of the spatial confinement reported in the literature are: (i) modified dependence (from T^7 to T^3) on the temperature of the linewidth of characteristic transitions of Eu^{3+} in Y_2O_3 [66, 67], (ii) increased lifetime (from 220 ns to 27 μs) of a multiplet of Eu^{3+} in Y_2O_3 nanocrystals as compared to the bulk counterpart [41], (iii) anomalous thermalization effect induced by excitation of Er^{3+} in Y_2O_2S nanocrystals [39] and ascribed to size-dependent elimination of direct phonon relaxation.

Surface defects are another contribution stemming from spatial confinement. In bulk crystals, the spatially localized *f*-states of RE ions are so weakly coupled to the embedding lattice that *f*-*f* transitions can yield very sharp emission lines (linewidth of the order of a few cm^{-1} at liquid He temperature). In disordered structures inhomogeneous line broadening causes line width to increase to 100 cm^{-1}. The same can be observed in nanophosphors where *f*-states of surface RE ions (i.e., ions at defect sites) also contribute to the spectra and increase line width due to inhomogeneous broadening [68]. Furthermore, in the case of RE elements embedded in nano-phosphors, the decay patterns of the luminescent RE ions can deviate from the standard exponential law of the bulk crystal at the beginning of the decay pattern. The decay patterns of RE luminescence in nano-phosphors vary considerably with the size of the nano-particles: at each temperature over a wide range (from 77 K to just above room temperature) they are found to be size-dependent with increasingly greater deviation from the exponential decay law for decreasing size. Two reasons are mainly responsible for quenching of the luminescence and shortening of the decay-pattern in RE-doped nanophosphors: first, the increasing role played by the surfaces as the nano-particle size is reduced and, second, the presence of surface molecular groups (such as O-H) that may enhance the decay process due to their stiff vibrations that rob the RE ion of its excitation energy (quenching of the luminescence/shortening of the decay-pattern).

2.4 Conclusions

Great progress in size, shape, and phase control of nano-phosphors achieved by using various processing methods has broadened the fields of applications of phosphors and the interest in their understanding at the fundamental level and the changes in their response at the nanoscale level.

In this review paper, fundamentals for luminescent materials and the main radiative and non-radiative mechanisms of bulk and nanoscale materials, that is spectroscopy of phospors and nanoposhors, have been are presented and discussed with a focus on the rare earths as luminescent ions. Scaling down dimensions to the nanometer-scale change the crystal field environment of an optical center as well as its peak position, emission intensity and lifetime. Basically, localized centers inserted in nanophosphors may differ from their counterpart in bulk in what we could call the "dynamical" characteristics, i.e., probability of radiative transitions, probability of non-radiative transitions, lifetime of excited states, thermal line broadening and thermal line shifting. Hence, these parameters and processes are key quantities to reveal the effects of the spatial confinement on the spectroscopy of luminescence rare earth-centers in nano-systems. Also, quantum confinement has significant effects through electron-phonon interactions as modified from the phonon density of states which becomes discrete with a lower cutoff frequency in the low energy side. Another interesting properties of doped nanophoshors is the size-dependent quantum efficiency.

References

1. Wiedemann, E. (1888). Über Fluorescenz und Phosphorescenz, I. Abhandlung (On fluorescence and phosphorescence, first paper). *Annalen der Physik, 34*, 446–463.
2. Blasse, G., & Grabmaier, B. C. (1994). *Luminescent materials*. Berlin: Springer.
3. Ronda, C. R. (2008). In C. R. Ronda (Ed.), *Luminescence from theory to applications* (pp. 1–34). Weinheim: Wiley-VCH Verlag GmbH & Co. KGaA.
4. Auzel, F. (2004). Upconversion and anti-Stokes processes with f and d ions in solids. *Chemical Reviews, 104*, 139–173.
5. Downing, E., Hesselink, L., Ralston, J., & Macfarlane, R. (1996). A three-color, solid state, three-dimensional display. *Science, 273*, 1185–1189.
6. Silversmith, A. J., Lenth, W., & Macfarlane, R. M. (1987). Green infrared-pumped erbium upconversion laser. *Applied Physics Letters, 51*, 1977–1979.
7. Sandrock, T., Scheife, H., Heumann, E., & Huber, G. (1997). High-power continuous-wave upconversion fiber laser at room temperature. *Optics Letters, 22*, 808–810.
8. Wang, F., & Liu, X. G. (2009). Recent advances in the chemistry of lanthanide-doped upconversion nanocrystals. *Chemical Society Reviews, 38*, 976–989.
9. Yi, G. S., Lu, H. C., Zhao, S. Y., Yue, G., Yang, W. J., & Chen, D. P. (2004). Synthesis, characterization, and biological application of size-controlled nanocrystalline NaYF4:Yb, Er infrared-to-visible up-conversion phosphors. *Nano Letters, 4*, 2191–2196.
10. Carlos, L. D., Ferreira, R. A., & Ribeiro, S. J. (2009). Lanthanide-containing light-emitting organic-inorganic hybrids: A bet on the future. *Advanced Materials, 21*, 509–534.

11. Sun, L. D., Dong, H., Zhang, P. Z., & Yan, C. H. (2015). Upconversion of rare earth nanomaterials. *Annual Review of Physical Chemistry, 66*, 619–642.
12. Feng, W., Han, C., & Li, F. (2013). Upconversion-nanophosphor-based functional nanocomposites. *Advanced Materials, 25*, 5287–5303.
13. Shen, J., Sun, L. D., & Yan, C. H. (2008). Luminescent rare earth nanomaterials for bioprobe applications. *Dalton Transactions, 42*, 5687–5697.
14. Wang, F., Banerjee, D., Liu, Y. S., Chen, X. Y., & Liu, X. G. (2010). Upconversion nanoparticles in biological labeling, imaging and therapy. *The Analyst, 135*, 1839–1854.
15. Zhou, J., Liu, Z., & Li, F. Y. (2012). Upconversion nanophosphors for small-animal imaging. *Chemical Society Reviews, 41*, 1323–1349.
16. Chen, J., & Zhao, J. X. (2012). Upconversion nanomaterials: Synthesis, mechanism, and applications in sensing. *Sensors, 12*, 2414–2435.
17. Zhang, P., Rogelj, S., Nguyen, K., & Wheeler, D. (2006). Design of a highly sensitive and specific nucleotide sensor based on photon upconverting particles. *Journal of the American Chemical Society, 128*, 12410–12411.
18. Hampl, J., Hall, M., Mufti, N. A., Yao, Y. M., Mac Queen, D. B., Wright, W. H., & Cooper, D. E. (2001). Upconverting phosphor reporters in immunochromatographic assays. *Analytical Biochemistry, 288*, 176–187.
19. VandeRijke, F., Zijlmans, H., Li, S., Vail, T., Raap, A. K., Niedala, R. S., & Tanke, H. J. (2001). Up-converting phosphor reporters for nucleic acid microarrays. *Nature Biotechnology, 19*, 273–276.
20. Lim, S. F., Riehn, R., Ryu, W. S., Khanarian, N., Tung, C. K., Tank, D., & Austin, R. H. (2006). In vivo and scanning electron microscopy imaging of upconverting nanophosphors in Caenorhabditis elegans. *Nano Letters, 6*, 169–174.
21. Chatterjee, D. K., Rufaihah, A. J., & Zhang, Y. (2006). Upconversion fluorescence imaging of cells and small animals using lanthanide doped nanocrystals. *Biomaterials, 29*, 937–943.
22. Gai, S. L., Yang, P. P., Li, C. X., Wang, W. X., Dai, Y. L., Niu, N., & Lin, J. (2010). Nanomorphology and charge generation in bulk heterojunctions based on low-bandgap dithiophene polymers with different bridging atoms. *Advanced Functional Materials, 20*, 1166–1172.
23. Huang, X. Y., Han, S. Y., Huang, W., & Liu, X. G. (2013). Enhancing solar cell efficiency: The search for luminescent materials as spectral converters. *Chemical Society Reviews, 42*, 173–201.
24. Idris, N. M., Jaykumar, M. K. G., Bansal, A., & Zhang, Y. (2015). Upconversion nanoparticles as versatile light nanotransducers for photoactivation applications. *Chemical Society Reviews, 44*, 1449–1478. and references therein.
25. Li, Z. X., Shi, F. B., Zhang, T., Wu, H. S., Sun, L. D., & Yan, C. H. (2011). Ytterbium stabilized ordered mesoporous titania for near-infrared photocatalysis. *Chemical Communications, 47*(28), 8109–8111.
26. de Wild, J., Meijerink, A., Rath, J. K., van Sark, W. G. J. H. M., & Schropp, R. E. I. (2011). Upconverter solar cells: Materials and applications. *Energy & Environmental Science, 4*, 4835–4848.
27. Jaque, D., & Vetrone, F. (2012). Luminescence nanothermometry. *Nanoscale, 4*(15), 4301–4326.
28. Zhang, C., Zhou, H. P., Liao, L. Y., Feng, W., Sun, W., Li, Z. X., Xu, C. H., Fang, C. J., Sun, L. D., Zhang, Y. W., & Yan, C. H. (2010). Luminescence modulation of ordered upconversion nanopatterns by a photochromic diarylethene: Rewritable optical storage with nondestructive readout. *Advanced Materials, 22*, 633–637.
29. Li, D., Ding, C. R., Song, G., Lu, S. Z., Zhang, Z., Shi, Y. N., Shen, H., Zhang, Y. L., Ouyang, H. Q., & Wang, H. (2010). Controlling the morphology of erbium-doped yttrium fluoride using acids as surface modifiers: Employing adsorbed chlorine ions to inhibit the quenching of upconversion fluorescence. *Journal of Physical Chemistry C, 114*, 21378–21384.

30. Suyver, J. F., Aebischer, A., Biner, D., Gerner, P., Grimm, J., Heer, S., Krämer, K. W., Reinhard, C., & Güdel, H. U. (2005). Novel materials doped with trivalent lanthanides and transition metal ions showing near-infrared to visible photon upconversion. *Optical Materials, 27*, 1111–1130.
31. Haase, M., & Schäfer, H. (2011). Upconverting nanoparticles. *Angewandte Chemie International Edition in English, 50*, 5808–5829.
32. Li, X. M., Zhang, F., & Zhao, D. Y. (2013). Highly efficient lanthanide upconverting nanomaterials: Progress and challenges. *Nano Today, 8*, 643–676.
33. Sun, L. D., Wang, Y. F., & Yan, C. H. (2014). Paradigms and challenges for bioapplication of rare earth upconversion luminescent nanoparticles: Small size and tunable emission/excitation spectra. *Accounts of Chemical Research, 47*, 1001–1009.
34. Chen, G. Y., Qiu, H. L., Prasad, P. N., & Chen, X. Y. (2014). Upconversion nanoparticles: Design, nanochemistry, and applications in theranostics. *Chemical Reviews, 114*, 5161–5214.
35. Gai, S. L., Li, C. X., Yang, P. P., & Lin, J. (2014). Recent progress in rare earth micro/nanocrystals: Soft chemical synthesis, luminescent properties, and biomedical applications. *Chemical Reviews, 114*, 2343–2389.
36. Bhargava, R. N., Gallagher, D., Hong, X., & Nurmikko, A. (1994). Optical properties of manganese-doped nanocrystals of ZnS. *Physical Review Letters, 72*, 416–419.
37. Huang, X., Han, S., Huang, W., & Liu, X. (2013). Enhancing solar cell efficiency: The search for luminescent materials as spectral converters. *Chemical Society Reviews, 42*, 173–201 and references therein.
38. Collins, J. (2016). Non-radiative processes in crystals and in nanocrystals. *ECS Journal of Solid State Science and Technology, 5*(1), R3170–R3184 and references therein.
39. Liu, G. K., Zhuang, H. Z., & Chen, X. Y. (2002). Restricted phonon relaxation and anomalous thermalization of rare-earth ions in nanocrystals. *Nano Letters, 2*, 535–539.
40. Meltzer, R. S., & Hong, K. S. (2000). Electron-phonon interactions in insulating nanoparticles: Eu_2O_3. *Physical Review B, 61*, 3396–3496.
41. Yang, H. S., Feofilov, S. P., William, D. K., Milora, J. C., Tissue, B. M., Meltzer, R. S., & Dennis, W. M. (1999). One phonon relaxation processes in Y_2O_3: Eu^{3+} nanocrystals. *Physica B, 263–264*, 476–478.
42. Tissue, B. M. (1998). Synthesis and luminescence of lanthanide ions in nanoscale insulating hosts. *Chemistry of Materials, 10*, 2837–2845.
43. Valeur, B., & Berberan-Santos, M. N. (2011). A brief history of fluorescence and phosphorescence before the emergence of quantum theory. *Journal of Chemical Education, 88*, 731–738.
44. Rakovich, Y. P., & Donegan, J. F. (2008). Anti-Stokes photoluminescence in semiconductor nanocrystal quantum dots. In A. L. Rogach (Ed.), *Semiconductor nanocrystal quantum dot* (pp. 257–275). Wien: Springer.
45. Görller-Walrand, C., & Binnemans, K. (1996). Rationalization of crystal field parametrization. In K. A. Gscheidner Jr., & L. Eyring (Eds.), *Handbook on the physics and chemistry of rare earths* (Vol. 23, pp. 121–283). Amsterdam: Elsevier Science B.V.
46. Soukka, T., Kuningas, K., Rantanen, T., Haaslahti, V., & Lövgren, T. (2005). Photochemical characterization of upconverting inorganic lanthanide phosphors as potential labels. *Journal of Fluorescence, 15*, 513–528.
47. Teng, X., Zhu, Y., Wei, W., Wang, S., Huang, J., Naccache, R., Hu, W., Yoong Tok, A. I., Han, Y., Zhang, Q., Fan, Q., Huang, W., Capobianco, J. A., & Huang, L. (2012). Lanthanide-doped $Na_xScF^{3+}{}_x$ nanocrystals: Crystal structure evolution and multicolor tuning. *Journal of the American Chemical Society, 134*(20), 8340–8343.
48. Ding, Y. J., Teng, X., Zhu, H., Wang, L. L., & Pei, W. B. (2013). Orthorhombic KSc_2F_7:Yb/Er nanorods: Controlled synthesis and strong red upconversion emission. *Nanoscale, 5*, 11928–11932.

49. Mahalingam, V., Hazra, C., Naccache, R., Vetrone, F., & Capobianco, J. A. (2013). Enhancing the color purity of the green upconversion emission from Er^{3+}/Yb^{3+}-doped $GdVO_4$ nanocrystals via tuning of the sensitizer concentration. *Journal of Materials Chemistry C, 1*, 6536–6540.
50. Wang, J., Deng, R., MacDonald, M. A., Chen, B., Yuan, J., Wang, F., Chi, D., Andy Hor, T. S., Zhang, P., Liu, G., Han, Y., & Liu, X. (2014). Enhancing multiphoton upconversion through energy clustering at sublattice level. *Nature Materials, 13*, 157–162.
51. Judd, B. R. (1962). Optical absorption intensities of rare-earth ions. *Physics Reviews, 127*, 750–761.
52. Chang, N. C., & Gruber, J. B. (1964). Spectra and energy levels of Eu^{3+} in Y_2O_3. *Journal of Chemical Physics, 41*, 3227–3234.
53. Buijs, M., Meyerink, A., & Blasse, G. (1987). Energy transfer between Eu^{3+} ions in a lattice with two different crystallographic sites: Y_2O_3:Eu^{3+}, Gd^2O^3: Eu^{3+} and Eu_2O_3. *Journal of Luminescence, 37*, 9–20.
54. Ofelt, G. S. (1962). Intensities of crystal spectra of rare-earth ions. *Journal of Chemical Physics, 37*, 511–520.
55. Dieke, G. H. (1968). *Spectra and energy levels of rare earth ions in crystals*. New York: Wiley Interscience.
56. Blasse, G. (1992). Vibronic transitions in rare earth spectroscopy. *International Reviews in Physical Chemistry, 11*, 71–100.
57. Di Bartolo, B., & Powell, R. C. (2014). *Crystal symmetry, lattice vibrations and optical spectroscopy of solids: A group theoretical approach*. Singapore: World Scintific Publishing Co. Ptc. Ltd. 596224.
58. Hizhnyakov, V., Boltrushko, V., Pae, K., & Vaikjarv, T. (2011). Zero-phonon lines: Novel manifestations of vibronic interactions in impurity centres of solid. *Condensed-Matter Spectroscopy-Optics and Spectroscopy, 111*, 377–385.
59. Auzel, F. (1980). In B. Di Bartolo (Ed.), *Radiationless processes*. New York: Plenum.
60. Dexter, D. L., & Schulman, J. H. (1954). Theory of concentration quenching in inorganic Phosphors. *The Journal of Chemical Physics, 22*, 1063–1070.
61. Smet, P. F., Avci, N., Loos, B., Van Haecke, J. E., & Poelman, D. (2007). Structure and photoluminescence of $(Ca,Eu)_2SiS_4$ powders. *Journal of Physics Condensed Matter, 19*, 246223 (12 pp).
62. Auzel, F. (2002). A fundamental self-generated quenching center for lanthanide-doped high-purity solids. *Journal of Luminescence, 100*, 125–130.
63. Tamura, A. (1995). Smoothed density of states of electrons and smoothed frequency spectrum of phonons for a mesoscopic system. *Physical Review B, 52*, 2688–2676.
64. Schuurmans, M. F. H., & van Dijk, J. M. F. (1984). On radiative and non-radiative decay times in the weak coupling limit. *Physica B, 123*, 131–155.
65. Cesaria, M., Collins, J., & Di Bartolo, B. (2016). On the efficient warm white-light emission from nano-sized Y_2O_3. *Journal of Luminescence, 169*, 574–580.
66. Chen, X., Di Bartolo, B., Barnes, N. P., & Walsh, B. M. (2004). Thermal tuning and broadening of the spectral lines of trivalent neodymium in laser crystals. *Physica Status Solidi B, 241*((8), 1957–1976.
67. Meltzer, R. S., Yen, W. M., Zheng, H., Feofilov, S. P., Dejneka, M. J., Tissue, B. M., & Yuan, H. B. (2000). Evidence for long-range interactions between rare-earth impurity ions in nanocrystals embedded in amorphous matrices with the two-level systems of the matrix. *Physical Review B, 64*, 100201-R (1–4).
68. Liu, G. K., Chen, X. Y., Zhuang, H. Z., Li, S., & Niedbala, R. S. (2003). Confinement of electron-phonon interaction on luminescence dynamics in nanophosphors of Er^{3+}: Y_2O_2S. *Journal of Solid State Chemistry, 171*, 123–132.

Chapter 3
Nanomaterials: Basic Concepts and Quantum Models

Maura Cesaria and Baldassare Di Bartolo

Abstract The term "nanosystem" refers to a system with at least one spatial dimension scaled down to the nanometer-scale (<100 nm) and includes zero-dimensional systems (such as metallic, semiconducting and ceramic nanoparticles), one-dimensional systems (such as nanowires, nanotubes and nanorods) and two-dimensional structures (thin films or plates).

The fascinating properties of materials at the nanoscale are continuing to attract the scientific interest in many research fields at both applicative and fundamental levels. The impossible to comprehensively review large number of reports and results available in the literature demonstrates the complexity in describing all the functionalities offered by the nanosystems as well as their properties as related to their fabrication approaches apart from departure from bulk form. To be able to gain insight into the potentialities and new future perspectives offered by the nanoworld and nanotechnology, knowledge and understanding of the physical fundamentals is a necessary starting point.

In this review paper, we consider and discuss the spectroscopy of nanomaterials by pointing out differences and breaking points as compared to the bulk counterparts, the importance of the surfaces, the characteristic length-scales (De Broglie wavelength, Fermi wavelength and exciton Bohr radius) that define various confinement regimes, the physics underlying the formalism to calculate the electronic dispersion of the low-dimensionality systems and the technological benefits on the excitonic binding energy implied by low-dimensionality (zero-, one- and two-dimensionaity). The presented discussion aims at laying a foundation to further studies for a reader new to the field of nanomaterials.

M. Cesaria
Department of Mathematics and Physics "Ennio De Giorgi", University of Salento, Lecce, Italy
e-mail: maura.cesaria@le.infn.it

B. Di Bartolo (✉)
Department of Physics, Boston College, Chestnut Hill, MA, USA
e-mail: dibartob@bc.edu

© Springer Science+Business Media Dordrecht 2017

B. Di Bartolo et al. (eds.), *Nano-Optics: Principles Enabling Basic Research and Applications*, NATO Science for Peace and Security Series B: Physics and Biophysics, DOI 10.1007/978-94-024-0850-8_3

3.1 Introduction

The nomenclature “nanomaterials” refers to materials with at least one dimension in the nanometer (1 nm = 10^{-9} m) range (from 1 to 100 nanometers) that are commonly classified as zero-dimensional (0D), one-dimensional (1D) and two-dimensional (2D) structures. In general, adding one more confinement direction can affects the system properties remarkably and involve different applications. In practice, the systems of interest in nano-science are more often composed of ensembles of nanostructures instead of single nanostructures because of processing. Hence, on one hand, fabrication processes able to control such dispersion are critical and, on the other hand, dispersion in size and shape must be taken into account to interpret the experimental findings and to design devices exploiting the properties at the nanoscale.

The first lecture (titled “There’s Plenty of Room at the Bottom”) regarding possible applications for nanomaterials dates to 1959 and was given by the Physics Nobel Laureate (in 1965) Richard Feynman at the annual American Physical Society meeting on the campus of Caltech [1]. The challenging concept expressed by Feynman in this seminal lecture was miniaturization as a possibility to collect/encode a huge amount of information in increasingly small spaces. Even if no practical way to accomplish all of this was known then, this idea laid the conceptual foundations for applications exploiting very short length-scales and new properties of materials. In this sense, Feynman challenged scientists to collaborate and explore increasingly small sizes and the applicative perspectives in this realm without thinking small about the solution of future challenging problems. In 1984 Feynman was invited to give an updated version of his landmark lecture at a weeklong seminar held at the Esalen Institute. His talk, titled “Tiny Machines”, while discussing the important technological advances since 1954, demonstrated that his first outlined picture of nanoworld was prescient and farsighted.

Over the years, there has been a growing widespread interest in nanomaterials (also termed mesoscopic systems) at both fundamental and applicative level leading to a new promising research field named “nanotechnology”. This term was first introduced in 1974 by Norio Taniguchi at the International Conference on Precision Engineering (ICPE) [2] and refers to processes enabling to engineer and fabricate objects with control on the nanoscale level.

The development of nanoscience demands fundamental knowledge of the physical properties at the nanoscale level and new technologies both in fabrication processes and investigation tools as well as interaction/exchange of various research fields including physics, chemistry, material science and engineering, just as previewed by Feynman. Efforts of fundamental research are presently focused on understanding the origin of the mechanical, chemical, reactivity, catalysis, thermal, electrical, optical, and magnetic properties of nanomaterials, which strongly differ from the ones of the large-scale structures of the same composition (i.e., bulk counterpart). Changes of the bulk properties involved by size-effects can be observed as regards

- chemical properties: reactivity and catalysis
- thermal properties: melting temperature and thermal conduction
- mechanical properties: adhesion and capillary forces
- optical properties: band-structure, absorption and scattering
- electrical properties: Coulomb charging, quantized conductance, quantum tunneling current
- magnetic properties: superparamagnetic effect

On the other hand, technology aims at engineering and controlling nanomaterial shape and size, size- and shape- dispersion and composition. Hence, since investigation of the nanoworld and nanotechnology takes advantage from imaging, measuring, modeling and manipulating matter at the nanometer scale, powerful observation tools, such as electron microscopes (scanning electron microscope (SEM) and transmission electron microscope (TEM)) [3–6], scanning-tunneling microscope (STM) [7, 8], atomic force microscope (AFM) [9] and scanning near-field optical spectroscopy (SNOM) [10, 11] were decisive in enabling and prompting progress in the nanoscale field. STM uses a sharp metal tip brought close (0.3÷1 nm) to the surface to be investigated to measure a detectable electron tunneling current. The "tunneling" of electrons (quantum tunneling effect) between the tip and the substance creates a current (flow of electrons) that lets acquire an image of the sample by changing the current intensity over the time. The reachable resolution (0.1 ÷1 nm) enables to observe even single atoms. Today's STMs allow not only to see atoms but also to measure the electronic density of states of single nanostructures as well as to move atoms around and arrange them in a specific designed configuration. Whereas STM works only on electrically conductive samples, AFM can image the surface of any (inorganic, organic, conductive and insulator) material with such a high resolution to define the position of individual atoms. The working principle of AFM can be summarized as follows. A tiny tip at the end of a flexible micro-cantilever is scanned in a raster pattern over the surface to be imaged by a piezoelectric actuator with sub-Angstrom accuracy. Cantilever deflection results from the tip-sample (short-range chemical binding, van-der Waals, electrostatic or magnetic) interaction which is kept to a fixed value by a feedback circuit in almost all operating modes of the AFM instrument. The cantilever deflection is carefully monitored by a laser beam reflected off the back of the cantilever towards a four-quadrant split position-sensitive photodetector acting as amplifier of the cantilever deflection.

Near-field optical probing (SNOM) enables to image and measure the optical properties of single semiconductor quantum nano-structures, even distributed at a spatial density of the order of $100/\mu m^2$, with spatial subnanometer resolutions, i.e., well beyond the diffraction limit of light. Conventional optical microscopes obtain the image of an object without being able to gain information about fine (subwavelength) features due to the diffraction limit [12], meaning that detail can't be greater than 0.61 λ/NA where λ is the probing wavelength and NA is the numerical aperture of the microscope optics. To overcome the limits of conventional microscopes the sample should be illuminated by a small aperture with diameter d much shorter than the probing wavelength and at a distance much shorter than d from the surface to be imaged (near-field region). SNOM technology exploits a tapered optical fiber with a metalized terminal part having diameter d of a few tens of nanometers and enables to collect local emission from the surface region just underneath the microscope aperture. Since a few hundreds of nW of laser light shine the sample, a high collection efficiency is a prerequisite to be able to detect very low level photoluminescence signals by SNOM spectroscopy.

Undoubtedly, nanotechnology would not exist without the above mentioned powerful tools and many fascinating properties of mesoscopic systems would never have been disclosed. In this respect, while the fundamental behavior of bulk materials is inherently determined by structure and composition, at the nanometer length-scale peculiar phenomena occur strictly related to either spatial confinement of the electronic as well as phonon wave-functions or surface-effects. For example, as it will be thoroughly discussed later, when at least one length (L) of a semiconductor system along a spatial direction is comparable with a characteristic length-scale (the so-called De Broglie electronic wavelength of the corresponding bulk material), then the energy spectrum becomes quantized along the confinement direction. Such phenomenon, termed "quantum size effect", involves a larger band gap and a blue shift of the absorption onset with decreasing size as compared to the bulk counterpart, meaning that the optical properties of a nanostructure are tunable as a function of both the confinement length L and number of the confinement directions (0D, 1D and 2D nanostructures).

A further consequence of low-dimensional confinement is the increase of the binding energy of an exciton (i.e., an electron-hole pair bounded by the Coulomb interaction) resulting from the enhanced Coulomb coupling between electrons and holes achievable by carrier localization in a nanostructure. This low-dimensional effect is a prerequisite to be able to observe excitonic effects in semiconductor mesoscopic systems working at room temperature and exploiting optical non-linearity in optoelectronic devices.

Moreover, discrete phonon density of states is characteristic of the nanoscale [13] and low-frequency acoustic phonon modes are cut off depending on the nanostructure size [14–17].

Since electron-phonon interaction determines the thermal conductivity and affects carrier dynamics in quantum-confined systems, different thermalization rules and processes (multiphonon processes [18, 19], Coulomb interactions and phonon-bottleneck effects [20, 21]) have to be considered in the nanoworld.

Turning to metal nanoclusters/nanoparticles, for which there is no band-gap, another nanoparticle phenomenon, known as *plasmon resonance*, is active that results from the resonant oscillation of surface conduction electrons stimulated by incident light with the plasma wavelength [22–29]. Scaling down to dimensions of a few nanometers, surface plasmon resonance disappears in metal nanostructures. In fact, as the spacings between adjacent energy levels (referred to as the Kubo gap) become comparable to the thermal energy $k_B T$ (where k_B is the Boltzmann constant and T the absolute temperature), a shift in conductive properties, from metallic to semiconducting and insulating, results for decreasing size [30–32].

Metallic nanoparticles have a huge potential in nanotechnology thanks to optimized synthesis approaches enabling control on their size, shape and degree of aggregation. Under resonant excitation, an enhancement of the local electric fields close to the metal nanoparticle surface occurs that is the basis of important applications such as surface-enhanced Raman Spectroscopy (SERS) [26, 33–35]. Such an active recent research field uses various chemical functional groups favoring bonds/interactions with antibodies, ligands, and drugs of interest in biotechnology, nano-medicine and diagnostic imaging. Furthermore, other applications exploit metal nanoparticles with modified composition and core-shell structure [36]. Metal nanoparticles are also interesting for sensor applications and enhanced catalytic activity involved by increased fraction of surface atoms and surface curvature with decreasing size as well as particle-particle and particle-chemical interactions [37].

Also, surface effects are responsible for fundamentally and technologically important experimental findings such as pre-melting (i.e., melting initiated at the surface) and size-dependent depression of the melting temperature (i.e., lowered melting temperature as compared to the bulk counterpart) [38–46].

A further amazing phenomenon peculiar of the nanoworld is the substantial increment of the thermal conductivity exhibited by nanofluids (very low concentrations of nanostructures dispersed in solvents with suitable relative thermal conductivity) which varies with size, shape, and material of the nanostructures as well as difference of thermal conductivity between bulk material and base fluid [47].

On the other hand, the evidence of quantum size effects makes the Planck's black-body model unable to describe the thermal emission spectrum at the nanoscale (at particle dimensions smaller than the radiation wavelength), where the Stefan–Boltzmann law yields strongly overestimated results [48–50]. Moreover, nanoworld also involves magnetic properties different as compared to the bulk materials [51, 52].

Extensive, even if not exhaustive because of continuous progress and developments, literature is available about nanoscience (nanomaterials and nanotechnology) and a comprehensive review of the existing scenario is outside the scope of this paper. The subject of quantum-confined structures is wide, in continuous progress and involves fundamental quantum mechanical models, applications in different fields and processing of materials and architectures at the nanoscale.

Instead, in the following we overview and critically discuss the spectroscopic properties of nanomaterials at the fundamental level with a focus on the underlying physics and breaking points with respect to the bulk counterparts. We aim at introducing a beginner reader to the fascinating world of low-dimensional systems from the fundamental standpoint, i.e., accounting for the understanding of low-dimensional physics as a tool for gaining insight in the potentialities and future perspectives of the nanoworld. In this perspective, we consider the surface-related effects (increased surface-to volume ratio (SVR) and the related modified chemistry and thermal stability of nanomaterials) and their technological importance while scaling down dimensions to the nanoscale level. Furthermore we overview and discuss basic concepts such as optical confinement characteristic lengths (De

Broglie wavelength, Fermi wavelength and exciton Bohr radius), the meaning of the classification-scheme of low-dimensionality systems (zero-, one- and two-dimensional systems), quantum mechanical models to mathematically describe the electronic structure and quantum size effects in semiconductor nanostructures as well as excitons in nanostructures, and the size-induced dramatic changes of the electronic dispersion of metallic nanostructures.

3.2 Basic Concepts Underlying Nanoscience

Nanoscience deals with phenomena and processing involving systems with nanometer dimensions (termed nanostructures or low-dimensional structures or mesoscopic structures). Drastic (from macroscopic to nanometer-scale) reduction in size results in physicochemical (electrical, mechanical, chemical, thermal, magnetic, and optical) properties that remarkably depart from the ones of the bulk counterpart whenever a characteristic dimension becomes comparable with some fundamental length-scale of the material. In this regime, the spatial confinement involves changes of the electronic dispersion, usually falling into the category of quantum-size effects, that can be detected in the optical response of nanostructures and lead to new device performances and technologies. Nanoscale spectroscopy is not only size-dependent but can also be controlled by the number of confinement dimensions (i.e. shape of the nanostructure) leading to atom-like spectrum when the confinement occurs in all directions (zero-dimensional structure or quantum dot or nanoparticle). At the nanoscale, tuning of size and shape also affects the surface energy: the smaller the nanostructure and the larger the number of edges, the larger is the contribution of the surface energy, a fact which decreases the cohesive energy. Since the cohesive energy dominates phenomena such as melting, evaporation and phase transition, scaling down to nanometer size involves modified thermal stability as compared to the bulk case. All of this has very interesting fundamental and applicative perspectives that will be examined.

In the present section of our paper we introduce the reader to nomenclature and basic concepts in nanoscience such as characteristic length-scales in solids (De Broglie wavelength, exciton Bohr radius and Fermi wavelength), their interplay with confinement effects leading to the classification of the mesoscopic systems as 2D-, 1D- and 0D-dimensional systems, size-induced (quantum-size effect) and surface-related effects.

3.2.1 Characteristic Scale-Lengths in Bulk Solids

The regimes where the properties of materials became size- and shape-dependent are defined by a set of characteristic length scales and their relationship with the physical dimensions of the system. In the following we briefly overview the basic concepts of the solid state physics enabling to introduce such relevant length scales (i.e., De Broglie wavelength and exciton Bohr radius in semiconductors, and Fermi wavelength in metals).

In the framework of the solid state theory a simple model, the so-called free-electron model, is often applied to solve for the Scroedinger equation of atomic and metal systems. Such a model assumes independent and free electrons obeying the Pauli-exclusion principle (Fermi-Dirac distribution function). To calculate systems described by a periodic potential (semiconductor crystals), it has been modified to include corrections to the pure free-electron model accounted for the band-structure and carriers' effective mass [53].

Based on the solid state physics [53], materials with ordered (periodic) lattice structure can exhibit electronic energy dispersion organized according alternate bands of allowed and forbidden energies. The highest fully occupied band and the lowest empty or partially occupied band are termed valence band and conduction band, respectively, and are separated by a continuum of electronic forbidden energies referred to as energy band-gap (Fig. 3.1). The width of such a band-gap (E_g) as compared to the thermal energy k_BT is of interest to classify materials as insulators or semiconductors. In fact, when in the ground state (zero absolute temperature, $T = 0$ K) E_g is much too wide to allow electrons from the valence band to be thermally promoted to the conduction band at room temperature ($k_BT \approx 25$ meV at $T = 300$ K) then a material is classified as insulator. In this case the valence band (and all bands of lower energy) is fully occupied, and the conduction band is empty (implying vanishing conductivity) even at temperature almost higher than the room temperature. Instead, a semiconductor material is insulator at $T = 0$ K but its energy band-gap smaller compared to insulators ($\sim$a few eV) enables the conduction band to be thermally populated around its minimum at room temperature, which implies electric conductivity. In addition to insulators and semiconductors, important materials in condensed matter physics are metals that do not have forbidden energy gap, i.e., at $T = 0$ K their conduction electrons occupy

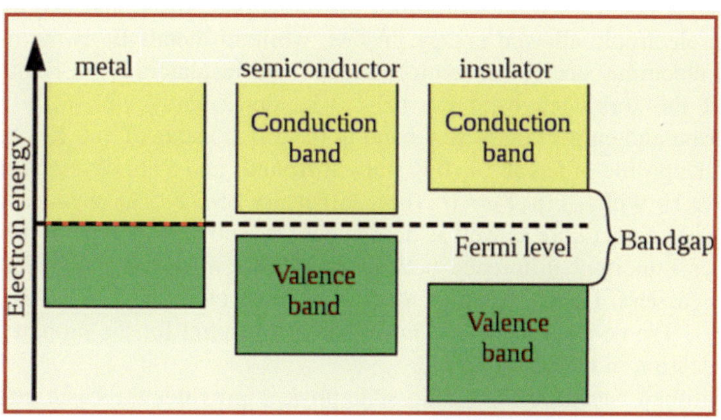

Fig. 3.1 Schematics of the band structure of a metal, semiconductor and insulator material

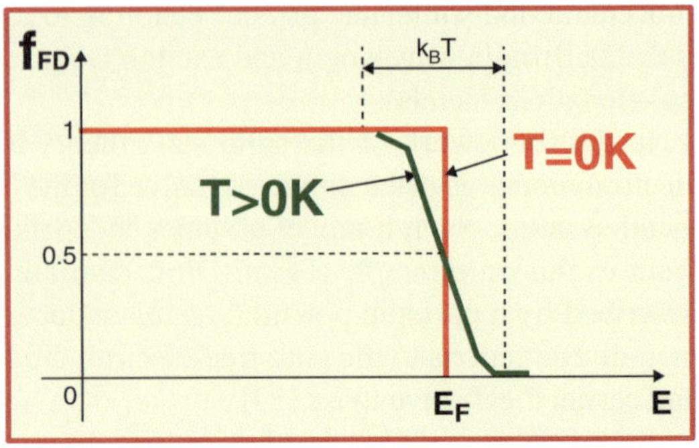

Fig. 3.2 Fermi-Dirac distribution function

a continuum of energy levels (the so-called "Fermi sphere") with maximum energy corresponding to the so-called "Fermi energy"(E_F). In metals the allowed electronic states with energy higher than E_F are depopulated at $T = 0$ K and can be thermally populated at any non-vanishing low temperature due to the absence of a forbidden energy gap. In general, E_F is defined as the highest energy level below which all electronic energy levels are filled and above which all electronic energy levels are empty at $T = 0$ K. In terms of statistics, E_F is defined by the Fermi-Dirac statistics which describes the probability density that fermions (like electrons) can occupy an allowed state at energy E as a function of the temperature (Fig. 3.2). Being T the absolute temperature, the Fermi-Dirac distribution function f_{FD} has the following mathematical expression:

$$f_{FD}(E, T) = \left(1 + e^{\frac{-(E-\mu)}{k_B T}}\right)^{-1}$$

where k_B is the Boltzmann's constant and μ is the so-called "total chemical potential" (tending to E_F at $T = 0$ K). Based on the Fermi-Dirac statistics, it results $f_{FD}(E_F, T = 0\ K) = 1/2$, i.e., the probability of occupation at E_F is 1/2. Notably, E_F is the ground state ($T = 0$ K) limit of the chemical potential μ and may be or may not be an electronic allowed energy. In fact, while in metals E_F is the maximum occupied electronic energy, in semiconductors and insulators E_F is located in the middle of the forbidden band-gap (Fig. 3.1), consistently with fully occupied valence band and empty conduction band in the ground state ($T = 0$ K). At $T > 0$ K the step-like profile of f_{FD} at $T = 0$ K smooths/rounds and a tail develops extending towards $E > E_F$ with $f_{FD}(\mu, T) = 1/2$ (Fig. 3.2) at any $T > 0$ K. The chemical potential μ shifts with temperature as well as doping content and type (*n*-type or *p*-type).

Another important difference between band-gap materials and metals is the nature of carriers, i.e. electrons in metals and both electrons and holes in semiconductors. The concept of hole is introduced to account for the properties of the missing electron in the valence band.

The outlined background of solid state physics provides the basic information necessary to introduce and discuss characteristic length scales in solids.

In semiconductors, the characteristic length scale of carriers (electrons and holes) is the De Broglie wavelength λ_{DB} given by

$$\lambda_{DB} = h/p$$

where $\mathbf{p} = m^* \mathbf{v}$ is the momentum of a carrier with effective mass m* (determined by the crystal lattice) and velocity **v** and $h = 6.62 \times 10^{-34}$ Js is the Planck's constant. In general, holes have larger effective mass and lower velocity than electrons in a given material. For example, in GaAs, the effective mass of electrons and holes and their thermal velocity are given by $m_e^* = 0.063\ m_o$, $m_h^* = 0.53\ m_o$, $\mathbf{v}_e = 4.4 \cdot 10^5$ m/s and $\mathbf{v}_h = 1.8 \cdot 10^5$ m/s. Based on the definition, after substituting $h = 6.62 \times 10^{-34}$ Js and 1 J = 1 kg m^2/s^2, the De Broglie wavelength can be expressed as $\lambda_{DB} = (727 \times 10^3\ nm)/(m^*/m_o)v$, where m^*/m_o is the carrier effective mass expressed in units of m_o ($m_o = 9.11 \times 10^{-31}$ Kg, mass of a free electron) and v is the value of the carrier velocity expressed in m/s. For example, in the case of electrons (holes) with isotropic effective mass $m_e^* = 0.1m_o$ ($m_h^* = 0.4m_o$) and thermal velocity $v_e = 10^5$ m/s, the electron (hole) De Broglie wavelength is equal to $\lambda_{DB,e} \approx 73$ nm ($\lambda_{DB,h} \approx 18$ nm).

In typical semiconductors with electron effective mass $m_e^* = (0.01–1)\ m_o$, a De Broglie wavelength of the order of ~730–73 Å (much larger than lattice constant) results at T = 300 K. For decreasing temperatures down to T = 4 K, λ_{DB} increases to 10^2–10^4 Å because of the lower electron thermal velocity. Furthermore, λ_{DB} can depend on the doping conditions due to changes in the carrier density.

In metals, the characteristic length-scale is termed Fermi wavelength (λ_F) and defined as De Broglie wavelength at the Fermi edge E_F (the highest occupied energy level at T = 0 K: any energy state below E_F is occupied and every level above E_F is empty). Based on its definition, the Fermi wavelength is expressed as follows

$$\lambda_F = h/mv_F$$

where $\mathbf{v}_F$ is the electron velocity at the Fermi energy E_F that, in terms of the electron density n at T = 0 K, is given by

$$E_F = (1/2)\, m{v_F}^2 = \left(3\pi^2\right)^{2/3}\left[h^2/\,(2m)\, n^{2/3}\right]$$

For increasing temperature above zero degrees Kelvin some electrons will be excited to higher states thus causing a blue-shift of the energy of the topmost filled level (the chemical potential energy μ) and increasing electron thermal velocity.

For example, in the case of copper (Cu) which is a monovalent element with atomic number Z = 29 and electronic configuration Cu:[Ar] $3d^{10}\ 4\ s^1$, the Fermi velocity is $v_F = 1.57 \times 10^6$ m/s. From the equations $E_F = (1/2)m{v_F}^2$ and $\lambda_F = h/mv_F$, after substituting $m \sim m_o$ [54], it results that $E_F \approx 7$ eV and $\lambda_F \approx 0.46$ nm.

Table 3.1 Fermi energy (E_F), Fermi velocity (v_F), effective mass normalized to the free electron mass (m^*/m_o) and Fermi wavelength (λ_F) of some common metals

Element	E_F (eV)	v_F (10^6 m/s)	m^*/m_o	λ_F (nm)
Li	4.74	1.29	1.28	0.44
Na	3.24	1.07	1.20	0.57
K	2.12	0.86	1.12	0.75
Cu	7	1.57	1.01	0.46
Ag	5.49	1.39	0.99	0.53
Au	5.53	1.4	1.10	0.47
Al	11.7	2.03	1.18	0.30
Zn	9.47	1.83	0.85	0.47

In metals, since the density of the conduction electrons ranges between approximately 10^{28} and 10^{29} electrons/m^3, a Fermi energy of the order of 2–10 eV results that depends on the atomic valency and the electronic density. As Table 3.1 shows, in many common metals, the electron effective mass is close to m_o and $\lambda_F \sim 0.3$–1 nm.

Another characteristic length scale in bulk solids is the average physical separation between the electron and hole, referred to as the "exciton Bohr radius", in a bound electron-hole (e-h) pair (termed "exciton"). The absorption of a photon by an interband transition in a bulk semiconductor or insulator can create an electron in the conduction band and hole in the valence band correlated through the Coulomb interaction. Such a state of an electron and hole bound by the electrostatic interaction is a neutral e-h complex called "exciton" [55–61]. The bulk exciton Bohr radius $a_{ex}{}^{bulk}$ is given by the following formula

$$a_{ex}{}^{bulk} = a_H \varepsilon_r \,(m_o/\mu_r)$$

where $a_H = 0.53$ Å is the Bohr radius of H atom, ε_r is the relative dielectric constant, m_o is the mass of a free electron and μ_r is the reduced mass of the e-h system ($\mu_r = m_e{}^* \, m_h{}^*/(m_e{}^* + m_h{}^*)$, being $m_e{}^*$ and $m_h{}^*$ the effective masses of the electron and hole, respectively).

Two kinds of excitons can be distinguished, termed Wannier–Mott excitons and Frenkel excitons (Fig. 3.3). The former ones, also called "free excitons", are delocalized (i.e., weak e-h coupling) states with binding energy $\sim$10 meV and mainly occur in semiconductors (large dielectric constant $\varepsilon \sim 10$). The latter ones, also called "tight bound excitons" are characterized by strong e-h coupling with binding energy of $\sim$0.1–1 eV (form in insulator and molecular crystals ($\varepsilon \sim 2$)), localization on specific sites and hopping motion from one site to another. Basically, differences in the binding energies implies that $a_{ex}{}^{bulk}$ of the Wannier–Mott excitons is larger than $a_{ex}{}^{bulk}$ associated to Frenkel excitons (Fig. 3.3).

Since in the following we will consider semiconductors, we will deal with Wannier–Mott excitons. The common technological important semiconductors such as Si, Ge, GaAs, GaP, InP, CdSe and InSb, have exciton Bohr radia of 4.9, 17.7, 14, 1.7, 9.5, 3 (5.07), and 69 nm, respectively.

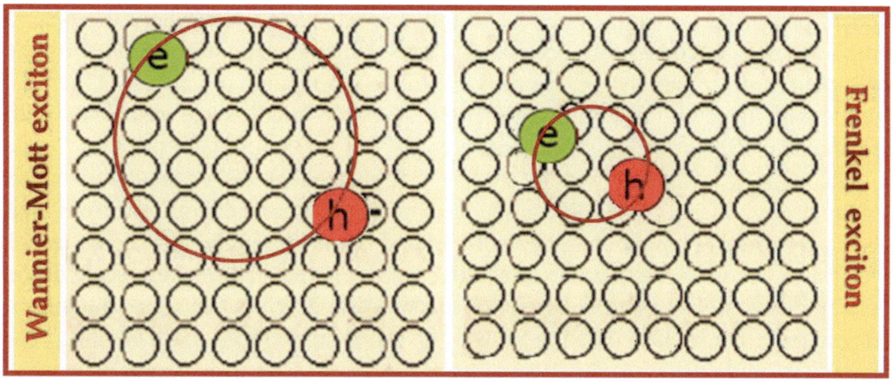

Fig. 3.3 Schematic representation of the two kinds of excitons

Definitively, to summarize, typical solid-state bulk systems can be characterized by their electronic dispersion (band-structure) and the Fermi-Dirac distribution function. Furthermore, characteristic length scales can be defined, namely a De Broglie wavelength of carriers (electrons and holes in semiconductors and electrons in metals) in the picture of the fermion gas and the exciton Bohr radius in semiconductors which is related to e-h pairs coupled by electrostatic interaction.

Once assessed that characteristic length scales exist in solid-state bulk systems and they are of the order of nanometer, it is straightforward to investigate their interplay with the size of the nanostructures in regard to fundamental properties of materials.

3.2.2 Quantum Confinement Effects and Classifications of the Nanostructures

The expression "quantum confinement effects" refers to changes in the electronic band structure and density of states (DOS) resulting from scaling the size of a system down to ultra-small length scales. The meaning of "ultra-small" is critical and can be specified once the interplay between characteristic length scales and size of the system is established. Notably, both optical and electric (transport regimes) properties of materials can be affected by reduction of their size to the nanometer level. On this small length scale electron transport is ruled by quantum mechanics (electron wave behaviour and quantum tunnelling are demanding) and novel phenomena (such as quantized conductance, Coulomb blockade and single electron tunnelling) [62–76] occur which are the subject of "nano-electronics" [77, 78]. Further length scales must be defined to account for the size-dependent electric properties that we will not consider in our discussion focused on the size-dependent optical properties of nanostructures. In this respect, the regime where the optical properties of semiconductors became size and shape dependent is ruled/set by the bulk exciton Bohr radius a_{ex}^{bulk} or the De Broglie wavelength of carriers in the bulk counterpart. Instead, the Fermi wavelength λ_F is the length-scale associated with confinement effects in metal nanostructures.

The most known and widely studied size-dependent effect is the co-called "quantum-size effect", i.e., the confinement-induced discrete energy spectrum, that occurs whenever a length characteristic (termed "confinement length" hereafter) of the system becomes comparable or smaller than a characteristic length-scale of the bulk counterpart. For example, a semiconductor sample with at least one dimension comparable to or smaller than the De Broglie wavelength of the corresponding bulk material exhibits discrete electron energy along the k-space direction associated to the confinement direction. On the other hand, while in bulk semiconductors, the exciton (Coulomb-bounded e-h complex) can move freely in all directions, an e-h pair forms and feels the system boundaries when a relevant length of a semiconductor is reduced to the same order as the bulk exciton Bohr radius a_{ex}^{bulk} (a few nanometers). In this situation, quantization of the exciton energy spectrum results.

Quantum size effects are also known for metal nanoparticles. However, since the typical De Broglie wavelengths λ_{DB} and exciton Bohr radius a_{ex}^{bulk} in semiconductors are larger than the Fermi wavelength λ_F in metals, then λ_F is the smallest length scale relevant to quantum-size effects and effects of spatial confinement are expected to be more easily observable in semiconductors than in metals. Indeed, being λ_F usually very short in metals (Table 3.1), then localization of the energy levels can be observed in metal nanostructures with size below 2–4 nm, that is whenever the level spacing exceeds the thermal energy ($\sim$25 meV at T = 300 K).

Commonly, nanostructures are classified based on their dimensionality, meaning the number of dimensions (one, two and three) scaled down to the nanometric-length. According to this criterion/standard, mesoscopic structures are termed two-dimensional or briefly 2D (confinement along one direction), one-dimensional or briefly 1D (confinement along two directions) and zero-dimensional or briefly 0D (confinement along three directions). Consistently with this classification, three-dimensional (3D) structure means bulk system. More precisely, the dimensionality of a mesoscopic system refers to the number of degrees of freedom in the particle momentum (Fig. 3.4). Hence, the nomenclature 2D-structure indicates confinement

Fig. 3.4 Classification of the nanostructures based on the number of confinement spatial directions

of the particle motion along one direction (z-direction in Fig. 3.4) and free motion along the other two independent directions (x and y directions in Fig. 3.4).The term 1D-structure means free movement along only one direction (x-direction in Fig. 3.4), i.e., confinement along two directions (y and z directions in Fig. 3.4). Confinement along all the directions defines 0D-structures and the nomenclature 3D structure refers to free motion along any direction. Generally 2D, 1D and 0D systems are also referred to as "quantum well" (QW), "quantum wire" (QWR) and "quantum dot" (QD), respectively. Sometimes 0D nanostructures are termed "quantum boxes" or "artificial atoms" too because electron confinement along all directions implies atom-like electron dispersion, i.e., spectrum consisting of discrete lines accessible to the system at characteristic energies.

In practice, the idea underlying the introduction of nanostructures was to create artificially, by combining different semiconductor materials, potential wells and barriers for confining carriers (electrons and holes) along one, two or three directions. The shape of the confining potential in conduction and valence bands is determined by the offset of band gap between different materials used as well as by the geometry of the structure. Nanostructures are the key systems for band gap engineering purposes in optolectronics: since the band gap energy of semiconductors determines their emission wavelength, new materials and combination of materials rather than alloys of semiconductors enable to carry out custom-designed emission energies.

3.2.3 *Surface to Volume Ratio and Applications*

In addition to quantum-size effects, another peculiarity of nanomaterials is the increasing importance of their surfaces with decreasing size that involves chemical and physical properties remarkably different from their macro counterparts. For example, while bulk gold does not react with many chemicals, gold nanoparticles can be used as highly reactive catalysts to speed up reactions due to their size-related reactivity [79–82]. This behavior can be ascribed to the increased surface to volume ratio (SVR) and role played by surface atoms for decreasing size, which is a peculiarity of the nanoworld.

Let there be a cube with sides L and a sphere with radius R (Fig. 3.5). For the cube, the volume and surface area is $V = L^3$ and $S = 6L^2$, respectively. For the sphere, the volume and surface area is $V = (4/3)\pi R^3$ and $S = 4\pi R^2$, respectively. As the characteristic size (L or R) increases, the corresponding increase in the surface area is squared and the increase in the volume is cubed. Hence, surface area and volume decrease for decreasing size with volume increasingly at a much faster rate than the surface area. Instead, since the SVR (S/R) scales with the inverse of the characteristic size (L or R), it increases very fast for decreasing size and the greater the characteristic size, the lower the SVR. In the case of characteristic length of the order of nanometer, a very large SVR would result, meaning that the role and weight of surfaces becomes important in the nanoworld. In fact, increased SVR involves higher contribution of the surface energy to the overall energy, i.e., decreased

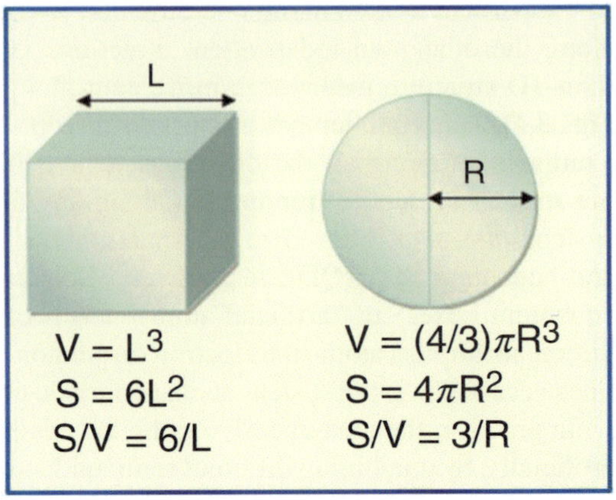

Fig. 3.5 Surface to volume ratio (S/V) of a cube and a sphere

cohesive energy [44, 83]. Moreover, the number of weakly coordinated surface atoms increases with decreasing size [44, 83, 84]. Such increased SVR and number of surface atoms in mesoscopic systems involve important phenomena peculiar of the nanoworld, such as increased surface chemical reactivity and depression of the melting point.

In practice, given a system of nanostructures (for example nanoparticles), the remarkably increased number of surface atoms would favor interactions (chemical reactions and bonds) between the surfaces of contiguous nanostructures as well as between the nanostructures and other chemicals. Therefore, nanoscale sizing maximizes possible reactivity because it maximizes the number of reaction sites. On the other hand, reduced coordination of the surface atoms (i.e., fewer neighboring bonds than inner atoms) reduces the cohesive energy that dominates the thermal stability (melting, evaporation and phase transition) [44, 85]. Since melting stabilizes the total energy by reducing the total surface energy, isolated and substrate-supported metallic, organic and semiconductor low-dimensional systems with relatively free surface fraction melt at temperatures lower than the melting point of the corresponding bulk material [41] and melting can be a surface-initiated process [86–89]. In this respect, the nomenclature "surface melting" or "pre-melting" is used because the mechanism can be modeled as the formation of a disordered "quasi-liquid" layer (where "quasi-liquid" indicates both liquid-like and solid-like) on the surface of an annealed solid system that thickens with increasing temperature and drives the complete melting of the nanostructure [90–92].

Surface melting is also responsible for collapse/breaking up of a continuous film, with thickness ranging from a few nanometers to hundreds of nanometers and deposited on an inert substrate, into patterns that spread, coalesce and decay into discrete droplets under high-temperature heating [93–96].

Such a dewetting phenomenon onsets at temperature lower than the one of bulk melting and is driven by the reduction of surface energy, optimization of shape and reduction of the heat for melting [96]. Therefore, pre-melting, melting and dewetting of low-dimensional systems are stabilizing processes prompted to minimize the surface energy in the presence of competition between surface and bulk free energy. Also, shape and sharp surface features (surface curvature, stepped surfaces and corners) favor reduced coordination of the surface atoms (i.e., fewer neighboring bonds than inner atoms) and surface diffusion, thus leading to increased surface-energy (reduced cohesive energy) [97–99] and reshaping. In this respect, "edge pre-melting" (i.e., pre-melting appearing at vertices and edges) is the first step of surface melting [100] and involves initial getting rounder and broaderly stepped features, due to mostly liquid-like atoms causing increasing mobility of vertex and edge atoms [97, 98, 101]. Another melting phenomenon peculiar of the nanoworld is "superheating", meaning that supported mesoscopic systems can melt below or above the melting point of the corresponding bulk crystal depending on the interface structure and matrix material [102–105]. To predict and explain the depression of the melting point at the nanoscale, a number of theoretical models have been developed, including the Thomson's and Pawlow's models [38, 106–109], the Lindemann criterion [110] and approaches based on the Lindemann melting rule [111–114], phenomenological approaches considering solid–liquid equilibrium (such as the homogeneous melting (HGM) model [38, 41, 115],the liquid skin melting (LSM) model and the liquid nucleation and growth model (LNGM) [111, 115–117], analytical models based on the increase of SVR and reduction of the cohesive energy [118–120], the liquid drop model [83, 121] and the bond-order-length-strength (BOLS) correlation mechanism [83].

Essentially two main classes of approaches can be identified to investigate the thermal stability of nanosystems: thermodynamic approaches, also termed top-down approaches, and bottom-up approaches.

Thermodynamic approaches apply basic concepts of classical thermodynamics (solid-liquid equilibrium, phase change and state variables, Gibb's free energy) to develop equations relating the melting point to the size and shape of a nanoscale-system.

Bottom-up approaches are based on computer modeling (mainly Molecular Dynamic (MD) simulations) and are becoming the mainstream method to investigate the melting behavior of nanoparticles/nanoclusters [91, 122–128]. Even if MD simulations can picture the melting dynamic over time scales of the order of picoseconds (i.e., cannot show the true melting dynamics of a system) and model systems containing a limited numbers of atoms, however they let a more accurate estimation of the melting point and melting processes (pre-melting, edge-melting and influence of shape) at the nanoscale. Indeed, the thermodynamic point of view (that predicts melting at a well-defined (equilibrium) temperature) contrasts with the experimental evidence that shows the lack of a precise melting temperature and, instead, the occurrence of a temperature interval over which melting phenomena can be observed [129, 130]. In the presence of melting point depression, a size-dependent broadening of the range of temperature in which melting occurs can be observed, being the broadening more evident for decreasing particle size.

On the other hand, based on the thermodynamic equilibrium between bulk solid state and bulk liquid state, the change of the melting temperature $\Delta T_m = T_{m,B} - T_{m,NP}$ (where $T_{m,B}$ and $T_{m,NP}$ are the bulk phase and size-dependent melting point, respectively) can be found to scale linearly as a function of the size R of the nanoparticle (subscript NP). In this calculation isolated, homogeneous and spherical nanoparticles are assumed. Indeed, the experimental findings indicate deviation from a linear trend for small enough ($R < 10$ nm) size and in the case of supported semiconductor nanoparticles [131–133]. Moreover, shape, substrate and free surface, crystalline orientation and dimensionality must be also considered to realistically model the size-dependent melting process. Therefore, in addition to the difficulty in determining the melting point from experiments, spreading use of MD simulations as compared to thermodynamic approaches is also favored by the conceptual limits of the thermodynamic approach. In this respect, the key question arising from the failing predictions of the conventional thermodynamics applied to the nanoworld is "May classic thermodynamics be applied to model the thermal stability of nanomaterials or, instead, nano-thermodynamics should be developed?"

Even if key issues still need to be understood about the thermal stability of nanomaterials, depression of melting point has important technological applications such as low-temperature (far below the bulk melting temperature) melting of nanostructure distributions resulting in a continuous dense film [130] and large area soldering applications leading to the packaging technology termed "wafer bonding" [134]. Wafer bonding is an emerging technology for fabricating complex three-dimensional structures at temperatures (a few hundreds of Celsius degrees) compatible with the ones required in the processing conditions leading to integration of multi-function microelectronic systems.

Annealing of arrays of nanorods can be used to obtain a dense continuous film going through several temperatures and structure (disordered phase, re-crystallization and formation of larger grains) regimes before the heated nanorods completely melt and collapse into a continuous phase [130, 134]. It is worth noticing that tilt angle change of the nanorods as well as tuning of their diameter and degree of package (reduced gaps between neighboring nanorods) can help their collapse and disappearing (as a result of coalescence) and decrease the annealing temperature requested to carry out a dense continuous film. At the applicative level all of this enables to form a continuous ordered film at much lower temperature as compared to the post-deposition annealing temperature of the counterpart bulk film.

The depression of the melting point can be also exploited to carry out wafer bonding, where two wafers, each consisting of a supported array of nanorods grown by an oblique angle deposition technique, are superposed with the nanorods layer in contact to act as an adhesive upon annealing. This two wafer structure is subsequently subjected to external pressure and heating (even at lower temperature than the collapse temperature of the nanorods) in order to favor first the formation of nano-structured bond at the nanorod/nanorod interface and then eliminate such a bonding interface and structure voids. The resulting sintering of the nanostructures and the formation of a dense bonding layer is not only size-dependent but depends on the bonding pressure too.

Since temperatures as low as a few hundreds Celsius degrees are involved in wafer bonding, this approach is an emerging technology with great promises in advancing integrated 3D chip technology.

3.3 Quantum Mechanical Models in Nanoscience

Low-dimensional systems are excellent examples of quantum mechanics in action [135]. Quantum confinement effects result essentially from changes in the electronic dispersion as compared to the bulk counterpart due to the influence of ultra-small (nanometric) length-scale constraints on the electron wavefunctions. Quantum size effects occur whenever a characteristic geometric-length of the system becomes comparable or smaller than a characteristic length-scale of the bulk counterpart. As a result of such "geometrical" constraints, electrons "feel" the presence of the structure boundaries and changes in its size as well as shape and adjust their energy spectrum that becomes discrete. Depending on the material (semiconductor, insulator or metal) and dimensionality of the nanostructure (2D, 1D or 0D nanostructure), an electronic dispersion with different expression results.

As previously discussed, the bulk characteristic length-scales are the De Broglie wavelength (λ_{DB}) (termed Fermi wavelength (λ_F) in metals) and the Bohr radius of an exciton ($a_{ex}{}^{bulk}$) in semiconductors. The basic building-block nanostructures are QWs (2D dimensional system), QWRs (1D dimensional system) and QDs (0D dimensional system). In any case, a quantum-mechanical treatment is demanding to be able to predict the physical behaviour of a low-dimensional structure, meaning that the Schrödinger equation must be solved in presence of a given quantum-confinement potential.

In general, in the Schrödinger description of a system the following equation must be solved

$$\left(-\frac{\hbar^2}{2m^*}\nabla^2 + \mathrm{V}(\mathbf{r},\mathrm{t})\right)\varphi(\mathbf{r},\mathrm{t}) = i\hbar\frac{\partial\varphi(\mathbf{r},\mathrm{t})}{\partial t}$$

where $\mathbf{r}$ and t are the spatial and temporal coordinates, respectively, $\nabla^2 = \frac{\partial^2}{\partial x^2} + \frac{\partial^2}{\partial y^2} + \frac{\partial^2}{\partial z^2}$ is the Laplacian operator, V($\mathbf{r}$,t) is the potential influencing the system's motion, m* is the (effective) mass of the system, $\hbar$ is the reduced Planck constant and $\varphi(\mathbf{r},t)$ is the wave function describing the system. The quantity $|\varphi(\mathbf{r},t)|^2$ is the probability of finding the system at a spatial location $\mathbf{r} = (x,y,z)$ and time t.

Under the assumption $V(\mathbf{r},t) = V(\mathbf{r})$ the dependence on time and spatial coordinates can be separated by writing the wave function as a product

$$\varphi(\mathbf{r},\ \mathrm{t}) = \psi(\mathbf{r})\,\phi(\mathrm{t}) = \psi(\mathbf{r})\ \exp(-\mathrm{i}2\pi\mathrm{Et/h})$$

which leads to the following time-independent Schrödinger equation

$$\left(-\frac{\hbar^2}{2m^*}\nabla^2 + \mathrm{V}(\mathbf{r})\right)\psi(\mathbf{r}) = \mathrm{E}\psi(\mathbf{r}) \tag{3.1}$$

where E is introduced as a constant to be meant as the energy of the system.

Standard quantum systems in solid state physics are a free (isolated) particle, a particle in a box with either infinite or finite walls and a particle moving under the influence of a periodic potential (such as an electron in a crystal). In all these cases the corresponding time-independent Schrödinger Eq. (3.1) can be solved analytically yielding energy spectrum and electronic wave functions.

A free electron model is the simplest way to interpret the electronic structure of metals. In this approximation the time-independent Schrödinger Eq. (3.1) can be solved for $V(\mathbf{r}) = 0$ and effective mass given by the free electron mass m_o. The resulting solutions have the form of plane-wave functions (i.e., $\psi_k(\mathbf{r}) = A \exp(ik \cdot r)$, where $\mathbf{k}$ is the wave vector constrained by the periodic boundary conditions) with energies $E = (\hbar k)^2/(2\, m_o)$.

Confining a particle along at least one spatial direction introduces quantized energies and wave function changing from a traveling wave to a standing wave [136]. In the most simple case of a particle spatially confined along the x-direction within a distance a (one-dimensional (1D) quantum box of width a) (Fig. 3.6), the effect of the 1D spatial confinement along the x-direction is introduced mathematically by the potential V(x) defined as $V(x) = 0$ if $0 < x < a$ (where a is the width of the confining well) and $V(x) = V_o$ elsewhere (where V_o is assumed finite and infinite in the case of a finite and infinite well, respectively) with boundary conditions requiring that the wave functions have nodes at the walls of the potential well. In the case of one-dimensional infinite well the solutions are oscillating functions vanishing at the walls (Fig. 3.6a) and the energy spectrum is discrete with eigenvalues labeled by one quantum number. In the case of a finite well having the same width a, the energies are slightly lower than the corresponding ones of the infinite well and leakage of the wavefunctions into the barrier occurs (quantum tunneling) (Fig. 3.6b) [136], which means that the wave functions are less well confined than in the corresponding infinite well. Notably, there is only a limited number of solutions, but there is always at least one.

Although the infinite-well model is an approximation, it is usually applied as a good starting point for describing the general effects of quantum confinement.

The results of the 1D quantum box can be extended to the two-dimensions (2D quantum box) and three-dimensions (3D quantum box) in a straightforward way. For example, if the particle with mass m is confined to a 2D box in the (x,y) plane of a Cartesian coordinate system, then the well potential V(x,y) is defined as $V(x,y) = 0$ if $0 < x < a$ and $0 < y < b$ (where a and b is the width of the confining well along x and y, respectively) and $V(x,y) = V_o$ elsewhere (where V_o is finite and infinite in the case of finite and infinite well, respectively). The solutions of the time-independent Schrödinger equation are still oscillating functions in each confinement direction and the energy spectrum is discrete with eigenvalues labeled by two quantum numbers, that is

$$E_{n_x,n_y} = \left(\frac{h^2}{8m^*}\right)\left[\left(\frac{n_x}{a}\right)^2 + \left(\frac{n_y}{b}\right)^2\right]$$

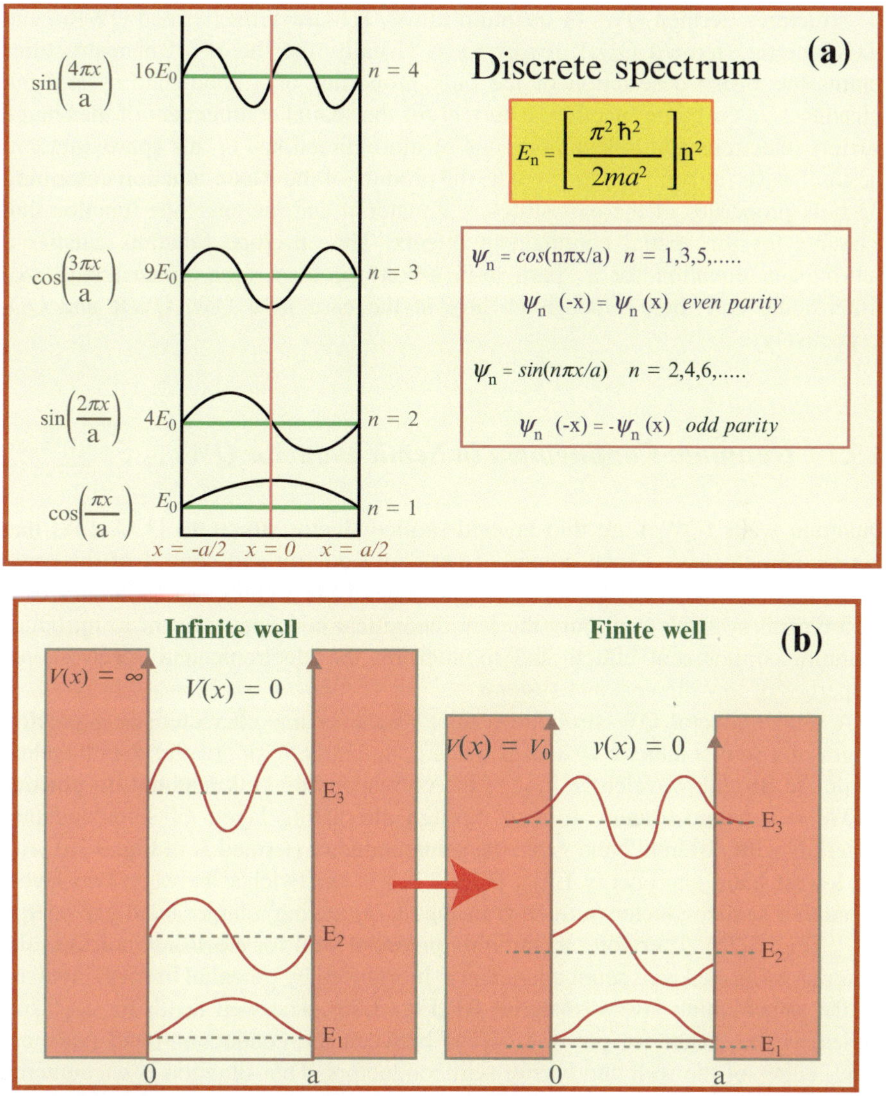

Fig. 3.6 (**a**) Quantized energy spectrum and wave functions of a one-dimensional infinite quantum box of width a. (**b**) Effect of the finite potential barrier versus the infinite one

where n_x and n_y are integers. The energy of the first allowed electron energy level in a typical 100 Å-wide GaAs quantum well is about 40 meV.

The standard problem of a N-dimensional box (N = 1,2,3) is the basis of the quantum-mechanical treatment of semiconductor quantum confined structures that, as already outlined, can be classified according to the number of confinement spatial directions. That is, confinement in one, two and three directions defines

2D-structures (termed QWs or quantum films), 1D-structures (termed QWRs) and 0D-structures (termed QDs), respectively. Usually, the theory of nanostructures retains the Bloch description of the bulk properties of a solid-state system but introduces an envelope function to correct for the spatial confinement of the charge carriers (electrons and holes) along one or more directions. In this approximation, the total wave function is expressed as the product of the Bloch function describing the bulk properties of the nanostructure's material and the envelope function that accounts for the spatial confinement effects. The envelope function satisfies a Schrödinger equation for a "particle-in a-box" problem (one-dimensional, two-dimensional and three-dimensional box in the case of a QW, QWR and QD, respectively).

3.3.1 Quantum Confinement in Semiconductor QWs

Quantum wells (QWs) are thin layered semiconductor structures [137, 138] that can be fabricated to a high degree of structural quality and control of thickness by modern epitaxial crystal growth approaches [139]. QWs can be assumed as prototypical systems to discuss the key theoretical concepts relevant to introduce quantum confinement effects and to interpret the electronic and spectroscopic properties of low-dimensional systems.

A semiconductor QW structure results whenever one characteristic geometric length of a semiconductor system is made comparable to or smaller than the electronic De Broglie wavelength λ_{DB} of the corresponding bulk material. In practice a QW is a heterostructure formed through alternating layers of semiconductor materials with different band gaps: the semiconductor (termed A in Fig. 3.7a) with the lowest band gap energy $E^{A}{}_{gap}$ (Fig. 3.7b) is sandwiched between two layers of another semiconductor (termed B in Fig. 3.7a) having a larger band gap energy $E^{B}{}_{gap}$ (Fig. 3.7b). Therefore, a confining potential well for electrons (holes) in the material A (termed well semiconductor) is introduced by potential barriers involved by the sandwiching layers (material B) (Fig. 3.7b). The well depth for electrons (holes) is the difference (i.e. the offset) between the conduction-band (valence-band) edges of the well and barrier semiconductors. This situation is encountered in practice for the following list of pairs of materials well/barrier: GaAs/GaAlAs, GaAs/GaInP, GaSb/GaAlSb, GaInAsP/InP, GaInAs/AlInAs, InP/GaInAs, Si/SiGe and GaSb/AlSb. In the case of a GaAs/AlGaAs QW, the quantum confinement is provided by the discontinuity in the band gap at the interfaces and the Al concentration is typically around 30 %, leading a band-gap discontinuity of 0.36 eV [140] such that electrons and holes see a confining barrier of 0.24 and 0.12 eV, respectively. The dispersion of a QW can be calculated by solving the Schrödinger equation for the electrons and holes in the potential wells created by the interface band discontinuities. The well width is the thickness of the intermediate layer (termed L_z in Fig. 3.7a, where z is the growth-direction of the tri-layer heterostructure) and represents the length to be made comparable with the bulk de Broglie wavelength of the

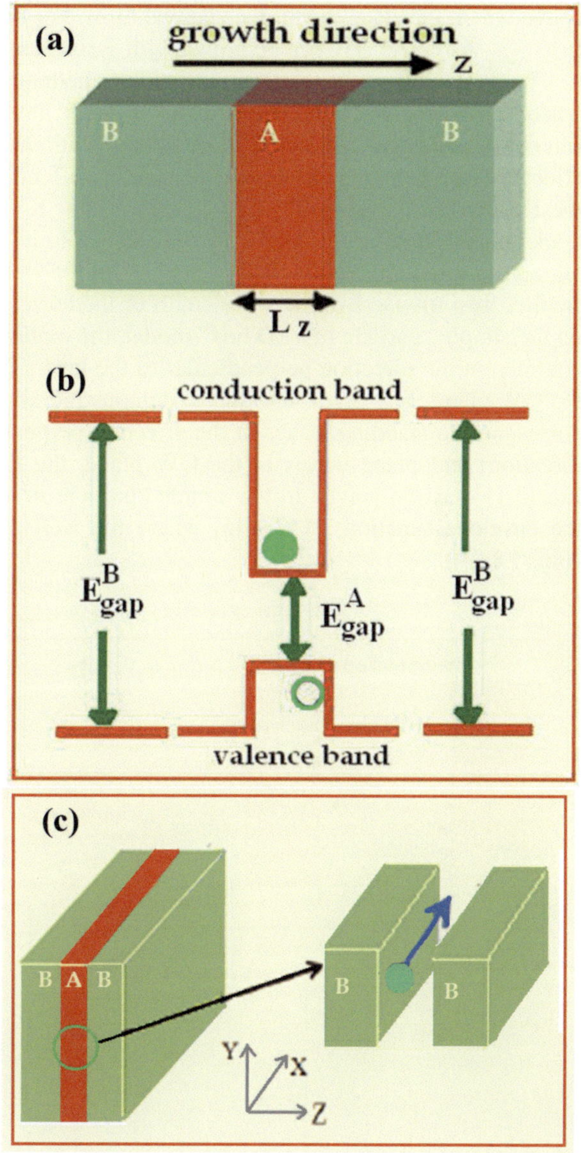

Fig. 3.7 A semiconductor QW structure A/B/A: (**a**) alternating layers of semiconductor materials B/A/B with (**b**) the lowest band gap material A sandwiched between two layers of a material B with a larger band-gap in such a way to form a potential well confining the electron motion (**c**) along the growth direction of the heterostructure

electrons or holes in the material A ($\lambda^A{}_{DB}$) to induce quantum size effects. Materials to be coupled to form a QW should satisfy the further request to be "lattice-matched", that is their lattice constants must be nearly identical, to reduce interface dislocations and defects. In practice relief of strain occurs above a critical thickness. Indeed, QWs can be routinely grown as strained layers on top of a lattice with a different cell constant (e.g., $In_xGa_{1-x}N$/GaN, In_xGa_{1-x} As/GaAs, and $Si_{1-x}Ge_x$/Si QWs) as long as the total thickness of the strained layer is less than a critical value above which interface misfit dislocations form. For example, a defect-free strained $In_xGa_{1-x}As/GaAs$ QW can be grown with x = 0.2 and L_z of nearly 10 nm.

The well width is the thickness of the intermediate layer (termed L_z in Fig. 3.7a, where z is the growth-direction of the tri-layer heterostructure) and is the characteristic geometric length to be compared with λ^A_{DB} to observe quantum-size effects [140, 141]. In GaAs, where the electron effective mass amounts to 0.067 m_0, the De Broglie wavelength $\lambda_{DB} = 42$ nm at 300 K results, meaning structures of thickness ≈ 10 nm are needed in order to be able to induce quantum-confinement effects at room temperature. If the thickness of the well material is comparable to or smaller than the De Broglie wavelength of the corresponding bulk material, based on the simple "particle in a 1D box" model, the motion of carriers is expected to be confined in the direction perpendicular to the layers (growth-direction) and free in the QW plane (Fig. 3.7c). In order to demonstrate that the allowed states in a QW correspond to standing waves in the direction perpendicular to the layers (growth-direction) and plane waves in the QW plane, the time-independent Schrödinger Eq. (3.1) must be solved in the simple "particle in a 1D box" approximation (of the envelope function) [142–145] where the well potential is the QW potential (Fig. 3.8a, b).

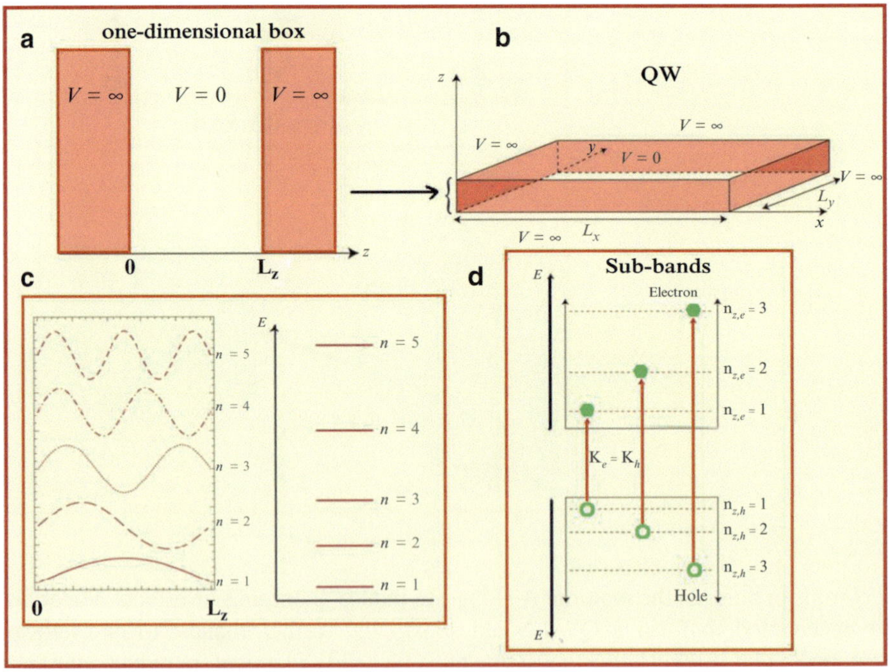

Fig. 3.8 Schematic diagram of (**a**) the potential barrier, (**b**) layer structure, (**c**) quantized eigenfunctions and eigenvalues and (**d**) band diagram with subband-structure (*dashed lines*) of a QW of thickness Lz and infinite barrier wall

Accounting for the structure geometry and presence of materials A and B (Fig. 3.7a), the QW potential can be written as follows:

$$V_{QW}\,(x,y,z) = V_A\,(x,y,z)\;F_A\,(z) + V_B\,(x,y,z)\;F_B\,(z)$$

where the functions $F_A(z)$ and $F_B(z)$ introduce the potential of the corresponding layer, that is $F_A(z)$ vanishes in the layer B and $F_B(z)$ vanishes in the layer A in such a way that $V_{QW}\,(x,y,z) = V_A\,(x,y,z)$ in the layer A and $V_{QW}\,(x,y,z) = V_B\,(x,y,z)$ in the layer B. Since the QW geometry decouples (x,y)-plane and z-direction, the QW potential can be decomposed as

$$V_{QW}\,(x,y,z) = V^{xy}{}_{QW}\,(x,y) + V_{QW}\,(z)$$

where $V_{QW}\,(z) = V^A\,F_A(z) + V^B\,F_B(z)$ is the confining potential seen by the carrier (electron and hole) along the direction z.

On the other hand, the translational invariance of the QW system in the (x,y)-plane enables to model the wavefunctions describing the motion in the QW plane as plane waves. Hence, being m* the carrier effective mass, a general solution of the time-independent Schrödinger equation associated with $V^{xy}{}_{QW}\,(x,y)$ and motion in the (x,y)-plane can be written as follows:

$$\varphi\left(\mathbf{r_{xy}}, z\right) = \frac{1}{\sqrt{S}}\left(e^{i\mathbf{k_{xy}}\cdot\mathbf{r_{xy}}}\right)\;\psi\,(z)$$

where z is the confinement direction (growth-direction), $\mathbf{r_{xy}} = (x, y)$ and $\mathbf{k_{xy}} = (k_x, k_y)$ is the position vector and wave vector in the QW-plane ((x,y)-plane), respectively, and the prefactor is a normalization constant. Therefore, being m* the carrier effective mass, the Schrödinger's equation to be solved for the particle of interest (electron or hole) in a QW is the following one-dimensional time-independent Schrödinger equation:

$$\left(-\frac{\hbar^2}{2m^*}\frac{\partial^2}{z^2} + V_{QW}\,(z)\right)\psi_n\,(z) = E_n\psi_n\,(z)$$

where $\psi_n(z)$ and E_n are the energy eigenfunction and eigenvalue (confinement energy) associated with the n-th solution and it is assumed that $L_z \leq \lambda^A{}_{DB} \ll L_x, L_y$. Rather than the continuity condition of the derivative of the wave functions at the interfaces, the conservation of both wave function $\psi_n(z)$ and particle flux $((1/m^*)(\partial\psi_n(z)/\partial z))$ across the QW walls is the boundary condition to be imposed to match the solutions in the well and the barrier materials. Essentially the equation to be solved is the time-independent Schrödinger equation of the standard problem of a particle confined in a one-dime1D box whose solutions are known to be standing waves (Fig. 3.6) labeled by one integer number.

In the case of an infinitely deep QW heterostructure, the solutions $\psi_n\,(z)$ of the time-independent Schrödinger equation are analytically known (oscillating sin and cos functions with nodes at the walls) and the energy spectrum is discrete (Fig. 3.8c) with eigenvalues

$$E_{n_z} = \left(-\frac{\hbar^2}{2m^*}\right)\left(\frac{n_z \pi}{L_z}\right)^2 \text{ where } n_z = 1, 2, 3, \ldots$$

Given the eigenvalue corresponding to $n_z = 1$

$$E_1 = \left(-\frac{\hbar^2}{2m^*}\right)\left(\frac{\pi}{L_z}\right)^2$$

the energy spectrum can be expressed as

$$E_n = E_1 {n_z}^2 \text{ where } n_z = 1, 2, 3, \ldots$$

Notably, the confinement energies E_{nz} are referred to the bottom of the well and the non vanishing energy E_1 is located above the bottom of the well.

Spacing between subsequent eigenvalues can be easily expressed as

$$E_{n_z+1} - E_{n_z} = \left(\frac{\hbar^2}{2m^*}\right)\left(\frac{\pi}{L_z}\right)^2 (2n_z + 1) \text{ where } n_z = 1, 2, 3, \ldots$$

Therefore, the QW confinement energies are quadratically spaced and their energy spacing increases for decreasing well width L_z and effective mass m* as well as for increasing quantum number n_z.

The solution of the QW problem for finite well height must be performed numerically and yields bound states whose wave functions are again standing oscillating functions inside the quantum well and exponentially decaying into the barriers (Fig. 3.6) with eigen-energies always somewhat lower than those of the infinite case [146]. In any case, the solution of the QW problem leads to discrete sequences of valence and conduction energy levels (termed sub-bands) (Fig. 3.8d). As the reduced translational symmetry of a QW structure lifts the bulk degeneracy of the heavy-hole (hh) and light-hole (lh) valence bands, each valence sub-band in Fig. 3.8d is a doublet of hh and lh subbands with the hh subband closer to the valence band bottom than the corresponding lh subband. Therefore, hh-to-conduction edge starts at a slightly lower energy and is more closely spaced than the lh-to-conduction edge.

Definitively, if $L_z \leq \lambda^A{}_{DB} \ll L_x, L_y$ and the effective mass is assumed isotropic, the total energy of a carrier in a QW is given by the formula

$$E^{tot}_{nz} = E_x + E_y + E_{nz} = \left(\frac{\hbar^2 {k_x}^2}{2m^*}\right) + \left(\frac{\hbar^2 {k_y}^2}{2m^*}\right) + \left(\frac{\hbar^2}{2m^*}\right)\left(\frac{n_z \pi}{L_z}\right)^2$$

where $n_z = 1, 2, 3, \ldots$

where the eigenvalues E_x and E_y are associated to plane waves solutions describing the free motion of a carrier along the x and y directions, respectively, and E_{nz} are the bounded energies of the carrier labeled by the integer index n_z. Since the confinement energies are inversely proportional to both the well width L_z and the carrier effective mass, narrow wells as well as lighter particles involve larger confinement energies. In a typical GaAs/$Al_{0.3}Ga_{0.7}As$ QW with $L_z = 10$ nm, the confinement energy for electrons and holes is 245 meV and 125 meV, respectively. The infinite well model predicts lowest quantized energies of $E_1 = 30$ meV and $E_2 = 113$ meV for the electrons [147, 148] and 11 meV (40 meV) and 44 meV (160 meV) for the first two bound states of the heavy (light) holes. Notably, the energy spacing of the electron levels should be larger than k_BT at 300 K (i.e., 25 meV) to make the quantum size effect readily observable at room temperature.

The above considerations result in the typical band structure depicted in Fig. 3.9: since the electronic dispersion is parabolic along the direction of free motion, the occurrence of discrete spectrum along the confinement direction (growth direction of the QW) implies a discrete sequence of parabolic bands ("sub-bands") with minimum energy given by the discrete confinement energy along z ($E_x = E_y = 0$). Basically, in a QW carriers (electrons and holes) are free to move in any direction parallel to the QW layers and a carrier in a given confined state can have additionally any amount of kinetic energy corresponding to its in-plane free motion that implies any energy greater than or equal to the confined- energy of that sub-band.

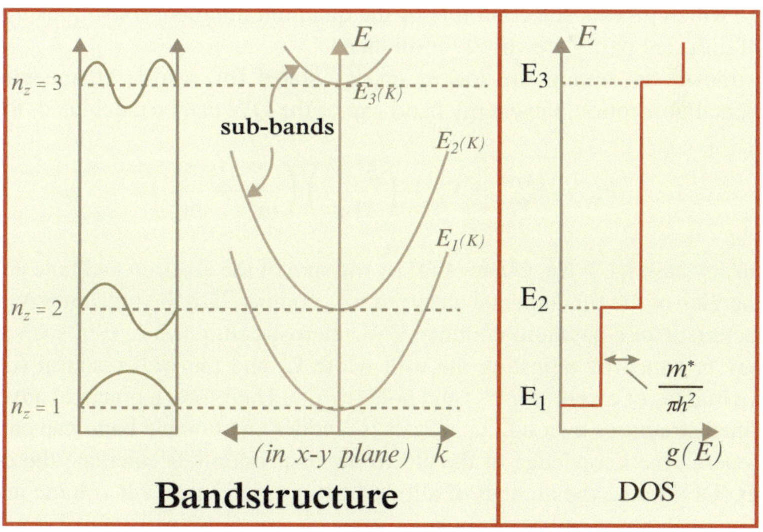

Fig. 3.9 Electronic dispertion of a QW: In a QW, as a result of the carrier confinement along the growth direction and its free motion in the QW plane, the electronic dispertion consists of parabolic subbands leading to a step-like DOS

The subband-structure that characterizes the energy spectrum of a QW determines its spectroscopic properties [149]. The optical transitions in QWs are termed "inter-subband transitions" and take place between electronic states that are confined in the z-direction and free in the well x–y plane.

In a bulk semiconductor, an optical transition is allowed between states having the same k-values in the case of a direct-gap material and momentum is conserved via a phonon interaction in the case of an indirect bandgap material. Turning to a QW, allowed interband transitions can occur according to the following relationship between exciting photon energy ($\hbar\omega$), energy gap of the bulk counterpart of the well material (E_g), energies of the valence (E_{nz}) and conduction (E_{mz}) subbands measured from the valence and conduction edges of the bulk material, respectively:

$$\hbar\omega = E_g + E_{nz} + E_{mz} = E_g + \left(\frac{\hbar^2\pi^2}{2m_h^* L_z^2}\right) n_z^2 + \left(\frac{\hbar^2\pi^2}{2m_e^* L_z^2}\right) m_z^2$$

under the selection rule that only transitions between states of the same quantum number in the valence and conduction bands are allowed. Indeed, in a QW, optical transitions must still conserve momentum in the plane of the QW, just as for bulk semiconductors. Instead of momentum conservation, there is an additional selection rule for the direction perpendicular to the layers resulting from optical absorption strength proportional to the overlap integral of the conduction and valence (envelope) wave functions. The transition rate is proportional to both the square of the overlap of the envelope wave functions and the joint density of states [150]. In the case of an infinitely deep well, the wave functions are orthogonal unless $n_z = m_z$, which gives a selection rule of the quantum number. Transitions between states of different parity are strictly forbidden.

For transitions among the lowest energy states ($m_z = n_z = 1$), meaning the fundamental absorption, the energy band gap of the QW can be calculated, that is:

$$\hbar\omega = E_g^{QW} = E_g + \left(\frac{\hbar^2\pi^2}{2L_z^2}\right)\left(\frac{1}{m_h^*} + \frac{1}{m_e^*}\right)$$

Being in any case $\hbar\omega > Eg$, a blue- shift by the sum of the electron and hole confinement energies of the fundamental absorption as compared to the bulk energy gap E_g is characteristic of a quantum-confined QW heterostructure. Moreover, such a blue-shift may be tuned by adjusting the well width L_z and the well material (i.e., the effective masses of electron (m^*_e) and hole (m^*_h)). Therefore, a practical advantage of confined structures over bulk materials is a wider and tunable band gap energy.

In general, the knowledge of the electronic dispersion lets calculate the density of states (DOS), i.e., the number of allowed energy states per unit volume and unit energy interval, based on the formula

$$g(E) \equiv \sum_n \int_{BZ} d\vec{k}\, \delta\left(E - E_n\left(\vec{k}\right)\right)$$

where E is a given allowed energy, g(E) dE is the total number of electronic states with energy ranging from E to E + dE per unit volume, the sum is extended to all the allowed values of the discrete band-index n, the integral sums all the wave vectors and averages on all allowed spin states.

If the band structure of a QW consisting of parabolic subbands is considered, then the corresponding DOS (per unit energy, per unit surface area) is given by the following expression

$$g^{QW}(E) = \frac{m^*}{\pi\hbar^2}\sum_{n_z}\theta(E - E_{nz})$$

where n_z is the quantum number associated with the confinement energy along the z-direction, $\theta(E–E_{nz})$ is the Heaviside unit step function defined by $\theta(E–E_{nz}) = 0$ if $E < E_{nz}$ and $\theta(E–E_{nz}) = 1$ if $E > E_{nz}$ and a factor 2 for spin has been included. Therefore, it results:

$$\begin{aligned} g^{QW}(E) &= 0 && \text{if } E < E_1 \\ g^{QW}(E) &= \tfrac{m^*}{\pi\hbar^2} && \text{if } E_1 < E < E_2 \\ g^{QW}(E) &= \tfrac{2m^*}{\pi\hbar^2} && \text{if } E_2 < E < E_3 \end{aligned}$$

Therefore, DOS associated to a QW for a given subband is a "step-like" function starting at the appropriate confinement energy [151]. Since in QWs, just as for bulk semiconductors, absorption transitions follow the DOS profile, the absorption coefficient (i.e., the change in the intensity of the photon flux per unit length) of a QW is a series of steps each corresponding to a QW subband [152].

To summarize, quantizing the carrier motion along the z-direction has three main consequences; (i) a blue-shift of the bulk band edge depending on QW's material and width, which enables effective band-gap engineering; (ii) an increase of the radiative recombination probability by keeping electrons and holes closer together due to the spatial confinment; and (iii) a DOS independent on energy (step-like profile) in contrast to that of the bulk counterpart.

The above approach (free-electron gas approximation) deals with absorption processes in which an exciting photon creates a free electron-hole pair without considering the effect of electron-electron correlation. The considered independent-electron approximation leads to an energy spectrum of carriers independent on the presence of other similar carriers which is taken into account only by populating the allowed energy states according the Pauli's exclusion principle. However, the electron-electron (electron-hole) interactions are important to a more accurate description of physical phenomena such as optical absorption below away from the band-edges. The influence of electron-electron correlation is addressed by excitonic effects in semiconductors that are introduced by defining the exciton as an electron-hole pair bounded by Coulomb interaction. As already discussed, the Wannier excitons, that are classified as weakly bounded excitons, form in semiconductor materials. Their large size makes them strongly sensitive to nanometer-scale variations of the surrounding, meaning to spatial confinement in nanostructures.

In bulk semiconductors the spectroscopy of (Wannier) excitons can be calculated based on the Schrödinger Eq. (3.1) solved within the approximation that treats electron and hole as two particles moving with the effective masses of the conduction (m_e^*) and valence (m_h^*) bands, respectively, and attracted to each other by the Coulomb interaction (according to a modified Bohr model of the hydrogen atom):

$$\left(-\frac{\hbar^2}{2m_e^*}\nabla_e^{\,2}-\frac{\hbar^2}{2m_h^*}\nabla_h^{\,2}-\frac{e^2}{4\pi\in|\mathbf{r_e}-\mathbf{r_h}|}\right)\varphi_{exc}(\mathbf{r_e},\mathbf{r_h})=E_{exc}\varphi_{exc}(\mathbf{r_e},\mathbf{r_h})$$

Upon the transformation of coordinates defined by $\mathbf{r}=\mathbf{r_e}-\mathbf{r_h}$ and the $\mathbf{R_{CM}}=(m_e^*\mathbf{r_e}+m_h^*\mathbf{r_h})/(m_e^*+m_h^*)$, the above two-body Schrödinger equation can be separated into an equation for the wave function $F(\mathbf{r})$ of the relative motion of the electron-hole system and an equation describing the motion of the center of mass (CM) by the wave function $G(\mathbf{R_{CM}})$:

$$\left(\frac{\hbar^2}{2m_r^*}k^2-\frac{e^2}{4\pi\in|\mathbf{r}|}\right)F(\boldsymbol{r})=E_rF(\boldsymbol{r})$$

$$\left(\frac{\hbar^2}{2\left(m_e^*+m_h^*\right)}K^2\right)G(\mathbf{R_{CM}})=E_{CM}G(\mathbf{R_{CM}})$$

where $1/m^*_r=(1/m^*_e+1/m^*_h)$ is the reduced mass of the electron-hole system, $\mathbf{K}=\mathbf{k_e}\text{-}\mathbf{k_h}$ is the wave vector of the CM motion and $\mathbf{k}=(m_e^*\mathbf{k_e}+m_h^*\mathbf{k_h})/(m_e^*+m_h^*)$ is the wave vector of the relative motion.

Since the Coulomb interaction depends on the relative coordinate, as a result, the CM motion can be pictured as a free motion (plane wave solution) of a particle with total mass $M=m^*_e+m^*_h$ and the equation of the relative motion is mathematically the Schrödinger equation of the hydrogen atom. Therefore the dispersion relationship of the exciton can be written as follows

$$E_{n,K\,ex}=E_g+E_{CM}+E_r=E_g+\frac{\hbar^2}{2M}K^2-\frac{R_y^*}{n^2}$$

where E_g is the band gap energy of the bulk semiconductor (i.e., the energy without the Coulomb interaction), E_{CM} is the CM energy (kinetic energy of the exciton) given by

$$E_{CM}=\frac{\hbar^2}{2M}K^2$$

E_r is the discrete spectrum energy of the electron-hole bound states labeled by a principal quantum number n = 1,2,3,… and expressed as follows

$$E_r = -\frac{R_y^*}{n^2}$$

and R_y^* is the exciton Rydberg given by

$$R_y^* = \left(\frac{m_r^*}{m_o \epsilon_r^2}\right) R_y$$

where ε_r is the relative dielectric constant of the semiconductor, m_o is the mass of the free electron and $R_y = 13.6$ eV is the atomic Rydberg constant. Therefore, the spectrum of the relative motion of an exciton is a Rydberg series similar to that of the hydrogen atom with an effective Rydberg constant modified by the reduced mass of the exciton and the dielectric relative constant of the material.

The exciton (envelope) functions solving the Schrödinger equation of the relative motion are the known hydrogen-like functions with the ground state solution given by

$$F(r) = F_{100}(r) = \frac{1}{\sqrt{\pi a_{ex}^3}} e^{-r/a_{ex}^{bulk}}$$

where

$$a_{ex}^{bulk} = a_H \varepsilon_r \frac{m_o}{m_r^*}$$

is termed "Bohr radius of exciton" and $a_H = 0.53$ Å is the Bohr radius of the hydrogen atom. In the case of Si, Ge and GaAs, a_{ex}^{bulk} amounts to 4.3, 11.5 and 12.4 nm, respectively. Physically, a_{ex}^{bulk} is the relative electron-hole separation in the pair coupled by the Coulomb interaction (exciton) and, as already mentioned, represents a characteristic confinement length-scale in semiconductors.

As compared to the independent-electron approximation (leading to the band-structure in Fig. 3.10a and the absorption coefficient in Fig. 3.10b), accounting for the electron-hole Coulomb correlation (i.e., excitonic effects) leads to a modified bandstructure that describes the system in terms of the electron-hole momentum and consisting of a sequence of discrete bands lying below the conduction band of the continuum of electronic states of the one-electron model (Fig. 3.11a). Such discrete excitonic bands approach the band-edge becoming closer and closer for increasing exciton energy.

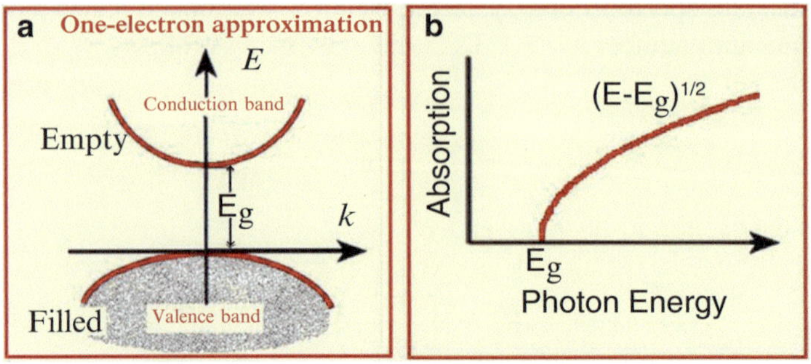

Fig. 3.10 Typical (**a**) bandstructure and (**b**) absorption coefficient of a bulk semiconductor in the framework of the independent-electron approximation

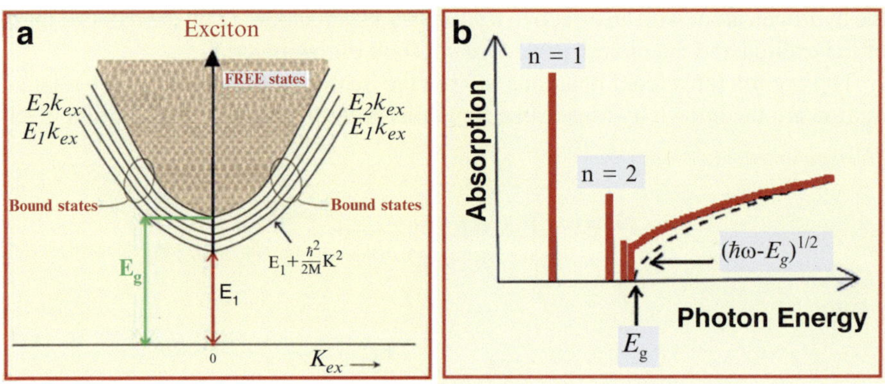

Fig. 3.11 Typical bulk semiconductor (**a**) bandstructure including excitonic contribution and (**b**) absorption coefficient with (*solid line*) and without (*dashed line*) excitonic effects

As for the spectroscopy in the presence of excitonic effects [138], while unbound electron-hole pairs can be continuously generated by interband absorption of photons with energy at least equal to the fundamental band-gap (E_g), for photon energy decreasing below E_g excitons can be created at discrete energies leading to sharp absorption peaks (Fig. 3.11b). Figure 3.11b sketches the main features of the absorption coefficient in the presence (solid line) and in absence (dashed line) of excitonic effects. It can be observed that as the exciting photon energy becomes lower than E_g, the (unbound electron-hole pairs) continuum absorption curve of a direct gap semiconductor with excitonic effects included doesn't decrease to zero according to the square-root trend of the one-electron model (Fig. 3.10b). Instead the absorption coefficient remains finite at the band edge and exhibits strong absorption peaks associated with the ground state ($n = 1$) and the first excited state ($n = 2$) of the exciton just below E_g (Fig. 3.11b). Basically, near the band edge, the absorption coefficient in the presence of excitonic effects (α_{ex}) can be related to the absorption coefficient without excitonic effects (α) as follows

$$\alpha_{ex} = \alpha \left(\frac{2\pi\sqrt{R_y^*}}{(\hbar\omega - E_g)^{1/2}} \right)$$

Finite enhanced absorption below E_g stemming from exciton formation is expressed by the Sommerfeld enhancement factor (i.e., the Coulomb-enhanced absorption coefficient divided by the absorption without Coulomb effects). In practice, in bulk semiconductors excitonic effects can be observed only in optical spectra at very low temperature in truly pure samples because of the low binding energy R_y^* [153]. Indeed, in technologically important bulk semiconductors, such as Si, Ge, and GaAs, the exciton binding energy R_y^* ranges from 3.8 to 14.7 eV, meaning it is lower than the thermal energy ($k_BT = 25$ meV at room temperature). As a consequence, excitonic resonances are not observed because exciton usually dissociates easily at room temperature and merging of excitonic resonances with interband transitions is also favored in poorer quality bulk samples.

Instead, the spatial confinement of carriers in QWs greatly enhances the binding energy of excitons and the oscillator strength in such a way that excitonic effects become observable at room temperature in the optical spectra of QWs as sharp resonances. Basically, exciton binding energy is related to the probability to have electron and hole in the same unit cell which is expressed by an overlap integral of the electron and hole wave functions in the QW. Spatial confinement is expected to strength the electron-hole interaction, thus enhancing the exciton binding energy through an increased overlap integral, and a decreased ground state electron-hole distance (Bohr radius of exciton) of the QW exciton (Fig. 3.12).

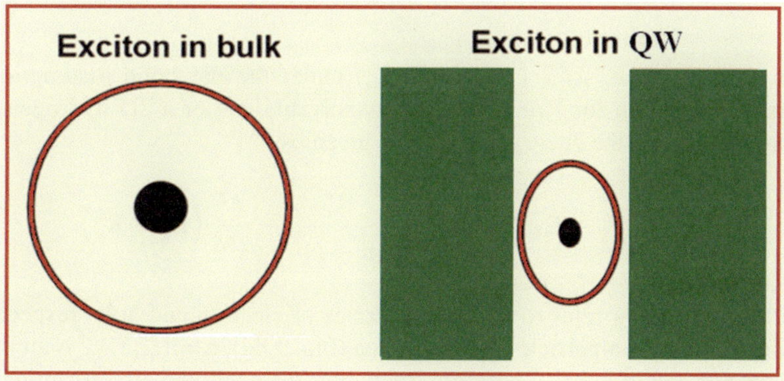

Fig. 3.12 Cartoon of the effects of the spatial confinement on an exciton in a QW: strengthed electron-hole interaction and decreased ground state electron-hole distance (excitonic Bohr radius)

To address the exciton problem in a QW, a time-independent Schrödinger equation must be solved with potential given by the sum of the QW confinement potentials of electron and hole ($V_e(z_e)$ and $V_h(z_h)$, respectively, being z-axis the growth direction of the QW structure) in the one-particle picture and the electron-hole Coulomb interaction:

$$\left(-\frac{\hbar^2}{2m_e^*}\nabla_e{}^2-\frac{\hbar^2}{2m_h^*}\nabla_h{}^2+V_e\left(Z_e\right)+V_h\left(Z_h\right)-\frac{e^2}{4\pi\in|\mathbf{r_e}-\mathbf{r_h}|}\right)\varphi_{exc}\left(\mathbf{r_e},\ \mathbf{r_h}\right)$$
$$=E_{exc}\varphi_{exc}\left(\mathbf{r_e},\ \mathbf{r_h}\right).$$

In general, such a problem doesn't have simple analytic solution because the parabolic approximation can be poor for holes, a QW system is a quasi 2D system and the effect of the well size and finite well depth must be considered [154, 155]. Under the assumption that the QW width is less or comparable to the bulk exciton Bohr diameter, then electron and hole are quantized independently of each other and a variational approach can be considered that introduces a class of trial functions with the following form

$$\varphi_{ex}=F\left(R\right)\ f\left(\rho\right)\ U_e\left(z_e\right)\ U_h\left(z_h\right)$$

where R is the exciton center of mass coordinate, ρ is the radius-vector of electron and hole relative motion in the QW-plane, F(R) describes the exciton center of mass motion, f(ρ) describes the relative electron-hole motion in the plane of the QW, $U_e(z_e)$ and $U_h(z_h)$ describe the electron and hole confined motion, respectively, and the following normalization is assumed:

$$\int dz_e|U_e\left(z_e\right)|^2=\int dz_h|U_h\left(z_h\right)|^2=\int_0^\infty 2\pi\rho d\rho|f\left(\rho\right)|^2=\int_0^\infty 2\pi RdR|F\left(R\right)|^2=1.$$

In the ideal 2D case, $|U_{e,h}\ (z_{e,h})|^2=\delta(z_{e,h})$ under the parabolic assumption for carriers, the equation for f(ρ) is the exactly solvable one of a 2D hydrogen atom problem. Therefore, the energy spectrum is given by

$$E_{ex}^{QW}=E_g+E_{nz,e}+E_{mz,h}-\frac{R_y^*}{\left(n-\frac{1}{2}\right)^2}+\frac{\hbar^2}{2M}\left(K_x^2+K_y^2\right)$$

where $E_{nz,e}$ and $E_{mz,h}$ refer to subband energies of electron and hole, respectively, resulting from the one-particle approximation (Fig. 3.8d), n = 1,2,3 ... is an integer and $K^2=K_x{}^2+K_y{}^2$ [156]. Therefore, including the excitonic effects implies that the eigen energies of the QW calculated in the single particle approximation (i.e., $E_g+E_{nz,e}+E_{mz,h}$) are shifted by the Rydberg energy $\left(-\frac{R_y^*}{\left(n-\frac{1}{2}\right)^2}\right)$ plus the kinetic energy $\left(\frac{\hbar^2}{2M}\left(K_x^2 + K_y^2\right)\right)$ of the exciton in the QW. As a general trend, while in bulk, the total energy of the exciton is simply the energy of the bang gap plus the exciton binding energy, in a nanostructure there are additional components due to the electron and hole confinement energies

By setting n = 1 (ground state), it can be observed

$$E_{bind,exc}^{QW} = 4R_y^* = 4\ E_{bind,exc}^{bulk}$$

that is, the binding energy of the exciton in an infinite QW is four times the binding energy of the corresponding bulk exciton. This fact lets observe excitonic effects at room temperature in QWs, in contrast to bulk semiconductors where they are only usually observed at low temperatures. Furthermore, in a QW, the electron-hole separation (Bohr radius of the exciton) decreases with respect to the bulk counterpart (Fig. 3.12). This is consistent with increasing binding energy in low-dimensional systems. In the 2D case,

$$E_{bind,exc}^{QW} = 4R_y^* = 4\ E_{bind,exc}^{bulk}$$

$$a_{exc}^{QW} = \frac{1}{2}\ a_{exc}^{bulk}$$

In the case of a realistic QW, a more complex expression of the exciton binding energy is obtained by the variational approach that, after minimization with respect to a variational parameter, yields that the exciton binding energy in a QW ranges from $E_{bind,exc}^{bulk}$ to $E_{bind,exc}^{QW}$. Moreover, it depends on the QW width and barrier heights for electrons and holes and increases for increasing exciton confinement: while for wide QWs the confinement increases for decreasing QW width, for very narrow QWs the opposite occurs due to leakage of the carrier wave function into the barriers.

In regard to excitons in QWs, they are characterized by two regimes depending on the relationship between the well thickness L_z and the bulk Bohr radius of the exciton ${a_{ex}}^{bulk}$: strong confinement if $L_z \sim {a_{ex}}^{bulk}$ and weak confinement if $L_z \gg {a_{ex}}^{bulk}$. Strong confinement regime (also termed "quasi-2D regime") means that the exciton binding energy is smaller than the confinement energy of the carriers (electrons and holes) and the exciton binding energy as well as the oscillator strength enhance on reducing the well thickness L_z. Weak-confinement regime (also termed "3D regime") means that the exciton binding energy is larger than the carrier quantization energy and the center-of-mass motion of the exciton is quantized as a whole and the oscillator strength is proportional to L_z [157].

The enhanced binding energy and increased oscillator strength of a exciton confined in a QW have critical impact on the spectroscopic properties, that can become dominated by the excitonic effects even at room temperature [154, 155, 158–162]. In this regard, Fig. 3.13c sketches the theoretical absorption spectrum

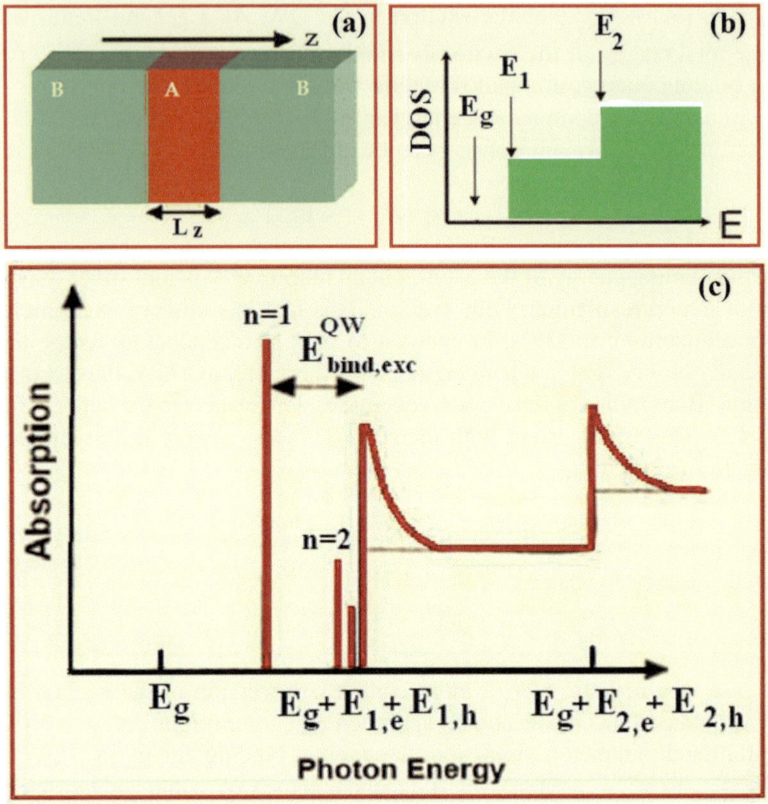

Fig. 3.13 (**a**) QW heterostructure and (**b**) its step-like DOS. (**c**) Sketched theoretical characteristic step-like absorption spectrum of a QW including excitonic effects (sharp intense absorption peaks just below the band-gap energy of the QW)

of a QW including excitonic effects: sharp intense absorption peaks just below the band-gap energy of the QW are associated to the exciton discrete spectrum and the characteristic step-like absorption stemming from the step-like profile of the DOS (Fig. 3.13b) results modified near the absorption edge, as observed in bulk materials while including excitonic influence (Fig. 3.11). The absorption profile of the QW above the exciton lines reflects the constant DOS in 2D, which differs from the increasing absorption of the bulk material resulting from the parabolic 3D DOS.

Indeed, the continuum absorption coefficient is increased over the value without excitonic effects (grey curve in Fig. 3.13c) by the Sommerfeld factor which expresses the influence of the electron-hole correlation on the unbound states. Excitonic features and their appearance can also be modulated by the well thickness

(L_z) [153]. Separate transitions are usually observed for the hhs anf lhs, as a consequence of their different effective masses and lower symmetry of the QW as compared to the bulk counterpart material.

3.3.2 *Quantum Confinement in Semiconductor QWRs*

Since the study by Esaki and Tsu [137] in the 1970s about semiconductor QWs and superlattices, progress in the processing technology let fabricate systems of lower dimensionality such as QWRs where the confinement of the carriers can be controlled through the size and shape of the QWR as well as through the selection of structure and barrier materials to produce various band offsets [163].

Our previous analysis dealing with QWs assumed carrier spatial confinement along one coordinate, where the confinement was provided by the offset between the band edges of two different semiconductors. The next logical step is to consider what happens if carriers are spatially confined along a further direction, leading to 2D confinement (that is one-dimensional free-motion) that defines the kind of nanostructure termed QWR. A semiconductor QWR structure results whenever two characteristic geometric lengths of a semiconductor system are comparable to or smaller than the electronic De Broglie wavelength of the corresponding bulk material. Hence, electron motion is free along the direction of the wire axis (termed x-direction hereafter) and confined along two direction (plane yz of the wire cross-section) (Fig. 3.14). Unlike the case of a QW, different potential profiles (e.g., rectangular and triangular) can be realized in the case of a QWR depending on the fabrication techniques and protocol [164–177]. The role of electron-hole Coulomb interaction also depends on the profile of the confining potential [178, 179].

Different analytical expressions of the confining potential lead to different classes of energy eigenfunctions and may require numerical solving of the Schrödinger equation. Anyway, quantum size effects exhibited by QWRs have general features that can be assessed in the most simple case of a rectangular QWR. Therefore, in the following we will develop the basic theoretical concepts for calculating the electronic and optical properties of QWRs in the case of a rectangular QWR, i.e., electrons (holes) are confined by a square potential well along each of two confinement independent directions (termed y and z hereafter).

In the case of 1D structures, to solve the time-independent Schrödinger equation (Eq. (3.1)) it is possible to decouple the motion along the axis of the QWR (x axis in Fig. 3.14) and the motion along the direction perpendicular (yz plane in Fig. 3.14a). Given the lengths L_x, L_y and L_z (Fig. 3.14), the 2D confinement potential of a rectangular QWR can be decomposed as follows

$$V_{QWR}(x, y, z) = V_1(x) + V_{2,3}(y, z)$$

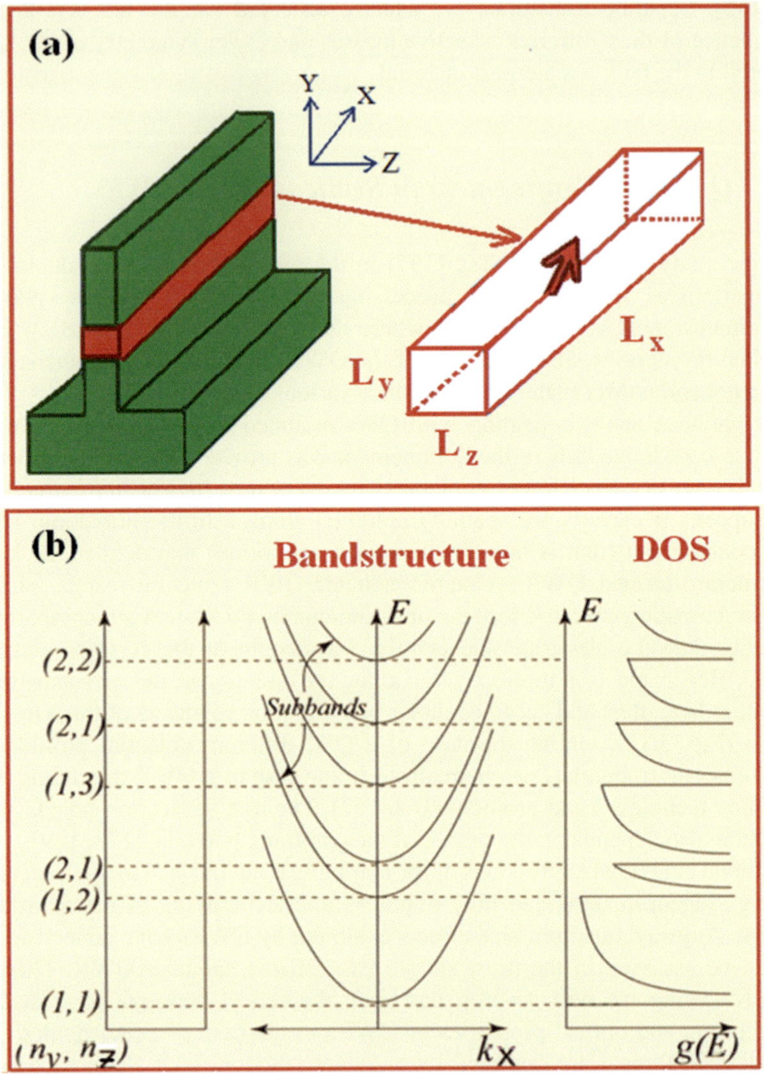

Fig. 3.14 (**a**) Schematics of a QWR heterostructure and corresponding confinement imposed to the electronic motion. (**b**) Bandstructure and DOS of a QWR without excitonic effects

where $V_1(x) = 0$ is the potential ruling the carrier free motion along the wire length (x-axis) and $V_{2,3}$ (y,z) vanishes for $0 < y < L_y$ and $0 < z < L_z$ and is infinite otherwise. Accordingly, the wave function can be written as the following product of wave functions

$$\psi(x, y, z) = \psi^1(x)\ \psi^{2,3}(y, z)$$

Substitution of the above general expression of potential and wave function in the time-independent Schrödinger Eq. (3.1) lets get the following independent Schrödinger equations:

$$\left(-\frac{\hbar^2}{2m_x^*}\frac{\partial^2}{\partial x^2}\right)\psi^1(x) = E^1\psi^1(x)$$

$$\left(-\frac{\hbar^2}{2m_y^*}\left(\frac{\partial^2}{\partial y^2}\right) - \frac{\hbar^2}{2m_z^*}\left(\frac{\partial^2}{\partial z^2}\right) + V_2(y,z)\right)\psi^2(y,z) = E^{2,3}\psi^{2,3}(y,z)$$

where m_x*, m_y* and m_z* is the carrier effective mass along the x, y and z-direction, respectively.

The first Schrödinger equation can be satisfied by a plane wave function

$$\psi^1(x) \approx e^{ik_x \cdot x}$$

and by energy eigenvalues

$$E^1 = \left(\frac{\hbar^2 k_x{}^2}{2m_x^*}\right)$$

where x is the position along the direction of the QWR axis (free-motion direction) and k_x is the wave vector along the x-axis.

The second Schrödinger equation can be easily solved by observing that in the case of a QWR with rectangular cross-section, geometry lets decouple the dependence on y and z, namely the potential $V_{2,3}$ can be written as $V_{2,3}$ (y,z) = V_2 (y) + V_3 (z), where $V_2(y) = 0$ if $0 < y < L_y$, $V_2(z) = 0$ if $0 < z < L_z$ and both V_2 and V_3 are infinite otherwise. This approach that separates the variables enables to express

$$\psi^2(y,z) = \psi^2(y)\,\psi^3(z)$$

and

$$E^{2,3} = E^2 + E^3$$

where

$$\left(-\frac{\hbar^2}{2m_y^*}\left(\frac{d^2}{dy^2}\right)\right)\psi^2(y) = E^2\psi^2(y)$$

$$\left(-\frac{\hbar^2}{2m_z^*}\left(\frac{d^2}{dz^2}\right)\right)\psi^3(z) = E^3\psi^3(z)$$

Since the obtained equations to solve are mathematically the Schrödinger equation of an infinite one-dimensional quantum well, eigenfunctions and eivgenvalues are already known. Therefore, under the assumption that L_y and L_z are comparable to the electron De Broglie wavelength in the wire material, it results the following quantized energy spectrum:

$$E^{2,3}=E^2+E^3=\left(\frac{\hbar^2}{2m_y^*}\right)\left(\frac{n_y\pi}{L_y}\right)^2+\left(\frac{\hbar^2}{2m_z^*}\right)\left(\frac{n_z\pi}{z}\right)^2 \text{ where } n_y, n_z = 1, 2, 3, \ldots$$

Therefore, the energy spectrum of a rectangular QWR in the free-electron approximation is given by

$$E_{QWR}^{tot} = E_g + \left(\frac{\hbar^2 {k_x}^2}{2m_x^*}\right) + \left(\frac{\hbar^2}{2m_y^*}\right)\left(\frac{n_y\pi}{L_y}\right)^2 + \left(\frac{\hbar^2}{2m_z^*}\right)\left(\frac{n_z\pi}{L_z}\right)^2$$
$$\text{where} \quad n_y, n_z = 1, 2, 3, \ldots$$

where E_g is the band gap energy of the bulk material and two quantum numbers, instead of one as in the case of a QW, are now introduced by 2D confinement. In analogy to a QW, a subband forms at each eigenvalue labeled by a couple of quantum numbers (n_y and n_z) with each subband having a free-particle dispersion (i.e., parabolic) as shown by the band structure in Fig. 3.14b. Given the energy dispersion, the resulting DOS has the following expression:

$$g^{QWR}(E) = \frac{1}{\pi}\sqrt{\frac{2m^*}{\hbar^2}}\sum_{n_y n_z}\frac{1}{\sqrt{E-E_{n_y n_z}}}\theta\left(E-E_{n_y n_z}\right)$$

where En_y,n_z are the confinement energies associated with the y and z directions and the θ function is the Heaviside unit step function The DOS of a QWR has a characteristic inverse square root dependence on the quantized confinement energies leading to the DOS profile shown in Fig. 3.14b. If only the first subband is filled, the system is purely 1D. As more subbands are filled, the system is quasi-1D and the DOS is found by summing over all subbands. The discontinuities in the DOS of a QWR are known as van Hove singularities.

More detailed calculation are reported in the literature to include the effects of the finite barrier height and the validity of the parabolic approximation to represent the valence subbands of a QWR [180, 181]. Notably, in contrast to a QW, where the reduced translational invariance lifts the bulk degeneration of the topmost valence band, the valence subbands of a QWR consist of a weighted linear combination of hh and lh states in such a way that distinction between lh and hh is unjustified in general [182–185]. A consequence of the peculiarities of the dispersion of a QWR is represented by the polarization selection rules accounting for the relative orientation of the polarization of the exciting field with respect to the axis of the QWR. The absorption strength of the hh exciton divided by the absorption strength of the lh exciton, depending on the polarization direction, gives indication of the anisotropy of the optical matrix elements [185, 186]. In regard to the allowed optical transitions of a QWR, the strength of hh (lh) transitions increases if the polarization of the exciting electric field is parallel (perpendicular) to the wire axis. Besides the valence mixing hh-lh, other quantities influencing the polarization anisotropy are the aspect ratio of rectangular QWRs [185] and the geometry of the cross-section (e.g., square, rectangular, triangular) of the QWR [187].

Because of larger exciton binding energies associated with confinement, as it occurs in QW structures, excitonic effects are expected to be very important in QWRs even at room temperature [188]. Indeed, to interpret the optical spectra of a QWR, the above calculated electron confinement energies must be modified as follows to include excitonic effects:

$$E_{QWR}^{tot} = E_g + \left(\frac{\hbar^2 k_x{}^2}{2m_x^*}\right) + \left(\frac{\hbar^2}{2m_y^*}\right)\left(\frac{n_y \pi}{L_y}\right)^2 + \left(\frac{\hbar^2}{2m_z^*}\right)\left(\frac{n_z \pi}{L_z}\right)^2 - E_{exc} \text{ where } n_y, n_z = 1, 2, 3, \ldots$$

where E_{ex} is the exciton binding energy. Since a carrier is confined in the wire plane and can move freely along the wire direction of the wire axis, the motion along the wire axis can be separated from the motion in the wire cross-section. The excitonic problem can be solved by introducing relative and center-of-mass coordinates that enable to decouple the starting two-body equation into two one-body equations, as described for a bulk exciton. The sum of the one-particle potentials due to the wire confinement taken at the center-of-mass coordinates is the potential to solve the center-of-mass motion. The relative motion along the wire direction is controlled by the Coulomb interaction.

As discussed in the case of a QW, the exciton binding energy depends on the probability that electron and hole can be found in the same unit cell expressed by the overlap integral of the electron and hole wave functions in the QW. For ideal QW, the binding energy of the ground-state exciton is four times the 3D effective Rydberg. This limit is reached when the well thickness is progressively reduced [136]. Turning from a QW to a QWR, the introduction of an additional confinement direction further enhances the exciton binding energy (by roughly 50 % and depending on the barrier height and degree of non-parabolicity of the valence band) [181]. Large exciton binding energies larger than the ones of QWs of comparable confinement length have been observed in semiconductor wires fabricated by different techniques [189–194]. In a QWR the exciton binding energy is strongly dependent on the spatial extension of the electron and hole wave-functions and theoretically can range from the bulk exciton binding energy to infinity [195–197]. This divergence would suggest that in the ideal 1D limit exciton binding energies can be increased much beyond the 2D limit of a QW [198, 199]. In realistic cases, the exciton binding energy in a QWR is finite and rarely is larger than four times the bulk exciton binding energy. Variational calculations were performed to predict the dependence of the excitonic ground state binding energy on the shape and height of the confining potential in a QWR as well as its scaling with size [200]. It was found a scaling governed by the extension of the single-particle wavefunctions in the plane perpendicular to the free motion wire direction and much less sensitive to the shape of the wire. Experimental results do not always confirm such a prediction about the dependence on the wire shape [201]. As a general rule, the influence of the excitonic effects on the optical properties of confined systems depends on the size of the confinement length, meaning that if such a length is much larger than the bulk Bohr radius of the exciton a_{ex}^{bulk} then a very weak influence of the confinement is expected (i. e., a weak confinement regime). In detail, for a wide wire ($L_y, L_z > 3\ a_{ex}$) in which the carriers are not confined (L_y, L_z larger than De Broglie wavelength) the excitons feel the boundaries of the wire structure and this involves discrete dispersion of the center of mass motion of the exciton [202, 203]. If $L_y, L_z < 3\ a_{ex}$ then carriers are confined with discrete spectrum (quantization-induced blue-shift of the electronic states) which dominates the optical properties of the QWR and the excitonic motion consists of a purely 1D-confined excitonic motion along the confinement direction and a relative motion along the wire direction controlled by the Coulomb interaction. The reduced dimensionality affects the excitonic wave functions in that the overlap integral of the electron-hole wave functions increases, hence involving enhanced exciton binding energy and oscillator strength of the optical transitions.

Laser operation from GaAs QWRs was first demonstrated in 1989 [204], but then the difficulty in fabricating QWR structures caused relatively slow progress that was further slowed following the discovery that 0D-systems form spontaneously during molecular beam epitaxy (MBE) growth in the Stranski–Krastanow regime [205].

3.3.3 *Quantum Confinement in Semiconductor QDs*

If carriers are spatially confined down to nanometric scale along all directions, which is the ultimate limit, then 3D confinement results that defines a QD heterostructure. Early synthesis of PbS QDs dates to more than 2000 years ago [206]. More recently, incorporation and tuning of the size and stoichiometry (CdS, CdSe, CuCl and CdSxSe1-x) of QDs in silicate glasses and colloidal solutions was a well assessed trick to change the color of glass [207–209]. In general, semiconductor QDs have been intensively investigated for their unique optical and electrical properties [210, 211] and applications as light emitting devices (LEDs and lasers), coupling QD-(semiconductor/photonic crystal) microcavity systems, solar cells, spintronic devices and biological markers. Fundamental and applicative interest in the potentialities of QDs prompted the development and optimization of processing and growth approaches such as molecular beam epitaxy (MBE) [212], metal-organic chemical-vapor deposition (MOCVD) [213], self organization approaches (QDs fabricated by Volmer-Weber [214] or Stranski-Krastanow [214, 215] growth) and "freestanding" (e.g., colloidal) QDs [216, 217].

Carriers spatially confined in all three coordinates cannot be described by plane-wave eigenstates. In this situation, whenever a characteristic geometric length of the system (e.g., the side for a cubic box or the radius of a spherical particle) is comparable or smaller than the electronic De Broglie wavelength of the corresponding bulk material, then quantum size effects result. To calculate the energy dispersion of a 0D-system (QD or quantum box) the same general theoretical guidelines developed in the case of QWs and QWRs can be applied. In a simplified model a QD can be assumed to be a box with side L_x, L_y and L_z along the direction x, y and z, respectively (Fig. 3.15a). Under the effective mass approximation of the single-electron model, the electronic states are described by the envelope function satisfying a time-independent Schrödinger equation with infinitely deep potential well along each independent spatial direction (x, y and z) resulting from the potential energy barriers formed at the QD's boundaries. Based on the same arguments applied in the case of a QWR, the 3D confinement potential of a rectangular QD (Fig. 3.15a) can be decomposed as follows

$$V_{QD}(x, y, z) = V_1(x) + V_2(y) + V_3(z)$$

where $V_i(x) = 0$ for $0 < x < L_j$ (where $(i,j) = (1,x), (2,y), (3,z)$) and is infinite otherwise. Accordingly to the method of separable variables, the wave function to be determined can be written as follows

$$\psi(x, y, z) = \psi^1(x)\ \psi^2(y)\ \psi^3(z)$$

Substitution of the above general expression of potential and wave function within the time-independent Schrödinger Eq. (3.1) lets get the following system of three decoupled time-independent Schrödinger equations:

$$\left(-\frac{\hbar^2}{2m_x^*}\left(\frac{d^2}{dx^2}\right)\right)\psi^1(x) = E^1\psi^1(x)$$

$$\left(-\frac{\hbar^2}{2m_y^*}\left(\frac{d^2}{dy^2}\right)\right)\psi^2(y) = E^2\psi^2(y)$$

$$\left(-\frac{\hbar^2}{2m_z^*}\left(\frac{d^2}{dz^2}\right)\right)\psi^3(z) = E^3\psi^3(z)$$

Following the same arguments of QWs and QWRs, the confined energies of a rectangular QD are given as follows

$$E_{n_x n_y n_z} = E^1 + E^2 + E^3 = \left(\frac{\hbar^2}{2m_x^*}\right)\left(\frac{n_x\pi}{L_x}\right)^2 + \left(\frac{\hbar^2}{2m_y^*}\right)\left(\frac{n_y\pi}{L_y}\right)^2 +$$
$$\left(\frac{\hbar^2}{2m_z^*}\right)\left(\frac{n_z\pi}{z}\right)^2 \quad (n_x,\ n_y, n_z = 1, 2, 3, \ldots)$$

Schematic representation of the quantum confinement effects of a QD shown in Fig. 3.15c) depicts the changes in the structure of both valence band and conduction

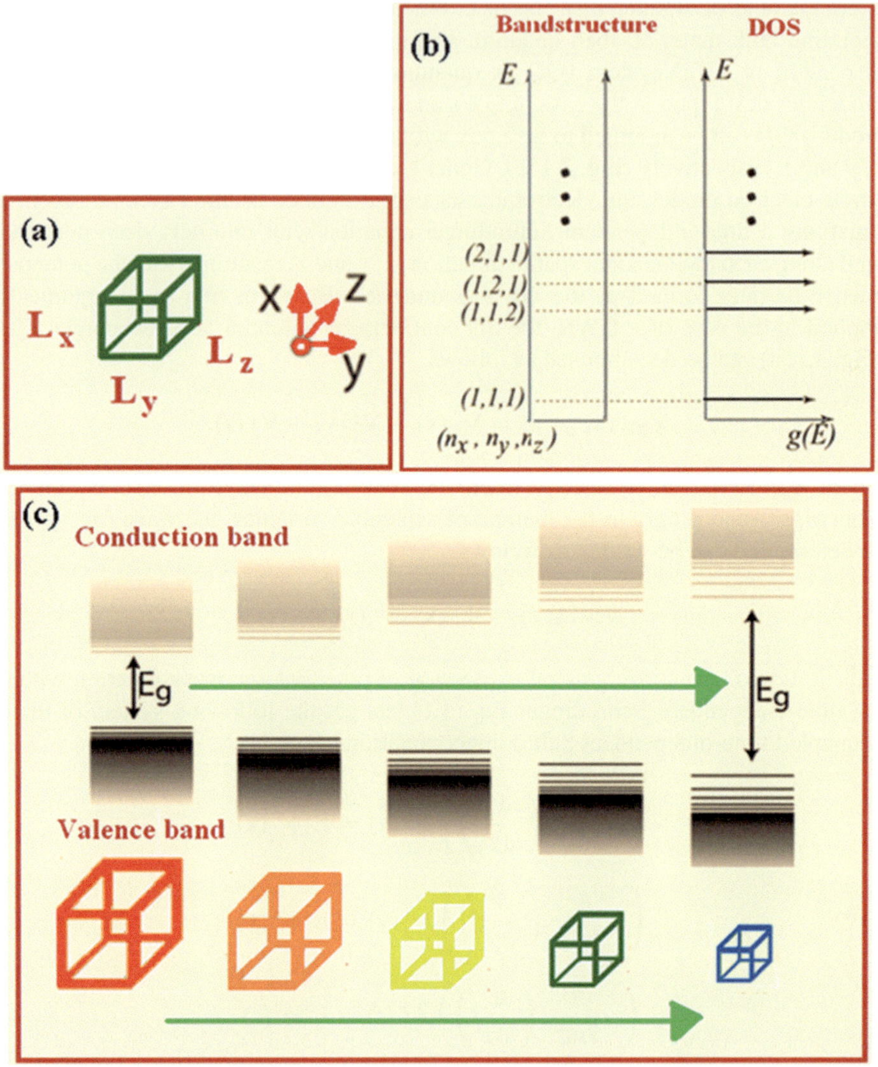

Fig. 3.15 (**a**) Cubic QD and (**b**) corresponding bandstructure and DOS. (**c**) Schematic representation of the quantum confinement effects of a QD: on the structure of both valence band and conduction band: discrete energy states at the band edges of the bulk material with energies and their relative spacing increasing with decreasing size

band: discrete energy states arise at the band edges of the bulk material and such allowed confined energies and their relative spacing increase with decreasing size of the QD [218]. Quantum confinement effects are observed when the QD's size is small enough that the energy level spacing of the confined energies exceeds k_BT (where k_B is Boltzmann's constant and T is the absolute temperature). Typically, the spacing range between QD's intraband energy levels is of 10–100 meV. The carrier effective mass is important in determining intraband energy spacings too. Notably, the calculated energy spectrum indicates that a QD exhibits distinct narrow optical line spectrum. Parameters involved in the mathematical expression of discrete energy spectrum of a QD indicate that an attractive property of a 0D-system is the possibility to tailor/tune its size, shape and composition for a desired application.

Therefore, the band structure of a QD is a sequence of discrete energy values (atom-like spectrum) (Fig. 3.15b) leading to a DOS given as a series of delta functions

$$g^{QD}(E) = 2\sum_{n_x n_y n_z} \delta\left(E - E_{n_x n_y n_z}\right)$$

where the factor 2 takes into account the spin degeneracy. This is why QDs are often termed "artificial atoms"(δ-function-like DOS) [219].

In contrast to a QW and a QWR, the DOS of a QD doesn't have at least a piece wise continuous profile, which has an important impact on spectroscopic properties. Indeed, the electron energy spectrum of a QD in the effective mass and free-electron approximations has the following expression:

$$E_{QD}^{tot} = E_g + \left(\frac{\hbar^2}{2m_x{}^*}\right)\left(\frac{n_x\pi}{L_x}\right)^2 + \left(\frac{\hbar^2}{2m_y{}^*}\right)\left(\frac{n_y\pi}{L_y}\right)^2 + \left(\frac{\hbar^2}{2m_z{}^*}\right)\left(\frac{n_z\pi}{z}\right)^2$$

$$\text{where } n_x,\ n_y, n_z = 1, 2, 3, \ldots$$

where E_g is the energy gap of the bulk material and n_x, n_y and n_z are three integer quantum numbers associated to the confinement directions. For example, the energy levels for a cubic QD ($L_x = L_y = L_z$) having side length L and electronic isotropic effective mass ($m_x{}^* = m_y{}^* = m_z{}^* = m_e{}^*$) are

$$E_{QD}^{tot} = E_g + \left(\frac{\hbar^2}{2m_e{}^*}\right)\left(\frac{\pi}{L}\right)^2\left(n_x^2 + n_y^2 + n_z^2\right) \text{ where } n_x,\ n_y, n_z = 1, 2, 3, \ldots$$

Energy direct transitions from the mth confined energy in the valence band to the nth confined energy in the conduction band are allowed under absorption of incident exciting photons with energy

$$\hbar\omega = E_g + E_n + E_m = E_g + \left(\frac{\hbar^2}{2m_e{}^*}\right)\left(\frac{n\pi}{L}\right)^2 + \left(\frac{\hbar^2}{2m_h{}^*}\right)\left(\frac{m\pi}{L}\right)^2$$

$$(n, m = 1, 2, 3, \ldots)$$

where $m_h{}^*$ is the hole effective mass. Hence, the band gap energy of a cubic QD is given by the above formula by setting $n = m = 1$:

$$E_g^{QD} = E_g + \left(\frac{\hbar^2}{2m_e{}^*}\right)\left(\frac{\pi}{L}\right)^2 + \left(\frac{\hbar^2}{2m_h{}^*}\right)\left(\frac{\pi}{L}\right)^2$$

Such a relationship indicates that the fundamental absorption edge of a QD is larger than the energy gap of the bulk counterpart and blue-shifts with decreasing size of the QD (Fig. 3.15c). The energy gap of a QD also depends on the semiconductor material and its stoichiometry. All of this allows tuning of the energy gap through changes in the QD size, material and composition (doping).

To summarize the key points in the case of a QD (0D system), quantization of the electron and hole spatial motion in any direction has the following implications: (i) quantized energy spectrum mimics the atomic one, (ii) the effective band-edge is blue-shifted; (iii) confinement forces the wave function of electron and hole to overlap, thus favoring the increase of the oscillator strength and probability of radiative transitions; and (iv) the discrete δ-function-like DOS produces narrow optical line spectra. Points (i) and (iv) justify the nomenclature "artificial atoms" adopted to describe the peculiar spectroscopy behavior of QD systems (Fig. 3.16a). Based on the delta-like DOS, the optical spectrum of a QD would be expected to consist of spectrally narrow emission characteristics. Instead, the conventional spectroscopic measurements of QDs reveal emission spectra with significantly broadened peaks [220–225]. This experimental evidence can be attributed to the fact that conventional procedures acquire the emission response of an ensemble of QDs rather than of a single QD. Hence, spectroscopy of single QDs is the starting point to understand and evaluate their intrinsic physical properties but, in practice, it is not possible to describe a QD ensemble/array by the spectroscopy of a single QD. In fact, fluctuations of size and shape ("regularly shaped" QDs require excellent growth conditions!) as well as compositional non-uniformities involve spectral inhomogeneous broadening of the spectral lines. Indeed, real samples consist of a distribution of QDs polydispersed in size and shape. If several families $i = 1, \ldots, N$ of QDs with dimension $R_i \pm \Delta R_i$ concur to the emission spectrum of a QD-based system, then the observed emission spectrum reflects the convolution of all the emission spectra resulting from each family of QDs composing the sample (Fig. 3.16b). Therefore, fabrication techniques are crucial to ensure structural quality of the samples avoiding strong inhomogeneous broadening of the absorption/emission features. Extremely high spatial resolution imaging tools can help in probing the optical response of a single QD and disclose its atom-like behavior [219, 222, 226].

Furthermore, spectroscopic properties of QDs cannot be correctly interpreted without taking into account the excitonic effects that often play a dominant role because of further increased binding energy of an exciton in a QD as compared to QWs and QWRs. Optical transitions in QDs can be described by an approximated expression (known as the Brus equation) that models the transition energy in spherical QDs with radius R and relative dielectric constant ε_r (QD with dielectric coefficient ε_{QD} surrounded by a medium of dielectric coefficient ε_m):

$$E_g^{QD} = E_g + \left(\frac{\hbar^2 \pi^2}{2R^2}\right)\left(\frac{1}{m_e{}^*} + \frac{1}{m_h{}^*}\right) - \left(\frac{1.8\, q^2}{4\pi R \epsilon_r\, \epsilon_o}\right)$$

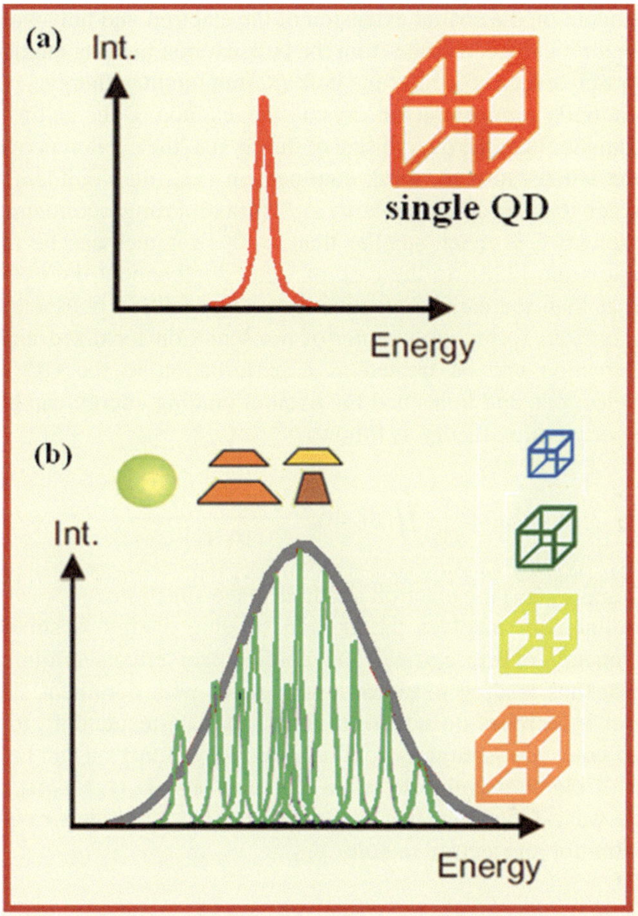

Fig. 3.16 (**a**) Emission spectrum of (a) a single QD and (**b**) a distribution of QDs with size and shape dispertion

In the Brus equation ε_o is the permittivity of vacuum and ε_r is the relative dielectric constant of the sherical QD that confines an exciton. The third term is related to the electron-hole Coulomb attraction and express the binding energy of the exciton which is influenced from the size R of the QD due to quantum confinement [227].

As described earlier, in bulk semiconductors, excitons are observable typically only at very low temperatures because room temperature thermal energy can easily dissociate the exciton. Instead, in quantum confined structures, a closer spacing between electron and hole enhances the excitonic binding energy (the exciton feels the confinement boundaries even if the confining well is larger than the exciton bulk Bohr radius) that turns out to be more enhanced and stable at room temperature the larger the number of confining directions. In a QD the exciton binding energy is strongly dependent on the spatial extension of the electron and hole wave-functions and, even if theoretically can range from the bulk exciton binding energy to infinity, in practice it rarely exceeds 4 times the bulk exciton binding energy.

Since in a bulk semiconductor crystal the exciton Bohr radius ($a_{ex}{}^{bulk}$) is significantly smaller than the overall size of the crystal, the exciton is free to migrate throughout the lattice. Instead, weak confinement (i.e., the confinement length L is slightly larger than or comparable to $a_{ex}{}^{bulk}$) and strong confinement (i.e., the confinement length L is much smaller than $a_{ex}{}^{bulk}$) regimes can be introduced at the mesoscopic scale [228]. In the case of weak confinement the exciton binding energy is larger than the carrier quantization energy and, in presence of quantum confinement, exciton forms with center-of-mass motion localized and quantized. Coulomb interaction can be treated as a perturbation to the QD confinement potential for electrons and holes and the exciton binding energy can be calculated based on the perturbation theory as follows

$$E_B \approx \frac{e^2}{\varepsilon_r} \iint d\mathbf{r}_e d\mathbf{r}_h \frac{|U_e(\mathbf{r}_e)\, U_h(\mathbf{r}_h)|^2}{|\mathbf{r}_e\text{-}\mathbf{r}_h|}$$

where $U_e(\mathbf{r_e})$ and $U_h(\mathbf{r_h})$ are electron and hole wave-functions.

Strong confinement involves the exciton binding energy is smaller than the confinement energy of the carriers and confinement effects dominate over the Coulomb ones, thus leading to electron-hole states with dominant single-particle character. That is, carriers do not form bond states corresponding to the exciton (electrons and holes are separately confined) and wave function can be represented as a product of electron and hole wave-functions $\psi = U_e(\mathbf{r_e})\ U_h(\mathbf{r_h})$, where the single-particle wave-functions are solutions of coupled Schrödinger equations with the QD potential for an electron (a hole) $V_{e,h}$

$$\left(-\frac{\hbar^2}{2m^*{}_{e,h}} \nabla^2_{e,h} + V_{e,h} - \frac{e^2}{\varepsilon_r} \int d\mathbf{r}_{h,e} \frac{|U_{h,e}(\mathbf{r}_{h,e})|^2}{|\mathbf{r}_e\text{-}\mathbf{r}_h|} \right) U_{e,h}(\mathbf{r}_{e,h}) = E_{e,h} U_{e,h}(\mathbf{r}_{h,e})$$

In this case, the exciton binding energy is defined by

$$E_B^{QD} = E_e^o + E_h^o - E_e - E_h$$

where E_e^o and E_h^o are energies of electron and holes that do not interact by Coulomb coupling.

In semiconductor QDs singly and multiply charged excitons were studied by photoluminescence spectroscopy applied to a single QD [229–232].

More generally, the bulk exciton Bohr radius ($a_{ex}{}^{bulk}$) is taken as the measure of quantum confinement in low-dimensional systems: given a solid state structure with characteristic length R:

(i) if $R \gg a_{ex}{}^{bulk}$, then confinement effects are generally negligible;
(ii) if $R > a_{ex}{}^{bulk}$ or $R \sim a_{ex}{}^{bulk}$, then the weak confinement regime occurs;
(iii) if $R < a_{ex}{}^{bulk}$, then we the strong confinement regime occurs.

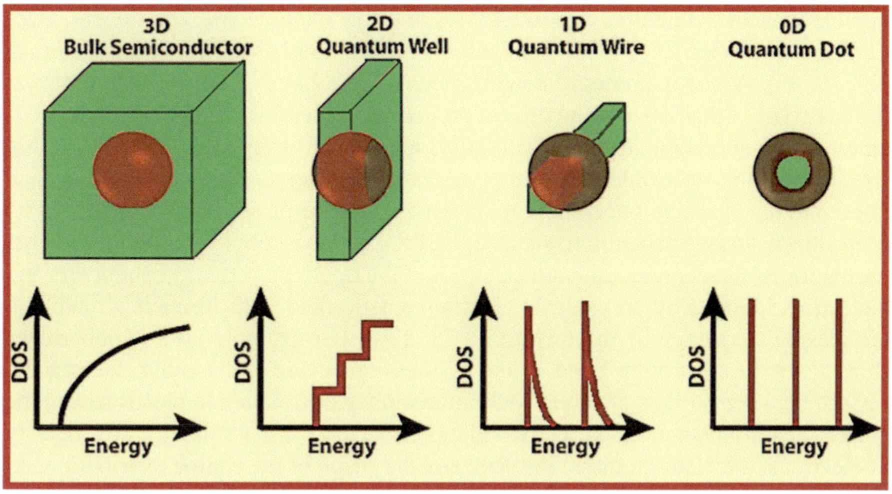

Fig. 3.17 Schematic overview of the DOS of semiconductor nanostructures (2D, 1D and 0D structures) in the excitonic strong confinement regime

A schematic overview of the DOS of semiconductor nanostructures (2D, 1D and 0D structures) in the strong confinement regime is shown in Fig. 3.17 [233]. For comparison the case of the bulk semiconductor (3D case) is also depicted and the exciton Bohr diameter is represented by a dark sphere.

Another important role in interpreting the optical properties of QDs is played by the electronic quantum states associated with the surface (called surface states) due to the high surface-to-volume ratio of mesoscopic systems [234, 235]. High density of surface sites associated with high surface-to-volume ratio may cause an enhanced or reduced transfer rate of photogenerated charge carriers (electron or hole or exciton) by their trapping and behaving as reducing (electron) or oxidizing (hole) agents [236, 237]. Capping or passivation (i.e., saturation of surface dangling bonds) of the QD's surface lets provide photostable QDs whenever surface states are detrimental because of their quenching radiative recombination and reducing the quantum yields.

3.4 Quantum Effects in Metal Nanoparticles

As already pointed out, the Fermi wavelength is the smallest length scale associated with confinement effects in metal nanostructures, that is properties become size-dependent when the geometrical size of the metal structure is comparable to or smaller than the Fermi wavelength. Since the conduction electron density is larger in metals than in semiconductors, the Fermi wavelength is of the order of a few nanometers in metals and tens of nanometers in semiconductors. A consequence of very short Fermi wavelength in metals is that confinement effects may be observed in metal nanostructures only at size-scales of a few nanometers. The optical, electromagnetic properties of metal nanoparticles remarkably differ from bulk materials [23, 29, 35]. At the nanometric length-scale, for size larger than a few nanometers (tipically larger than 10 nm) the optical properties of metallic nanoparticles are governed by the so called surface plasmon resonance that disappears with decreasing size. Indeed, the electronic structure of metal nanoclusters/nanoparticles exhibits an important nanoparticle phenomenon known as "surface plasmon resonance" (SPR) or localized surface plasmon resonance (LSPR), that is a coherent collective charge oscillation of the conduction band electrons at the interface between conductors and dielectrics induced by an external electromagnetic field with frequency matching the plasma frequency of the metal [22–28, 238]. For example gold nanoparticles show a strong absorption band in the visible region due to coupling through the surface between surface plasmon oscillation modes of conduction electrons and the applied electromagnetic field. The oscillation frequency is determined by the density of electrons, the electron mass, the size, and the shape of the charge distribution. At the fundamental level, the free electrons of a metal are modeled like a gas of free charge carriers (a so-called plasma) and can be excited to sustain propagating plasma waves, that is longitudinal electromagnetic charge density waves, whose quanta are termed "plasmons". Plasmons exist in the form of both bulk plasmons in the volume of a plasma and surface plasmons bound to the interface between a plasma and a dielectric. When the size of a metal nanoparticle is in the range of the penetration depth of an electromagnetic field into the metal (e.g., ~20 nm for silver in the optical spectral range), there is no distintion between surface and bulk plasmons.

The first example of this nanoscale phenomenon dates to some glassblowers from imperial Rome made by embedding colloidal metal nanoparticles within glassy works. Another well-known fine example is the famous Lycurgus Cup (fourth century A.D.): such a chalice has a dark greenish tint under reflected lighting and appears red once illuminated from behind. Some of the beautiful bright colors in medieval stained glass windows are a metal-nanoparticle phenomenon that fascinated scientists long before studies of semiconductors and was explained later when Faraday (1857) proposed metallic gold in its colloidal form as origin of colorful emission and Mie (1908) [239] provided the theoretical foundations for understanding the physical behavior and optical responses of metal nanoparticles by solving Maxwell's equation for the absorption and scattering of electromagnetic radiation by spherical particles ($2R << \lambda$, being λ the wavelength of the light irradiating the particle with radius R). On the basis of Mie theory, the plasmon band of metal particles involves electric-field induced dipolar oscillations of the free electrons in the conduction band that occupy energy states near the Fermi level.

Essentially, bright colors can be observed from colloidal solutions of metallic nanoparticles with their color emission tunable by changing shape, size, composition, orientation and local dielectric environment of the nanostructures [240, 241]. In what concerns the comparison between the optical properties of semiconductor QD [242] and metal nanoparticles [243] as a function of size and shape, it can be observed that while the optical spectra of semiconductor QDs are very sensitive to changes in size (due to the already discussed confinement energies which scale as the inverse of the squared radius), instead just a slight change and a drastic change of the optical response can be observed in the case of metal nanoparticles versus size and shape changes, respectively. Therefore, metal nanoparticles are more sensitive to anisotropy than to size tuning. Introducing an anisotropy (for example a shape transition from sphere to rod by femtosecond laser irradiation of colloidal metal nanoparticles) results in the splitting of the SPR into two modes. Also, an increase in the refractive index of the dielectric surrounding a metal nanoparticle leads to a red-shift of the surface plsmon resonance peak as well to an increase in the plasmon band intensity and bandwidth. The shift if the SPR peak caused by changes in the local environment (such as adsorbed species or a core-shell structure) within a distance d around the nanoparticle can be calculated by $\Delta\lambda_{max} = n\ \Delta n[1-\exp(-2d/l_d)]$, where n is the bulk refractive-index response of the nanoparticle(s); Δn is the change in refractive index induced by the surrounding and l_d is the characteristic exponential decay length of the electromagnetic field [244, 245]. In the case of a metal-core/shell structure of the nanoparticle, thickening of the shell embedding the nanoparticle can cause a blue-shift or a red-shift of the SPR peak depending on the shell composition [246–248].

Under resonant excitation of metal nanoparticles, an enhancement of the local electric fields close to the particle surface occurs. Enhancement and local SPR dependency on size and shape indicate that the smallest nanoparticles have a maximum SPR peak at shorter wavelengths [249, 250], too small nanoparticles exhibit reduced effective conductivity and light scattering [251], for nanoparticle size of the order of the exciting wavelength non radiative modes become favoured [251] and shapes with high curvature corners and edges strongly enhance the local field around these geometrical features [252, 253].

Scaling down to dimensions of a few nanometers removes the SPR from the optical properties of the metal nanostructures and enables to observe a shift in the conductive properties from metallic to semiconducting and insulating with progressively decreasing size. The origin of such a modified electronic structure of metals at very small scale-lengths is the discreteness of the electronic states due to confinement of the electron wavefunction. As the nanoparticle size decreases, the energy continuum of the bulk metal transforms into discrete energy levels (Fig. 3.18). The average spacing of successive quantum levels, δ, is termed the Kubo gap and is given by $\delta = 4E_F/3\ N$, where E_F is the Fermi energy of the bulk material and N is the nuclearity (number of atoms in the metal nanocluster), which is related to the number of valence electrons in the metal [30–32]. In this picture, metallic behavior is ascribed to gap between occupied and unoccupied electronic states at the Fermi level vanishing (bulk metal) or smaller than the thermal energy k_BT (metal nanoparticle) (Fig. 3.18). For scaling size from bulk to nano-scale metal the Kubo gap increases because the number of atoms decreases. Therefore, for decreasing size (and/or temperature) a threshold is reached when the Kubo gap equals the thermal energy k_BT. In this circumstance a transition/shift in conductive properties, from metallic to semiconducting/insulating behavior, results. Based on a well-known mechanism applied in semiconductor solid-state physics, when electrons can be thermally excited across the Kubo gap, a low temperature insulator becomes a semiconductor and a metal at higher temperatures. According to the principles of band theory, the overlapping between atomic orbitals of neighboring atoms

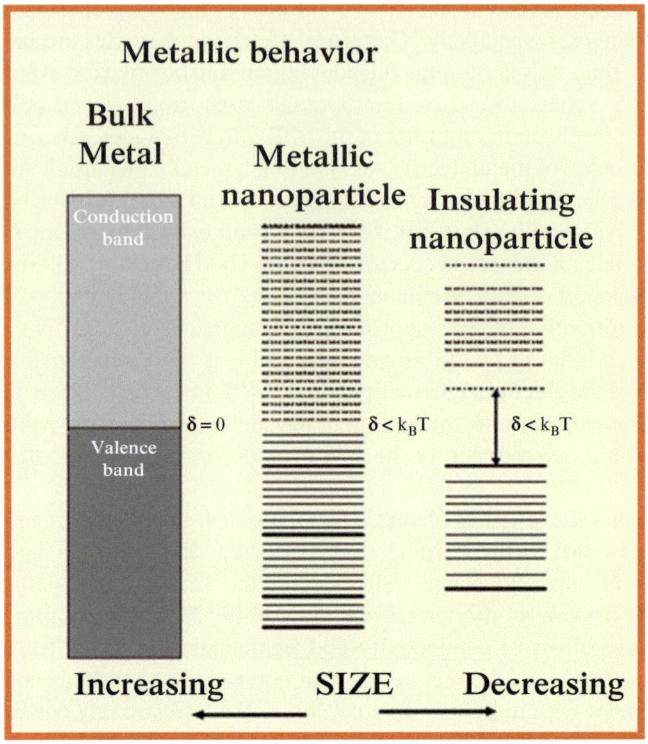

Fig. 3.18 Changes of the electronic structure of metals going from the bulk phase to the nanoscale (a few nanometers) where the plasmonic resonance disappears and discrete energy spectrum accounts for the metal-to-insulator conductivity transition

leads to bonding and antibonding states, namely valence and conduction bands in semiconductors, with a band width related to the bond strength. Hence, the insulating properties of nanometric metal systems can be ascribed to the small number of neighboring atoms. Indeed, in the case of systems consisting of a relatively poor number of atoms (e.g., nanoparticles with size of a few nanometers), weak non metallic interactions (e.g., van der Waal forces) hold together the atom constituents and no band may form.

3.5 Summary and Conclusions

Since the famous Feynman's lecture, titled "There's Plenty of Room at the Bottom", regarding the challenging applicative perspectives for nanomaterials and the first use of the term "nanotechnology" by Taniguchi, there has been a widespread interest devoted to nanomaterials at both fundamental and applicative level.

While the fundamental behavior of bulk materials is inherently determined by structure and composition, at the nanometer length-scale fascinating effects and phenomena occur strictly related to either spatial confinement of the electronic wavefunctions or surface-effects.

This review paper has introduced and discussed the fascinating fundamental properties of nanomaterials by treating the meaning of quantum confinement, the importance of the surface effects and their technological applications, the classification usually adopted to refer to the low-dimensional systems, the quantum models applied to solve for both electronic and excitonic dispertion of the mesoscopic semiconductor systems and the nanoscale behavior of the metal nanoparticles.

The developed discussion has pointed out the following main different behaviors of the nanoworld as compared to the 3D bulk world.

In regard to the confinement effects, when at least one of the material dimensions L is comparable with the De Broglie electronic wavelength of the corresponding bulk band-gap material, then the energy spectrum becomes quantized along the confinement direction. Such phenomenon, termed "quantum size effect" involves a larger band gap and a blue shift of the absorption onset with decreasing size as compared of the bulk counterpart as well as material optical properties tunable as a function of both the confinement dimension L and number of the confinement directions.

Moreover, whenever the length scale of the material is comparable to the wavelength of the phonons, discrete phonon density of states form too. Since electron-phonon interaction determines the thermal conductivity and affects carrier dynamics in quantum-confined systems, different thermalization rules and processes (multiphonon processes, Coulomb interactions and phonon-bottleneck effects) occur in nanomaterials as compared to the bulk counterpart due to discrete density of states.

In the case of metallic nanoclusters/nanoparticles, for which there is no band-gap between valence and conduction bands, another phenomenon known as *plasmon resonance* is active, involving localized enhancement of the field intensity with important applicative perspectives in bio-imaging and sensing. For metallic nanoclusters with dimensions smaller than 2 nm, the surface plasmon absorption disappears: the spacings between adjacent energy levels (referred to as the *Kubo* gap) become comparable to the thermal energy, k_BT, resulting in a shift in the conductive properties from metallic to semiconducting and insulating with decreasing size.

Among the technological important behaviors of nanomaterials, pre-melting and lowered melting temperature as compared to the bulk counterpart are very recently active research topics which are ascribed to surface effects such as the increased fraction of surface atoms and surface curvature with decreasing size. Lowered melting temperature may occur for isolated and un-supported nanomaterials. In fact, nanocrystals embedded in a matrix can melt below or above the melting point of the corresponding bulk crystal depending on the interface structure between embedded nanocrystals and the matrix. If the interfaces are coherent or semi-coherent, an enhancement of the melting point is present. Otherwise, there is a depression of the melting point.

Amazing thermal behavior is also exhibited by nanofluids, which can exhibit substantial increment of the thermal conductivity even when the concentrations of suspended nanoparticles are very low, and the more enhanced the smaller nanomaterial size. The thermal conductivity of nanofluids varies with the size, shape, and material of nanoparticles as well as difference of thermal conductivity between nanomaterial and base fluid.

On the other hand, the evidence of quantum size effects make the Plank's black-body model unable to describe the thermal emission spectrum at the nanoscale (at particle dimensions smaller than the radiation wavelength), where the Stefan–Boltzmann law yields strongly overestimated results.

Furthermore, the catalysis and magnetic properties of nanomaterials strongly differ as compared to the bulk materials. As can example, superparamagnetism is a form of magnetism peculiar of ferromagnetic or ferromagnetic nanomaterials.

Efforts of science and technology are presently focused on understanding the origin of the mechanical, chemical, reactivity, catalysis, thermal, electrical, optical, magnetic properties of nanomaterials, which strongly differ from the bulk counterpart, as well as being able to engineer and control shape and size, size- and shape- dispersion as well as composition at the nanoscale.

The huge amount of studies and papers about nanomaterials published in the last decades and still presently demonstrates that the interest in this research field is growing more and more. What do you expect?. Very likely further fundamental surprising properties as well as challenging opportunities and novel applications made possible by improved nanofabrication approaches and nanostructuring.

References

1. Feynman, R. (1992). There's plenty of room at the bottom. *Journal Microelectromechanical Systems, 1*, 60–66.
2. Taniguchi, N. (1974). On the basic concept of 'Nano-Technology, Proc. Intl. Conf. Prod. Eng. Tokyo, Part II, Japan Society of Precision Engineering.
3. McMullan, D. (1995). Scanning electron microscopy 1928–1965. *Scanning, 17*(3), 175–185.
4. Smith, K. C. A., & Oatley, C. W. (1955). The scanning electron microscope and its fields of application. *British Journal of Applied Physics, 6*(11), 391–399.
5. Zaefferer, S. (2011). A critical review of orientation microscopy in SEM and TEM, Crystal Research and Technology. *New Developments in Electron Diffraction, 46*, 607–628.
6. Williams, D. B., & Carter, C. B. (2009). *Transmission electron microscopy, a textbook for materials science* (2nd ed.). New York: Springer-Verlag.
7. Binnig, G., Rohrer, H., Gerber, C., & Weibel, E. (1982). Surface studies by scanning tunneling microscopy. *Physical Review Letters, 49*, 57–61.
8. Giaever, I. (1960). Energy gap in superconductors measured by electron tunneling. *Physical Review Letters, 5*, 147–148.
9. Binnig, G., Quate, C. F., & Gerber, C. (1986). Atomic force microscope. *Physical Review Letters, 56*, 930–933.
10. Synge, E. H. (1928). A suggested method for extending microscopic resolution into the ultra-microscopic region. *Philosophical Magazine, 6*, 356–362.

11. Betzig, E., Trautman, J. K., Harris, T. D., S: Weiner, J., & Kostelak, R. L. (1991). Breaking the diffraction barrier: Optical microscopy on a nanometric scale. *Science, 251*, 1468–1470.
12. Abbe, E. (1873). Beiträge zur theorie des mikroskops und der mikroskopischen wahrnehmung. *Archiv für mikroskopische Anatomie, 9*, 413–468.
13. Stroscio, M. A., & Dutta, M. (2001). *Phonons in nanostructures*. New York: Cambridge University Press.
14. Liu, G. K., Zhuang, H. Z., & Chen, X. Y. (2002). Restricted phonon relaxation and anomalous thermalization of rare-earth ions in nanocrystals. *Nano Letters, 2*, 535–539.
15. Meltzer, R. S., & Hong, K. S. (2000). Electron-phonon interactions in insulating nanoparticles: Eu_2O_3. *Physical Review B, 61*, 3396–3496.
16. Tamura, A. (1995). Smoothed density of states of electrons and smoothed frequency spectrum of phonons for a mesoscopic system. *Physical Review B, 52*, 2688–2676.
17. Collins, J. (2016). Non-radiative processes in crystals and in nanocrystals. *ECS Journal of Solid State Science and Technology, 5*(1), R3170–R3184 and references therein.
18. Heitz, R., Grundmann, M., Ledentsov, N. N., Eckey, L., Veit, M., Bimberg, D., Ustinov, V. M., Egorov, A. Y., Zhukov, A. E., Kopev, P. S., & Alferov, Z. I. (1996). Multiphonon-relaxation processes in self-organized InAs/GaAs quantum dots. *Applied Physics Letters, 68*, 361–363.
19. Raymond, S., Fafard, S., Charbonneau, S., Leon, R., Leonard, D., Petroff, P. M., & Merz, J. L. (1995). Photocarrier recombination in $Al_yIn_{1-y}As/Al_xGa_{1-x}As$ self-assembled quantum dots. *Physical Review B, 52*, 17238–17242.
20. Efros, A. L., Kharchenko, V. A., & Rosen, M. (1995). Breaking the phonon bottleneck in nanometer quantum dots: Role of Auger like processes. *Solid State Communications, 93*, 281–284.
21. Benisty, H., Sotomayor-Torres, C. M., & Weisbuch, C. (1991). Intrinsic mechanism for the poor luminescence properties of quantum-box systems. *Physical Review B, 44*, 10945–10948.
22. Otto, A. (1968). Excitation of nonradiative surface plasma waves in silver by the method of frustrated total reflection. *Zeitschrift fur Physik A: Hadrons and Nuclei, 216*, 398–410.
23. Novotny, L., & Hecht, B. (2006). *Principles of nano-optics* (pp. 378–393). Cambridge: Cambridge University Press.
24. Fan, X., Zheng, W., & Singh, D. J. (2014). Light scattering and surface plasmons on small spherical particles (review). *Light: Science & Applications, 3*, e179. doi:10.1038/lsa.2014.60.
25. Homol, J., Yee, S. S., & Gauglitz, G. (1999). Surface plasmon resonance sensors: Review. *Sensors and Actuators B: Chemical, 54*, 3–15.
26. Willets, K. A., & Van Duyne, R. P. (2007). Localized surface plasmon resonance spectroscopy and sensing. *Annual Review of Physical Chemistry, 58*, 267–297.
27. Bohren, C. F., & Huffman, D. R. (1983). *Absorption and scattering of light by small particles*. New York: Wiley.
28. Kreibig, U., & Vollmer, M. (1995). *Optical properties of metal clusters*. Berlin: Springer.
29. Warnes, W. L., Dereux, A., & Bobesen, T. W. (2003). Surface plasmon subwavelength optics. *Nature, 424*, 824–830.
30. Ghosh, S. K. (2011). Kubo gap as a factor governing the emergence of new physicochemical characteristics of the small metallic particulate. *Assam University Journal of Science & Technology: Physics & Science Technology, 7*, 114–121.
31. Kubo, R. (1962). Electronic properties of metallic fine particles. I. *Journal of the Physical Society of Japan, 17*, 975–986.
32. Kubo, R., Kawabata, A., & Kobayashi, S. (1984). Electronic properties of small particles. *Annual Review of Materials Science, 14*, 49–66.
33. Campion, A., & Kambhampati, P. (1998). Surface-enhanced Raman scattering. *Chemical Society Reviews, 27*, 241–250.
34. Sharma, B., Cardinal, M. F., Kleinman, S. L., Greeneltch, N. G., Frontiera, R. R., Blaber, M. G., Schatz, G. C., & Van Duyne, R. P. (2013). High-performance SERS substrates: Advances and challenges. *MRS Bulletin, 38*, 615–624.

35. Nie, S., & Enmory, S. R. (1997). Probing single molecules and single nanoparticles by surface-enhanced. *Science, 275*, 1102–1106.
36. Jankiewicz, B. J., Jamiola, D., Choma, J., & Jaroniec, M. (2012). Silica–metal core–shell nanostructures. *Advances in Colloid and Interface Science, 170*, 28–47.
37. Biener, J., Wittstock, A., Baumann, T. F., Weissmüller, J., Bäumer, M., & Hamza, A. V. (2009). Surface chemistry in nanoscale materials. *Materials, 2*, 2404–2428.
38. Pawlow, P. (1909). The dependency of the melting point on the surface energy of a solid body. *Zeitschrift für Physikalische Chemie, 65*, 545–548.
39. Takagi, M. (1954). Electron-diffraction study of liquid-solid transition of thin metal films. *Journal of the Physical Society of Japan, 9*, 359–363.
40. Goldstein, A. N., Echer, C. M., & Alivisatos, A. P. (1992). Melting in semiconductor nanocrystals. *Science, 256*, 1425–1427.
41. Buffat, P., & Borel, J.-P. (1976). Size effect on the melting temperature of gold particles. *Physical Review A: Atomic, Molecular, and Optical Physics, 13*, 2287–2298.
42. Sar, D. K., Nayak, P., & Nanda, K. K. (2008). Thermodynamic model for the size-dependent melting of prism-shaped nanoparticles. *Physics Letters A, 372*, 4627–4629.
43. Qi, W. H. (2005). Size effect on melting temperature of nanosolids. *Physica B, 368*, 46–50.
44. Sun, C. Q., Wang, Y., Tay, B. K., Li, S., Huang, H., & Zhang, Y. B. (2002). Correlation between the melting point of a nanosolid and the cohesive energy of a surface atom. *The Journal of Physical Chemistry. B, 106*, 10701–10705.
45. Bachels, T., Guntherodt, H. J., & Schafer, R. (2000). Melting of isolated tin nanoparticles. *Physical Review Letters, 85*, 1250–1253.
46. Zhang, M., Efremov, M. Y., Schiettekatte, F., Olson, E. A., Kwan, A. T., Lai, S. L., Wisleder, T., Greene, J. E., & Allen, L. H. (2000). Size-dependent melting point depression of nanostructures: Nanocalorimetric measurements. *Physical Review B, 62*, 10548–10557.
47. Wang, X.-Q., & Mujumdar, A. S. (2007). Heat transfer characteristics of nanofluids: A review. *International Journal of Thermal Sciences, 46*, 1–19.
48. Planck, M. (1901). Ueber das Gesetz der Energieverteilung in Normalspectrum. *Annalen der Physik, 4*, 553–563.
49. Martynenko, Y. W., & Ognev, L. I. (2005). Thermal radiation from nanoparticles. *Technical Physics, 50*, 1522–1254.
50. Wuttke, C., & Rauschenbenbeutel, A. (2013). Thermalization via heat radiation of an individual object thinner than the thermal wavelength. *Physical Review Letters, 111*(1–5), 024301.
51. Kolhatkar, A. G., Jamison, A. C., Litvinov, D., Willson, R. C., & Lee, T. R. (2013). Tuning the magnetic properties of nanoparticles. *International Journal of Molecular Sciences, 14*, 15977–16009.
52. Issa, B., Obaidat, I. M., Albiss, B. A., & Haik, Y. (2013). Magnetic nanoparticles: Surface effects and properties related to biomedicine applications. *International Journal of Molecular Sciences, 14*, 21266–21305.
53. Ashcroft, N. W., & Mermin, N. D. (1976). *Solid state physics*. New York: Holt/Rineheart/Winston.
54. Kasap, S. O. (2002). *From principles of electronic materials and devices* (2nd ed.). Boston: McGraw-Hill.
55. Frenkel, Y. I. (1936). On the solid body model of heavy nuclei. *Phys. Z. Soviet Union, 9*, 158–186.
56. Wannier, G. H. (1937). The structure of electronic excitation levels in insulating crystals. *Physics Review, 52*, 191–197.
57. Mott, N. F. (1938). Conduction in polar crystals. II. The conduction band and ultra-violet absorption of alkali-halide crystals. *Transactions of the Faraday Society, 34*, 500–506.
58. Knox, R. S. (1963). *Theory of excitons*. New York: Academic Press.
59. Bassani, F., & Pastori Parravicini, G. (1975). *Electronic states and optical transitions in solids*. Oxford: Pergamon Press.

60. Koch, W., Kira, M., Khitrova, G., & Gibbs, H. M. (2006). Semiconductor excitons in new light. *Nature Materials, 5*, 523–531.
61. Haug, H., & Koch, S. W. (2006). In H. Haug & S. Koch (Eds.), *Quantum theory of the optical and electronic properties of semiconductors* (4th ed.). Singapore: World Scientific.
62. Nakajima, A., Futatsugi, T., Kosemura, K., Fukano, T., & Yokoyama, N. (1999). Si single-electron tunneling transistor with nanoscale floating dot stacked on a Coulomb island by self-aligned process. *Journal of Vacuum Science & Technology, 17*, 2163–2171.
63. Leobandung, E., Guo, L., Wang, Y., & Chou, S. Y. (1995). Observation of quantum effects and Coulomb blockade in silicon quantum-dot transistors at temperatures over 100 K. *Applied Physics Letters, 67*, 938–940.
64. Pikus, F. G., & Likharev, K. K. (1997). Nanoscale field-effect transistors: An ultimate size analysis. *Applied Physics Letters, 71*, 3661–3663.
65. Zhang, S. K., Zhu, H. J., Lu, F., Jiang, Z. M., & Wang, X. (1998). Coulomb charging effect in self-assembled Ge quantum dots studied by admittance spectroscopy. *Physical Review Letters, 80*, 3340–3343.
66. Ishikuro, H., Fujii, T., Saraya, T., Hashiguchi, G., Hiramoto, T., & Ikoma, T. (1996). Coulomb blockade oscillations at room temperature in a Si quantum wire metal-oxide-semiconductor field-effect transistor fabricated by anisotropic etching on a silicon-on-insulator substrate. *Applied Physics Letters, 68*, 3585–3587.
67. Hanna, A. E., & Tinkham, M. (1991). Variation of the Coulomb staircase in a two-junction system by fractional electron charge. *Physical Review B, 44*, 5919–5922.
68. Grabert, H. (1991). Single charge tunneling: A brief introduction. *Zeitschrift fur Physik B: Condensed Matter, 85*, 319–325.
69. Kim, K. (1998). Visible light emissions and single-electron tunneling from silicon quantum dots embedded in Si-rich SiO_2 deposited in plasma phase. *Physical Review B, 57*, 13072–13076.
70. Wilkins, R., Ben-Jacob, E., & Jaklevic, R. C. (1989). *Physical Review Letters, 63*, 801–804.
71. Markovich, G., Leff, D. V., Chung, S. W., Soyez, H. M., Dunn, B., & Heath, J. R. (1997). Parallel fabrication and single-electron charging of devices based on ordered, two-dimensional phases of organically functionalized metal nanocrystals. *Applied Physics Letters, 70*, 3107–3109.
72. Wang, B., Xiao, X., Huang, X., Sheng, P., & Hou, J. G. (2000). Single-electron tunneling study of two-dimensional gold clusters. *Applied Physics Letters, 77*, 1179–1181.
73. Klein, D. L., McEuen, P. L., Bowen Katari, J. E., Roth, R., & Alivisatos, A. P. (1996). An approach to electrical studies of single nanocrystals. *Applied Physics Letters, 68*, 2574–2276.
74. Chen, W., Ahmed, H., & Nakazoto, K. (1995). Coulomb blockade at 77 K in nanoscale metallic islands in a lateral nanostructure. *Applied Physics Letters, 66*, 3383–3385.
75. Meirav, U., Kastner, M. A., & Wind, S. J. (1990). Single-electron charging and periodic conductance resonances in GaAs nanostructures. *Physical Review Letters, 65*, 771–774.
76. van Wees, B. J., van Houten, H., Beenakker, C. W. J., Williamson, J. G., Kouwenhoven, L. P., van der Marel, D., & Foxon, C. T. (1988). Quantized conductance of point contacts in a two-dimensional electron gas. *Physical Review Letters, 60*, 848–850.
77. Datta, S. (1997). *Electronic transport in mesoscopic systems*. Cambridge: Cambridge University Press.
78. Lu, W., & Lieber, C. M. (2007). Nanoelectronics from the bottom up. *Nature Materials, 6*, 841–850.
79. Bond, G. C., & Thomson, D. T. (1999). Catalysis by gold. *Catalysis Reviews: Science and Engineering, 41*(3–4), 319–388.
80. Bond, G. C. (2002). Gold: A relatively new catalyst. *Catalysis Today, 72*, 5–9.
81. Bone, W. A., & Wheeler, R. V. (1906). The combination of hydrogen and oxygen in contact with hot surfaces. *Philosophical Transactions of the Royal Society A, 206*, 1–67.
82. Klabunde, K. J., Stark, J. V., Koper, O., Mohs, C., Park, D. G., Decker, S., Jiang, Y., Lagadic, I., & Zhang, D. (1996). Nanocrystals as stoichiometric reagents with unique surface chemistry. *The Journal of Physical Chemistry, 100*, 12142–12153.

83. Sun, C. Q. (2007). Size dependence of nanostructures: Impact of bond order deficiency. *Progress in Solid State Chemistry, 35*, 1–159.
84. Klabunde, K. J., & Richards, R. M. (Eds.). (2009). *Nanoscale materials in chemistry* (2nd ed.). Hoboken: Wiley.
85. Sun, C. Q., Bai, H. L., Li, S., Tay, B. K., & Jiang, E. Y. (2004). Size-effect on the electronic structure and the thermal stability of a gold nanosolid. *Acta Materialia, 52*, 501–505.
86. Sakai, H. (1996). Surface-induced melting of small particles. *Surface Science, 351*, 285–291.
87. Stranski, I. N. (1942). Über das Verhalten nichtpolarer Kristalle dicht unterhalb des Schmelzpunktes und beim Schmelzpunkt selbst. *Zeitschrift für Physik, 119*, 22–34.
88. Frenken, J. W. M., Maree, P. M. J., & Van Der Veen, J. F. (1986). Observation of surface-initiated melting. *Physical Review B, 34*, 7506–7516.
89. Tartaglino, U., Zykova-Timan, T., Ercolessi, F., & Tosatti, E. (2005). Material surfaces and nanosystems close to the melting temperature (Proceedings of the IV International Conference High Temperature Capillarity). *Journal of Materials Science, 40*, 2141–2147.
90. Mei, Q. S., & Lu, K. (2007). Melting and superheating of crystalline solids: From bulk to nanocrystals. *Progress in Materials Science, 52*, 1175–1262.
91. Sang, L. V., Hoang, V. V., & Hang, N. T. T. (2013). Molecular dynamics simulation of melting of fcc Lennard-Jones nanoparticles. *European Physical Journal: D, 67*(64), 1–8.
92. Oxtoby, D. W. (1990). New perspectives on freezing and melting. *Nature, 347*, 725–730.
93. Bischof, J., Scherer, D., Herminghaus, S., & Leiderer, P. (1996). Dewetting modes of thin metallic films: Nucleation of holes and spinodal dewetting. *Physical Review Letters, 77*, 1536–1539.
94. Herminghaus, S., Jacobs, K., Mecke, K., Bischof, J., Fery, A., Ibn-Elhaj, M., & Schlagowski, S. (1998). Spinodal dewetting in liquid crystal and liquid metal films. *Science, 282*, 916–919.
95. Habenicht, A., Olapinski, M., Burmeister, F., Leiderer, P., & Boneberg, J. (2005). Jumping, nanodroplets. *Science, 309*, 2043–2045.
96. Gromov, D. G., & Gavrilov, S. A. (2009). Manifestation of the heterogeneous mechanism upon melting of low-dimensional systems. *Physics of the Solid State, 51*, 2135–2144 and references therein.
97. Wang, Y. T., Teitel, S., & Dellago, C. (2005). Melting of icosahedral gold nanoclusters from molecular dynamics simulations. *The Journal of Chemical Physics, 122*, 214722–214738.
98. Zhao, Y., & Yakobson, B. I. (2003). What is the ground-state structure of the thinnest Si nanowires? *Physical Review Letters, 91*(1–4), 035501.
99. Guisbiers, G., Van Overschelde, O., & Wautelet, M. (2008). Theoretical investigation of size and shape effects on the melting temperature and energy bandgap of TiO_2 nanostructures. *Applied Physics Letters, 92*(1–3), 103121.
100. Dash, J. G. (1999). History of the search for continuous melting. *Reviews of Modern Physics, 71*, 1737–1743 and references therein.
101. Wang, Y., Teitel, S., & Dellago, C. (2004). Melting and equilibrium shape of icosahedral gold nanoparticles. *Chemical Physics Letters, 394*, 257–261.
102. Chattopadhyuy, K., & Goswami, R. (1991). Melting and superheating of metals and alloys. *Progress in Materials Science, 42*, 287–300.
103. Jiang, Q., Zhang, Z., & Li, J. C. (2000). *Acta Materialia, 48*, 4791–4795 and references therein.
104. Shirinyan, A. S., Gusak, A. M., & Wautelet, M. (2005). Melting and superheating of nanowires. *Acta Materialia, 55*, 5025.
105. Saka, H., Nishikawa, Y., & Imura, T. (1988). Melting temperature of In particles embedded in an Al matrix. *Philosophical Magazine A, 57*, 895–906.
106. Thomson, J. J. (1888). *Application of dynamics to physics and chemistry*. London: McMillan.
107. Pawlow, P. (1910). Über den Dampfdruck der Körner einer festen Substanz. *Zeitschrift für Physikalische Chemie, 68*, 316–322.
108. Pawlow, P. N. (1910). Ueber den Einfluss der Oberfläche einer festen Phase auf die latente Wärme und die Temperatur des Schmelzens. *Zeitschrift für Chemie und Industrie der Kolloide, 7*, 37–39.

109. Barybin, A., & Shapovalov, V. (2011). Modification of Pawlow's thermodynamical model for the melting of small single-component particles. *Journal of Applied Physics, 109*(1–9), 034303.
110. Lindemann, F. A. (1910). The calculation of molecular vibration frequencies. *Zeitschrift für Physik, 11*, 609–612.
111. Couchman, R., & Jesser, W. A. (1977). Thermodynamic theory of size dependence of melting temperature in metals. *Nature, 269*, 481–483.
112. Couchman, P. R., & Ryan, C. L. (1978). The Lindmann hypothesis and the size-dependence of melting temperature. *Philosophical Magazine A, 37*, 369–373.
113. Couchman, P. R. (1979). The Lindemann hypothesis and the size dependence of melting temperatures. II. *Philosophical Magazine A, 40*, 637–643.
114. Hoshino, K., & Shimamura, S. (1979). A simple model for the melting of fine particles. *Philosophical Magazine A, 40*, 137–141.
115. Nanda, K. K. (2009). Size-dependent melting of nanoparticles: Hundred years of thermodynamic model. *Pramana, 72*, 617–628.
116. Hanszen, K. J. (1960). Theoretische Untersuchungen über den Schmelzpunkt Ein Beitrag zur Thermodynamik der Grenzflächen. *Zeitschrift für Physik, 157*, 523–553.
117. Beaglehole, D. (1991). Surface melting of small particles, and the effects of surface impurities. *Journal of Crystal Growth, 112*, 663–669.
118. Wautelet, M. (1990). Size effect on the melting (or disordering) temperature of small particles. *Solid State Communications, 74*, 1237–1239.
119. Qi, W. H., Wang, M. P., Zhou, M., Shen, X. Q., & Zhang, X. F. (2006). Modeling cohesive energy and melting temperature of nanocrystals. *Journal of Physics and Chemistry of Solids, 67*, 851–855.
120. Yang, C. C., & Li, S. (2007). Investigation of cohesive energy effects on size-dependent physical and chemical properties of nanocrystals. *Physical Review B, 75*(1–5), 165413.
121. Nanda, K. K., Sahu, S. N., & Behera, S. N. (2002). Liquid-drop model for the size-dependent melting of low-dimensional systems. *Physical Review A, 66*(1–8), 013208.
122. Ercolessi, F., Andreoni, W., & Tosatti, E. (1991). Melting of small gold particles: Mechanism and size effect. *Physical Review Letters, 66*, 911–914.
123. Celestini, F., Pellenq, R. J.-M., Bordarier, P., & Rousseau, B. (1996). Melting of Lennard-Jones clusters in confined geometries. *Zeitschrift für Physik D Atoms, 37*, 49–53.
124. Delogu, F. (2005). Structural and energetic properties of unsupported Cu nanoparticles from room temperature to the melting point: Molecular dynamics simulations. *Physical Review B, 72*, 205418.
125. Alavi, S., & Thompson, D. L. (2006). Molecular dynamics simulation of the melting of aluminum nanoparticles. *The Journal of Physical Chemistry. A, 110*, 1518–1523.
126. Wang, Y., & Dellago, C. (2003). Structural and morphological transitions in gold nanorods: A computer simulation study. *The Journal of Physical Chemistry. B, 107*, 9214–9219.
127. Zhang, Y., Wen, Y.-H., Zheng, J.-C., & Zhu, Z. (2009). Energetic and structural evolution of gold nanowire under heating process: A molecular dynamics study. *Physics Letters A, 373*, 3454–3458.
128. Wen, Y.-H., Zhang, Y., Zheng, J.-C., Zhu, Z.-Z., & Sun, S.-G. (2009). Orientation-dependent structural transition and melting of Au nanowires. *Journal of Physical Chemistry C, 113*(48), 20611–20617.
129. Cross, M. W., & Varhue, W. J. (2008). Radiative melting of crystalline ruthenium oxide nanorods. *Nanotechnology, 19*, 435705 (5 pp).
130. Karabacak, T., DeLuca, J. S., Wang, P.-I., Ten Eyck, G. A., Ye, D., Wang, G., & Lu, T. (2006). Low temperature melting of copper nanorod arrays. *Journal of Applied Physics, 99*(1–6), 064304.
131. Olson, E. A., Yu Efremov, M., Zhang, M., Zhang, Z., & Allen, L. H. (2005). Size-dependent melting of Bi nanoparticles. *Journal of Applied Physics, 97*, 034304 (9 pages).
132. Coombes, C. J. (1972). The melting of small particles of lead and indium. *Journal of Physics F: Metal Physics, 2*, 441–449.

133. Wronski, C. R. M. (1967). The size dependence of the melting point of small particles of tin. *Journal of Applied Physics, 18*, 1731–1737.
134. Wang, P.-I., Lee, S. H., Parker, T. C., Frey, M. D., Karabacak, T., Lu, J.-Q., & Lu, T.-M. (2009). Low temperature wafer bonding by copper nanorod array. *Electrochemical and Solid-State Letters, 124*, H138–H141.
135. Weisbuch, C., & Vinter, B. (2014). *Quantum semiconductor structures: Fundamentals and applications*. Boston: Academic Press.
136. Bastard, G. (1990). *Wave mechanics applied to semiconductor heterostructures*. New York: Wiley.
137. Esaki, L., & Tsu, R. (1970). Superlattice and negative differential conductivity in semiconductors. *IBM Journal of Research and Development, 14*, 61–65.
138. Nag, B. R. (2000). *Physics of quantum well devices*. Dordrecht: Kluwer Academic Publisher.
139. Cho, A. Y. (1991). Advances in molecular beam epitaxy (MBE). *Journal of Crystal Growth, 111*, 1–13.
140. Esaki, L., & Chang, L. L. (1974). New transport phenomenon in a semiconductor superlattice. *Physical Review Letters, 33*, 495–498.
141. Dingle, R., Wiegmann, W., & Henry, C. H. (1974). Quantum states of confined carriers in very thin $Al_xGa_{1-x}As$-GaAs-$Al_xGa_{1-x}As$ heterostructures. *Physical Review Letters, 33*, 827–830.
142. Bastard, G. (1981). Superlattice band structure in the envelope-function approximation. *Physical Review B, 24*, 5693–5697.
143. Bastard, G. (1982). Theoretical investigations of superlattice band structure in the envelope-function approximation. *Physical Review B, 25*, 7584–7597.
144. Altarelli, M. (1986). In G. Allan, G. Bastard, N. Boccara, M. Lannoo, & M. Voods (Eds.), *Heterojunctions and semiconductor superlattices*. Berlin/New York: Springer-Verlag.
145. White, S. R., & Sham, L. J. (1981). Electronic properties of flat-band semiconductor heterostructures. *Physical Review Letters, 47*, 879–882.
146. Weisbuch, C. (1987). Fundamental properties of III-V semiconductor two-dimensional quantized structures: The basis for optical and electronic device applications. In R. Dingle (Ed.), *Semiconductors and Semimetals* (Vol. 24). New York: Academic Press.
147. Dean, P. J. (1977). III–V compound semiconductors. In J. I. Pankove (Ed.), *Electroluminescence* (pp. 63–132). Berlin/Heidelberg: Springer.
148. Nakamura, S., Pearton, S., & Fasol, G. (2000). *The blue laser diode* (2nd ed.). Berlin/Heidelberg: Springer.
149. Bugajski, M., & Reginski, K. (1996). Optical properties of semiconductor quantum wells. *Opto-Electronics Review, 4*, 83–100.
150. Fox, M. (2001). *Optical properties of solids*. Oxford: Clarendon.
151. Ando, T., Fowler, A. D., & Stern, F. (1982). Electronic properties of two-dimensional systems. *Reviews of Modern Physics, 54*, 437–672.
152. Han, J., Crawford, M. H., Shul, R. J., Figiel, J. J., Banas, M., Zhang, L., Song, Y. K., Zhou, H., & Nurmikko, A. V. (1998). AlGaN/GaN quantum well ultraviolet light emitting diodes. *Applied Physics Letters, 73*, 1688–1690.
153. Sturge, M. D. (1962). Optical Absorption of Gallium Arsenide between 0.6 and 2.75 eV. *Physics Review, 127*, 768–773: *Erratum Physics Review, 129*, 2835 (1963).
154. Greene, R. L., Bajaj, K. K., & E: Phelps, D. (1984). Energy levels of Wannier excitons in $GaAs-Ga_{1-x}Al_xAs$ quantum-well structures. *Physical Review B, 29*, 1807–1812.
155. Bastard, G., Mendez, E. E., Chang, L. L., & Esaki, L. (1982). Exciton binding energy in quantum wells. *Physical Review B, 26*, 1974–1979.
156. Shinada, M., & Sugano, S. (1966). Interband optical transitions in extremely anisotropic semiconductors. I. Bound and unbound exciton absorption. *Journal of the Physical Society of Japan, 21*, 1936–1946.
157. Iotti, R. C., & Andreani, L. C. (1997). Crossover from strong to weak confinement for excitons in shallow or narrow quantum wells. *Physical Review B, 56*, 3922–3932.

158. Chemia, D. S., & Miller, D. A. B. (1985). Excitonic nonlinear-optical effects in semiconductor quantum-well structures. *Journal of the Optical Society of America B: Optical Physics, 2*, 1155–1173.
159. Loehr, J. P., & Singh, J. (1990). Nonvariational numerical calculations of excitonic properties in quantum wells in the presence of strain, electric fields, and free carriers. *Physical Review B, 42*, 7154–7162.
160. Miller, R. C., Kleinman, D. A., Nordland, W. A., Jr., & Gossard, A. C. (1980). Luminescence studies of optically pumped quantum wells in GaAs-$Al_xGa_{1-x}As$ multilayer structures. *Physical Review B, 22*, 863–871.
161. Sanders, G. D., & Chang, Y. C. (1985). Optical properties in modulation-doped GaAs-$Ga_{1-x}Al_xAs$ quantum wells. *Physical Review B, 31*, 6892–6895.
162. Miller, D. A. B., Chemia, D. S., Eilenberger, D. J., Smith, P. W., Gossard, A. C., & Tsang, W. T. (1982). Large room-temperature optical nonlinearity in GaAs/Ga_{1-x} Al_xAs multiple quantum well structures. *Applied Physics Letters, 41*, 679–681.
163. Law, M., Goldberger, J., & Yang, P. (2004). Semiconductor nanowires and nanotubes. *Annual Review of Materials Research, 34*, 83–122. -(first published online as a Review in Advance, 2004 DOI:10.1146/annurev.matsci.34.040203.112300).
164. Petroff, P. M., Gossard, A. C., Logan, R. A., & Wiegmann, W. (1983). Toward quantum well wires: Fabrication and optical properties. *Applied Physics Letters, 41*, 635–637.
165. Stern, M. B., Craighead, H. G., Liao, P. F., & Mankiewich, P. M. (1984). Fabrication of 20-nm structures in GaAs. *Applied Physics Letters, 45*, 410–412.
166. Kash, K., Scherer, A., Worlock, J. M., Craighead, H. G., & Tamargo, M. C. (1986). Optical spectroscopy of ultrasmall structures etched from quantum wells. *Applied Physics Letters, 49*, 1043–1045.
167. Grambow, P., Demel, T., Heitmann, D., Kohl, M., Schiile, R., & Ploog, K. (1989). Preparation of One-Dimensional Single and Multi-Layered quantum wire by ultrafine deep mesa etching techniques. *Microelectronic Engineering, 9*, 357–360.
168. Kohl, M., Heitmann, D., Grambow, P., & Ploog, K. (1989). One-dimensional magneto-excitons in GaAs/$Al_xGa_{1-x}As$ quantum wires. *Physical Review Letters, 63*, 2124–2127.
169. Kapon, E., Tamargo, M. C., & Hwang, D. M. (1987). Molecular beam epitaxy of GaAs/AlGaAs superlattice heterostructures on nonplanar substrates. *Applied Physics Letters, 50*, 347–349.
170. Turco, F. S., Simhony, S., Kash, K., Hwang, D. M., Ravi, T. S., Kapon, E., & Tamargo, M. C. (1990). Molecular beam epitaxial growth of GaAs/AlAs and GaAs/AlGaAs quantum wells on sub-micron-period corrugated substrates. *Journal of Crystal Growth, 104*, 766–772.
171. Hoenk, M. E., Nieh, C. W., Chen, H. Z., & Vahala, K. J. (1989). Compositional modulation in $Al_xGa_{1-x}As$ epilayers grown by molecular beam epitaxy on the (111) facets of grooves in a nonplanar substrate. *Applied Physics Letters, 55*, 53–55.
172. Bhat, R., Kapon, E., Hwang, D. M., Koza, M. A., & Yun, C. P. (1988). Patterned quantum well heterostructures grown by OMCVD on non-planar substrates: Applications to extremely narrow SQW lasers. *Journal of.Crystal Growth, 93*, 850–856.
173. Kapon, E., Walther, M., Christen, J., Grundmann, M., Caneau, C., Hwang, D. M., Colas, E., Bhat, R., Song, G. H., & Bimberg, D. (1992). Quantum wire heterostructure for optoelectronic applications. *Superlattices and Microstructures, 12*, 491–499.
174. Kash, K., Bhat, R., Mahoney, D. D., Lin, P. S. D., Scherer, A., Worlock, J. M., Van der Gaag, B. P., Koza, M., & Grabbe, P. (1989). Strain-induced confinement of carriers to quantum wires and dots within an InGaAs-InP quantum well. *Applied Physics Letters, 55*, 681–683.
175. Gershoni, D., Weiner, J. S., Chu, S. N. G., Baraff, G. A., Vandenberg, J. M., Pfeiffer, L. N., West, K., Logan, R. A., & Tanbun-Ek, T. (1990). Optical transitions in quantum wires with strain-induced lateral confinement. *Physical Review Letters, 65*, 1631–1634.
176. Petroff, P. M., Gossard, A. C., & Wiegmann, W. (1984). Structure of AlAs-GaAs interfaces grown on (100) vicinal surfaces by molecular beam epitaxy. *Applied Physics Letters, 45*, 620–623.

177. Pryor, C. (1991). Electronic structure and optical properties of serpentine superlattice quantum-wire arrays. *Physical Review B, 44*, 12912–12917.
178. Rossi, F., & Molinari, E. (1996). Linear and nonlinear optical properties of realistic quantum-wire structures: The dominant role of Coulomb correlation. *Physical Review B, 53*, 16462–16473.
179. Chang, K., & Xia, J. B. (1998). Quantum-confined Stark effects of exciton states in V-shaped GaAs/$Al_xGa_{1-x}As$ quantum wires. *Physical Review B, 58*, 2031–2037.
180. Bockelmann, U., & Bastard, G. (1991). Interband optical transitions in semiconductor quantum wires: Selection rules and absorption spectra. *Europhysics Letters, 15*, 215–220.
181. Suemune, I., & Coldren, L. A. (1988). Band-mixing effects and excitonic optical properties in GaAs quantum wire structures-comparison with the quantum wells. *IEEE Journal of Quantum Electronics, QE-24*, 1778–1790.
182. Einevoll, G. T., & Chang, Y. C. (1989). Effective bond-orbital model for acceptor states in semiconductors and quantum dots. *Physical Review B, 40*, 9683–9697.
183. Citrin, D. S., & Chang, Y. C. (1990). Subband structures of semiconductor quantum wires from the effective-bond orbital model. *Journal of Applied Physics, 68*, 161–168.
184. Citrin, D. S., & Chang, Y. C. (1991). Theory of optical anisotropy in quantum-well-wire arrays with two-dimensional quantum confinement. *Physical Review B, 43*, 11703–11719.
185. McIntyre, C. R., & Sham, L. H. (1992). Theory of luminescence polarization anisotropy in quantum wires. *Physical Review B, 45*, 9443–9446.
186. Sercel, P. C., & Vahala, K. J. (1990). Analytical technique for determining the polarization dependence of optical matrix elements in quantum wires with band-coupling effects or determining the polarization dependence of optical matrix elements in quantum wires with band-coupling effects. *Applied Physics Letters, 57*, 545–547.
187. Sercel, P. C., & Vahala, K. J. (1991). Polarization dependence of optical absorption and emission in quantum wires. *Physical Review B, 44*, 5681–5691.
188. Oberli, D. Y., Vouilloz, F., Dupertuis, M.-A., Fall, C., & Kapon, E. (1995). Optical spectroscopy of semiconductor quantum wires. *Il Nuovo Cimento D, 17*(11–12), 1641–1650.
189. Nagamune, Y., Arakawa, Y., Tsukamoto, S., Nishioka, M., Sasaki, S., & Miura, N. (1992). Photoluminescence spectra and anisotropic energy shift of GaAs quantum wires in high magnetic fields. *Physical Review Letters, 69*, 2963–2966.
190. Wegscheider, W., Pfeiffer, L. N., Dignam, M. M., Pinczuk, A., West, K. W., McCall, S. L., & Hull, R. (1993). Lasing from excitons in quantum wires. *Physical Review Letters, 71*, 4071–4074.
191. Rinaldi, R., Cingolani, R., Lepore, M., Ferrara, M., Catalano, I. M., Rossi, F., Rota, L., Molinari, E., Lugli, P., Marti, U., Martin, D., Morier-Gemoud, F., Ruterana, P., & Reinhart, F. K. (1994). Exciton binding energy in GaAs V-shaped quantum wires. *Physical Review Letters, 73*, 2899–2902.
192. Someya, T., Akiyama, H., & Sakaki, H. (1995). Laterally squeezed excitonic wave function in quantum wires. *Physical Review Letters, 74*, 3664–3666.
193. Someya, T., Akiyama, H., & Sakaki, H. (1996). Enhanced binding energy of one-dimensional excitons in quantum wires. *Physical Review Letters, 76*, 2965–2967.
194. Weman, H., Potemski, M., Lazzouni, M. E., Miller, M. S., & Merz, J. L. (1996). Magneto-optical determination of exciton binding energies in quantum-wire superlattices. *Physical Review B, 53*, 6959–6962 and references therein.
195. Loudon, R. (1959). One-dimensional hydrogen atom. *American Journal of Physics, 27*, 649–655.
196. Elliot, R. J., & Loudon, R. (1959). Theory of fine structure on the absorption edge in semiconductors. *Journal of Physics and Chemistry of Solids, 8*, 382–388.
197. Elliot, R. J., & Loudon, R. (1960). Theory of the absorption edge in semiconductors in a high magnetic field. *Journal of Physics and Chemistry of Solids, 15*, 196–207.
198. Ogawa, T., & Takagahara, T. (1991). Interband absorption spectra and Sommerfeld factors of a one-dimensional electron-hole system. *Physical Review B, 43*, 14325–14328.

199. Ogawa, T., & Takagahara, T. (1991). Optical absorption and Sommerfeld factors of one-dimensional semiconductors: An exact treatment of excitonic effects. *Physical Review B, 44*, 8138–8156.
200. Chang, Y. C., Chang, L. L., & Esaki, L. (1985). A new one-dimensional quantum well structure. *Applied Physics Letters, 47*, 1324–1326.
201. Rossi, F., Goldoni, G., & Molinari, E. (1997). Shape-independent scaling of excitonic confinement in realistic quantum wires. *Physical Review Letters, 78*, 3527–3530.
202. Lage, H., Heitmann, D., Cingolani, R., Granbow, P., & Ploog, K. (1991). Center-of-mass quantization of excitons in GaAs quantum-well wires. *Physical Review B, 44*, 6550–6553.
203. Rinaldi, R., Cingolani, R., Ferrara, M., Lage, H., Teitmann, D., & Ploog, K. (1992). Emission properties of quantum-well wires under stationary conditions. *Physical Review B, 47*, 7275.
204. Kapon, E., Hwang, D. M., & Bhat, R. (1989). Stimulated emission in semiconductor quantum wire heterostructures. *Physical Review Letters, 63*, 430–433.
205. Bimberg, D., Grundmann, M., & Ledentsov, N. N. (1998). *Quantum dot heterostructures*. Chichester: Wiley.
206. Walter, P., Welcomme, E., Hallegot, P., Zaluzec, N. J., Deeb, C., Castaing, J., Veyssiere, P., Breniaux, R., Leveque, J. L., & Tsoucaris, G. (2006). Early use of PbS nanotechnology for an ancient hair dyeing formula. *Nano Letters, 6*, 2215–2219.
207. Onushchenko, A. A. (1981). Quantum size effect in 3-dimensional microscopic semiconductor crystals. *JETP Letters, 34*, 345–349.
208. Efros, A. L. (1982). Interband absorption of light in a semiconductor sphere. *Soviet Physics – Semiconductors, 16*, 772–775.
209. Rossetti, R., Ellison, J. L., Gibson, J. M., & Brus, L. E. (1984). Size effects in the excited electronic states of small colloidal CdS crystallites. *The Journal of Chemical Physics, 80*, 4464–4469.
210. Yoffe, A. D. (2001). Semiconductor quantum dots and related systems: Electronic, optical, luminescence and related properties of low dimensional systems. *Advances in Physics, 50*, 1–208.
211. Bryan, J. D., & Gamelin, D. R. (2005). Doped semiconductor nanocrystals: Synthesis, characterization, physical properties, and applications. *Progress in Inorganic Chemistry, 54*, 47–126.
212. Jiang, W. H., Xu, H. Z., Xu, B., Ye, X. L., Wu, J., Ding, D., Liang, J. B., & Wang, Z. G. (2000). Annealing effect on the surface morphology and photoluminescence of InGaAs/GaAs quantum dots grown by molecular beam epitaxy. *Journal of Crystal Growth, 212*, 356–359.
213. Sellin, R. L., Ribbat, C., Grundmann, M., Ledentsov, N. N., & Bimberg, D. (2001). Close-to-ideal device characteristics of high-power InGaAs/GaAs quantum dot lasers. *Applied Physics Letters, 78*, 1207–1209.
214. Tsatsulnikov, A. F., Kovsh, A. R., Zhukov, A. E., Shernyakov, Y. M., Musikhin, Y. G., Ustinov, V. M., Bert, N. A., Kopev, P. S., Alferov, Z. I., Mintairov, A. M., Merz, J. L., Ledentsov, N. N., & Bimberg, D. (2000). Volmer–Weber and Stranski–Krastanov InAs-(Al,Ga) As quantum dots emitting at 1.3 μm. *Journal of Applied Physics, 88*, 6272–6275.
215. Ledentsov, N. N., Shchukin, V. A., Grundmann, M., Kirstaedter, N., Böhrer, J., Schmidt, O., Bimberg, D., Ustinov, V. M., Egorov, A. Y., Zhukov, A. E., Kopev, P. S., Zaitsev, S. V., Gordeev, N. Y., Alferov, Z. I., Borovkov, A. I., Kosogov, A. O., Ruvimov, S. S., Werner, P., Gosele, U., & Heydenreich, J. (1996). Direct formation of vertically coupled quantum dots in Stranski-Krastanow growth. *Physical Review B, 54*, 8743–8750.
216. Xia, Y., Xia, X., & Peng, H.-C. (2015). Shape-controlled synthesis of colloidal metal nanocrystals: Thermodynamic versus kinetic products. *Journal of the American Chemical Society, 137*, 7947–7966.
217. Overbeek, J. T. G. (1982). Monodisperse colloidal systems, fascinating and useful. *Advances in Colloid and Interface Science, 15*, 251–277.
218. Reed, M. A., Randall, J. N., Aggarwal, R. J., Matyi, R. J., Moore, T. M., & Wetsel, A. E. (1988). Observation of discrete electronic states in a zero-dimensional semiconductor nanostructure. *Physical Review Letters, 60*, 535–537.

219. Alivisatos, A. P. (1996). Semiconductor clusters, nanocrystals, and quantum dots. *Science, 271*, 933–937.
220. Banin, U., Bruchez, M., Alivisatos, A. P., Ha, T., Weiss, S., & Chemla, D. S. (1999). Evidence for a thermal contribution to emission intermittency in single CdSe/CdS core/shell nanocrystals. *The Journal of Chemical Physics, 110*, 1195–1201.
221. Empedocles, S. A., Neuhauser, R., Shimizu, K., & Bawendi, M. G. (1999). Photoluminescence from single semiconductor nanostructures. *Advanced Materials, 11*, 1243–1256.
222. Chavez-Pirson, A., Temmyo, J., Kamada, H., Gotoh, H., & Ando, H. (1998). Near-field optical spectroscopy and imaging of single InGaAs/AlGaAs quantum dots. *Applied Physics Letters, 72*, 3494–3496.
223. Nirmal, M., Dabbousi, B. O., Bawendi, M. G., Macklin, J. J., Trautman, J. K., Harris, T. D., & Brus, L. E. (1996). Fluorescence intermittency in single cadmium selenide nanocrystals. *Nature, 383*, 802–804.
224. Nötzel, R., Niu, Z., Ramsteiner, M., Schönherr, H.-P., Tranpert, A., Däweritz, L., & Ploog, K. H. (1998). Uniform quantum-dot arrays formed by natural self-faceting on patterned substrates. *Nature, 392*, 56–59.
225. Tersoff, J., Teichert, C., & Lagally, M. G. (1996). Self-organization in growth of quantum dot superlattices. *Physical Review Letters, 76*, 1675–1678.
226. Gammon, D., Snow, E. S., Shanabrook, B. V., Katzer, D. S., & Park, D. (1996). Homogeneous linewidths in the optical spectrum of a single Gallium Arsenide Quantum Dot. *Science, 273*, 87–90.
227. Brus, L. E. (1986). Electronic wave functions in semiconductor clusters: Experiment and theory. *The Journal of Physical Chemisty, 90*(12), 2555–2560; Brus, L. E. (1984). Electron–electron and electronhole interactions in small semiconductor crystallites: The size dependence of the lowest excited electronic state. *The Journal of Chemical Physics, 80*, 4403–4409.
228. Ronda, C. R. (2008). In R. Ronda (Ed.), *Luminescence: From theory to applications*. Weinheim: Wiley-VCH Verlag GmbH&Co.KGaA.
229. Hartmann, A., Ducommun, Y., Kapon, E., Hohenester, U., & Molinari, E. (2000). Few-particle effects in semiconductor quantum dots: Observation of multicharged excıtons. *Physical Review Letters, 84*, 5648–5651.
230. Warburton, R. J., Schaflein, C., Haft, D., Bickel, F., Lorke, A., Karrai, K., Garcia, J. M., Schoenfeld, W., & Petroff, P. M. (2000). Optical emission from a charge-tunable quantum ring. *Nature, 405*, 926–929.
231. Finley, J. J., Ashmore, A. D., Lemaître, A., Mowbray, D. J., Skolnick, M. S., Itskevich, I. E., Maksym, P. A., Hopkinson, M., & Krauss, T. F. (2001). Charged and neutral exciton complexes in individual self-assembled In (Ga) As quantum dots. *Physical Review B, 63*, 073307–073310.
232. Findeis, F., Baier, M., Zrenner, A., Bichler, M., Abstreiter, G., Hohenester, U., & Molinari, E. (2001). Optical excitations of a self-assembled artificial ion. *Physics Review, 63*(1–4), 121309.
233. Alivisatos, A. P. (1996). Perspectives on the physical chemistry of semiconductor nanocrystals. *The Journal of Physical Chemistry, 100*, 13226–13239.
234. Wang, Y., & Herron, N. (1991). Nanometer-sized semiconductor clusters–materials synthesis, quantum size effects, and photophysical properties. *The Journal of Physical Chemistry, 95*, 525–532.
235. Bang, J., Yang, H., & Holloway, P. H. (2006). Enhanced and stable green emission of ZnO nanoparticles by surface segregation of Mg. *Nanotechnology, 17*, 973–978.
236. Ma, Y., Wang, X., Jia, Y., Chen, X., Han, H., & Li, C. (2014). Titanium dioxide-based nanomaterials for photocatalytic fuel generations. *Chemical Reviews, 114*, 9987–10043.
237. Kamat, P. V. (2014). Quantum dot solar cells. Semiconductor nanocrystals as light harvesters. *Journal of Physical Chemistry C, 112*, 18737–18753.
238. Homola, J., Yeea, S. S., & Gauglitzb, G. (1999). Surface plasmon resonance sensors: Review. *Sensors and Actuators B: Chemical, 54*, 3–15.

239. Mie, G. (1908). Beiträge zur optik trüber medien, speziell kolloidaler metallösungen. *Annalen der Physik, 330*, 377–445.
240. Kelly, K. L., Coronado, E., Zhao, L. L., & Schatz, G. C. (2003). The optical properties of metal nanoparticles: The influence of size, shape, and dielectric environment. *The Journal of Physical Chemistry. B, 107*, 668–677.
241. Sosa, I. O., Noguez, C., & Barrera, R. C. (2003). Optical properties of metal nanoparticles with arbitrary shapes. *The Journal of Physical Chemistry. B, 107*, 6269–6275.
242. Dabbousi, B. O., Rodriguez-Viejo, J., Mikulec, F. V., Heine, J. R., Mattoussi, H., Ober, R., Jensen, K. F., & Bawendi, M. G. (1997). (CdSe) ZnS core − shell quantum dots: Synthesis and characterization of a size series of highly luminescent nanocrystallites. *The Journal of Physical Chemistry. B, 101*, 9463–9475.
243. Anker, J. N., Paige Hall, W., Lyandres, O., Shah, N. C., Zhao, J., & Van Duyne, R. P. (2008). Biosensing with plasmonic nanosensors. *Nature Materials, 7*, 442–453.
244. Haes, A. J., & Van Duyne, R. P. (2002). A nanoscale optical biosensor: Sensitivity and selectivity of an approach based on the localized surface plasmon resonance of triangular silver nanoparticles. *Journal of the American Chemical Society, 124*, 10596–10604.
245. Jung, L. S., Campbell, C. T., Chinowsky, T. M., Mar, M. N., & Yee, S. S. (1998). Quantitative interpretation of the response of surface plasmon resonance sensors to adsorbed films. *Langmuir, 14*, 5636–5648.
246. Oldenburg, S. J., Averitt, R. D., Westcott, S. L., & Halas, N. J. (1998). Nanoengineering of optical resonances. *Chemical Physics Letters, 288*, 243–247.
247. Ma, Y., Zhou, J., Zou, W., Jia, Z., Petti, L., & Mormile, P. (2014). Localized surface plasmon resonance and surface enhanced Raman scattering responces of Au@Ag core-shall nanorods with different thickness of Ag shell. *Journal of Nanoscience and Nanotechnology, 14*, 4245–4250.
248. Csapo, E., Oszko, A., Varga, E., et al. (2012). Synthesis and characterization of Ag/Au alloy and core (Ag)-shell(Au) nanoparticles. *Colloids and Surfaces A, Physicochemical and Enginnering Aspects, 415*, 281–287.
249. El-Brolossy, T. A., Abdallah, T., Mohamed, M. B., et al. (2008). Shape and size dependence on the surface plasmon resonance of gold nanoparticles studied by photoacoustic techniques. *The European Physical Journal: Special Topics, 153*, 361–364.
250. Ringe, E., Zhang, J., Langille, M. R., et al. (2010). Effect of size, shape, composition, and support film on localized surface plasmon resonance frequency: A single particle approach applied to silver bipyramids and gold and silver nanocubes. In *Proceedings of the material research society fall meeting*, 1208.
251. Moskovits, M. (2005). Surface enhanced Raman spectroscopy: A brief retrospective. *Journal of Raman Spectroscopy, 36*, 485–496.
252. Moore, C. B., Rison, W., Mathis, J., & Aulich, G. (2000). Lightning rod improvement studies. *Journal of Applied Meteorology, 39*, 593–609.
253. Lee, J., Hua, B., Park, S., et al. (2014). Tailoring surface plasmons of high-density gold nanostar assemblies on metal films for surface-enhanced Raman spectroscopy. *Nanoscale, 6*, 616–623.

Chapter 4
Non-radiative Processes in Nanocrystals

J. M. Collins

Abstract This paper discusses non-radiative processes relevant to the luminescence characteristics of optically active ions doped into insulators or large-gap semiconductors, with particular attention to how these processes are affected as the particle size is reduced from bulk single crystals to as small as a few nanometers. The non-radiative processes discussed in this article are thermal line broadening and thermal line shifting, relaxation via phonons between excited electronic states, and vibronic emission and absorption. One prominent effect of confinement in the systems of interest is a reduction in the phonon density of states. Thus, we focus on how these non-radiative processes are altered due to the change in the phonon density of states as particle size decreases.

4.1 Introduction

Inorganic insulators doped with rare-earth ions and transition metal ions represent an important class of luminescent materials for many applications, including phosphors for lighting, scintillators, solid-state laser materials, bio-markers for imaging, and nanothermometry. Following excitation by radiation, the optical ions usually undergo some degree of non-radiative relaxation, releasing part or all of its energy to the lattice. During the non-radiative relaxation, all or part of the electronic energy initially stored in the optically active ion is converted into phonons.

The specific non-radiative processes addressed in this work are thermal line broadening, thermal line shifting, decay via a phonon from one electronic level to another, vibronic transitions, and phonon-assisted energy transfer. Generally speaking, the two main effects of going from the bulk to the nano are: (1) an increase in the surface to volume ratio, and (2) a reduction in the phonon density of states. In this paper we focus on the second point. Most non-radiative processes that play a significant role in the luminescent properties of these systems involve phonons, and most of those are determined in part by the phonon density of states of the

J.M. Collins (✉)
Department of Physics and Astronomy, Wheaton College, Norton, MA 02766, USA
e-mail: jcollins@wheatonma.edu

© Springer Science+Business Media Dordrecht 2017
B. Di Bartolo et al. (eds.), *Nano-Optics: Principles Enabling Basic Research and Applications*, NATO Science for Peace and Security Series B: Physics and Biophysics, DOI 10.1007/978-94-024-0850-8_4

lattice. One goal of this paper is to present results that demonstrate how the reduced density of states in nanoparticles affect the aforementioned processes, and under what conditions will such affects be noticeable.

Before doing so, however, we present some detailed theory of the non-radiative processes in solids, including various forms of the adiabatic approximation, the non-adiabatic operators that drive the transitions, and prediction about the rates of non-radiative transitions. The electron-phonon coupling is a central idea to the theory. We consider primarily systems where this coupling is weak, such as f-f transitions of rare earth ions.

To set up the problem, we first address the following question: How is the energy stored in the electronic system of the optically active ion converted into phonons? To answer this question, it is useful to first review the notion of an adiabatic process.

4.2 Adiabatic Processes

In this section, we introduce the *Adiabatic Approximation*, which allows separation of the motion of the nuclei from that of the electrons. The idea of an *adiabatic* process is fundamental to Quantum Mechanics and Thermodynamics, and so a brief review of an adiabatic process, is useful. For interested readers, the original proof of the adiabatic theorem by Max Born is given in Ref. [1].

Suppose we have a system with a Hamiltonian having the time dependence shown in Fig. 4.1. At $t < t_i$, the Hamiltonian is constant at H_i, and at $t > t_f$, the Hamiltonian is constant at H_f, and during a time $\Delta t = t_f - t_i$ the Hamiltonian is time-dependent. Suppose also that the system has a characteristic time T. (For example, if the system were a harmonic oscillator, T would represent the period of the oscillator.) In an adiabatic process, we demand that the following three conditions hold.

1. Suppose there exists another state $\left|m\right\rangle$ nearby to $\left|n\right\rangle$ with an energy ε_m. The variation in the Hamiltonian during the time Δt must be less than or on the order of $\varepsilon_n - \varepsilon_m$: $\left\langle H_f - H_i\right\rangle <\sim |\varepsilon_n - \varepsilon_m|$.
2. The time over which the Hamiltonian varies must be much greater than the characteristic time of the system: $\Delta t \gg T$

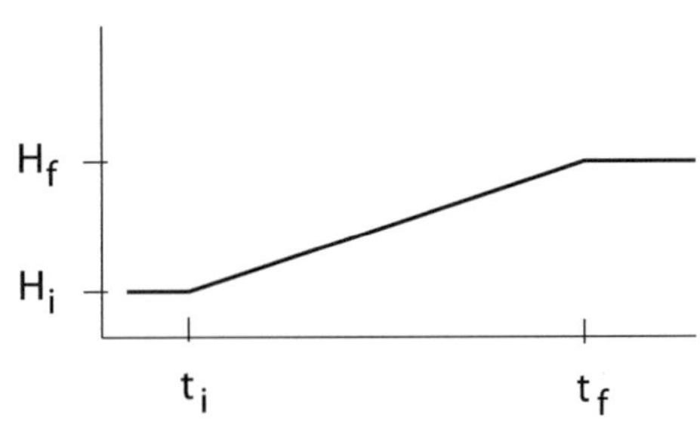

Fig. 4.1 Generic diagram of a time-dependent Hamiltonian between t_{initial} and t_{final}. In this diagram, the Hamiltonian varies linearly with time, but a linear dependence is not required for the system to vary adiabatically

3. The initial state $\left|n\right\rangle$ of the system is nondegenerate.

Consider s system for which the above conditions hold, and that at $t < t_i$ such a system is in the stationary state $\left|n\right\rangle$ with an energy ε_n, i.e., $H_i\left|n\right\rangle = \varepsilon_n\left|n\right\rangle$. The Adiabatic Approximation states that, although the time dependence of the Hamiltonian causes the system to evolve during the time Δt, the system will remain in the state $\left|n\right\rangle$ throughout. At $t = t_f$ the energy of the system will be ε'_n, (i.e., $H_f\left|n\right\rangle = \varepsilon'_n\left|n\right\rangle$), where in general $\varepsilon'_n \neq \varepsilon_n$. The wavefunction represented by the state $\left|n\right\rangle$ will also have changed during the time Δt. That is, if we define $\langle x \mid n\rangle = \psi_n$, then in general $\psi_n^i \neq \psi_n^f$.

As an elementary example, consider the case of an electron trapped in an infinite, one-dimensional well. For a particle of mass m and a box with potential energy is zero region between $x = 0$ and $x = L$, the energies and wavefunctions of the system are given by

$$\varepsilon_n = \frac{h^2\pi^2}{8\ \mathrm{mL}^2}n^2 \text{ and } \psi(x) = \sqrt{\frac{2}{L}}\, sin\left(\frac{n\pi x}{L}\right), \tag{4.1}$$

where $n = 1, 2, 3, \ldots$ is the quantum number of the state of the system. The characteristic time of the particle in the *nth* state is associated with the round trip time, T_n, between $x = 0$ and $x = L$, and is given by $T_n \approx h/\varepsilon_n$.

Suppose now that the wall of the box at $x = L$ is moved slowly (adiabatically) in a time $\Delta t \gg T_n$ to a final position $x = 2\ L$. During that time, the size of the box can be described by the function $l(t)$ such that $l(t_i) = L$ and $l(t_i + \Delta t) = 2\ L$. Figure 4.2a shows the particle's wavefunction (in the $n = 1$ state) at t_i and at t_f. Note that as the wall is moved, the particle remains in the $n = 1$ state, as long as the wall is moved slowly.

In this scenario, the system consists of a slow subsystem (the expanding box) and a fast subsystem (the particle). As the wall moves, the particle responds nearly instantaneously to the new position of the wall, and always remains in the *nth* energy state; a the transition from the *nth* to the state $(n + 1)th$ state (or any other state) will not occur. During the time Δt, the instantaneous energy and wavefunction of the system are given by:

$$\varepsilon_n(l) = \frac{h^2\pi^2}{8\mathrm{m}(l(t))^2}n^2 \text{ and } \psi\ (x,\ l) = \sqrt{\frac{2}{l(t)}}\, sin\left(\frac{n\pi x}{l(t)}\right) \tag{4.2}$$

As the wall moves, the wavefunctions and energies have a dependence on l, the width of the box, and hence on time. Because the form of the energies and wavefunctions are unaltered, this dependence on the length of the box is a *parametric* one.

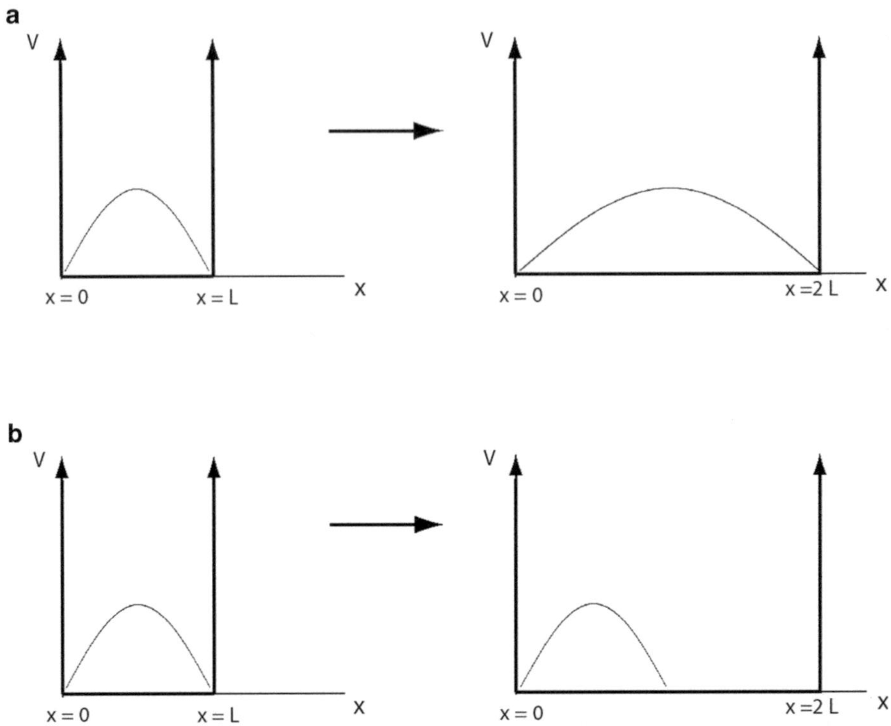

Fig. 4.2 (**a**) In an adiabatic process the wall moves from x = L to x = 2L, the particle remains in the same state, and the wavefunction adapts adiabatically. (**b**) If the wall is moved suddenly, the initial wavefunction is unchanged immediately after the wall is moved. In each case above the n = 1 wavefunction is shown

For the sake of comparison, it is useful to consider briefly the case where the wall moves suddenly (i.e. non-adiabatically) from $x = L$ to $x = 2L$ in a time $\Delta t \ll T_n$. This case is shown in Fig. 4.2b. In this case, the particle does not have time to react to the movement of the wall. Immediately after the wall is moved, the wavefunction of the particle is unaltered, and the particle is now in a superposition state consisting of many stationary states of the box of length $2L$. The motion of the wall in this case is decidedly non-adiabatic.

4.3 Ion in a Solid: The Adiabatic Approximation

An optically active of ion embedded in a solid consists of two subsystems: the electrons and the nuclei. Because the ratio of the nuclear mass to the electronic mass is on the order of 10^5, the nuclei move more slowly than the electrons by a factor of 10^{-2}–10^{-3}. Thus, the electrons and the nuclei constitute a fast subsystem and a slow subsystem, respectively, and an adiabatic approximation is indeed justified.

Treatments in the literature on the adiabatic approximation as applied to molecules or to ions in a solid are plentiful, [e.g. 2–9]. Though different authors each present valid treatments of the adiabatic approximation, there are inconsistencies among them, particularly in the nomenclature. For an excellent overview of the distinctions between some of the treatments, see the article by Azumi and Matsuzaki [6].

4.3.1 The Full Hamiltonian

The Hamiltonian of a molecular system can be expressed as

$$H_{total} = T_e(r) + T_N(R) + U\,(r, R) + V(R), \tag{4.3}$$

where

$$T_e(r) = \sum\nolimits_i \frac{p_i^2}{2\mathrm{m}} = -\frac{\hbar^2}{2\mathrm{m}} \sum\nolimits_i \vec{\nabla}_i^2 \tag{4.4}$$

$$T_N(R) = \sum\nolimits_i \frac{P_\alpha^2}{2M_\alpha} = -\frac{\hbar^2}{2} \sum\nolimits_\alpha \frac{\vec{\nabla}_\alpha^2}{M_\alpha} \tag{4.5}$$

$$U\,(r, R) = \sum\nolimits_i \sum\nolimits_{j>i} \frac{e^2}{\left|\vec{r}_i - \vec{r}_j\right|} + \sum\nolimits_i \sum\nolimits_\alpha \frac{Z_\alpha e^2}{\left|\vec{r}_i - \vec{R}_\alpha\right|} \tag{4.6}$$

$$V(R) = \sum\nolimits_\alpha \sum\nolimits_{\beta>\alpha} \frac{Z_\alpha Z_\beta e^2}{\left|\vec{R}_\alpha - \vec{R}_\beta\right|} \tag{4.7}$$

In the equations above, $\vec{r}_i$ and $\vec{r}_j$ are the positions of the electrons, and $\vec{R}_\alpha$ and $\vec{R}_\beta$ are the positions of the nuclei. $T_e\,(r)$ and $T_N\,(R)$ are the kinetic energy operators for the electrons and nuclei, respectively. $U(r,R)$ contains the electron-electron and electron-nuclear interactions, and $V(R)$ is represents the repulsive interaction between the nuclei. Z_α and M_α are the atomic number and mass of the αth nucleus, respectively, and m is the mass of the electron. To simplify the notation, for the remainder of the article we will use the symbols r and R instead of $\vec{r}$ and $\vec{R}$.

The eigenstates and eigenvalues of the system are given by the solutions to the time-independent Schrodinger equation:

$$(H_{total} - E_{total})\,\Psi\,(r, R) = 0 \tag{4.8}$$

Because of the electron-nuclear interaction term, the Hamiltonian H_{total} does not allow for separation of variables, i.e., $\Psi(r, R)$ cannot be separated into the product of electronic and nuclear wavefunctions.

4.3.2 *The Born-Oppenheimer Adiabatic Approximation*

Let us define an electronic Hamiltonian consisting all the terms containing the electronic coordinates,

$$H_{elec} = T_e(r) + U(r, R) \ . \tag{4.9}$$

As constructed, the potential energy term in H_{elec} contains both electronic and nuclear positions, so the eigenfunctions of H_{elec} will be functions of both r and R.

Let us assume that at any moment the nuclear positions can be considered fixed, that is, the electrons are "unaware" of the nuclear motion. This assumption is justified by the fact that the nuclei move much more slowly than the electrons, so that the electrons can react instantaneously to any change in position of the nuclei. (Note that the nuclear motion in this system is the analog to the moving wall in the previous section.) In such an assumption, R is treated as a parameter and r as the variable. The Schrodinger equation for H_{elec} becomes

$$(T_e(r) + U(r, R) - \epsilon_k(R))\, \phi_k(r, R) = 0. \tag{4.10}$$

The wavefunctions $\phi_k(r, R)$ and the eigenvalues $\epsilon_k(R)$ correspond to the *kth* electronic state at a particular set of nuclear coordinates. As the nuclear coordinates change, the system remains in the *kth* electronic state as the wavefunction and energy eigenvalues adjust adiabatically. The set of wavefunctions $\phi_k(r, R)$ may be chosen to form a complete, orthonormal set, i.e., they obey the following relationship:

$$\langle \phi_i(r, R)\, \phi_j |\, (r, R) \rangle = \delta_{ij}. \tag{4.11}$$

Combining Eqs. (4.3), (4.8), and (4.9), we may write:

$$(T_N(R) + V(R) + H_{elec} - E_{total})\, \Psi(r, R) = 0 \tag{4.12}$$

We see that the Hamiltonian is a sum of terms that depend on the nuclear and electronic coordinates separately. Since the $\phi_k(r, R)$ form a complete set, $\Psi(r, R)$ may be expanded as follows:

$$\Psi(r, R) = \sum\nolimits_k \phi_k(r, R)\, \chi_k(R), \tag{4.13}$$

where the $\chi_k(R)$ are expansion coefficients that are functions of the nuclear positions.

Inserting (4.13) into (4.12), and writing the electronic state in Dirac notation, we obtain:

$$(T_N(R) + V(R) + H_{elec} - E_{total}) \sum_k \chi_k(R) \left| \phi_k(r, R) \right\rangle = 0 \tag{4.14}$$

Operating on Eq. (4.14) on the left with $\left\langle \phi_i(r, R) \right|$, and using (4.5), (4.9), (4.10) and (4.11), we get:

$$\begin{aligned}
&[T_N(R) + V(R) + \epsilon_i(R) - E_{total}]\, \chi_i(R) \\
&\quad - \sum_k \sum_\alpha \frac{\hbar^2}{2M_\alpha} \chi_k(R) \left\langle \phi_i(r, R) \right| \vec{\nabla}_\alpha^2 \left| \phi_k(r, R) \right\rangle \\
&\quad - \sum_k \sum_\alpha \frac{\hbar^2}{M_\alpha} \left\langle \phi_i(r, R) \right| \vec{\nabla}_\alpha \left| \phi_k(r, R) \right\rangle \vec{\nabla}_\alpha \chi_k(R) = 0
\end{aligned} \tag{4.15}$$

In arriving at (4.15) we integrated over all electronic coordinates, i.e. we imposed the orthonormality condition defined by Eq. (4.11), so $\epsilon_i(R)$ does not depend on the electronic coordinates. We also used the following:

$$\begin{aligned}
\vec{\nabla}_\alpha^2 \phi(r, R)\, \chi(R) &= \vec{\nabla}_\alpha \cdot \left[\vec{\nabla}_\alpha (\phi \chi) \right] \\
&= \vec{\nabla}_\alpha \cdot \left(\phi \vec{\nabla}_\alpha \chi + \chi \vec{\nabla}_\alpha \phi \right) \\
&= \phi \vec{\nabla}_\alpha^2 \chi + \chi \vec{\nabla}_\alpha^2 \phi + 2 \left(\vec{\nabla}_\alpha \phi \right) \cdot \left(\vec{\nabla}_\alpha \chi \right)
\end{aligned}$$

The term in brackets [] on the left side of Eq. (4.15) depends only on the nuclear positions. The last two terms in Eq. (4.15) contain derivatives of the electronic wavefunctions with respect to the nuclear positions, and are responsible for coupling the nuclear and electronic subsystems.

We now make the assumption that the last two terms on the right side of Eq. (4.15) are negligible, simplifying Eq. (4.15) to

$$[T_N(R) + V(R) + \epsilon_i(R) - E_{total}]\, \chi_i(R) = 0 \tag{4.16}$$

Equation (4.16) is essentially the Schrodinger equation for the nuclear wavefunctions, $\chi_i(R)$. Note that $\epsilon_i(R)$ comes from the solution to the electronic Schrodinger Eq. (4.10) and plays the role of a potential energy in (4.16).

Neglecting the last two terms in (4.15) has allowed for the separation of the electronic and nuclear wavefunctions whose solutions are found by first solving

Eq. (4.10) and then (4.16). Consequently, the wavefunction of the system can be written as the product of an electronic wavefunction and a nuclear wavefunction:

$$\Psi_k(r,R) = \phi_k(r,R)\,\chi_k(R) \tag{4.17}$$

Equations (4.10), (4.13) and (4.17) comprise the *Born-Oppenheimer Adiabatic Approximation*. In this approximation, $\epsilon_i(R)$ corresponds to an average electronic potential in which the nuclei move, and includes, as seen from Eq. (4.10), the kinetic energy of the electrons, the electron-electron interaction, and the electron-nuclear interaction. If the nuclear positions change, the electrons, with their much smaller masses, react immediately to the new position of the nuclei, producing a new $\epsilon_i(R)$, that is, the system behaves adiabatically in response to the motion of the nuclei.

The validity of the Born-Oppenheimer Adiabatic Approximation rests on the last two terms in (4.15) to be considered negligible, that is, on terms of the form $\vec{\nabla}_\alpha\phi_k(r,R)$ and $\vec{\nabla}^2_\alpha\phi_k(r,R)$, and on the nuclear mass, M_α. Neglecting the terms containing $\vec{\nabla}_\alpha\phi_k(r,R)$ and $\vec{\nabla}^2_\alpha\phi_k(r,R)$ assumes a weak dependence of the electronic wavefunction, $\phi_k(r,R)$, on the nuclear coordinates. Specifically, it assumes that the slope and the curvature of $\phi_k(r,R)$ in R-space are small. This is reasonable for many systems, especially if the amplitudes of the vibrations of the nuclei are small. Also, the large nuclear mass M_α in the denominator contributes to making the last two terms in (4.15) small perturbations to the Hamiltonian in (4.16).

4.3.3 The Born-Huang Adiabatic Approximation

In this section we relax slightly the constraints of the Born-Oppenheimer approximation. Recall that in adiabatic process the motion of the slow subsystem will not induce a change in the state of the fast subsystem. Thus, if the fast system is perturbed by an interaction with slow system, only matrix elements that connect two different electronic states can induce the system to transition to a different electronic state. In examining Eq. (4.15), one notices that only the off diagonal components of matrix elements in the last two terms of Eq. (4.15) connect different electronic states. The diagonal components, on the other hand, connect each electronic state to itself, and so can be incorporated into the Hamiltonian in (4.16).

Following this thought, we first consider the diagonal component of the last term in Eq. (4.15):

$$-\frac{\hbar^2}{M_\alpha}\left\langle\phi_k(r,R)\middle|\vec{\nabla}_\alpha\middle|\phi_k(r,R)\right\rangle\vec{\nabla}_\alpha\chi_k(R) \tag{4.18}$$

where $\left\langle\phi_i(r,R)\middle|\vec{\nabla}_\alpha|\phi_k(r,R)\right\rangle$ represents an integral over electronic coordinates, r, and can be rewritten as

$$-\frac{\hbar^2}{M_\alpha}\left[\int \phi_k\left(r,R\right)\overrightarrow{\nabla}_\alpha \phi_k\left(r,R\right)d\overrightarrow{r}\right]\overrightarrow{\nabla}_\alpha \chi_k(R) \tag{4.19}$$

Inside the brackets [] in the above expression, we may rearrange the integration over $\overrightarrow{r}$ with the derivative with respect to the nuclear coordinates to obtain

$$-\frac{\hbar^2}{M_\alpha}\overrightarrow{\nabla}_\alpha\left[\int \phi_k\left(r,R\right)\phi_k\left(r,R\right)d\overrightarrow{r}\right]\overrightarrow{\nabla}_\alpha \chi_k(R) \tag{4.20}$$

Imposing orthonormality of the electronic states (at all R), (4.20) is equal to zero, since $\overrightarrow{\nabla}_\alpha(1) = 0$. Thus, the diagonal matrix elements of the last term in (4.15) are identically zero.

We now consider the diagonal elements of the second term in Eq. (4.15), namely

$$-\frac{\hbar^2}{2M_\alpha}\chi_k(R)\langle \phi_k\left(r,R\right)\left|\overrightarrow{\nabla}^2{}_\alpha\right|\phi_k\left(r,R\right)\rangle. \tag{4.21}$$

We may choose $\phi_k(r,R)$ to the real, in which case the integral form of (4.21) appears as

$$-\frac{\hbar^2}{2M_\alpha}\chi_k(R)\int \phi_k\left(r,R\right)\overrightarrow{\nabla}^2{}_\alpha \phi_k\left(r,R\right)d\overrightarrow{r}. \tag{4.22}$$

To evaluate (4.22), consider the identity

$$\begin{aligned} 0 &= \nabla_\alpha^2\left[\int \phi_k\left(r,R\right)\phi_k\left(r,R\right)d\overrightarrow{r}\right] \\ &= \int \nabla_\alpha^2\left[\phi_k\left(r,R\right)\phi_k\left(r,R\right)\right]d\overrightarrow{r} \\ &= 2\int\left(\overrightarrow{\nabla}_\alpha \phi_k\left(r,R\right)\right)^2 d\overrightarrow{r} + 2\int \phi_k\left(r,R\right)\nabla_\alpha^2 \phi_k\left(r,R\right)d\overrightarrow{r} \end{aligned} \tag{4.23}$$

where in the first line above we have use the orthonormality condition (4.11). Using (4.23), (4.22) can be written as

$$\frac{\hbar^2}{2M_\alpha}\chi_k(R)\int\left(\overrightarrow{\nabla}_\alpha \phi_k\left(r,R\right)\right)^2 d\overrightarrow{r}. \tag{4.24}$$

Thus, the diagonal elements of the last two terms in (4.15) are reduced to (4.24). In carrying out the integration over $\overrightarrow{r}$ in (4.24) results in a term that is a function of R only, and so may be incorporated into the left side of Eq. (4.16), which becomes

$$\left[T_N(R) + V'(R) + \epsilon_k(R) - E'_{total}\right]\chi_k(R) = 0, \tag{4.25}$$

where

$$V'(R) = V(R) + \frac{\hbar^2}{2M_\alpha}\int \left(\vec{\nabla}_\alpha \phi_k(r,R)\right)^2 d\vec{r}. \qquad (4.26)$$

Equation (4.25) can be solved to obtain the nuclear wavefunctions, $\chi_k(R)$. As previously done, the separation of the electronic and nuclear equations allows us to write the states of the system as

$$\Psi_k(r,R) = \phi_k(r,R)\,\chi_k(R). \qquad (4.27)$$

where the $\chi_k(R)$ in (4.27) are different than those in (4.17). Equations (4.10), (4.25), and (4.27) define the *Adiabatic Approximation*, sometimes referred to as the *Born-Huang Adiabatic Approximation*. Equation (4.25) represents a slight improvement over (4.16).

4.3.4 The Crude Adiabatic Approximation

It will prove useful to consider one additional form of the adiabatic approximation. We begin by rewriting (4.9) as follows:

$$\begin{aligned} H_{elec} &= T_e(r) + U(r,R_0) + \Delta U(r,R) \\ &= H^0_{elec}(r,R_0) + \Delta U(r,R) \end{aligned} \qquad (4.28)$$

where R_0 indicates the nuclei are at fixed positions, and $\Delta U(r,R)$ is the interaction due to the displacements of the nuclei from those positions. Neglecting $\Delta U(r,R)$, and using the same arguments as in Sect. 4.3.2, the electronic wavefunctions can be found by solving the Schrodinger equation

$$\left(T_e(r) + U(r,R_0) - \epsilon_i^0(R_0)\right)\phi_k^0(r,R_0) = 0, \qquad (4.29)$$

which is analogous to Eq. (4.10). In (4.29), the $\phi_k^0(r,R_0)$ are electronic wavefunctions with the nuclei at fixed positions, and are not identical to the analogous terms in (4.10). These $\phi_k^0(r,R_0)$ also form a complete set of orthonormal wavefunctions, so

$$\Psi_k(r,R) = \sum_k \phi_k^0(r,R_0)\,\chi_k(R), \qquad (4.30)$$

with the understanding that the $\chi_k(R)$ in (4.26) are not the same as those in (4.13). The states defined in (4.26) are called the *crude* Born-Oppenheimer states.

Following arguments similar to those in Sect. 4.2.2, the Schrodinger equation analogous to (4.16), is

$$\left[T_N(R) + V(R) + \Delta U(r,R) + \epsilon_k^0(R_0) - E^0_{total}\right]\chi_k(R) = 0 \qquad (4.31)$$

Because $\Delta U(r,R)$ depends on both the electronic and nuclear coordinates, (4.31) as written cannot be solved for all values of r.

The Schrodinger equation for the entire system is

$$\left[T_N(R) + V(R) + \Delta U\left(r,R\right) + \epsilon_i^0\left(R_0\right) - E_{total}^0\right]\sum_i \phi_i^0\left(r,R\right)\rangle\chi(R) = 0 \quad (4.32)$$

Operating on (4.27) with $\phi_k^0\left(r,R_0\right)\Big|$ we obtain

$$\begin{aligned}&\langle\phi_k^0\left(r,R_0\right)\left|\left[T_N(R) + V(R) + \Delta U\left(r,R\right) + \epsilon_k^0\left(R_0\right) - E_{total}^0\right]\right|\sum_i\phi_i^0\left(r,R_0\right)\rangle\chi_k(R)\\&= \left\{T_N(R) + V(R) + \epsilon_k^0\left(R_0\right) - E_{total}^0 + \langle\phi_k^0\left(r,R_0\right)\left|\Delta U\left(r,R_0\right)\right|\phi_k^0\left(r,R_0\right)\rangle\right\}\chi_k(R)\\&+ \sum_{i\neq k}\langle\phi_k^0\left(r,R_0\right)\left|\Delta U\left(r,R\right)\right|\phi_i^0\left(r,R_0\right)\rangle\chi_k(R) = 0\end{aligned} \quad (4.33)$$

Note that we have separated out the diagonal matrix element of $\Delta U(r,R)$, which does not connect different electronic states. All terms in the brackets {} depend only on R. Also, because R_0 represents fixed coordinates, T_{N} does not act on the electronic wavefunctions, as it did in the previous two sections.

Neglecting the off-diagonal terms in (4.33), which couple the electronic to the nuclear motion, results in:

$$\left\{T_N(R) + V(R) + \epsilon_k^0\left(R_0\right) - E_{total}^0 + \langle\phi_k^0\left(r,R_0\right)\left|\Delta U\left(r,R\right)\right|\phi_k^0\left(r,R_0\right)\rangle\right\}\chi_k(R) = 0\,. \quad (4.34)$$

Equation (4.34) can be solved to find the nuclear wavefunctions, $\chi_k(R)$. Finally, the separation of the electronic and nuclear wavefunctions into functions of r and R, respectively, allows the wavefunction of the system to be written as

$$\Psi\left(r,R\right) = \phi_k^0\left(r,R\right)\chi_k(R) \quad (4.35)$$

Of course, the $\chi_k(R)$ in (4.35) are different from those in (4.17) and (4.27). Equations (4.29), (4.30) and (4.35) define the *Crude Adiabatic Approximation*. The defining characteristic of this approximation is that the electronic wavefunction are determined with the nuclei fixed at fixed positions.

In each of the three forms of the adiabatic approximations presented, the total wavefunction is a product of an electronic wavefunction and a nuclear wavefunction, thus separating the fast and slow subsystems. The next section is devoted to finding expressions for the coupling between the electronic and nuclear subsystems, known as the electron-phonon coupling. Before proceeding, we make the following observations.

1. In deriving the Crude Adiabatic Approximation, we neglected the off diagonal matrix elements between electronic states of the potential energy term in the Hamiltonian, $\Delta U(r,R)$. On the other hand, in the adiabatic approximation in

Sects. 4.2.2 and 4.2.3, the neglected terms arose from the kinetic energy term, T_N, operating on the electronic wavefunctions. Thus, the different formalisms presented in the previous sections neglect different terms in the Hamiltonian in order to separate the fast and slow subsystems. It should not be surprising, then, that these different formalisms will give rise to different forms for the electron-phonon coupling.

2. The Crude Adiabatic Approximation assumes that the positions of the nuclei are fixed. If we take this position to be that at which the ground state of the ion is at equilibrium, this approximation is only valid in systems for which the higher electronic states have equilibrium positions close to that of the ground state. Such is the case for ions in electronic states that are weakly coupled to the lattice. Examples are the 4f states of the lanthanides ions and also the ^2E level of the Cr^{3+}.
3. The neglected terms in the Born-Oppenheimer and Born-Huang Adiabatic Approximations are valid starting points for finding the electron-phonon coupling in weakly- and strongly-coupled systems. Transitions among most *d*-states in transition metal ions and *f-f* or *f-d* transition in rare earth ions can be treated with electron-phonon couplings found from the neglected terms in the Born-Oppenheimer or Born-Huang formalisms.

4.4 Electron-Phonon Coupling

A breakdown in the adiabatic approximation leads to an interaction that couples the electronic motion to the nuclear motion, allowing from the conversion of electron energy to be converted into the kinetic and potential energy associated with of the vibration of the nuclei. This interaction is often called the electron-phonon interaction, and is formally included in the quantum mechanical transition matrix elements as an electron-phonon coupling operator. In this section, we present the two most commonly used expressions for the electron-phonon coupling operator for an ion in a solid. These operators are used to explain a variety of phonon-related transitions between different electronic states, including vibronic absorption and emission, one-phonon transitions, multiphonon relaxation, and Raman scattering.

4.4.1 Type A Electron-Phonon Coupling

In the Crude Adiabatic Approximation, we neglected the off-diagonal terms of $\Delta U(r, R)$, which takes into account the change in potential energy of the system due to the motion of the nuclei. To find the electron-phonon coupling within this approximation, we first write $\Delta U(r, R)$ in a more useful form.

In the presence of an acoustic wave, the optically active ion is displaced from its equilibrium positions, changing the energy of the system. This change in energy is due only to a *relative* change in the distance between the optically active ion and its neighbors. We denote the distance between the optically active ion and its neighbor (labeled α) as $R_{ion,\alpha}$, and the equilibrium distance between them as $R_{ion,\alpha,0}$. For the systems considered in this work, the wavefunctions are highly localized to the central ion, and so we need concern ourselves mainly with the perturbation of the optically active ion by the motion of the surrounding ions. The change in the potential energy between the optically active ion and its neighbors due to the vibrations, which we write as $\Delta U_{ion}(r, R)$, can be expressed in terms of a Taylor expansion:

$$\begin{aligned}\Delta U_{ion}(r,R) &= U_{ion}(r,R) - U_{ion}(r,R_0) \\ &= \sum\nolimits_{\alpha}\left[\frac{\partial U_{ion}}{\partial R_{\alpha}}\right]_0 (R_{ion,\alpha} - R_{ion,\alpha,0}) + \\ &\frac{1}{2}\sum\nolimits_{\alpha,\beta}\left[\frac{\partial^2 U_{ion}}{\partial R_{ion,\beta}\partial R_{ion,\alpha}}\right]_0 (R_{ion,\alpha} - R_{ion,\alpha,0})\left(R_{ion,\alpha} - R_{ion,\beta,0}\right) + \cdots \\ &= \Delta U_{ion}^{(1)} + \Delta U_{ion}^{(2)} + \cdots\end{aligned} \tag{4.36}$$

where $\Delta U_{ion}^{(1)}$ and $\Delta U_{ion}^{(2)}$ are given by

$$\Delta U_{ion}^{(1)} = \sum\nolimits_{\alpha}\left[\frac{\partial U_{ion}}{\partial R_{\alpha}}\right]_0 (R_{ion,\alpha} - R_{ion,\alpha,0}) \tag{4.37}$$

$$\Delta U_{ion}^{(2)} = \frac{1}{2}\sum\nolimits_{\alpha,\beta}\left[\frac{\partial^2 U_{ion}}{\partial R_{ion,\beta}\partial R_{ion,\alpha}}\right]_0 (R_{ion,\alpha} - R_{ion,\alpha,0})\left(R_{ion,\alpha} - R_{ion,\beta,0}\right) + \cdots \tag{4.38}$$

In order to gain physical insight into these terms, we consider the simple case of a linear solid, shown in Fig. 4.3. The atoms nuclei are separated from their neighbors by a distance a, and they are free to vibrate longitudinally. The displacement of the *lth* ion from its equilibrium position is given by q_l. In addition to the displacements, Fig. 4.3 also shows a representation of the acoustic wave in the crystal. In the long wavelength approximation, the strain, ε_l, at the *lth* site can be written as [4]:

$$\varepsilon_l = \frac{\partial q_l}{\partial x} \approx \frac{q_{l+1} - q_l}{a} = \frac{(x_{l+1} - x_l) - a}{a}. \tag{4.39}$$

Thus, the strain is proportional to the change in distance between the optically active ion and a neighboring ion. In the case of an ion in a solid, $(x_{l+1} - x_l) - a$ in (4.39) is replaced by $R_{ion,\alpha} - R_{ion,\alpha,0}$. Comparing (4.39) and (4.36),

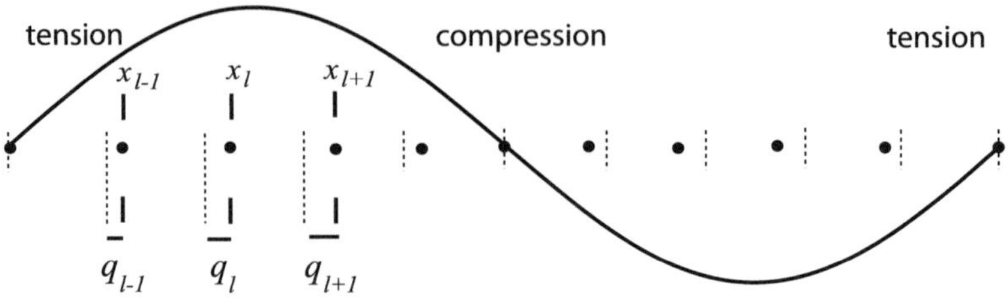

Fig. 4.3 Representation of a one-dimensional solid in the presence of a longitudinal acoustic wave. Equilibrium positions of the atoms are given by the dashed lines, separated by a distance a, and actual positions are labeled by x_l. The sine wave is a representation of the displacement of the ions from equilibrium as a function of the horizontal position, with regions of tension and compression indicated. In the long wavelength approximation, the strain in the lattice is proportional to $(q_{l+1} - q_l)$ (Figure 4.3 is adapted from Ref. [4])

the change in potential energy experienced by the optically active ion due to the displacements of the ions in the solid can be written as the sum of powers of the strain.

$$\Delta U_{ion}(r, R) = \sum\nolimits_{\alpha} V_{1,\alpha}\varepsilon_{\alpha} + \sum\nolimits_{\alpha} V_{2,\alpha}{\varepsilon_{\alpha}}^{2} + \cdots \tag{4.40}$$

For a linear system of like ions experiencing a longitudinal wave, as shown in Fig. 4.3, the form of the strain given in (4.39) is particularly simple. For a three dimensional crystal consisting of different ion types, transverse and longitudinal waves, and going beyond the long wavelength limit, the situation is very much more complex. We shall, however, make the assumption that we can write the $\Delta U_{ion}(r, R)$ in the form given in (4.40). We will utilize Eq. (4.40) later in this paper.

In a perturbation treatment, the first term given in (4.36) can be used to find the first order correction to the electronic wavefunction:

$$\phi_i(r, R) = \phi_i^0(r, R_0) + \sum\nolimits_{j \neq i} \frac{\left\langle \phi_j(r, R_0) \left| \Delta U_{ion}^{(1)} \right| \phi_i(r, R_0) \right\rangle}{\epsilon_j - \epsilon_i} \phi_j(r, R_0) \tag{4.41}$$

The states in the Crude Adiabatic Approximation corrected to the first order are:

$$\Psi_{i,k}^{(1)}(r, R) = \left[\phi_i^0(r, R_0) + \sum\nolimits_{j \neq i} \frac{\left\langle \phi_j(r, R_0) \middle| \Delta U_{ion}^{(1)} \middle| \phi_i(r, R_0) \right\rangle}{\epsilon_j - \epsilon_i} \phi_i(r, R_0) \right] \chi_k(R) \tag{4.42}$$

The sum in (4.42) contains terms of the form $\left\langle \phi_j(r, R_0) \middle| \Delta U_{ion}^{(1)} \middle| \phi_i(r, R_0) \right\rangle$, which mix the various electronic states of the system, indicating that the displacements of the nuclei from equilibrium cause transitions from one electronic state to another. Thus, $\Delta U_{ion}^{(1)}$ is an example of an electron-phonon coupling operator. It is interesting to note that the wavefunctions in (4.42) are the product of a purely electronic wavefunction, in brackets [], with nuclear wavefunction. That is, when $\Delta U_{ion}^{(1)}$ is used as a perturbation to the electronic wavefunction only, it mixes various electronic states, but does not couple the electronic wavefunction to the vibrational wavefunction. The electronic and the nuclear states are not truly coupled in the usual sense, and the corrected wavefunction (4.42) is still considered "adiabatic". This type of the coupling is known as *electron-phonon coupling of type A* [6].

4.4.2 Type B Electron-Phonon Coupling

In the Born Oppenheimer Adiabatic Approximation (Sect. 4.2.2), the states of the system are given by (4.27). We define a non-adiabatic Hamiltonian, H_{NA}, by the neglected, off-diagonal matrix elements of the last two terms in (4.15) in the following manner.

$$\begin{aligned} H_{NA}\Psi_{i,k}(r,R) = &-\frac{\hbar^2}{2}\sum_k\sum_\alpha \frac{1}{M_\alpha}\chi_{i,k}(R)\overrightarrow{\nabla}_\alpha^2\phi_i(r,R) \\ &-\hbar^2\sum_k\sum_\alpha \frac{1}{M_\alpha}\left(\overrightarrow{\nabla}_\alpha\phi_i(r,R)\right)\cdot\left(\overrightarrow{\nabla}_\alpha\chi_{i,k}(R)\right) \end{aligned} \tag{4.43}$$

At this point, it is typical to assume that the first term on the right in (4.43) is negligible compared with the second. Under that assumption, (4.38) becomes

$$H'_{NA}\Psi_{i,k}(r,R) = -\hbar^2\sum_k\sum_\alpha\frac{1}{M_\alpha}\left(\overrightarrow{\nabla}_\alpha\phi_i(r,R)\right)\cdot\left(\overrightarrow{\nabla}_\alpha\chi_{i,k}\right). \tag{4.44}$$

Treating H'_{NA} as a perturbation, the first order, non-adiabatic wavefunction is given by:

$$\begin{aligned} \Psi_{i,k}^{(1)}(r,R) = \Psi_{i,k}(r,R) - \sum_{j\neq i}\sum_\alpha\frac{\hbar^2}{M_\alpha}&\frac{\left\langle\Psi_{j,k}(r,R)\middle|\left(\overrightarrow{\nabla}_\alpha\phi_i(r,R)\right)\cdot\left(\overrightarrow{\nabla}_\alpha\chi_{i,k}\right)\middle|\Psi_k(r,R)\right\rangle}{E_j-E_i} \\ &\times\Psi_{j,k}(r,R) \end{aligned} \tag{4.45}$$

The wavefunction $\Psi_{i,k}^{(1)}$ in (4.45) cannot be expressed as a product of an electronic wavefunction and a nuclear wavefunction, so it is a true non-adiabatic wavefunction. The electron-phonon coupling operator expressed in (4.44) leads to a breakdown of the adiabatic approximation, coupling the electronic and nuclear motions and allowing the nuclear motion to cause transitions between electronic states. H'_{NA} is sometimes referred to as the *electron-phonon coupling of type B*[6].

4.5 Representation of Eigenstates and Operators

In order to apply the previous treatments to phonon-related processes in solids, we must first represent the vibrational states of the lattice and electron-phonon coupling operators in the appropriate forms. To do so, we make the following assumptions:

1. *The nuclei vibrate in a harmonic potential.* In this so-called "harmonic approximation", each normal coordinates Q_k, is associated with the *kth* vibrational mode of the solid and oscillates with a frequency ω_k. Each mode acts as a harmonic oscillator, with the excitation of the *kth* oscillator corresponding to the number of phonons, n_k, in that mode. The energy of the lattice is:

$$E = E_0 + \sum\nolimits_k \left(n_k + \frac{1}{2}\right) \hbar\omega_k \tag{4.46}$$

where E_0 is the energy of the lattice with the nuclei at their equilibrium positions.

2. *The normal modes of the solid act independently, with no communication between them.* In this assumption, the eigenstates of the lattice vibrations, $\chi(Q_k)$, are products of the states of the normal modes:

$$\chi\left(Q_k\right) = \prod\nolimits_k \left|n_k\right\rangle = \left|n_1\right\rangle\left|\, n_2\right\rangle\left|n_3\right\rangle \ldots \left|\, n_k\right\rangle \ldots \left|\, n_{3N-6}\right\rangle, \tag{4.47}$$

where N is the number of atoms in the solid.

It should be noted that neither of the assumptions above are strictly valid. Experiments on phonon decay times have shown that phonons generally do decay into other, lower energy modes [e.g. 10, 11]. Also, the assumption of a harmonic approximation is only valid for very small amplitudes of vibration. As the amplitudes increase (i.e., as temperature increases), the restoring force becomes more non-linear. Frenkel noted early on that such non-linear effects cause a breakdown of the adiabatic separation between the electronic and nuclear subsystems, thus allowing lattice vibrations to cause electronic transitions [12]. Despite these assumptions, working with the states as described in (4.47) does lead to results that adequately explain the behavior of many systems across a range of temperatures.

In Sect. 4.3 we derived the electron-phonon coupling operators (types A and B) in terms of the nuclear coordinates. Within the constraints of the above assumptions, we now present the form of those electron-phonon coupling operators in terms of the normal coordinates.

The electron-phonon coupling of type A, in terms of the strain operator, is given in Eq. (4.39). It is convenient to express the strain in terms of the phonon annihilation and creation operators, a and $a^{\dagger}$, respectively. We do not derive this expression here, but simply present the result. The reader is referred to the text by Henderson and Imbusch [4] for the details of the derivation.

Referring to the example of a linear solid shown in Fig. 4.3, it can be shown that the displacement of the lth ion from its equilibrium position, q_l, can be written in terms of the generalized position has the following form [4]:

$$q_l = \left(\frac{1}{N}\right)^{1/2} \sum_k Q_k exp\left(-i\kappa a\right) \tag{4.48}$$

where a is the spacing between atoms, k is the wave vector associated with the kth mode, N is the number of atoms in the linear chain. Using (4.48), the local strain at the site of the lth ion, as approximated in (4.39), due to the kth normal mode can be expressed in terms of a_k and $a_k^{\dagger}$ as follows:

$$\varepsilon_k = -i\left(\frac{\hbar\omega_k}{2MNv_k^2}\right)^{1/2} \left(a_k^{\dagger} - a_k\right) \tag{4.49}$$

In (4.49) v_k is the velocity of the sound associated with the kth mode. Again, the reader is referred to [4] for the full derivation of (4.49). The operator for the electron-phonon coupling of type A is obtained by inserting (4.49) into the expansion similar to (4.40), except the sum is over all the normal coordinates. Keeping only the first two terms, the result is:

$$\Delta U_{ion} \approx \Delta U_{ion}^{(1)} + \Delta U_{ion}^{(2)} = V_{1,k}\varepsilon_k + V_{2,k}\varepsilon_k^2 \tag{4.50}$$

Recall that (4.48) applies to a linear chain of atoms in the long wavelength approximation. For practical reasons, however, it is usually assumed that the simplified version of the strain operator in (4.49) has the same form for all normal modes in a three-dimensional crystal with different masses, whether those modes correspond to transverse or longitudinal waves.

The electron phonon coupling of type B is given by the non-adiabatic Hamiltonian, H_{NA}', defined by Eq. (4.44). To express (4.44) in terms of the normal coordinates, recall that $i\hbar\vec{\nabla}_{\alpha} = P_{\alpha}$. The kinetic energy of the lattice in Cartesian coordinates and in normal coordinates is:

$$T_N = \sum_{\alpha} \frac{P_{\alpha}^2}{2M_{\alpha}} = \sum_k \frac{P_k^2}{2M} = -\hbar^2 \sum_k \frac{1}{2M}\left(\frac{\partial}{\partial Q_k}\right)^2 \tag{4.51}$$

where the first sum is over all nuclei, the second and third sums are over all normal modes, and M is a properly weighted mass. Using the term on the far right in (4.51), and re-deriving Eq. (4.44), it is readily seen that the electron-phonon coupling operator of type B operating on the adiabatic wavefunction of the system is expressed as:

$$H'_{NA}\Psi_i(r,Q) = -\hbar^2 \sum\nolimits_k \frac{1}{M}\left(\frac{\partial \phi_i(r,Q)}{\partial Q_k}\right)\cdot\left(\frac{\partial \chi_{i,}(Q)}{\partial Q_k}\right) \tag{4.52}$$

This defines the non-adiabatic operator written in terms of the normal coordinates of the lattice. We note that calculating the first term in parentheses in (4.52) is very difficult, requiring detailed knowledge of the electronic wavefunction. On the other hand, the last term in the sum in (4.52) is readily calculated in the harmonic approximation, since it contains the first derivatives of the standard harmonic oscillator wavefunctions. The matrix elements containing this term are frequently calculated in determining non-radiative transition rates between electronic levels using a single configurational coordinate model [13,14, and references therein].

4.6 Thermal Broadening and Shifting of Sharp Spectral Lines

4.6.1 Thermal Broadening of Spectral Lines

The broadening of a spectral line can be caused by several interactions, among which are the following:

1. Strain Broadening – These are site-to-site variations in the crystalline field at the ion due to strains in the crystal. This is a static interaction and is present at even low temperatures.
2. Lifetime Broadening: This category includes all processes that affect the lifetime (τ) of the ion in its excited state, thereby changing the linewidth (ΔE) through the uncertainty relation: $\Delta E\tau \geq \hbar/2$. The processes are radiative decay, nonradiative decay, and vibronic transitions. Even for allowed transitions, the broadening due to this term is less than the strain broadening observed in single crystals.
3. Direct processes: These processes involve a transition from one level to another via the absorption or emission of a phonon. This term has been found to be of secondary importance in most systems, and so will not be discussed here.
4. Raman Scattering: This occurs via the emission of a phonon at one frequency and the absorption of another phonon at a different frequency, and where the initial and final electronic states are the same, and the intermediate state is virtual. It is a second order process, and is found to be the dominant contributor to the shift in several systems [e.g. 15–18]. This process is shown in Fig. 4.4.

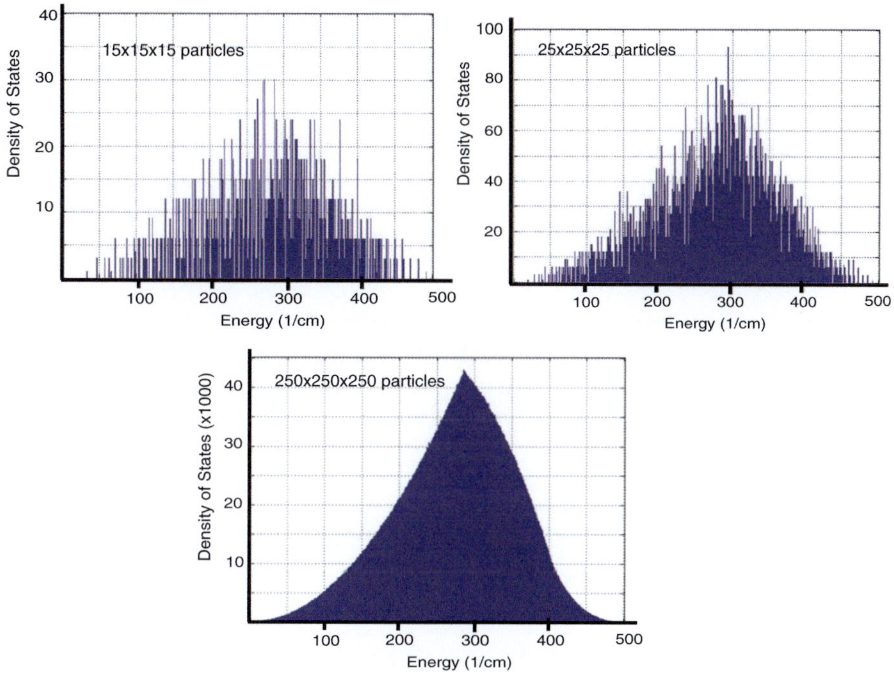

Fig. 4.4 The phonon density of states vs. phonon energy of cubic nanoparticles with (*left*) $15 \times 15 \times 15$ atoms, (*center*) $25 \times 25 \times 25$ atoms, and (*bottom*)$250 \times 250 \times 250$ atoms. The velocity of sound was set to 3400 m/s

To investigate this interaction, we utilize the electron-phonon coupling of type A, as given by Eqs. (4.49) and (4.50). For weak electron-phonon coupling, Eqs. (4.49) and (4.50) can be used to determine the interaction term to the second order. The result is:

$$\Delta U_{ion} \approx \Delta U_{ion}^{(1)} + \Delta U_{ion}^{(2)}$$
$$= V_1 \sum_k \left(\frac{\hbar \omega_k}{2Mv^2} \right)^{1/2} \left(a_k - a_k^\dagger \right) + V_2 \frac{\hbar}{2Mv^2} \sum_k \sqrt{\omega_k \omega_{k'}} \left(a_k - a_k^\dagger \right) \left(a_{k'} - a_{k'}^\dagger \right). \tag{4.53}$$

where we have assumed V_1 and V_2 are independent of the phonon mode k. The states of the system are products of an electronic part and a nuclear part.

$$|\Psi\rangle = |\phi\rangle \otimes \left| n_1 n_2 n_3 \ldots n_k \ldots \right\rangle = \left| \phi,\ n_1 n_2 n_3 \ldots n_k \ldots \right\rangle \tag{4.54}$$

Since Raman scattering is a second order process, the contributing terms derive from: (4.1) the first order term in (4.53) with the first order correction to the initial and final states, and (4.2) the second order term in (4.5) with the zeroth order states. The relevant matrix element for the Raman process is:

$$\left\langle \Psi_f \middle| \Delta U_{ion} \middle| \Psi_i \right\rangle$$
$$= \sum_j \frac{\left\langle \phi_i^0, n_k - 1, n_{k'} + 1 \middle| \Delta U_{ion}^{(1)} \middle| \phi_j^0 n_k - 1, n_{k'} \right\rangle \left\langle \phi_j^0, n_k - 1, n_{k'} \middle| \Delta U_{ion}^{(1)} \middle| \phi_i^0, n_k, n_{k'} \right\rangle}{\epsilon_i - \left(\epsilon_j - \hbar\omega_k\right)}$$
$$+ \sum_j \frac{\left\langle \phi_i^0, n_k - 1, n_{k'} + 1 \middle| \Delta U_{ion}^{(1)} \middle| \phi_j^0 n_k - 1, n_{k'} \right\rangle \left\langle \phi_j^0, n_k, n_{k'} + 1 \middle| \Delta U_{ion}^{(1)} \middle| \phi_i^0, n_k, n_{k'} \right\rangle}{\epsilon_i - \left(\epsilon_j - \hbar\omega_{k'}\right)}$$
$$+ \left\langle \phi_i^0, n_k - 1, n_{k'} + 1 \middle| \Delta U_{ion}^{(2)} \middle| \phi_i^0, n_k, n_{k'} \right\rangle \tag{4.55}$$

Using (4.53) to replace for $\Delta U_{ion}^{(1)}$ and $\Delta U_{ion}^{(2)}$ in (4.55), recalling that $\boldsymbol{a}\left|n\right\rangle = \sqrt{n}\left|n-1\right\rangle$ and $\boldsymbol{a}^{\dagger}\left|n\right\rangle = \sqrt{n+1}\left|n+1\right\rangle$, and assuming that $\omega_k, \omega_{k'} \ll \epsilon_j$ for all intermediate states, j, (4.55) becomes:

$$\left\langle \Psi_f \middle| \Delta U_{ion} \middle| \Psi_i \right\rangle = \alpha' \sqrt{\omega_k \omega_{k'} n_k \left(n_{k'} + 1\right)} \tag{4.56}$$

where

$$\alpha' = \frac{\hbar}{Mv^2}\left[\sum_{j \neq i} \frac{\left|\left\langle \phi_j \middle| \mathrm{V}_1 \middle| \phi_i \right\rangle\right|^2}{\epsilon_i - \epsilon_j} + \left\langle \phi_i \middle| \mathrm{V}_2 \middle| \phi_i \right\rangle\right]. \tag{4.57}$$

The transition probability per unit time due to modes k and k' is

$$\left(W_{Raman}\right)_{kk'} = \frac{2\pi}{\hbar^2}\left|\alpha'\right|^2 \omega_k \omega_{k'} n_k \left(n_{k'} + 1\right) \varrho\left(\omega_f\right), \tag{4.58}$$

where $\varrho(\omega_f)$ represents the density of final states of the phonon field. To find the total transition rate for these Raman processes, we must integrate over all phonon modes k and k'. For sharp lines, where the width of the line is much less than the Debye frequency, we estimate $\varrho(\omega_f)$ as

$$\varrho\left(\omega_f\right) \simeq \varrho\left(\omega_k\right) \varrho\left(\omega_{k'}\right) \delta\left(\omega_k - \omega_{k'}\right) d\omega_k d\omega_{k'} = \varrho^2\left(\omega_k\right) d\omega_k. \tag{4.59}$$

Inserting (4.59) into (4.58) and integrating over all phonon modes, we obtain:

$$W_{Raman} = \frac{2\pi}{\hbar^2} \left|\alpha'\right|^2 \int \omega_k^2 n_k \left(n_k + 1\right) \varrho^2 \left(\omega_k\right) d\omega_k \tag{4.60}$$

Recalling the expression for the phonon occupation number of the *kth* mode,

$$n_k = \frac{1}{e^{\hbar\omega_k/kT} - 1}, \tag{4.61}$$

the total transition probability becomes

$$W_{Raman} = \frac{2\pi}{\hbar^2} \left|\alpha'\right|^2 \int \omega_k^2 \varrho^2 \left(\omega_k\right) \frac{e^{\hbar\omega_k/kT}}{\left(e^{\hbar\omega_k/kT} - 1\right)^2} d\omega_k \, . \tag{4.62}$$

In the Debye approximation, $\varrho\left(\omega\right) = 3V\omega^2/2\pi^2 v_s^3$, (4.61) takes the following form:

$$W_{Raman} = \frac{2\pi}{\hbar^2} \left|\alpha'\right|^2 \frac{9V^2}{4\pi^4 v_s^6} \int_0^{\omega_D} \frac{\omega_k^6 \, e^{\hbar\omega/kT}}{\left(e^{\hbar\omega_k/kT} - 1\right)^2} d\omega_k, \tag{4.63}$$

where ω_D is the Debye frequency. It is convenient to rewrite (4.63) in terms of the unitless parameter $x = \hbar\omega/kT$ and the Debye temperature $T_D = \hbar\omega_D$. The result is:

$$W_{Raman} = \overline{\alpha} \left(\frac{T}{T_D}\right)^7 \int_0^{T_D/T} \frac{x^6 e^x}{\left(e^x - 1\right)^2} dx \, . \tag{4.64}$$

where $\overline{\alpha} = \frac{9V^2}{2\pi\hbar^2} \left(\frac{\omega_D}{v_s}\right)^6 \left|\alpha'\right|^2$, and is referred to as the electron-phonon coupling constant.

The temperature dependence of (4.64) is contained in the T^7 term outside the integral and in the upper limit of the integral, T_D/T. In the limit as $T \to 0$, the upper limit goes to infinity, and the integral is simply a constant, so the contribution of Raman processes to the linewidth goes as T^7, which goes to zero as $T \to 0$. At $T \gg T_D$ we may use the approximation $e^x \sim 1 + x$, so the integral goes roughly as x^5, and the (4.64) goes as T^2.

4.6.2 *Thermal Shifting of Spectral Lines*

Any interaction of a system an external agent will, in general, affect the energies of the states of the system. As temperature increases, the interaction of the phonons with the electrons also increases, leading to the thermal shift of the energy level. The contribution of the electron-phonon interaction is a second-order effect, and so contains the matrix element of $\Delta U_{ion}^{(2)}$ between the zeroth order states, and of $\Delta U_{ion}^{(1)}$

between the first order states. The correction to the energy of the level due to the electron-phonon interaction is given by:

$$\Delta\epsilon_{i,k} = \left\langle \phi_i^0,\ n_k \,\middle|\, \Delta U_{ion}^{(2)} \middle| \phi_i^0,\ n_k \right\rangle + \sum_j \frac{\left\langle \phi_i^0,\ n_k \middle| \Delta U_{ion}^{(1)} \middle| \phi_j^0,\ n_k - 1 \right\rangle \left\langle \phi_j^0, n_k - 1 \middle| \Delta U_{ion}^{(1)} \middle| \phi_i^0, n_k \right\rangle}{\epsilon_i - \left(\epsilon_j - \hbar\omega_{k'}\right)} + \sum_j \frac{\left\langle \phi_i^0, n_k \middle| \Delta U_{ion}^{(1)} \middle| \phi_j^0, n_k + 1 \right\rangle \left\langle \phi_j, n_k + 1 \middle| \Delta U_{ion}^{(1)} \middle| \phi_i^0, n_k \right\rangle}{\epsilon_i - \left(\epsilon_j + \hbar\omega_k\right)} \tag{4.65}$$

Using (4.53) to rewrite $\Delta U_{ion}^{(1)}$ and $\Delta U_{ion}^{(2)}$ in terms of the creation and annihilation operators, assuming that $\left|\epsilon_i - \epsilon_j\right| \gg \hbar\omega_k$, and using the identities $\boldsymbol{a}\left|n\right\rangle = \sqrt{n}\left|n-1\right\rangle$ and $\boldsymbol{a}^\dagger\left|n\right\rangle = \sqrt{n+1}\left|n+1\right\rangle$, (4.65) reduces to

$$\Delta\epsilon_{i,k} = \frac{\hbar}{Mv_s^2}\left[\left\langle \phi_i^0 \middle| \mathrm{V}_2 \middle| \phi_i^0 \right\rangle + \sum_j \frac{\left|\left\langle \phi_i^0 \middle| \mathrm{V}_1 \middle| \phi_j^0 \right\rangle\right|^2}{\epsilon_i - \epsilon_j}\right] \omega_k \left(2n_k + 1\right) \tag{4.66}$$

The thermal shift of the line is due only to the terms containing n_k. The total thermal shift is found by summing over all k. For *large particles*, this sum can be approximated by an integral, so the total shift is given by:

$$\Delta\epsilon_i = \frac{\hbar}{Mv_s^2}\left[\left\langle \phi_i^0 \middle| \mathrm{V}_2 \middle| \phi_i^0 \right\rangle + \sum_j \frac{\left|\left\langle \phi_i^0 \middle| \mathrm{V}_1 \middle| \phi_j^0 \right\rangle\right|^2}{\epsilon_i - \epsilon_j}\right] 2\int \rho\left(\omega_k\right) \omega_k n_k d\omega_k \tag{4.67}$$

Assuming the Debye distribution $\left(\varrho\left(\omega\right) = 3V\omega^2/2\pi^2 v_s^3\right)$, using the equilibrium value of n_k as defined in (4.61), setting the upper limit of the integral in (4.67) to the Debye frequency, and making the substitution $x = \hbar\omega/kT$, the thermal shift of a an energy level becomes:

$$\delta E_{ther} = \Delta V \int_0^{T_D/T} \frac{x^3}{e^x - 1} dx \tag{4.68}$$

where

$$\Delta V = \frac{2\hbar}{Mv_s^2}\left(\frac{kT}{\hbar}\right)^4 \left[\left\langle \phi_i^0 \left|\mathrm{V}_2\right| \phi_i^0 \right\rangle + \sum_j \frac{\left\langle \phi_i^0 \left|\mathrm{V}_1\right| \phi_j^0 \right\rangle^2}{\epsilon_i - \epsilon_j}\right] \tag{4.69}$$

The temperature dependence of the thermal shift of a spectral line is determined by the T^4 term contained in ΔV and by the upper limit of the integral. As $T \to 0$, the integral approaches a constant value, and the line shift goes as T^4. At high T the integrand goes roughly as x^2, the integral goes as T^{-3}, and so the line shift becomes linear with T.

4.7 The Phonon Density of States in Nanoparticles

Consider a simple cubic solid with side length L and atomic spacing a. The wavelengths of the phonons vary from roughly twice the atomic spacing to twice the side length of the particle. The energy of a phonon in such a solid is given by

$$\varepsilon_{ph} = \frac{hv_s}{\lambda} = \frac{hv_s}{2L}n, \tag{4.70}$$

where v_s is the velocity of sound, $n = \left(n_x{}^2 + n_y{}^2 + n_z{}^2\right)^{1/2}$, and n_x, n_y, n_z are integers ranging from 1 to L/a. Note that the maximum phonon energy is determined by the interatomic spacing, and so is independent of the particle size, while the low frequency phonons increase in energy as the particle size, L, decreases. Thus, many low frequency phonons that exist in the bulk are no longer supported in a nanoparticle.

Calculated phonon densities of states (DOS) of cubic nanoparticles $15 \times 15 \times 15$ atoms on a side (L ~ 3 nm), $25 \times 25 \times 25$ atoms on a side (L ~ 5 nm), and that of a nanoparticle $250 \times 250 \times 250$ atoms on a side (L ~ 50 nm) have been calculated using the speed of sound equal to 3400 m/s and an interatomic spacing of 0.2 nm. The modes were accumulated in 1000 bins, which for the diagrams show were each approximately 0.5 cm^{-1} in width. The results are shown in Fig. 4.4. We note the following:

1. For the $250 \times 250 \times 250$ atoms system, the DOS exhibits a ε^2 dependence, as expected from the Debye theory, out to a frequency at which the DOS reaches a maximum. At higher energies, the DOS is decidedly un-Debye-like, decreasing smoothly to zero. This behavior is due to the finite size of the crystal.
2. For the smaller particles, the DOS is a discrete function at lower energies, with a large energy gap between zero energy and the first mode. For 250 nm particles, the DOS appears nearly continuous at all energies, and is very similar in appearance to the DOS in a bulk cubic particle. Figure 4.5, shows the DOS at low phonon energies.
3. The results in Figs. 4.4 and 4.5 are for cubic crystals, but the discreteness of the DOS at low energies is a common feature to all very small particles. The exact shape of the DOS function, however, depends on the particle shape.

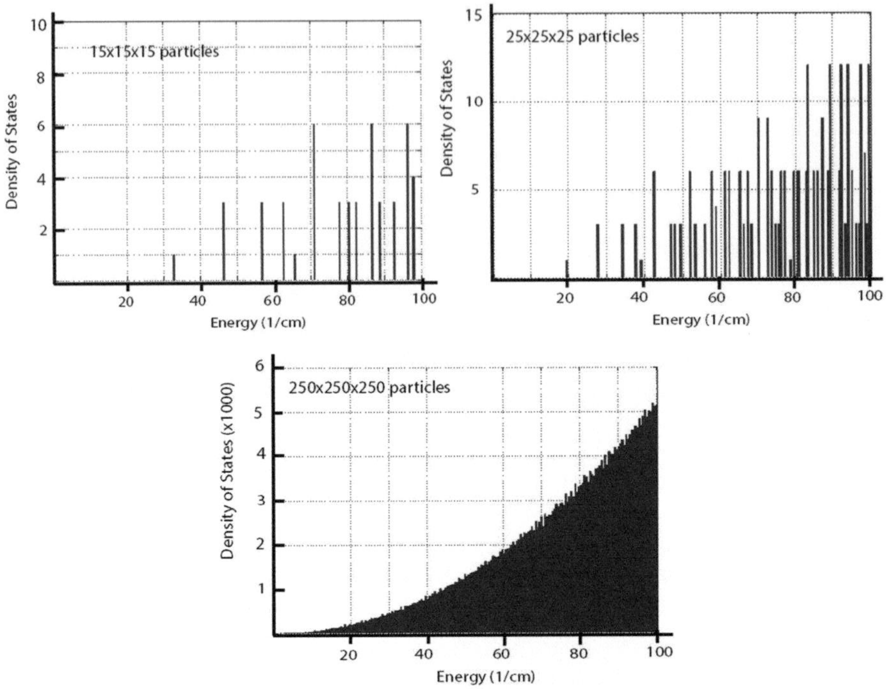

Fig. 4.5 The phonon density of states shown in Fig. 4.4 in the energy range from 0 to 100 cm^{-1}

The discreteness seen in the small particle is due to the fact that in going from bulk to nano, the total number of phonon modes decreases drastically. This decrease can be shown by noting that the total number of phonon modes is simply *3 N-6*, where N is the number of atoms in the particle, and can be estimated as $N \sim \left(\frac{L}{a}\right)^3$. For a bulk crystal with $L = 0.3$ cm and $a = 0.2$ nm, $3N \sim 4.5 \times 10^{21}$, whereas when $L = 3$ nm, $3N \sim 4.5 \times 10^3$. Thus, going from a particle size of 0.3 cm to 3 nm the total number of allowed phonon modes decreases by 18 orders of magnitude! As a result, the phonon spectrum is no longer a continuous function of energy.

Given the importance that phonon-related processes play in the luminescence behavior of ions in solids, the change in the phonon DOS as one moves from the bulk to the nano regime is likely have observable experimental effects. In the following sections, we take note of experiments that have, and in some cases have not, revealed such effects.

4.8 Effect of the Phonon DOS on the Establishment of Thermal Equilibrium

Following absorption of a photon, a luminescent ion generally relaxes to a state of quasi-thermal equilibrium. The time it takes to reach this quasi-thermal equilibrium is generally on the order of picoseconds. The relaxation can be within a single electronic state or among different electronic states, the latter of which would require the breakdown of the adiabatic approximation. Even relaxation within the same electronic state requires the participation of all phonon modes and the mixing of those modes for thermal equilibrium to be established. For small particles, where the low frequency modes are discrete and well separated from one another, we may expect the establishment thermal equilibrium following excitation to be inhibited.

Experimental evidence of this effect has been observed by G. Liu et al. [19, 20], who conducted emission and excitation experiments on nanoparticles of Er-doped Y_2O_3 with radii of $\sim$400 nm and 25 nm. Figure 4.6 shows an emission spectrum at 3 K of the $^4S_{3/2} \rightarrow {}^4I_{11/2}$ transition of Er in Y_2O_3 following excitation with a pulsed laser at energy levels into the $^4F_{7/2}$ manifold. In the 400 nm particles, the emission originates from only the lowest energy level of the $^4S_{3/2}$ manifold. In the 25 nm nanoparticles, however, anomalous hot emission bands are observed. Excitation into

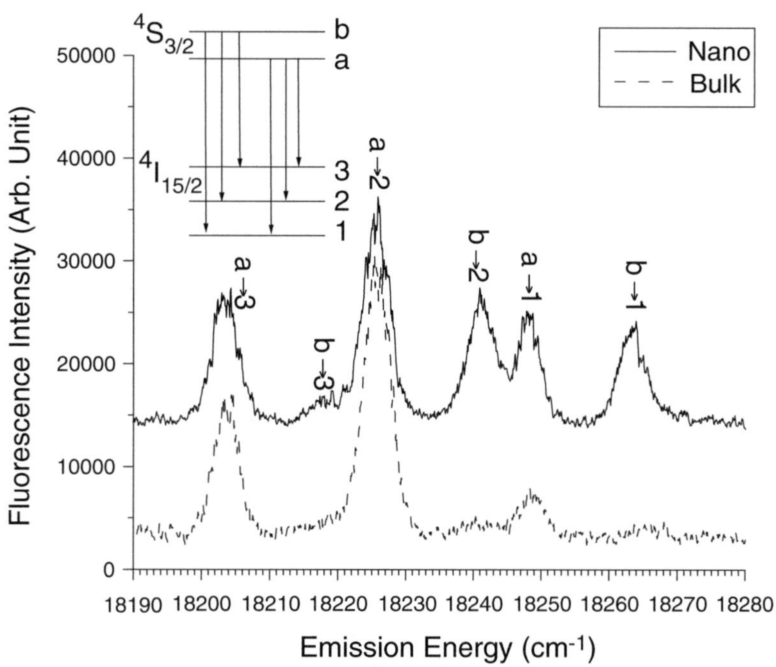

Fig. 4.6 The luminescence of the $^4S_{3/2}$ to $^4I_{15/2}$ transition in bulk (*dotted line*) and diameter nanocrystals (*solid line*) of Er:Y_2O_2S at 2.6 K [19]

the $^4F_{7/2}$ levels is followed by fast relaxation to the $^4S_{3/2}$, resulting in the population of the upper crystal field level of that manifold. In the 400 nm particles there is fast relaxation from level (b) of the $^4S_{3/2}$ manifold to its lower level (a) (See Fig. 4.6). In the 25 nm particles, however, there is no available mode to accept a phonon of that low frequency (~25 cm^{-1}). Consequently, the one-phonon decay process at that energy does not occur in the nanoparticle, and level (b) remains populated long enough to emit a photon.

This experiment demonstrates the effect of the discreteness of the phonon DOS in small particles. However, it also hints that observing such effects may be difficult; the discreteness of the phonon DOS can be masked by second-order processes and/or by the mixing of phonons due to anharmonic contributions to the potential, even at low temperatures.

4.9 Thermal Broadening and Shifting of Sharp Spectral Lines in Nanoparticles

In this section, we focus on the changes in the density of states of nanoparticles of different sizes affects the shift thermal broadening and shifting of sharp spectral lines.

4.9.1 Broadening of a Spectral Line in a Nanoparticle

In Sect. 4.6, the broadening of a spectral line was found to depend on the density of phonon states and on the phonon occupation number of each state. Earlier, we estimated the phonon density of states using Debye approximation, and the sum over all phonon states was carried out by integration. For nanoparticles, the density of phonon states is a discrete function, and the sum over states can be carried out directly. We begin our discussion of the thermal broadening of a spectral line in a nanoparticle with Eq. (4.58):

$$(W_{Raman})_{kk'} = \frac{2\pi}{\hbar^2}\left|\alpha'\right|^2 \omega_k \omega_{k'} n_k \left(n_{k'} + 1\right) \varrho\left(\omega_f\right), \tag{4.71}$$

where α' is given by (4.57).

The total transition rate is found by summing over all final states of the lattice, subject to the condition that energy must be conserved. In nanoparticles, the density of phonon states, $\varrho_{nano}(\omega)$, depends on the size and shape of the sample. Examples are shown in Fig. 4.4. In describing the density of phonon states, it is important to recall that each phonon mode represents a resonance peak with a particular line shape ($f(\omega)$) and line width ($\Delta\omega$). The density of states may be written as:

$$\varrho_{nano}(\omega_i) = g(\omega_i) f(\omega_i) \tag{4.72}$$

where $g(\omega_i)$ is the degeneracy modes at frequency ω_i. Though the line shape is more correctly represented as a Lorentzian, we shall simplify the shape to the "top hat" function, that is:

$$f(\omega) = \begin{cases} 1/\Delta\omega \text{ for } \omega - \Delta\omega \le \omega \le \omega + \Delta\omega \\ 0 \text{ elsewhere} \end{cases} \tag{4.73}$$

We also assume that line width of each resonance is $\Delta\omega$, independent of ω. For sharp spectral lines, we further assume that the main contribution to the broadening occurs when $|\omega_k - \omega_{k'}| \le \Delta\omega/2$, that is, the phonons in the scattering process are of nearly the same frequency. In such a scheme, we may approximate the density of final states as

$$\varrho(\omega_f) = \varrho_{nano}(\omega_i)\, \varrho_{nano}(\omega_i{}')\, \Delta\omega\Delta\omega'\delta(\omega_i - \omega_i{}') \tag{4.74}$$

Summing (4.71) over phonon frequencies, and using (4.72), (4.73) and (4.74), the total transition rate of the Raman process is

$$\begin{aligned} W_{Raman} &= \frac{2\pi}{\hbar^2}|\alpha'|^2 \sum_i g^2(\omega_i)\, \omega_i^2 n_{\omega_i}(n_{\omega_i}+1) f^2(\omega)\, \Delta\omega \\ &= \frac{2\pi}{\hbar^2}|\alpha'|^2 \frac{1}{\Delta\omega} \sum_i g^2(\omega_i)\, \omega_i{}^2 n_{\omega_i}(n_{\omega_i}+1) \end{aligned} \tag{4.75}$$

The degeneracy term $g(\omega)$ includes all modes within a range $\omega \pm \Delta\omega/2$. In Fig. 4.4, the energy axes are broken up into 1000 bins, each of energy ~0.5 cm^{-1}. $g(\omega_i)$ is given by the number of modes in the *ith* bin, where ω_i is the central frequency of the bin. The sum in (4.78), which carries the temperature dependence of the broadening, was carried out for four particle sizes from T=1 K to T=700 K. The results are shown in Fig. 4.7.

We make the following observations regarding these results.

1. The strongest temperature dependence of the line broadening for all particle sizes occurs at temperatures below 10 K. For the 3 nm (15 × 15 × 15 atoms) nanoparticle, the temperature dependence is strong between 1 K and 10 K.
2. Above 300 K the curvature of the lines in Fig. 4.7 are independent of particle size, thus thermal dependence of the broadening should be the same for particles of all sizes.
3. Figure 4.7 also shows that above ~10 K the expected broadening to be larger for smaller particle sizes. However, this effect depends on the details of the calculation (e.g. the binning of the data and the "top hat" line shape function), and should not be taken too seriously.

To understand the strong temperature dependence of the broadening at low temperatures, it is useful to consider not just the phonon density of states, but also the phonon occupancies. Figure 4.8 shows the product $\varrho_{nano}(\epsilon)n(\epsilon)$ for the 3 nm particle at T = 10 K, 100 K, and 500 K. At 10 K, only a few modes are occupied, and even then the phonon occupation numbers are very small, much less than 1. At 1 K, the occupancy of the lowest mode in the 3 nm particle is $\sim 10^{-23}$. Thus, in a nanoparticle there are essentially no phonons at 1 K, so there is no broadening. This helps explain the steep slope for the 3 nm particle in Fig. 4.7 at temperatures less that a few degrees K.

Using hole-burning experiments, Meltzer and Hong [21] examined the broadening of the $^7F_0 \rightarrow {}^5D_0$ transition of Eu_2O_3 spherical nanoparticles of different diameters (5.4, 7.6, and 11.6 nm) at temperatures between 4 K and 10 K. They observed a T^n dependence, where $3 < n < 4$, for the thermal broadening of the line. This dependence is much smaller than that shown in Fig. 4.7 for the 5 nm particles, and was also much smaller than their own calculated predictions. In contradiction to the results in Fig. 4.7, the authors observed the thermal broadening increasing as the particle size decreases. To explain this they refer to calculations that posit an inverse relation between the electron-phonon coupling and particle size [22].

Erdem et al. [23] measured the linewidth of the $^2E \rightarrow {}^4T_2$ transition of Cr-doped YAG nanoparticles at temperatures from 30 K to 300 K as a function of particle size (Fig. 4.9). The results were fit to Eq. (4.64) assuming Debye temperature of 550 K and where the electron-phonon coupling parameter, α', was allowed to vary. The fits to Eq. (4.64) are reasonably good, showing that even in nanoparticles as small as 28 nm behave similar to the bulk crystal. This is consistent with the behavior shown in Fig. 4.7. Excellent fits to Eq. (4.64) were also reported by Bilir et al. on the temperature dependence of the linewidth of Nd nanoparticles ranging in diameters from 16 to 250 nm [24].

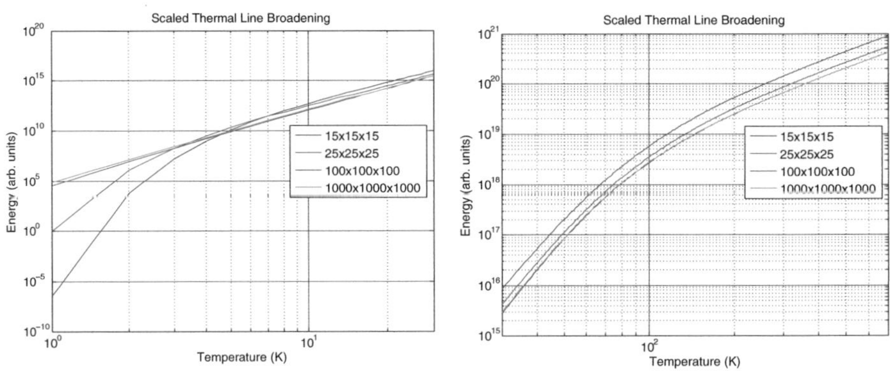

Fig. 4.7 Temperature dependence on the thermal broadening of a spectral line (given by the sum in Eq. (4.74)) for cubic nanoparticles (15 × 15 × 15-black, 25 × 25 × 25-blue, 100 × 100 × 100-red, 1000 × 1000 × 1000-green) for temperatures ranges 1–30 K (*left*) and 30–700 K (*right*)

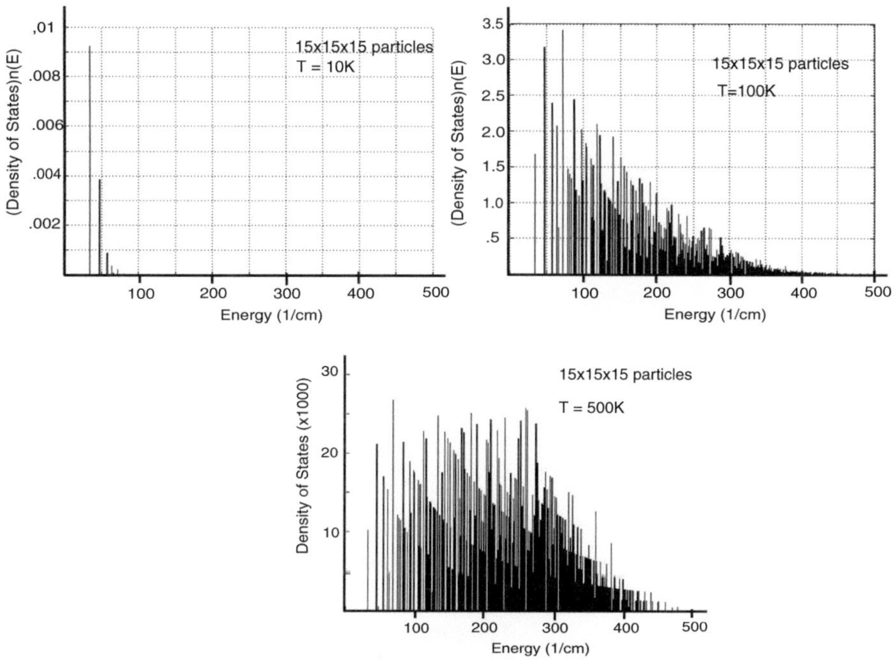

Fig. 4.8 Graphs showing the number of phonons as a function of phonon energy, i.e. $\varrho(\epsilon)n(\epsilon)$ *vs.* ϵ, for cubic nanoparticls$15 \times 15 \times 15$ atoms (size ~3 nm) at T = 10, 100, and 500 K

The examples above show an unresolved conflict in the behavior of the electron-phonon coupling as the particle size changes. Whereas Meltzer et al. [21] suggest an increase in the coupling as the particle size decreases, the data from Erdem et al. [23] and Bilir et al. [24] suggest a decrease in the coupling as particle size decreases. And finally Suyver et al. [25] conclude that electron-phonon coupling is independent of particle size. These works indicate that a definitive answer as to how the electron-phonon coupling changes with particle size remains elusive.

4.9.2 *Shifting of a Sharp Spectral Line in a Nanoparticle*

The theoretical treatment of the thermal shift of the energy of the spectral line in a nanoparticle begins with Eq. (4.66) the shift due to a particular phonon mode, *k*.

$$\Delta\epsilon_{i,k} = \frac{\hbar}{Mv_s^2}\left[\left\langle\phi_i^0 \,\middle|\, \mathrm{V}_2 \middle|\phi_i^0\right\rangle + \sum\nolimits_j \frac{\left|\left\langle\phi_i^0 \middle| \mathrm{V}_1 \,\middle|\, \phi_j^0\right\rangle\right|^2}{\epsilon_i - \epsilon_j}\right]\omega_k\,(2n_k + 1) \qquad (4.76)$$

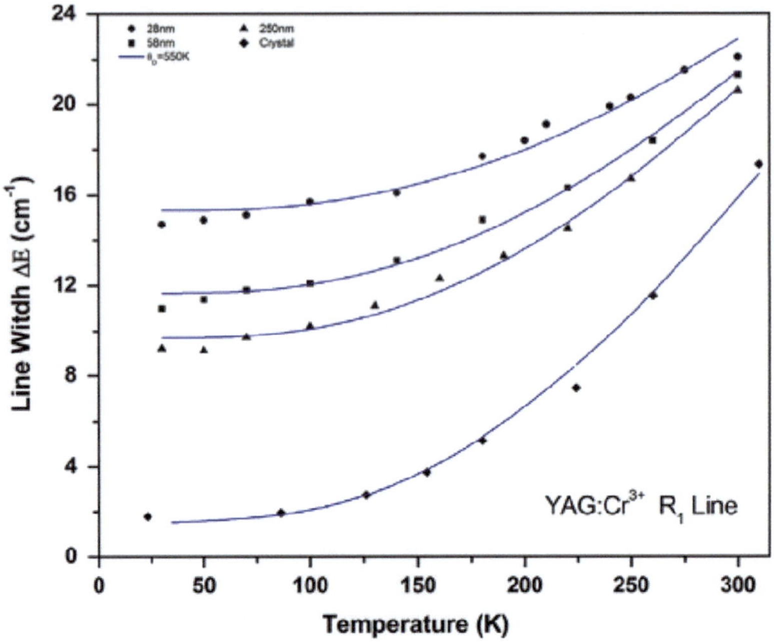

Fig. 4.9 Linewidth vs. temperature of the R_1 line in nanopowders (●28 nm, ■58 nm, ▲250 nm), and in a single crystal (♦) of Cr-doped YAG. The solid lines are fits to Eq. (4.64) [23]

To find the total thermal shift of the particle, we neglect the contribution of spontaneous emission to (4.79) (i.e., $(2n_k + 1)$ becomes $(2n_k)$, and sum (4.79) over all phonon modes. Referring to (4.75) and (4.76), the total thermal shift can be written as:

$$\Delta\epsilon_i = \frac{2\hbar}{Mv_s^2}\left[\left\langle \phi_i^0 \middle| V_2 \middle| \phi_i^0 \right\rangle + \sum_j \frac{\left|\left\langle \phi_i^0 \middle| V_1 \middle| \phi_j^0 \right\rangle\right|^2}{\epsilon_i - \epsilon_j}\right] \sum_k g(\omega_k)\,\omega_k n_k \qquad (4.77)$$

The sum in (4.77) carries the temperature dependence of the thermal line shift. Figure 4.10 shows the fit the sum in Eq. (4.77) for cubic nanoparticles ($15 \times 15 \times 15$, $25 \times 25 \times 25$, $100 \times 100 \times 100$, and $1000 \times 1000 \times 1000$ atoms) at temperatures between 1 K and 700 K. We see that, as with the line broadening, the main differences between the shift in large and small particles occurs at very low temperatures. At high temperatures, the thermal shift is nearly independent of particle size.

Erdem et al. studied the lineshift of the ${}^2E \rightarrow {}^4T_2$ emission line of Cr^{3+} in YAG in nanoparticles and in a bulk crystal [23]. The results showed that the thermal lineshift the lineshift decreases as the particle size decreases, which is consistent with Fig. 4.10. In that work, the lineshift was fit to Eq. (4.68) assuming a Debye

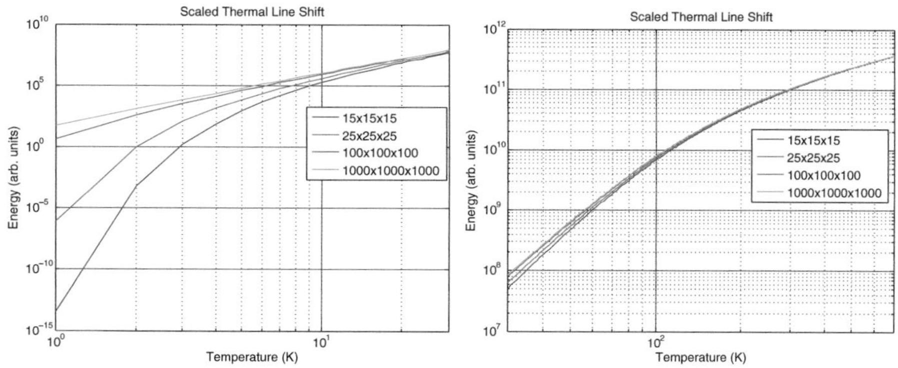

Fig. 4.10 Temperature dependence on the thermal shift of a spectral line (given by the sum in Eq. (4.78) for cubic nanoparticles (15 × 15 × 15-black, 25 × 25 × 25-blue, 100 × 100 × 100-red, 1000 × 1000 × 1000-*green*) for temperatures ranges 1–30 K (*left*) and 30–700 K (*right*)

temperature of 550 K, and it was found that, as with the thermal line broadening data, the electron-phonon coupling parameter decreased with particle size. Given the lack of a theoretical explanation of the electron-phonon coupling decreasing with particle size, more work is required to say definitively why the shift is less in smaller particles.

4.10 Vibronic Transitions

Following excitation, the decay of an isolated ion can occur via (4.1) emission of a photon, (4.2) emission of phonon(s), or (4.3) the emission of a photon concurrent with the absorption or emission of one or more phonons. This third process is referred to as a vibronic transition. In this section we discuss vibronic lines that appear as sidebands to a zero-phonon line in weakly-coupled systems [26–28].

4.10.1 Vibronic Sidebands of Sharp Lines: Theory

In the limit of weak electron-phonon coupling, the narrow zero-phonon line is accompanied by sidebands, which result from the modulation of the zero-phonon transition by the vibrations of the solid. In Fig. 4.11, one immediately notices the rich structure contained in that sideband, structure that contains information of the phonon density of states. To explain such a structure, we must consider the interaction of the ion with the different phonon modes of the crystal.

The transition rate of a vibronic transition involving the emission of a photon and of a phonon in the *kth* mode is governed by terms having the following form:

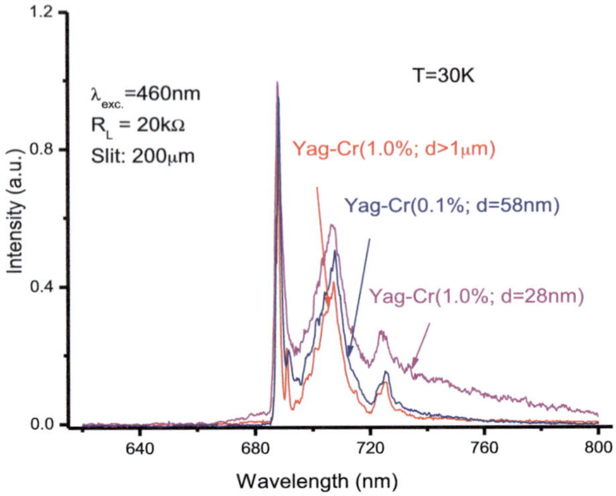

Fig. 4.11 Emission spectra at 30Kof Cr:YAG particles of 28 nm, 58 nm, and >1000 nm. Notice the broadening of the spectral lines and an enhanced background signal as the particle size decreases

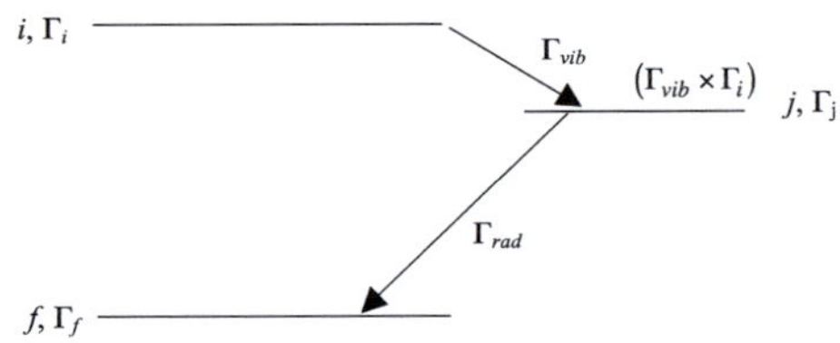

Fig. 4.12 The vibronic emission process with states and transition operators labeled according to irreducible representations

$$W_k \propto \left[\left| \langle \phi_f n_{k+1} \left| O_{ph} \right| \phi_j n_k \rangle \langle \phi_j n_k \left| O_{rad} \right| \phi_i n_k \rangle \right|^2 + \left| \langle \phi_f n_{k+1} \left| O_{rad} \right| \phi_j n_k \rangle \langle \phi_j n_{k+1} \left| O_{ph} \right| \phi_i n_k \rangle \right|^2 \right] \tag{4.78}$$

where O_{rad} and O_{ph} in Eq. (4.78) represent the appropriate radiative and nonradiative operators and ϕ_i, ϕ_j, and ϕ_f are the wavefunctions of the initial, intermediate and final electronic states, respectively. The first term in Eq. (4.78) represents a process whereby a photon is created in the transition from the initial electronic state to an intermediate electronic state, and then a phonon is emitted in the transition to the final electronic state. The second term reverses the order of these transitions. Figure 4.12 shows a schematic drawing of a vibronic transition that represents by the second term in Eq. (4.78).

Each of the electronic wavefunctions and the operators have a certain symmetry, and using group theory one can associate them with certain irreducible representations. We make the following definitions.

Γ_i: the irreducible representation of the initial electronic state of the transition
Γ_f: the irreducible representation of the final electronic state of the transition
Γ_r: the irreducible representation of the radiative operator (We will assume that this is the electric dipole operator.)
Γ_v: the irreducible representation of the vibrational mode involved in the transition

Thus, for the vibronic transition shown in Fig. 4.12 to occur the direct product $\Gamma_i \times \Gamma_{vib} \times \Gamma_r$ must contain Γ_f:

$$\Gamma_f \in \Gamma_i \times \Gamma_v \times \Gamma_r \tag{4.79}$$

We note that (82) is merely a selection rule, and can only be used to determine if a particular transition can occur; it cannot be used to determine the strength of a transition.

4.10.2 *Vibronic Sidebands of Sharp Lines: The Case of Mgo:V^{2+}*

Consider the case of a vibronic spectrum in emission at low temperature of MgO:V^{2+}, shown in Fig. 4.13 [28]. In MgO the V^{2+} ion sits in a site of octahedral symmetry, surrounded by six oxygen ions. Because the site has inversion symmetry, electric dipole transitions between two electronic states within the d^3 configuration are forbidden. As a result, the purely radiative transitions (accounting for the zero-phonon line) are driven by the magnetic dipole operator. Odd vibrations of the local complex destroy this inversion symmetry, so that the vibronic transitions involving such vibrations are electric dipole allowed.

We now examine the relationship between theses vibronic transitions in MgO:V^{2+} and the density of phonon states of the MgO lattice. First, we observe that the normal vibrational modes of the site symmetry of the octahedral group O_h are either purely even or purely odd. The representation of the final state (Γ_f) of the V^{2+} ion is known to be even. Since the electric dipole operator (Γ_{rad}) is odd, then a transition from the intermediate state via the electric dipole, according to Eq. (4.79), will be allowed only if the intermediate state is odd. The initial (excited) electronic state of V^{2+} is also even, so that only odd vibrations will be involved in the transition from Γ_i to Γ_j. Thus, Eq. (4.79) reduces to a statement of the parity selection rule.

It can be shown that of the phonons modes featured most prominently in the density of states of MgO, most of them can induce the octahedral complex to oscillate in one or more of its odd vibrational modes [29]. As a result, nearly all of the crystal phonon modes are able to participate in the vibronic transitions. The phonon spectrum of the MgO crystal (obtained by neutron scattering data [30]) is shown in Fig. 4.13a. The similarity of the shape of the low temperature vibronic sideband (Fig. 4.13b) to that of the phonon spectrum is striking, and suggests that the vibronic sideband can be closely related to the phonon spectrum of the lattice. That these two spectra show similarities and the fact that nearly all phonon modes of the MgO crystal can cause local vibrations to participate in the transition is not coincidental. However, proving that there is a one-to-one correspondence between the peaks (and valleys) of the two spectra is not trivial, since that would require calculating the transition probabilities for each of the 3N-6 normal modes of the crystal. Even if such a calculation could be done, it is no guarantee that such a calculation would be able to reproduce the observed vibronic spectrum. Generally, the shape of the vibronic spectrum will not exactly mimic that of the density of phonon states. It is, however, a practical way of gaining insight into the phonon density of states for some crystals.

In nanocrystals, where the confinement on the density of phonon states is most severe, one would expect that changes to the density of states would be obvious in the vibronic spectrum of the nanoparticle. In fact, such a result would represent the most direct experimental evidence of the reduced density of states in nanoparticles. Unfortunately, there is a significant amount of broadening of the zero-phonon line in small nanocrystals, due to the fact that zero phonon lines from various sites (due to the proximity of the surface) are shifted in energy with respect to one another. The sum of the contributions from various sites overlaps with a large portion of the phonon sidebands of the zero phonon line from the "normal" site. An example of this is shown in the vibronic spectra of Cr-doped YAG nanoparticles shown in Fig. 4.11. Perhaps due to the fact that this overlap is most prominent near

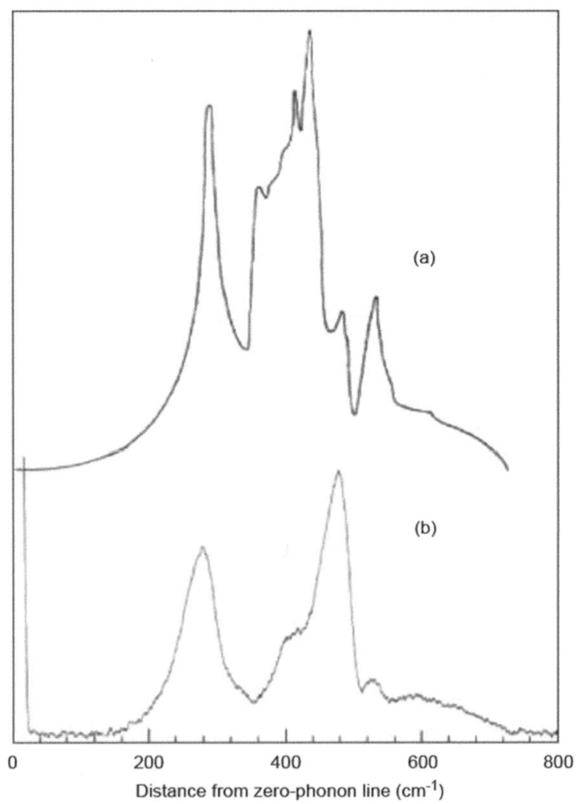

Fig. 4.13 (**a**) The density of states of MgO as determined from neutron scattering shown with (**b**) the vibronic sideband of $MgO:V^{2+}$ [29]

in the low energy range of the sidebands, where the most obvious changes (i.e., discreteness of the density of states and absence of the very low energy modes) to the density of states occur, there is no reported vibronic spectrum that clearly shows the vibronic spectrum changing with particle size. The difficulty in observing this is also complicated by the fact that the emission from nanoparticles is often very weak, probably because of the large number of surface states.

4.11 Conclusions

This paper first presented a detailed discussion of the Adiabatic Approximation(s), and the breakdown of that approximation, which allows for the existence of non-radiative processes. The electron-phonon coupling terms for the different adiabatic approximations were then discussed. This electron-phonon coupling was then used to determine the thermal broadening and shifting of sharp spectral lines of optically active ions in bulk solids. Integral to the broadening and shifting of spectral lines, and indeed to most non-radiative processes, is the phonon density of states in the system under investigation. Given that one goal of the paper was to examine how non-radiative processes depend on particle size, we then investigated how the phonon density of states depended on particle size. This investigation consisted of calculations of the phonon density of states for cubic nanoparticle, where it was found that for very small particles, the phonon density of states becomes very different than for bulk particles. The most obvious change in the phonon density of states between macro and nano systems occurs at the low energy end of the spectrum. After presenting this extensive background, the question of how non-radiative processes in doped insulators are altered as the size of the particles change from macroscopic to nano-sized was considered.

The fact that the electronic states of optically active ions in insulators are highly localized to the site of the ion, the general theory of non-radiative transitions is largely unaltered as the particle size changes. Using the calculated densities of states for cubic nanoparticles, we examined the thermal broadening and shifting of spectral lines for various particle sizes over a wide temperature range. Initial results hint that the effects of particle size on the broadening and shifting of lines are most likely to be observed only at low temperatures and in very small particles. Even in particles on the order of 50 nm, one is unlikely to be able to discern any contribution to these processes due to confinement effects of the phonon density of states. Also discussed were how the reduced phonon density of states inhibits the systems ability to reach thermal equilibrium and should changes in the vibronic sidebands in weakly-coupled systems.

Acknowledgements The Author would like to that Ryan Clair of Wheaton College for the calculations of the density of states of the nanoparticles, and also to Professor Rino Di Bartolo of Boston College for the opportunity to lecture at the Erice 2015 School. The author also acknowledges that the material presented here was supported in part by the Nation Science Foundation under grant number 1105907.

References

1. Born, M., & Fock, V. (1928). *Zeitschrift für Physik, 51*, 165–180.
2. Born, M., & Oppenheimer, R. (1927). *Annals of Physics, 84*, 457.
3. Born, M., & Huang, K. (1956). *Dynamical theory of crystal lattices*. Oxford: Oxford University Press.
4. Henderson, B., & Imbusch, G. (2006). *Optical spectroscopy of inorganic solids*. Oxford: Oxford University Press.
5. Demtröder, W. (2005). *Molecular physics: Theoretical principles and experimental methods*. Weinheim: Wiley Press.
6. Azumi, T., & Matsuzaki, K. (1977). *Photochemistry and Photobiology, 25*, 315–326.
7. Ballhausen, C., & Hansen, A. (1972). *Annual Review of Physical Chemistry, 23*, 15.
8. Di Bartolo, B. (1976). Interaction of radiation with ions in solids. In B. Di Bartolo (Ed.), *Luminescence of inorganic solids*. New York: Plenum Press.
9. Herzberg, G. (1966). *Electronic spectra of polyatomic molecules*. Amsterdam: van Nostrand Press.
10. Wang, X., Dennis, W. M., & Yen, W. M. (1992). *Physical Review B, 46*(13), 8168–8172.
11. Tobert, W. M., Dennis, W. M., & Yen, W. M. (1990). *Physical Review Letters, 65*(5), 607–609.
12. Frenkel, J. (1931). *Physical Review, 37*, 17–44.
13. Struck, C. W., & Fonger, W. H. (1976). *Journal of Luminescence, 14*, 253–279.
14. Struck, C. W., & Fonger, W. H. (1975). *Journal of Luminescence, 10*, 1–30.
15. McCumber, D. E., & Sturge, M. D. (1963). *Journal of Applied Physics, 34*, 1682–1684.
16. Di Bartolo, B., & Peccei, R. (1965). *Physical Review, 137*(6A), 1770–1776.
17. Ellens, A., Andres, H., Meijerink, A., & Blasse, G. (1997). *Physical Review B, 55*(1), 173–179.
18. Ellens, A., Andres, H., Meijerink, A., & Blasse, G. (1997). *Physical Review B, 55*(1), 180–186.
19. Liu, G. K., Chen, X. Y., Zhuang, H. Z., Li, S., & Niedbala, R. S. (2003). *Journal of Solid State Chemistry, 171*, 123–132.
20. Liu, G. K., Zhuang, H. Z., & Chen, X. Y. (2002). *Nano Letters, 2*, 535.
21. Meltzer, R., & Hong, K. (2000). *Physical Review B, 61*(5), 3396–3403.
22. Takagahara, T. (1996). *Journal of Luminescence, 70*, 129–143.
23. Erdem, M., Ozen, G., Yahsi, U., & DiBartolo, B. (2015). *Journal of Luminescence, 158*, 464–468.
24. Bilir, G., Ozen, G., Collins, J., & Di Bartolo, B. (2014). *Applied Physics A, 115*(1), 263–273.
25. Suyver, J., Kelly, J., & Meijerink, A. (2003). *Journal of Luminescence, 104*, 187–196.
26. Miyakawa, T., & Dexter, D. L. (1970). *Physical Review B, 1*, 2961–2969.
27. Auzel, F. (1978). Multiphonon interaction of excited luminescent centers in the weak coupling limit: Non-radiative decay and multiphonon sidebands. In B. Di Bartolo (Ed.), *Luminescence of inorganic solids*. New York: Plenum Press.
28. Wall, W., Di Bartolo, B., Collins, J., & Orucu, H. (2009). *Journal of Luminescence, 129*(12), 1782–1785.
29. Wall, W. (1974). Ph.D. thesis, Boston College.
30. Sangster, M. J., Peckham, G., & Saunderson, D. H. (1970). *Journal of Physics C, 3*, 1026.

Chapter 5
3D Optical Laser Lithography

Martin Wegener

Abstract We very briefly review our recent progress regarding three-dimensional (3D) optical laser lithography or "3D direct laser writing (DLW)". We focus on approaches inspired by 3D stimulated-emission depletion (STED) to break the diffraction barrier, on 3D dip-in DLW to obtain very tall structures with fine features, and on galvo-scanner based 3D DLW to increase the writing speed by two orders of magnitude.

5.1 Introduction

In the field of nanophotonics, many interesting new concepts and devices have been suggested. This includes, for example, three-dimensional complete-photonic-band-gap materials, metamaterials, and transformation optics. To bring these ideas to live, one generally needs to fabricate complex three-dimensional (3D) nanostructures.

The basic idea underlying 3D optical laser lithography or 3D "direct laser writing (DLW)" is to focus a (pulsed) laser tightly into the volume of a photoresist. To concentrate the excitation to the focal volume and to avoid exposure of the periphery, one needs some sort of a super-linear behavior of the photoresist versus the incident light intensity. This nonlinearity can, e.g., originate from two-photon absorption of light or from a chemical process. In a negative-tone photoresist, the local exposure then leads to a cross-linking of a liquid monomer to a solid polymer. In a subsequent development process, the insufficiently cross-linked regions are washed out, whereas the exposed solidified material remains. In this fashion, by computer-controlled scanning of the focus with respect to the photoresist, one can fabricate arbitrary complex three-dimensional micro- and nanostructures. Such instrument can be seen as a 3D printer with (sub-) micrometer spatial resolution. After all, "lithography" means nothing but "printing". 3D DLW instrumentation is, e.g., commercially available by Nanoscribe GmbH (see www.nanoscribe.de).

M. Wegener (✉)
Institute of Applied Physics and Institute of Nanotechnology, Karlsruhe Institute of Technology (KIT), 76128 Karlsruhe, Germany
e-mail: martin.wegener@kit.edu

© Springer Science+Business Media Dordrecht 2017

B. Di Bartolo et al. (eds.), *Nano-Optics: Principles Enabling Basic Research and Applications*, NATO Science for Peace and Security Series B: Physics and Biophysics, DOI 10.1007/978-94-024-0850-8_5

In what follows, we will briefly describe our recent progress in regard to (i) spatial resolution [1], (ii) accessible height of the structures [2], and (iii) writing speed [3].

5.2 Spatial Resolution Beyond the Abbe Barrier

I learned in school that lithography is limited by the Abbe diffraction barrier. However, this statement should be taken with some caution. Figure 5.1 illustrates a tight focus of light by iso-surfaces of the squared intensity (left) and by a corresponding cut through the focal plane (right). A simple but useful way of thinking about many negative-tone photoresists is the threshold model: If the local intensity or dose exceeds a certain polymerization threshold, the photoresist is cross-linked sufficiently and remains after the development step. Regions below this threshold are washed out. Effectively, the light field is digitized. If the threshold is moved closer and closer to the maximum of the spatial profile, the resulting volume

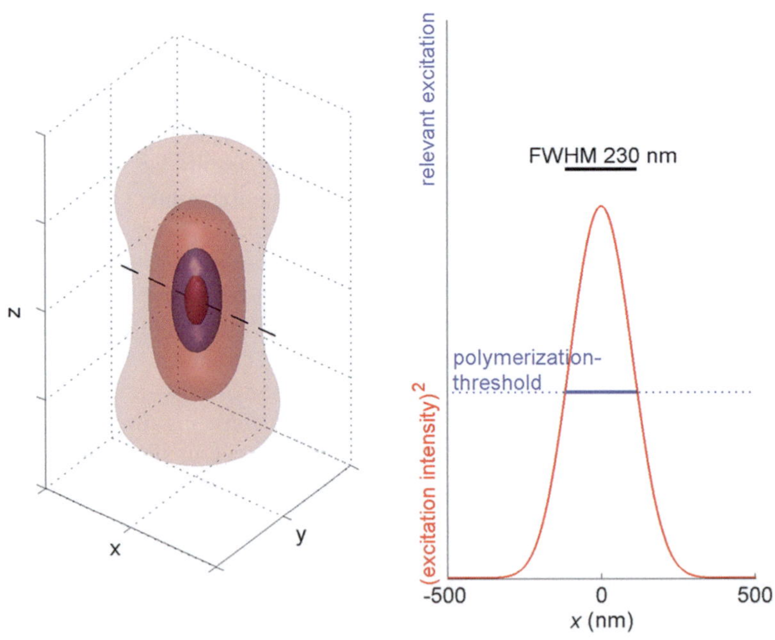

Fig. 5.1 Laser light at 800 nm wavelength and with circular polarization is tightly focused by a lens with numerical aperture NA = 1.4. The optical axis is along z. This leads to the iso-surfaces of the squared intensity (relevant for two-photon absorption) shown on the *left-hand side*. A cut along the *dashed line* is depicted on the *right-hand side*. For the shown polymerization threshold of the photoresist, the region indicated in *blue* solidifies and remains after the development step (Figure provided by Joachim Fischer, also see Ref. [1])

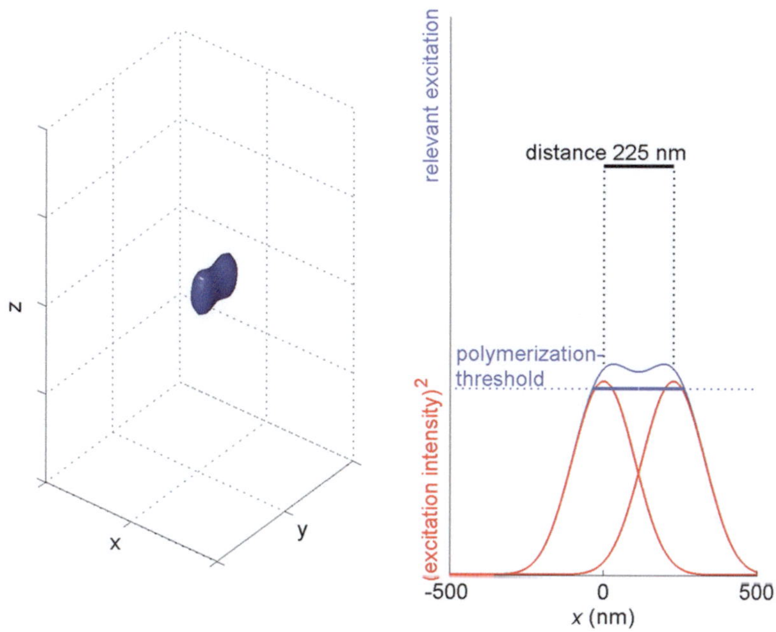

Fig. 5.2 Illustration of the two-photon Sparrow criterion in 3D laser lithography (compare Fig. 5.1) (Figure provided by Joachim Fischer, also see Ref. [1])

element or "voxel" (in analogy to the picture element or "pixel") becomes smaller and smaller. In principle, it can become as small as 1 nm. Hence, diffraction does not impose any fundamental limit on the accessible linewidth in lithography. In practice, linewidths around 100 nm can be achieved for the parameters given in the caption of Fig. 5.1.

However, diffraction does impose a limit on the accessible resolution, i.e., on the minimum separation between adjacent features (see Fig. 5.2). Suppose the photoresist just adds up successive exposures. In this case, the dose minimum between two adjacent exposures becomes less and less pronounced when moving them closer and closer together. For the parameters chosen, this minimum even completely disappears at around 200 nm separation, corresponding to the lateral two-photon Sparrow criterion [1]. At this point, it is no longer possible to write two separated voxels.

Applications of such ordinary 3D DLW include 3D photonic quasi-crystals [4], 3D helical metamaterials [5], and 3D carpet invisibility cloaks [6].

In 2014, Stefan W. Hell obtained the Nobel Prize in Chemistry for breaking the diffraction barrier in microscopy by introducing the concept of stimulated-emission depletion (STED). In a nutshell [1], a second specially shaped laser focus de-excites or depletes the photoresist in the periphery of the writing focus. In particular, this second laser focus has a zero of its intensity where the writing laser has its maximum

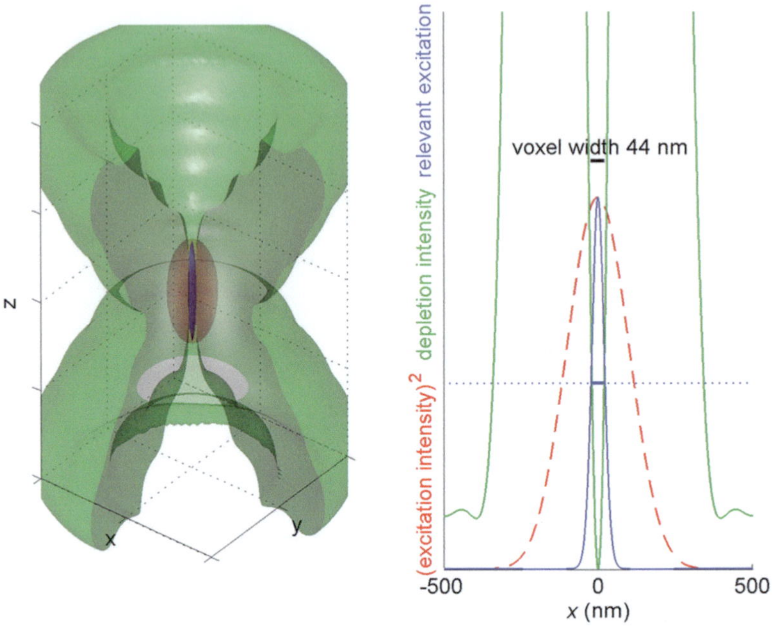

Fig. 5.3 Illustration of stimulated-emission depletion (STED) 3D DLW (compare Fig. 5.1). The depletion focus is shown in *green* (Figure provided by Joachim Fischer, also see Ref. [1])

(see Fig. 5.3). When successively increasing the intensity of this depletion beam, the photoresist is eventually depleted at all locations except for the immediate vicinity of this zero. In this fashion, the effectively exposed region becomes arbitrarily small.

We have given an extensive review [1] of the corresponding state-of-the-art. Application examples of 3D STED DLW include visible-frequency 3D invisibility cloaking [7, 8], 3D photonic crystals with a complete 3D photonic band gap at visible frequencies [9], and advanced 3D gold helical architectures [10].

5.3 Dip-in Direct Laser Writing

Originally, we used an oil-immersion approach as illustrated in Fig. 5.4a. To write tall structures with respect to the glass substrate, one needs to move the microscope lens closer to the substrate. Due to the finite free working distance of the lens, the lens eventually bumps into the substrate, imposing a limitation to the maximum height of structures. Typically, this limit lies around 100 μm. Dip-in DLW illustrated in Fig. 5.4b avoids this limitation. Here, the liquid photoresist itself is used as the

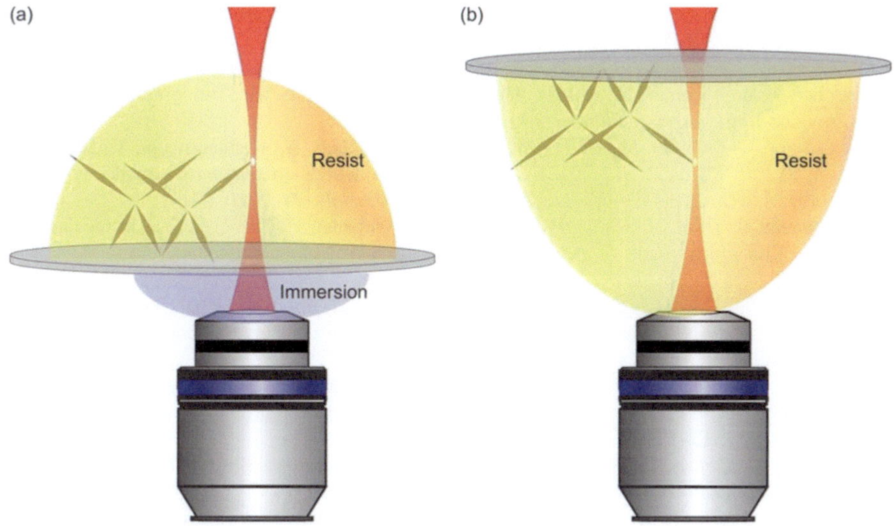

Fig. 5.4 Comparison of regular 3D DLW (**a**) and dip-in 3D DLW (**b**). For clarity, the components are not shown to scale (Figure provided by Michael Thiel, also see Ref. [2])

immersion liquid. (Obviously, the photoresist refractive index needs to be adjusted accordingly.) After anchoring the structure to the substrate (which no longer needs to be transparent), one can pull the structure up and write arbitrarily tall structures. Importantly, in sharp contrast to regular 3D DLW, the focus does not change versus height because light always experiences the same optical path.

Application examples of dip-in 3D DLW [2] include pentamode mechanical metamaterials [11] and mechanical cloaks [3].

5.4 Galvo-Scanner Direct Laser Writing

Originally, we scanned the photoresist by using high-precision 3D piezoelectric stages while fixing the focus. In this fashion, the writing speed is limited to about 100 μm/s in practice. More recently, Nanoscribe GmbH introduced scanning of the focus by using galvo-scanners (motorized mirrors) while fixing the photoresist, allowing for focus speeds as large as several m/s. Due to limitations regarding speed of data transfer, practically accessible velocities are currently around several cm/s, i.e., more than 100 times faster than ordinary 3D DLW.

A recent application example of galvo-scanner dip-in 3D DLW is the mechanical counterpart of invisibility cloaking [3], where overall volumes as large as a few mm^3 are achieved at sub-micrometer precision.

References

1. Fischer, J., & Wegener, M. (2013). Three-dimensional optical laser lithography beyond the diffraction limit. *Laser & Photonics Reviews, 7*, 22–44.
2. Bückmann, T., Stenger, N., Kadic, M., Kaschke, J., Frölich, A., Kennerknecht, T., Eberl, C., Thiel, M., & Wegener, M. (2012). Tailored 3D mechanical metamaterials made by dip-in direct-laser-writing optical lithography. *Advanced Materials, 24*, 2710–2714.
3. Bückmann, T., Thiel, M., Kadic, M., Schittny, R., & Wegener, M. (2014). An elasto-mechanical unfeelability cloak made of pentamode metamaterials. *Nature Communications, 5*, 4130–4134.
4. Ledermann, A., Cademartiri, L., Hermatschweiler, M., Toninelli, C., Ozin, G. A., Wiersma, D. S., Wegener, M., & von Freymann, G. (2006). Three-dimensional silicon inverse photonic quasicrystals for infrared wavelengths. *Nature Materials, 5*, 942–945.
5. Gansel, J. K., Thiel, M., Rill, M. S., Decker, M., Bade, K., Saile, V., von Freymann, G., Linden, S., & Wegener, M. (2009). Gold helix photonic metamaterial as broadband circular polarizer. *Science, 325*, 1513–1515.
6. Ergin, T., Stenger, N., Brenner, P., Pendry, J. B., & Wegener, M. (2010). Three-dimensional invisibility cloak at optical wavelengths. *Science, 328*, 337–339.
7. Fischer, J., Ergin, T., & Wegener, M. (2011). Three-dimensional polarization-independent visible-frequency carpet invisibility cloak. *Optics Letters, 36*, 2059–2061.
8. Ergin, T., Fischer, J., & Wegener, M. (2011). Optical phase cloaking of 700-nm light waves in the far field by a three-dimensional carpet cloak. *Physical Review Letters, 107*, 173901.
9. Frölich, A., Fischer, J., Zebrowski, T., Busch, K., & Wegener, M. (2013). Titania woodpiles with complete three-dimensional photonic bandgaps in the visible. *Advanced Materials, 25*, 3588–3592.
10. Kaschke, J., Blume, L., Wu, L., Thiel, M., Bade, K., Yang, Z., & Wegener, M. (2015, in press). A helical metamaterial for broadband circular polarization conversion. *Advanced Optical Materials, 3*, 3411–1417. doi:10.1002/adom.201500194.
11. Kadic, M., Bückmann, T., Stenger, N., Thiel, M., & Wegener, M. (2012). On the practicability of pentamode mechanical metamaterials. *Applied Physics Letters, 100*, 191901.

Chapter 6
Nanostructures and Nanocrystals with Radiation Induced Color Centers: Optical Properties and Applications

Rosa M. Montereali and Aleksandr P. Voitovich

Abstract Optical properties and some applications of radiation-induced point defects (color centers) in dielectric nanocrystals and thin films are presented. The recent developments of novel high-spatial resolution X-ray imaging detectors based on the photoluminescence of optically active defects in lithium fluoride thin films are described. Opportunities to locally change and control the material optical properties at sub-micron scale through the formation of electronic point defects are discussed. Comparison between the optical response of radiation-induced defects in bulk crystal and in thin films is highlighted. It is shown that in nanocrystals and in the near-surface layers of lithium fluoride crystals, some color centers possess optical properties different from those of the defects in bulk crystal. Photoluminescence and photoluminescence excitation spectra and Huang-Rhys parameters are presented for radiation-induced defects in lithium fluoride nanocrystals. Some dosimetric characteristics of nanocrystals and crystals (microcrystals) of the same composition for doped insulating materials are compared.

6.1 Introduction

Nano-sized particles and structures play an important role in environment, biomedicine, manufacturing technologies and diagnostic techniques. Evolution and health of living organisms depend essentially on nano-sized biological agents.

R.M. Montereali
ENEA C.R. Frascati, Fusion and Technologies for Nuclear Safety and Security Department, Photonics Micro- and Nanostructures Lab, FSN-TECFIS-MNF, 45 Via E. Fermi, Frascati, 00044 Rome, Italy

A.P. Voitovich (✉)
Institute of Physics, National Academy of Sciences, 68 Nezalezhnosti Av., 220072 Minsk, Belarus
e-mail: voitovich@imaph.bas-net.by; avoitovich@gmail.com

© Springer Science+Business Media Dordrecht 2017
B. Di Bartolo et al. (eds.), *Nano-Optics: Principles Enabling Basic Research and Applications*, NATO Science for Peace and Security Series B: Physics and Biophysics, DOI 10.1007/978-94-024-0850-8_6

The basic and applied research on nano-objects and nanotechnologies is rapidly developing and these investigations include a great variety of nanostructures and phenomena.

Impurities and intrinsic electronic defects in solids are nanoscale objects. Their dimensions are close to lattice spacing and lie typically in the range 0.5–2.0 nm for crystalline inorganic solids. These peculiar features open up many opportunities for the application of materials containing such defects. In [1] it was suggested that advanced technologies could be used for fabrication of miniaturized optical devices and luminescent structures with high spatial resolution: photomask templates in lithography, waveguides, etc. A review of the first exciting research achievements is given in [2]. Recently diamond containing the nitrogen-vacancy defects is even considered as a candidate for realization of a quantum register in a solid [3]. The investigated defects generally possess specific spectral features, like optical absorption bands, which often give a distinctive color to the crystals. Therefore, they are also known as color centers (CCs).

Intrinsic point defects can be produced in different kinds of crystalline materials by extreme ultraviolet, X or gamma rays, electrons, ions or neutrons. The defect dimensions combined with the peculiar properties of the hosting matrix are promising for use as efficient storage media with a high recording density as well as for direct writing of nanostructures and for radiation imaging at nanoscale.

Moreover, insulating crystals have been successfully applied for dosimetry of ionizing radiations. Their thermoluminescence (TL) is proportional to the absorbed radiation doses. Nanocrystals (NCs) could be considered as promising candidates for dosimetry detectors of improved performances.

6.2 Radiation Detectors, Photonic Devices and Imaging on Nanometric Scale

6.2.1 Radiation-Induced Luminescent Point Defects in LiF Thin Layers: General Overview and Applications

Luminescence properties of point defects in insulating materials are successfully used for solid state light sources and radiation detectors. Nowadays there is a great interest in technological research about broad-band luminescent inorganic and organic materials and novel integrated device configurations [4] are under study for applications in the field of photonics. The area of growth and characterization of thin films has seen a considerable expansion due to the need for a substantial shrink in the scale and costs of optical devices. The miniaturization of solid-state light sources, as well as lasers and amplifiers, generally requires the use of thin emitting layers [2] and photoluminescence (PL) represents a very powerful tool for their investigation [5].

Among dielectric materials, lithium fluoride (LiF) has several peculiar characteristics. It is almost non-hygroscopic (0.13 g/100 g H_2O solubility) in comparison with other alkali halides and its Knupp hardness is relatively high (99 kg/mm^2). It is optically transparent from the near-ultraviolet to the near-infrared spectral ranges, extending from $\cong$ 120 nm to $\cong$ 7 μm and its refractive index ($\sim$1.39 in the visible spectral range) is one of the lowest of any solid material.

Polycrystalline LiF films have been grown by physical deposition methods, mainly thermal and electron-beam evaporation, on different amorphous and crystalline substrates, such as glass and silica [6], silicon [7] as well as plastic ones [8] with controlled thickness, extending from a few nanometers to several microns.

Point defects in LiF are well known for their application in tunable lasers and dosimeters. As in LiF crystals, radiation-induced CCs in LiF thin films are stable at RT and some of them efficiently emit broad-band photoluminescence in the visible spectral interval under optical excitation [2]. Among the different types of CCs produced by ionizing radiation in LiF, the interest of this chapter is mainly focused on F_2 and F_3^+ centers, which consist of two electrons bound to two and three close anion vacancies, respectively. These CCs have almost overlapping broad absorption bands centered at around 450 nm, known as M band [9]; therefore they can be simultaneously excited with a single pump wavelength in the blue spectral interval. On the other hand, under optical excitation at wavelengths close to 450 nm, they exhibit two different broad emission bands in the green (F_3^+) and red (F_2) spectral regions, peaked at 541 nm and 678 nm, respectively. Both this kind of electronic defects are laser active even at RT and their PL properties were investigated for the realization of optically pumped solid-state lasers [10, 11], as well as for the development of miniaturized light-emitting devices, like single mode [12, 13] and multi-mode active waveguides [14, 15] or distributed feedback (DFB) lasers [16, 17] based on irradiated LiF crystals. The optical amplification properties of F_2 CCs were investigated also in confining structures based on LiF films, like simple layered guiding structures [18] and fully dielectric vertical optical microcavities [19] operating in the red spectral interval.

LiF is also a well-known dosimeter material in pure [20] and doped [21] form. Radiation detectors based on microcrystalline dispersion of LiF in a polymeric matrix have been introduced for gamma and electron high-dose dosimetry [22], mainly based on TL. The visible PL of F_2 and F_3^+ CCs in polycrystalline thick LiF films grown by thermal evaporation on radiation-hard silica substrates was also proposed for gamma dosimetry at high doses [23].

On this basis, in the last decade many efforts have been devoted to the development of novel high-spatial resolution imaging luminescent film detectors for Extreme Ultra-Violet (EUV) radiation and soft X-rays [24, 25] generated by powerful table-top laser plasma sources, as well as for X-rays up to 10 keV [26, 27] produced by compact X-ray tubes, and even in large-scale facilities like synchrotrons [28] and Free Electron Lasers (FEL) [29].

These solid-state imaging detectors are based on the optical reading of PL from stable, visible-emitting CCs in LiF, locally produced by irradiation with X-rays, after impinging and interacting with the object under X-ray investigation. The aggregate electronic defects suitable for high-spatial resolution imaging applications are again the F_2 and F_3^+ ones. After exposure to X-rays, the latent images stored in the LiF thin layers by local formation of these optically active defects are read with conventional and advanced optical fluorescence microscopes operating at visible wavelengths. The considerable values, close to unity, of photoemission efficiencies for both types of defects [10], whose intense PL falls in an extended visible spectral range, where the efficiency of common electronic detectors is maximum, allow achieving a good enough radiation sensitivity.

Among the main advantages of the LiF thin-film radiation imaging detectors, there are an intrinsic very high spatial resolution ($<$250 nm), related to the atomic scale of the radiation-induced defects utilized as single luminescent units. This means that the resolution is below few nanometers as the lattice constant of LiF is the shortest among all alkali halides, i.e. 0.2013 nm. In practice, the spatial resolution is essentially limited by the optical microscope and the technique utilized for PL detection. By using advanced fluorescence optical microscopes as readout instruments, like a Confocal Laser Scanning Microscope (CLSM) operating in far-field, a sub-micrometric [25, 26] spatial resolution of 250 nm was generally obtained. Reading resolutions below 100 nm were demonstrated to be achievable with more sophisticated near-field microscopy techniques and Scanning Near-field Optical Microscope (SNOM) fluorescence images with spatial resolution below 80 nm were obtained on LiF thin film X-ray imaging detectors. Such detectors have a wide dynamic range ($> 10^3$) and a large field of view, as they can be grown on large areas, greater than 10 cm^2 [27]; they are easy to handle, being insensitive to ambient light, as their band gap ($\sim$14 eV) is the largest among common insulating solids, and no additional process is needed after X-ray exposure.

These high-spatial resolution thin-film radiation detectors offer a great versatility, since the material is sensitive to almost any kind of radiation, including charged particles. Recently, their use has been proposed and successfully extended to low-energy proton beam advanced diagnostics [30]. Accurate accumulated two-dimensional (2D) dose distribution with high spatial resolution over large areas and a wide dynamic range, covering at least three orders of magnitude of doses, together with nondestructive optical fluorescence readout, were successfully demonstrated for the first time by using low-energy proton beams, with the main challenge of application in clinical hadron therapy.

Some properties of F_2 and $F_3{}^+$ CCs in LiF thin films will be discussed in the following sections to highlight the relevance of the radiative properties of these defects hosted in a polycrystalline LiF material and/or in confining micro- and nanostructures for photonic devices and radiation imaging detector development.

6.2.2 *Photonics and Radiation Imaging: Patterning of Luminescent Point Defects on Nanometric Scale*

CCs in LiF thin films are currently offering great opportunities of research and development in the field of miniaturized light sources and novel luminescent radiation detectors [31]. LiF films have the advantages of compatibility with amorphous and crystalline substrates, the possibility to grow multi-layered structures of different materials [2] and functionalize their emission properties, like in colored LiF film-based fully dielectric optical microcavities [19, 32], as well as the control of dopants concentrations and interfaces by versatile preparation techniques [33].

The recent progress in laser technologies, electron and particle beam methods, and novel advanced photon sources have led to a wide range of opportunities to locally change and control the LiF optical properties at sub-micron scale through the formation of electronic point defects. Combining these opportunities with LiF thin film growth capabilities, interesting developments are under way.

The local modification of the LiF material properties in a thin surface layer through the formation and stabilization of CCs is dependent on the nature and energy of the utilized radiation, on the irradiation parameters (dose, dose-rate, etc.) and on conditions of sample preparation [34]. From the practical point of view, one of the main aspects of the radiation-matter interactions to carefully take into account is the penetration depth d in solids. For low energy radiation depth d can be comparable with the typical thickness t of the films: as it is sketched in Fig. 6.1, the surface colored layer can be host in a bulk crystal, at the LiF-air interface, or in a LiF thin film; in this example the penetration depth is comparable or less than the thickness of the irradiated LiF film and a two layers structure is obtained. However, even in the case of energetic radiation, that is $d > t$, the use of thin films allows to directly fabricate thin colored layers, as the thickness itself of the irradiated LiF film limits the depth of the colored regions. In this way a two-dimensional (2-D) colored thin layer can be physically realized. In the following, only examples of low-energy radiation, with limited coloration depth, will be considered.

Low energy radiation can create high concentrations of stable primary and aggregate electronic defects, which locally change optical properties of LiF in thin irradiated surface layers. Among them, spectral-dependent modifications of refractive index [12] and absorption (gain) properties [13] have been simultaneously achieved in LiF within volume of size comparable with optical wavelengths, by direct writing with low energy (12 keV) electron beams (for a review see [2] and reference therein), whose typical penetration depth in LiF is about 1.5 μm.

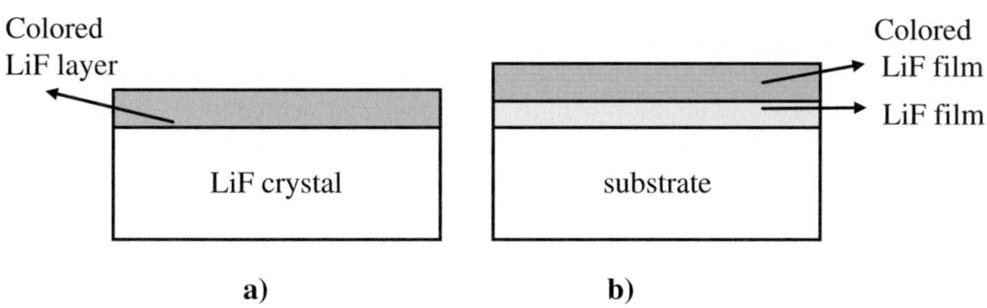

Fig. 6.1 Sketch of a single colored surface layer hosted in a LiF single crystal matrix (**a**) and of the surface colored layer in a LiF polycrystalline film of thickness t grown on a substrate (**b**). In this case the depth of the coloration $d < t$, and a two layers structure is obtained

Among light particles, low-energy electrons have the ability to interact mainly with the electrons in the solid; on the contrary, ions can substantially modify the host matrix by atom displacement, depending on their mass and energy, and other phenomena, besides point defects generation, in this case take place and become eventually predominant. While energy dissipation is almost constant up to the end of the track for low-energy electrons [35], in the case of light ions the energy transfer along the particle tracks is not uniform, being strongly dependent on the velocity of the incident ion [36]. The irradiation involves the simultaneous generation of isolated F centers and of more complex defects formed by F-aggregation, such as F_2 and F_3 or those obtained by their ionization, whose spatial distribution along the depth can be monitored by photoluminescence microscopy [35, 37, 38].

A careful choice of energy and particle beams could allow a better control of the nature and concentrations of formed defects and their depth distribution, as well as direct patterning of sub-micrometric structures [37, 39] by using advanced lithographic techniques based on direct writing with focused particle beams.

Using a Scanning Electron Microscope (SEM) equipped with a lithography apparatus, which allows controlling of the electron beam deflection, several fluorescent lattices of colored dots were made on the surface of a 2.4 μm thick polycrystalline LiF film grown on glass substrate maintained at 250 °C during the deposition [40]. The nominally 100 nm diameter dots of CCs were directly written on the LiF layer with 2 keV electron energy, a 100 pA current and a dose of 450 $\mu C/cm^2$. The dots were periodically separated by a distance of 1 μm, large enough to allow the fluorescence detection of a single light-emitting dot, with a good signal to noise ratio and high contrast. The observed dimensions (a convolution between their real size and that of the probe aperture of the used SNOM) were of the order of 150 nm. The good optical quality and the quite smooth surface of LiF films thermally evaporated in the selected growth conditions, with typical grain size of 150–250 nm and root mean square roughness less than 20 nm, cannot exclude a broadening due to defect migration, like the diffusion of fluorine vacancies from irradiated volume to the surrounding regions [37]. Moreover, lateral spreading of the electrons by scattering processes and charge effects could further enlarge the interaction volume and could decrease the final spatial resolution.

The continuous shrinking in size of the photonic structures requires to explore also short wavelength radiation, like EUV and soft X-rays, whose photon energies extend from about 20 eV to 2 keV. They are characterized by a limited penetration in solids and their attenuation length (the depth at which the intensity of a monochromatic beam impinging normally on the material surface is reduced by a factor *e*) varies from few tens of nanometers to few micrometers in LiF [41], as shown in the plot of Fig. 6.2a; these values are compatible with the typical thickness of thermally evaporated LiF films.

Direct writing with a focused monochromatic soft X-ray microprobe of fixed energy from 400 to 700 eV allows the realization of low-dimensional, nominally down to one hundred of nanometers, fluorescent arrays of lines and dots, based on active F_2 and F_3^+ CCs, in LiF crystals and thin films [42, 43]. In Fig. 6.2b the CLSM fluorescence image of a colored regular array of light-emitting lines, directly written

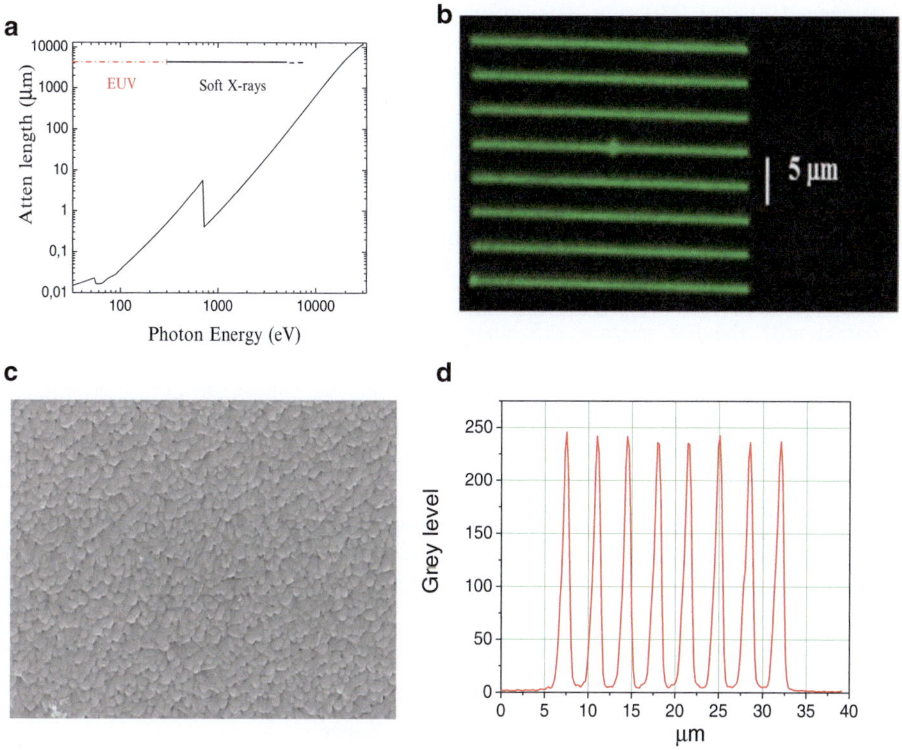

Fig. 6.2 Attenuation length of EUV and soft X-rays in LiF (**a**). Fluorescent grating based on F_2 and F_3^+ CCs directly written by a 450 eV X-ray beam on a 0.3 μm thick LiF film grown on Si substrate (**b**). SEM image (10×10) μm^2 of the surface of a 3.1 μm thick LiF film grown on glass substrate (**c**). PL intensity profile of light-emitting grating of Fig.2b along a vertical line (**d**)

on a 0.3 μm thick LiF film on Si substrate, is shown together with its intensity profile (Fig. 6.2d). It consists of 8 lines regularly spaced by 3.5 μm, whose measured full-width at half-maximum is about 800 nm, a typical value easily achievable and detectable in standard operating microscope conditions. The fluorescence image, stored in the LiF film grown on a reflective substrate, was obtained by a CLSM (Nikon Eclipse 80i-C1) equipped with an Argon laser at 457.9 nm. Irradiation with blue light excites the visible PL of the F_2 and F_3^+ lattice defects locally created in the areas previously exposed to the incoming focused X-ray beam, which appear bright in the picture of Fig. 6.2b.

Also in these cases, the SNOM fluorescence image of the thinnest lines [22] directly written on a 3.1 μm thick LiF film grown on a glass substrate by using the X-ray microprobe, with a beam diameter of 100 nm at 640 eV, provides a minimum dimension of about 170 nm. The surface morphology of the used LiF film was investigated by SEM and its image, reported in Fig. 6.2c, shows a good uniformity of the grains, which have a quite regular shape, whose estimated dimensions are between 100 and 250 nm. The influence of the polycrystalline nature of LiF films and of their properties, like density, grain size and surface morphology, on these complex phenomena requires further investigations. Although the final spatial resolution is virtually unlimited, being in principle related to the physical extension of the emitting defects (below few nanometers), the diffusion lengths of vacancies and electrons can play a peculiar role in self-assembling phenomena of defects aggregation and clustering.

High-resolved, permanent, visible-emitting regular patterns were efficiently created on large areas in LiF crystals and films by EUV light from a laser-plasma source by masking the incoming radiation with metallic meshes placed in close contact with the LiF surface [25, 44]. The soft X-ray imaging in contact mode has been successfully extended to thin biological objects [24], even for *in vivo* samples, whose soft X-ray absorption is dependent on composition and thickness. The transmitted radiation creates CCs, whose local densities are proportional to the received dose and then the fluorescence spatial intensities are proportional to the sample X-ray transmission. The soft X-ray radiography, stored at the surface of LiF, can be read by an optical microscope operating in fluorescence mode, through the visible PL of F_2 and $F_3{}^+$ active defects under simultaneous optical pumping in the blue, at around 450 nm.

6.2.3 LiF Thin Film Microstructure and Enhanced Photoluminescence of Point Defects

LiF films, generally grown by thermal evaporation on amorphous substrates, like glass, silica and plastics, as well as on some crystalline materials, like LiF, NaF, silicon and metallic thin layers, are polycrystalline in a wide range of thickness extending from few nanometers to several microns. Their structural, morphological and optical properties are strongly dependent on the main deposition parameters, that is the substrate temperature during growth and the total film thickness, as well as on the deposition rate, the evaporation geometry, and, obviously, the choice of the substrate [2]. Moreover, the LiF film refractive index is strongly related to the density of the deposited film, which is in turn dependent on the film structure. Higher substrate temperatures increase the mobility of the film molecules and thus favor the formation of more tightly packed microcrystals. In LiF films thermally evaporated on glass substrates the extinction coefficient is below 10^{-3}, and the real part is always lower than that of the bulk material. Rising the substrate temperature during film deposition induces an increase of the refractive index of the film, which is generally lowered by increasing the film thickness. The choice of an oriented substrate, like a polished LiF single crystal, raises the refractive index close to the bulk value.

Even in LiF thin films grown on optically transparent substrates, the direct use of optical absorption spectra to individuate the presence of different kinds of point defects is often precluded by the presence of interference fringes due to the

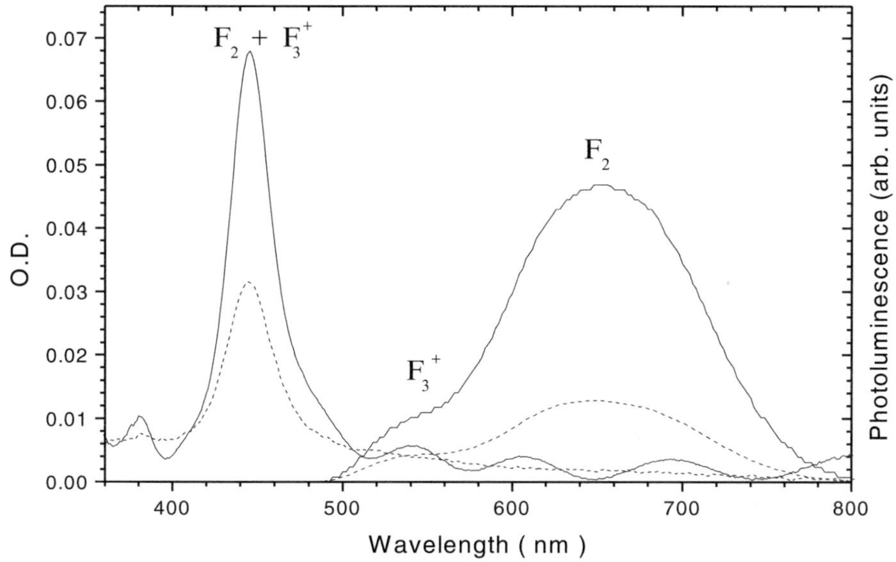

Fig. 6.3 RT optical absorption (left axis) and photoluminescence (right axis) spectra of a surface-colored LiF crystal (dashed line) and a colored LiF film of 1.8 μm thickness thermally evaporated on a glass substrate kept at 300 °C during the growth (solid line) irradiated by 12 keV electron beam in the same conditions. The used pumping wavelength is 458 nm

refractive index difference between film and substrate [2, 6]. PL measurements are more sensitive and in the case of short penetrating radiation ($d < t$) they easily allow a direct comparison of formation efficiency of several active CCs in the polycrystalline LiF film matrix with respect to single crystal, as well as for LiF films of similar thickness grown in different conditions [45].

Figure 6.3 shows the RT optical absorption and PL spectra of a LiF crystal and a LiF film of thickness 1.8 μm thermally evaporated on a glass substrate kept at 300 °C during the growth, uniformly irradiated by a 12 keV electron beam in the same conditions. The F_2 and F_3^+ CCs formation efficiencies appear to be much larger in the LiF film with respect to a LiF crystal. On the left, the absorption spectrum of a LiF crystal colored by 12 keV electrons, corrected for the absorption of the non-irradiated part of the sample, is compared with that of a film roughly corrected by the interference oscillations due to differences between the refractive indexes in the LiF film and the substrate. The irradiation parameters are the same for the two samples and the electron penetration depth about 1.5 μm is below the thickness of the used film, as sketched in Fig. 6.1. By comparing the absorption spectra of the colored LiF film with the bulk one, it is evident that the intensity of the M absorption band is 2–3 times greater in films. By pumping these samples with the 458 nm line of an Argon laser, the F_3^+ and F_2 emissions can be simultaneously observed. On the right of Fig. 6.3, the emission spectrum of a LiF film of $t = 1.8$ μm irradiated by 12 keV electrons is reported together with the LiF crystal one. These

PL spectra clearly indicate that the active centers concentration is 2–5 times greater in films than in a bulk. It seems that the films evaporated at higher temperatures reach a higher defect density [45].

This behavior should be mainly related to the enlarged surface-to-volume ratio in polycrystalline LiF films, where the increased density of grain boundaries, which could act as a source of vacancies, strongly influences the CCs stabilization properties and increases the formation efficiency of visible active CCs, eventually reducing the final densities of larger complex reabsorbing defects.

6.2.4 Toward Control of Point Defects in Layered Thin Films for Luminescent Photonic Devices and Radiation Detectors

The sensitivity of X-ray LiF film based detectors relies mainly on CCs formation efficiency during irradiation. It has been already observed that the formation and stabilization processes of CCs are strongly dependent on the characteristics of the LiF film hosting material.

Figure 6.4a shows the typical transmittance and reflectance spectra of a LiF film about 1 μm thick thermally evaporated on a glass substrate. The measured values and the clear presence of interference fringes are a signature of excellent transparency, presence of sharp edges and good uniformity of the grown LiF films. By proper theoretical approach, one can usually estimate a few physical parameters, such as film thickness, surface roughness, material inhomogeneity along the growth axis, refractive index and extinction coefficient spectral behaviour, and so on. In Fig. 6.4b the photograph of a LiF film thermally evaporated on an untreated Si (100) substrate is presented. Also in this case the optical reflectance spectrum (transmittance measurements are precluded by the opacity of Si at optical wavelengths) confirms the good transparency of the LiF film.

A direct comparison of the optical response of EUV induced CCs between a 2.2 μm thick LiF film, thermally evaporated on a silicon substrate kept at 250 °C, and a LiF crystal, irradiated in the same conditions, has been performed by CLSM fluorescence imaging on uniform irradiated areas [46]. Their images are reported in Figs. 6.4c, d. The photo-induced light emission coming from the LiF film is significantly larger than the one coming from the LiF crystal. The higher sensitivity of the film, corresponding to a factor of (8 ± 2), is comparable with the previsions of a theoretical model that takes into account light confinement effects in the investigated planar microstructures (see [31] and references therein). In them a key role is clearly played by the presence of the reflective Si substrate, apart the peculiar features of the film, that makes it morphologically different from the crystal and favouring the formation of higher densities of optically active CCs. Just the reflective properties of Si allow recovering a relevant part of emitted light that would be lost in the opposite direction to the observation one. Moreover, this reflected part can constructively

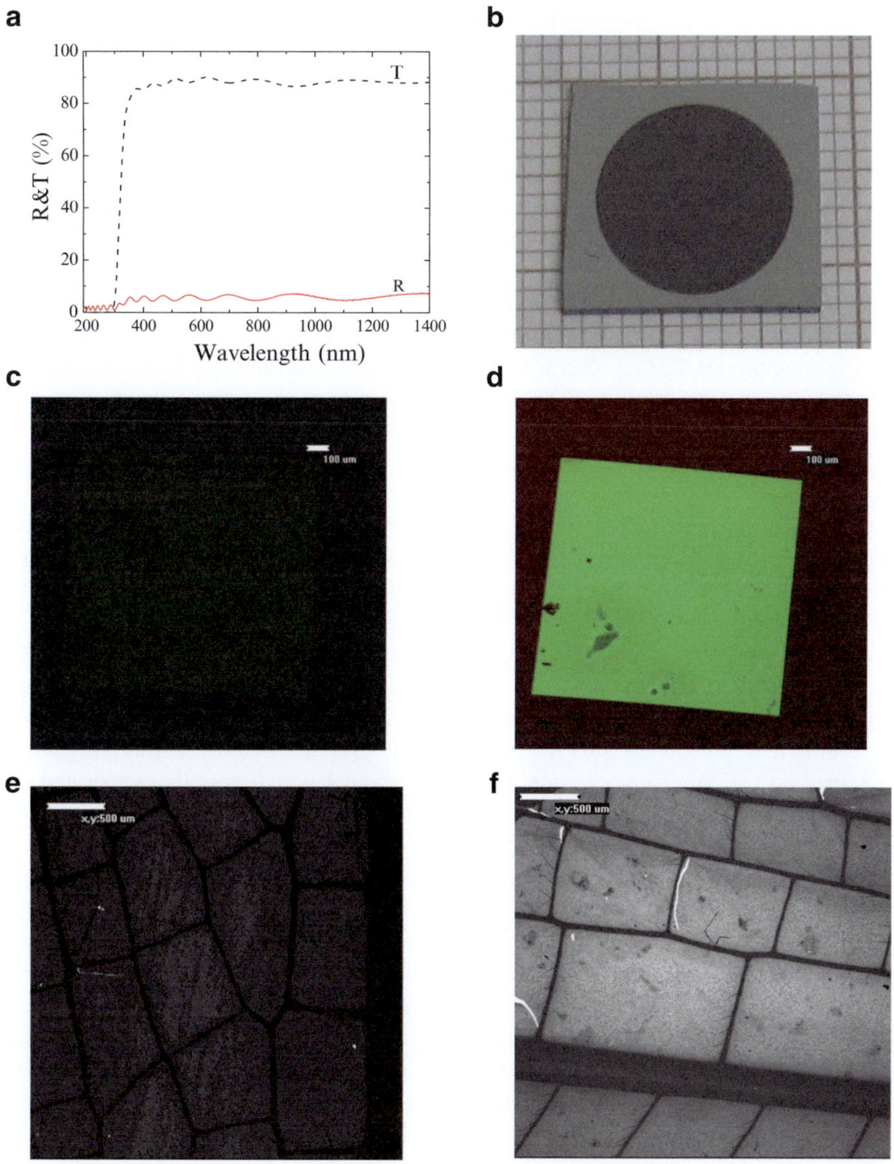

Fig. 6.4 Optical transmittance and reflectance spectra of a LiF film about 1 μm thick thermally evaporated on a glass substrate (**a**). Photograph of a LiF film thermally evaporated on a Si (100) substrate (**b**). CLSM fluorescence images of EUV uniform irradiated areas, masked by a commercially available 100 nm thick Si_3N_4 windows of typical size (1×1) mm^2, stored in a LiF crystal (**c**) and in a LiF film of thickness 2.2 μm, thermally evaporated on a silicon substrate (**d**), irradiated in identical conditions (50.000 shots). Bar length: 100 μm. CLSM fluorescence images of soft X-ray contact micro-radiographies of a dried dragonfly wing stored on a 1 μm thick LiF film thermally evaporated on glass (**e**) and silicon (**f**); substrates obtained in the same irradiation conditions (single-shot, Cu target). Bar length: 500 μm

interfere with the straightforward emitted one, and this is true for all orders of multireflection at all emission angles, provided the LiF thickness is suitable.

Indeed, the planar structure constituted by the LiF film over the Si substrate forms an elementary half microcavity and the thin surface layer containing CCs is located in proximity of a cavity mode. In quantum-optics language, the CCs in the LiF film are located in a thin layer where the vacuum-field modes density is increased with respect to a LiF crystal. Because spontaneous emission is stimulated by vacuum-field photons, the high vacuum-field modes density induces the CCs in the LiF film to perform their optical cycle faster than in the LiF crystal, resulting in an increased time-averaged photon emission amount.

In the case of the LiF samples in Fig. 6.4c, d, the colored region was assumed as a homogeneous surface thin film, where the CCs densities are constant along a depth of 30 nm, superimposed to the uncolored part of the sample itself, as in the bi-layer structure sketched in Fig. 6.1. For a geometrical mounting as in our CLSM, with optical pumping at 458 nm perpendicular to the colored surface, and emitted light collection in the same direction, numerical simulations provide an increase of 6.6 times for a red wavelength at 615 nm, while only an increase of 1.4 is expected for a green wavelength at 515 nm. From the experimental point of view, it is very difficult to separate the effects of light confinement from those related to the film microstructure, especially for very thin surface colored layers, whose thickness is comparable with the film surface roughness. In the case of EUV irradiation, the PL enhancement was already observed also for LiF films deposited on glass substrate, where the substrate optical effects can be neglected.

Figure 6.4e, f show the soft X-ray micro-radiographies of a dried dragonfly wing obtained on LiF thin films of equal thickness, about 1 μm, thermally evaporated on glass and Si (100) substrates, respectively. They were simultaneously exposed under vacuum with a single-shot of a Nd-YAG laser-plasma source with Cu target, whose penetration depth is comparable and/or higher than the total film thickness, at a soft X-ray fluence of $\sim 350\ \mu J \cdot cm^{-2}$/shot. In the integrated PL images, the F_2 and F_3^+ signal intensities can be tentatively assumed to be proportional to the concentrations of active CCs, which are directly related to the X-ray transparency of the exposed sample; thus, the more absorbing parts of the biological object appear as darker. Also in this case, with the same acquisition parameters, the higher sensitivity of the LiF film on Si substrate with respect to the LiF film on glass is noticeable, and the higher optical contrast allows identifying smaller details.

6.3 Nanocrystals with Radiation Defects: Characteristics and Applications to Dosimetry

The formation and optical properties of radiation-induced color centers in micro- and nano-sized crystals have been studied for many years (see, e.g. [47, 48, 49, 50, 51]). Changes in the concentration ratio of different kinds of defects and in the spectral properties of the same type of centers have been found moving from microcrystalline to nanocrystalline host. These differences are associated with a higher concentration of surface color centers in nanocrystals, where the surface area to volume ratio is larger. The conditions for surface centers formation and stabilization are found to be different from those in the bulk.

Hereinafter, for radiation surface color centers (SCCs) we will use the following notation: ${F_{Sn}}^{-}$, ${F_{Sn}}^{+}$ and F_{Sn}, where the subscript n is the number of anion vacancies in SCCs; a superscript "+" or "–," or no superscript at all indicates deficiency, excess or balance of electrons with respect to the number of vacancies; subscript S refers to SCCs.

6.3.1 Near-Surface Layer Radiation Induced Color Centers in Lithium Fluoride Nanocrystals

In this section we consider the characteristics of the optical absorption, and also PL and photoluminescence excitation (PLE) spectra for the SCCs in LiF. These centers were created in LiF nanocrystals (NCs) by gamma rays from a ^{60}Co source. NCs were produced by mechanical fragmentation of a nominally pure LiF monocrystal. For experimental convenience they have been pressed into pellets. NCs and their aggregations, imaged with electron microscope, are presented in Fig. 6.5. It shows that manufactured particles dimensions are in the range of $d \leq 1$ μm. CCs have been also investigated in crystal plates cut out from the same LiF monocrystal.

The samples were irradiated at room temperature (RT), and at the liquid nitrogen temperature (LNT). Reactions of the surface centers separately with electrons and with anion vacancies were investigated. The numbers of anion vacancies and

Fig. 6.5 Electron microscope image of LiF NCs

electrons entering into the centers composition were established and it has been found that F_{S1}, F_{S1}^{-}, F_{S2}^{+}, F_{S2}, F_{S2}^{-}, F_{S3}^{+} and F_{S3} types of the surface centers are formed [50]. After irradiation, samples were kept at RT until the completion of all processes of aggregation.

Optical absorption, PL and PLE spectra were measured in pellets and in crystalline plates. They are shown in Fig. 6.6 for SCCs and CCs. It can be seen that the PL and PLE spectra of SCCs which contain more than one anion vacancy

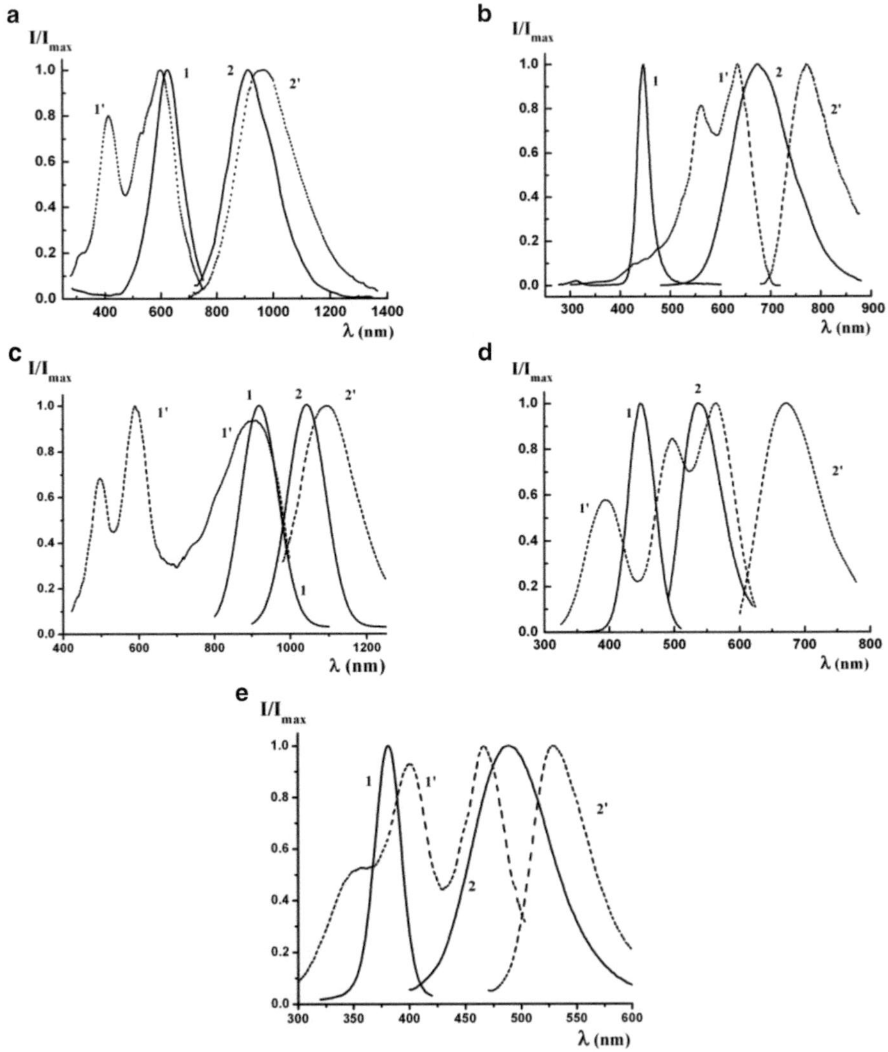

Fig. 6.6 PLE (1) and PL (2) spectra for CCS (1, 2) and for SCCs (1′, 2′) of the following types: F_2^{+} and F_{S2}^{+} (**a**), F_2 and F_{S2} (**b**), F_2^{-} and F_{S2}^{-} (**c**), F_3^{+} and F_{S3}^{+} (**d**), $F_3(R_2)$ and F_{S3} (**e**)

differ from the corresponding spectra of the bulk centers of the same composition. One of the main difference is that the PLE spectra for SCCs have two (for F_{S2} centers) or three (for $F_{S2}{}^{+}$, $F_{S2}{}^{-}$, $F_{S3}{}^{+}$ and F_{S3} centers) bands with not strongly variation of intensities. A possible explanation of the PLE spectra splitting is the crystalline electric field asymmetry in the near surface layer. Differences in the absorption spectra for the simplest surface (F_{S1}) and bulk (F_1) defects have not been found.

After the full post-radiation annealing of pellets at RT the following aggregated SCCs remain in appreciable concentrations in NCs: $F_{S2}{}^{+}$, $F_{S3}{}^{+}$ and F_{S3}. Absorption of these centers leads to substantial optical density of an initially transparent pellet within the spectral range of 310–650 nm [51] and correlates with their PLE spectra (Fig. 6.6a, d, e). Let's note that, after annealing, the optical density of the crystalline plate, which has the same thickness and irradiation dose as the pellet, is 4–8 times lower than the one of the pellet within the indicated spectral range.

PLE spectra for SCCs were fitted as the sum of Gaussian bands. The peak wavelengths λ_m and the full widths at half maximum $\Delta\lambda$ of the PLE components were determined. The obtained values are presented in Table 6.1 together with those of PL spectra of SCCs. For comparison, the values for PL and PLE spectra of CCs in bulk crystal are also given in Table 6.1.

Figure 6.6 and Table 6.1 clearly show that peak wavelengths of the PLE and PL spectra as well as the wavelength shift among them for the $F_{S3}{}^{+}$ and F_{S3} defects vary greatly in comparison with those of the $F_3{}^{+}$ and F_3 defects, respectively. This behavior indicates significant differences in the electric field and in the electron density distribution in defects of the same composition located in the near-surface layers and in the bulk crystal. We suppose that the last condition could reduce the Huang-Rhys parameter for the $F_{S3}{}^{+}$ (F_{S3}) SCCs in comparison with the one for the $F_3{}^{+}$ (F_3) defects in the bulk crystal. It has been shown [52] that the values of the Huang-Rhys parameter are as follows: 5.6 (3.6) for the $F_{S3}{}^{+}$ (F_{S3}) SCCs and 7.0 (4.9) for the $F_3{}^{+}$ (F_3) defects in the bulk crystal. It is well known [53, 54] that an optical transition is characterized by a large Huang-Rhys factor when the equilibrium position of the ions in the crystal lattice is strongly modified as a result of the electron transition. Such position shifting requires the electronic transition to be accompanied by a large change in the electrons density distribution. The more diffuse the electronic distribution, the less its influence on the lattice, and the Huang-Rhys parameter becomes small.

In near-surface layers of LiF crystals, the activation energy of the defects diffusion is significantly higher than in the bulk of the crystals [55, 56]. The activation energy for anion vacancies in these layers is equal to 1.0 eV, whereas it is 0.6 eV in the crystal bulk. Consequently, the potential barrier separating the fluorine ion at the lattice site from the vacant site (vacancy) in a near-surface layer is substantially greater than that in the bulk of crystal. In contrast to the bulk of crystal, there is no diffusion of defects consisting of two vacancies and one electron in near-surface layers at room temperature. As a result, the concentration of these defects remains constant for a long time if the sample is stored at room temperature.

Table 6.1 Values of peak wavelengths λ_m and full widths at half maximum $\Delta\lambda$ for absorption and luminescence bands of different CCs and SCCs in LiF

CCs and SCCs types	F_2^+		F_2		F_2^-		F_3^+		F_3		F_{S2}^+		F_{S2}		F_{S2}^-		F_{S3}^+		F_{S3}	
Maxima, widths of bands, nm	λ_m	$\Delta\lambda$	λ_m	$\Delta\lambda$	λ_m	$\Delta\lambda$	λ_m	$\Delta\lambda$	λ_m	$\Delta\lambda$	λ_m	$\Delta\lambda$	λ_m	$\Delta\lambda$	λ_m	$\Delta\lambda$	λ_m	$\Delta\lambda$	λ_m	$\Delta\lambda$
The first absorption band	624	118	444	25.5	921	116	448	48	381	28	600	100	632	82	907	210	564	68	463	52
The second absorption band	–	–	–	–	–	–	–	–	–	–	510	60	559	68	587	88	493	54	399	58
The third absorption band	–	–	–	–	–	–	–	–	–	–	414	74	–	–	497	75	393	68	354	68
Luminescence band	913	193	678	133	1042	120	538	73	488	82	960	244	771	110	1099	184	671	108	529	60

6.3.2 Nanocrystals for Radiation Dosimetry

PL and PLE spectra and also data on the electron-phonon interaction in surface and bulk CCs of the same composition have been presented in the previous section. There are differences in the following characteristics between SCCs and CCs: (i) in the position of the energy levels and in the peak and width values of PL and PLE bands; (ii) in the magnitudes of the transition probabilities; (iii) in the parameters of the electron-phonon interaction. The depths of the electrons and holes traps and the defects concentrations will change moving from the crystal volume to its near-surface layer. Thus, if such a transition occurs, considerable changes in the properties of radiation-induced defects have to take place.

All the differences listed above will be dominant in NCs due to their extended surface. They will change the characteristics of thermoluminescence (TL) of irradiated NCs compared to the bulk crystal. These effects stimulate research on the possibilities of using NCs for radiation dosimetry as the TL based techniques are widely used for measurements of ionizing radiation doses.

Figure 6.7 shows measurements of TL intensities as a function of the annealing time for LiF: Mg, Ti crystals and NCs. The change of the annealing temperature over time is also reported by the broken line 1 in Fig. 6.7. The used crystals are commercially available thermoluminescent detectors for dosimetry. NCs were prepared by mechanical fragmentation of these crystals. Weights of used NCs are not controlled, therefore it is not possible to compare the values of TL intensities in Fig. 6.7a, b. PL and PLE spectra of LiF: Mg, Ti crystals and NCs (not reported in this paper) are similar those shown in Fig. 6.6.

Time (temperature) dependencies of TL intensities I for crystals and NCs differ greatly. In Fig. 6.7a, a main band of TL is observed for crystals at annealing temperature T = 240 °C. This glow peak is commonly used for the dosimetry of ionizing radiation. There are two main bands in the TL signal for NCs (Fig. 6.7b): one peaked at $T \approx 150$ °C and the second one at $T = 215$ °C. The large width is typical for these TL bands. This feature is due to the large range of local crystal field variations in NCs and, as a result, variations of the positions of the traps energy levels. Increasing of the field variations is caused by strains in NCs arising during their manufacturing.

Currently further studies and efforts are still being made to improve the TL characteristics of dosimetric materials: sensitivity, linear response range, fading, reusability. NCs could be considered as potential candidates for dosimeters with improved parameters. In Table 6.2 the range of radiation exposures with linear TL response are shown for several nano- and microcrystals of doped insulating materials used as radiation detectors. Typical dimensions of the used NCs were few tens of nanometers. It can be derived that the range of linear response is much larger for the nanocrystals in comparison with the microcrystals of same composition. This behavior of the NCs is useful also for measurements of the heavy charged particles doses. Heavy ion beams are widely used in radiotherapy and diagnostics. Along with a wider range of linear response, many NCs exhibit negligible fading and good

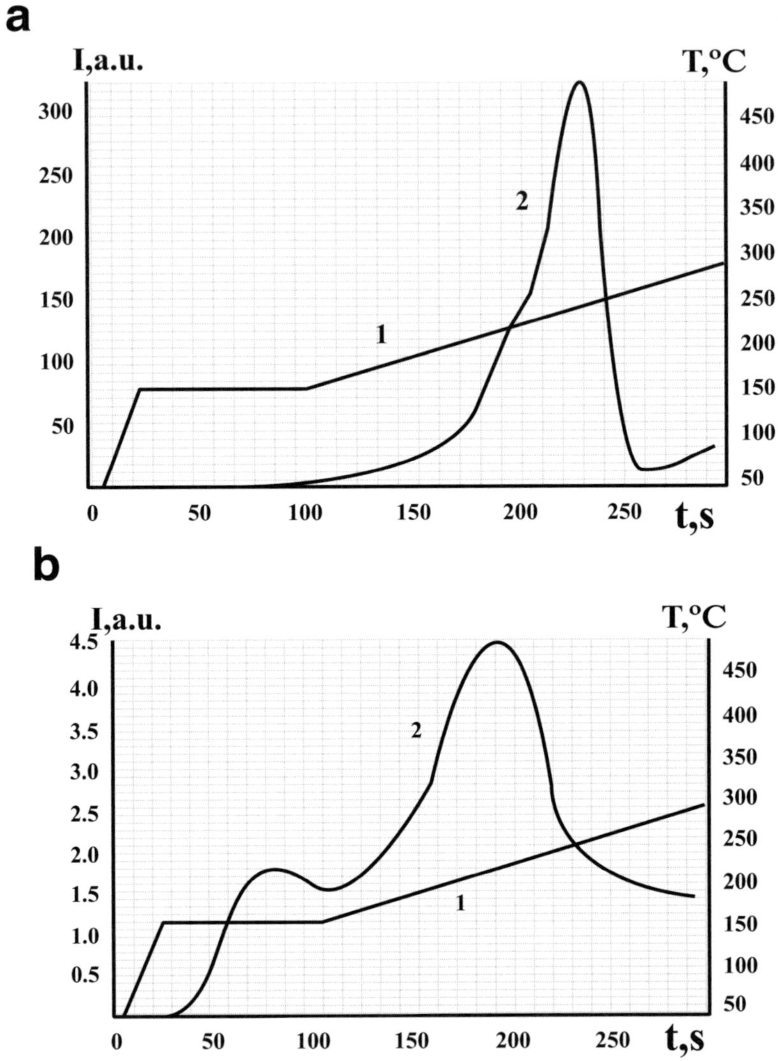

Fig. 6.7 Annealing temperature (*1*) and TL intensities (*2*) versus annealing time for LiF: Mg, Ti crystals (**a**) and NCs (**b**)

reusability. Thus, the NCs demonstrate good opportunities for use as detectors of ionizing radiation, including heavy charged particles, at high doses.

Increasing the range of linear response at the higher doses using NCs can be explained on the basis of the model [65] taking into consideration the particle tracks in crystals. According to this model an ionized zone formed in the TL material by a swift charged particle is located inside the track of the particle. At low fluences of the particles the TL signal is proportional to the number of tracks. In this case

Table 6.2 Ranges of TL linear response for nano- and microcrystals of several doped insulating materials used for ionizing radiation detection

Chemical composition	Radiation	Range of linear response: Nanocrystals	Crystals or micro-crystals
LiF:Mg, Cu, P [57, 58]	γ–rays, ^{137}Cs	0.1–10^3 Gy	0.1–20 Gy
	48 MeV, $^7Li^+$	10^{11}–10^{13} ions/cm^2	–
MgB_4O_7:Dy [59]	γ–rays, ^{60}Co	$10-5 \cdot 10^3$ Gy	$10^{-5}-10$ Gy
$CaSO_4$: Dy [58]	75 MeV, C^+	10^9–10^{13} ions/cm^2	10^9–10^{11} ions/cm^2
$K_3Na(SO_4)$:Eu [60, 61]	γ–rays, ^{60}Co	$0.1–7 \cdot 10^4$ Gy	0.1–10 Gy
$BaSO_4$:Eu [62]	γ–rays, ^{60}Co	$0.1–7 \cdot 10^3$ Gy	0.1–10 Gy
$Ba_{0.97}Ca_{0.03}SO_4$:Eu [63]	48 MeV, $^7Li^+$	10^9–10^{13} ions/cm^2	10^9–10^{10} ions/cm^2
	75 MeV, C^+	$10^9 \sim 10^{13}$ ions/cm^2	
	90 MeV, O^+	$10^9 \sim 10^{12}$ ions/cm^2	
$K_2Ca_2(SO_4)$:Eu [58]	48 MeV, $^7Li^+$	10^9–10^{13} ions/cm^2	10^9–10^{11} ions/cm^2
	75 MeV, C^+	10^9–10^{12} ions/cm^2	
	90 MeV, O^+	10^9–10^{12} ions/cm^2	
ZnS:Cu [64]	β–rays, ^{90}Sr, ^{90}Y	$1.5 \cdot 10^{-3}–5 \cdot 10^3$ Gy	–

the recombination of electrons with holes, which leads to TL, occurs entirely within the track. At very high fluencies the tracks overlap each other and the effective ionization is reduced that leads to saturation of TL. The NCs have sizes of tens of nanometers. Even at high doses, the possibility of the neighboring track to be in different NCs is very high due to their small size. Then the TL response is linear even at high doses.

6.4 Conclusions

The peculiarities of the solid-state radiation imaging LiF detectors, such as high spatial resolution over a large field of view, a wide dynamic range, simplicity of use and efficiency of the PL reading technique, can be exploited for X-ray microradiography and microscopy in different configurations, also for in vivo observation of biological samples. Its characteristics should be particularly suitable for X-ray lensless imaging techniques, including contact microscopy as well as absorption and phase contrast tomography with parallel beams. Improvements of LiF film detector performances and optimization of the reading technique are currently under development.

Different multilayer systems based on LiF film and nanomaterials are under development, as well as many irradiation methods and processes are under investigation in order to allow better understanding and control of point defects in LiF thin layers and nanocrystals.

Recently, low-energy ion sputtering allowed simultaneous periodic surface nanostructuring and optical activation of LiF crystal surface [66], through the stable formation of visible emitting F_2 and F_3^+ defects, thus providing the opportunity to conceive and realize functionalized insulating substrates.

Thin LiF films irradiated by swift heavy ions seem to show a variable F_2 to F_3^+ ratio as a function of the irradiation dose, following the simultaneous radiation induced modification of the mean grain size [67]. All these results are promising for the future of this radiation-sensitive material, grown by simple and well-assessed film deposition methods on different substrates.

SCCs with previously unobserved absorption and luminescence characteristics have been found in NCs produced by mechanical fragmentation of the LiF crystal [48, 49]. The absorption and photoluminescence spectra of SCCs are shifted toward longer wavelengths as compared with those for CCs of the same type in the bulk crystal. The PLE spectra for SCCs have two (for F_{S2} centers) or three (for F_{S2}^+, F_{S2}^-, F_{S3}^+ and F_{S3} centers) bands with not strongly various intensities. Optical transitions in the F_{S3}^+ (F_{S3}) SCCs and in the F_3^+ (F_3) CCs are characterized by different Huang-Rhys parameters. The observed behavior suggests that the electron density distribution function of these defects in the near-surface layers is more diffuse compared to the bulk crystal. In near-surface layers of LiF crystals, the activation energy of the defects diffusion is significantly higher than in the bulk of the crystals [55, 56].

For many doped insulating materials used in TL radiation dosimetry, NCs have a much larger range of linear response to dose compared with crystals and micro-crystals of the same composition. Simultaneously, some of them exhibit negligible fading and good reusability. Thus, the NCs demonstrate good opportunities for use as detectors at high doses of ionizing radiation, including heavy charged particles. Characteristics improvements of NCs for detecting ionizing radiation are currently under study.

Acknowledgments One of the authors (R.M.M.) is indebted to F. Bonfigli, E. Nichelatti and M.A. Vincenti for valuable discussions and with M. Piccinini for helpful suggestions and a critical reading of the manuscript. The second author (A.P.V.) thanks L.P. Runets for useful discussions.

References

1. Voitovich, A. P., Scavarda do Carmo, L. C., Kalinov, V. S., & Saltanov, A. V. (1990). *Doklady Akademii Nauk BSSR, 34*(1), 21–23 (in Russian).
2. Montereali, R. M. (2002). In H. S. Nalwa (Ed.), *Handbook of thin film materials* (vol. 3, chap. 7, p. 399). New York: Academic.
3. Doherty, M. W., Manson, N. B., Delaney, P., Jelezko, F., Wrachtrup, J., & Hollenberg, L. C. L. (2013). *Physics Reports, 528*(1), 1–45.
4. Montereali, R. M. (2009). In B. Di Bartolo & O. Forte (Eds.), Proceeding of the international school of atomic and molecular spectroscopy, 24th Course "*Frontier Developments in Optics and Spectroscopy*". ISBN 978-1-936036-00-4.
5. Montereali, R. M. (1997). *Journal of Luminescene, 72–74*, 4.
6. Montereali, R. M., Baldacchini, G., & Scavarda do Carmo, L. C. (1991). *Thin Solid Films, 201*, 106.

7. Cosset, F., Celerier, A., Barelaud, B., & Vareille, J. C. (1997). *Thin Solid Films, 303*, 191.
8. Di Lazzaro, P., Bollanti, S., Flora, F., Mezi, L., Murra, D., Torre, A., Bonfigli, F., Montereali, R. M., Vincenti, M. A. (2013). Proceeding of the SPIE 8677 (86770T, pp. 1–7).
9. Nahum, J., & Wiegand, D. A. (1967). *Physics Review, 154*, 817.
10. Basiev, T. T., Mirov, S. B., & Osiko, V. V. (1988). *IEEE Journal of Quantum Electronics, 24*, 1052.
11. Ter-Mikirtychev, V. V., & Tsuboi, T. T. (1996). *Progress in Quantum Electronics, 20*(3), 219.
12. Montereali, R. M., Mancini, A., Righini, G. C., & Pelli, S. (1998). *Optics Communication, 153*, 223.
13. Montereali, R. M., Piccinini, M., & Burattini, E. (2001). *Applied Physics Letters, 78*(26), 4082.
14. Mussi, V., Somma, F., Moretti, P., Mugnier, J., Jacquier, B., Montereali, R. M., & Nichelatti, E. (2003). *Applied Physics Letters, 82*(22), 3886.
15. Chiamenti, I., Bonfigli, F., Gomes, A. S. L., Michelotti, F., Montereali, R. M., & Kalinowski, H. J. (2014). *Journal of Applied Physics, 115*(2), 023108-1-7.
16. Kurobori, T., Kawamura, K., Hirano, M., & Hosono, H. (2003). *Journal of Physics. Condensed Matter, 15*, L399.
17. Kurobori, T., Obayashi, Y., Kurashima, M., Hirose, Y., Sakai, T., Aoshima, S., Kojima, T., & Okuda, S. (2008). *Nuclear Instruments and Methods in Physics Research B, 266*, 2762.
18. Montereali, R. M. (2003). In B. Di Bartolo (Ed.), *Spectroscopy of systems with spatially confined structures* (NATO Science Series, pp. 617–632). Dordrecht: Kluwer Academic Publisher.
19. Bonfigli, F., Jacquier, B., Menchini, F., Montereali, R. M., Moretti, P., Nichelatti, E., Piccinini, M., Rigneault, H., & Somma, F. (2003). In B. Di Bartolo (Ed.), *Spectroscopy of systems with spatially confined structures* (NATO Science Series, pp. 697–703). Dordrecht: Kluwer Academic Publisher.
20. McLaughlin, W. L., Miller, A., Ellis, S. C., Lucas, A. C., & Kapsar, B. M. (1980). *Nuclear Instruments and Methods in Physics Research B, 175*, 17.
21. Lakshmanan, R., Madhusoodanan, U., Natarajan, A., & Panigrahy, B. S. (1996). *Physica Status Solidi A, 153*, 265.
22. Kovacs, A., Baranyai, M., McLaughlin, W. L., Miller, S. D., Miller, A., Fuochi, P. G., Lavalle, M., & Slezsak, I. (2000). *Radiation Physics and chemistry, 57*, 691.
23. Montecchi, M., Baccaro, S., Nichelatti, E., Bonfigli, F., Marolo, T., & Montereali, R. M. (2002). In M. Barone, E. Borchi, J. Huston, C. Leroy, P.G. Rancoita, P. Riboni, & R. Ruchiti (Eds.), *Proceeding of 7th international conference on advanced technology and particle physics* (World Scientific, pp. 819–825).
24. Baldacchini, G., Bonfigli, F., Faenov, A., Flora, F., Montereali, R. M., Pace, A., Pikuz, T., & Reale, L. (2003). *Journal of Nanoscience and Nanotechnology, 3*(6), 483–486.
25. Baldacchini, G., Bollanti, S., Bonfigli, F., Flora, F., Di Lazzaro, P., Lai, A., Marolo, T., Montereali, R. M., Murra, D., Faenov, A., Pikuz, T., Nichelatti, E., Tomassetti, G., Reale, A., Reale, L., Ritucci, A., Limongi, T., Palladino, L., Francucci, M., Martellucci, S., & Petrocelli, G. (2005). *The Review of Scientific Instruments, 76*, 113104–1.
26. Almaviva, S., Pelliccia, D., Franzini, I., Cedola, A., Bonfigli, F., Montereali, R. M., & Lagomarsino, S. (2006). *Applied Physics Letters, 89*, 054102.
27. Kurobori, T., & Matoba, A. (2014). *Japanese Journal of Applied Physics, 53*, 02BD 14.
28. Heidari Bateni, S., Bonfigli, F., Cecilia, A., Baumbach, T., Pelliccia, D., Somma, F., Vincenti, M. A., & Montereali, R. M. (2013). *Nuclear Instruments and Methods A, 720*, 109.
29. Pikuz, T. A., Faenov, A. Y., Fukuda, Y., Kando, M., Bolton, P., Mitrofanov, A., Vinogradov, A. V., Nagasono, M., Ohashi, H., Yabashi, M., Tono, K., Senba, Y., Togashi, T., & Ishikawa, T. (2013). *Applied Optics, 52*(3), 509.
30. Piccinini, M., Ambrosini, F., Ampollini, A., Picardi, L., Ronsivalle, C., Bonfigli, F., Libera, S., Nichelatti, E., Vincenti, M. A., & Montereali, R. M. (2015). *Applied Physics Letters, 106*(261108), 1.
31. Montereali, R. M., Bonfigli, F., Piccinini, M., Nichelatti, E., & Vincenti, M. A. (2016). *Journal of Luminescence, 170*, 761.

32. Belarouci, A., Menchini, F., Rigneault, H., Jacquier, B., Montereali, R. M., Somma, F., Moretti, P., & Cathelinaud, M. (2001). *Optical Materials, 16*, 63.
33. Goncharova, O., Kalinov, V., & Voitovich, A. (2004). *Radiation Measurements, 38*, 775.
34. Itoh, N., & Stoneham, A. M. (2001). *Materials modification by electronic excitation*. Cambridge: University Press.
35. Montereali, R. M., Bigotta, S., Giammatteo, M., Piccinini, M., Picozzi, P., & Santucci, S. (2000). *Radiation Effects and Defects in Solids, 156*, 135.
36. Perez, A., Davenas, J., & Dupuy, C. H. S. (1976). *Nuclear Instruments and Methods, 132*, 219.
37. Martin, J., Bischoff, L., & Wannemacker, R. (2001). *Optics Communication, 188*, 119.
38. Piccinini, M., Ambrosini, F., Ampollini, A., Carpanese, M., Picardi, L., Ronsivalle, C., Bonfigli, F., Libera, S., Vincenti, M. A., & Montereali, R. M. (2014). *Nuclear Instruments and Methods B, 326*, 72.
39. Caine, E. J., & Miller, S. D. (1998). *Journal of Vacuum Science and Technology B, 16*(6), 3232.
40. Adam, P., Benrezzak, S., Bijeon, J. L., Royer, P., Guy, S., Jacquier, B., Moretti, P., Montereali, R. M., Piccinini, M., Menchini, F., Somma, F., Seassal, C., & Rigneault, H. (2001). *Optics Express, 9*(7), 353.
41. http://www-cxro.lbl.gov/optical_constants/filter2.html
42. Montereali, R. M., Bonfigli, F., Dietler, G., Gregoratti, L., Kiskinova, M., Larciprete, R., Marolo, T., & Sekatskii, S. K. (2005). *Physica Status Solidi C, 2*(1), 298.
43. Larciprete, R., Gregoratti, L., Danailov, M., Montereali, R. M., Bonfigli, F., & Kiskinova, M. (2002). *Applied Physics Letters, 80*, 3862.
44. Baldacchini, G., Bonfigli, F., Flora, F., Montereali, R. M., Murra, D., Nichelatti, E., Faenov, A., & Pikuz, T. (2002). *Applied Physics Letters, 80*, 4810.
45. Baldacchini, G., Cremona, M., d'Auria, G., Martelli, S., Montereali, R. M., Montecchi, M., Burattini, E., Grilli, A., & Raco, A. (1996). *Nuclear Instruments and Methods B, 116*, 447.
46. Montereali, R. M., Almaviva, S., Bonfigli, F., Faenov, A., Flora, F., Franzini, I., Nichelatti, E., Pikuz, T., Vincenti, M. A., & Baldacchini, G. (2007). In A. Sepenguzel, G. Badenes, G. C. Righini (Eds.), *Photonic materials, devices, and applications II* (Proceedings of SPIE, vol. 6593B, pp. 1–8). Bellingham: SPIE.
47. Moharil, S. V., & Deshmukh, B. T. (1977). *Radiation Effects and Defects in Solids, 34*, 189–202.
48. Ichimura, N., Kondo, H., Harada, Y., & Hashimoto, S. (2000). *Journal of Luminescence, 87–89*, 586–588.
49. Zielasek, V., Hildebrandt, T., & Henzler, M. (2000). *Physical Review B, 62*(4), 2912–2919.
50. Voitovich, A. P., Kalinov, V. S., Korzhik, M. V., Martynovich, E. F., Runets, L. P., & Stupak, A. P. (2013). *Radiation Effects and Defects in Solids, 168*(2), 130–136.
51. Voitovich, A. P., Kalinov, V. S., Stupak, A. P., Novikov, A. N., & Runets, L. P. (2015). *Journal of Luminescence, 157*, 28–34.
52. Voitovich, A. P., Kalinov, V. S., Mudryi, A. V., Pavlovskii, V. N., Runets, L. P., & Svitsiankou, I. E. (2016). *Journal of Luminescence, 172*, 147–153.
53. Stoneham, A. M. (2001). *Theory of defects in solids: Electronic structure of defects in insulators and semiconductors*. Oxford: Oxford University Press.
54. Solé, J. G., Bausá, L. E., & Jaque, D. (2005). *An introduction to the optical spectroscopy of inorganic solids*. Chichester: John Wiley & Sons Ltd.
55. Voitovich, A. P., Kalinov, V. S., Martynovich, E. F., Stupak, A. P., & Runets, L. P. (2015). *Physics of the Solid State, 57*(9), 1752–1758.
56. Voitovich, A. P., Kalinov, V. S., Martynovich, E. F., Novikov, A. N., & Stupak, A. P. (2012). *Physics of the Solid State, 54*(9), 1768–1775.
57. Salah, N., Sahare, P. D., & Rupasov, A. A. (2007). *Journal of Luminescence, 124*, 357–364.
58. Salah, N. (2011). *Radiation Physics and Chemistry, 80*, 1–10.
59. Lochab, S. P., Pandey, A., Sahare, P. D., Chauhan, R. S., Salah, N., & Ranjan, R. (2007). *Physica Status Solidi A: Applications and Material Science, 204*(7), 2416–2425.
60. Dhoble, S. J., Moharil, S. V., Dhopte, S. M., Muthal, P. L., & Kondawar, V. K. (1993). *Physica Status Solidi A: Applications and Material Science, 135*, 289–297.

61. Sahare, P. D., Ranjan, R., Salah, N., & Lochab, S. P. (2007). *Journal of Physics D: Applied Physics, 40*, 759–764.
62. Salah, N., Habib, S., Khan, Z. H., Al-Hamedi, S., & Lochab, S. P. (2009). *Journal of Luminescence, 129*, 192–196.
63. Lochab, S. P., Kanjilal, D., Salah, N., Habib, S. S., Lochab, J., Ranjan, R., Aleynikov, V. E., Rupasov, A. A., & Pandey, A. (2008). *Journal of Applied Physics, 104*(033520), 1–4.
64. Yazici, A. N., Öztaş, M., & Bedir, M. (2007). *Optical Mater., 29*, 1091–1096.
65. Horowitz, Y. S., Avila, O., & Rodriquez-Villafuerte, M. (2001). *Nuclear Instruments and Methods B, 184*, 85–112.
66. Mussi, V., Granone, F., Marolo, T., Montereali, R. M., Boragno, C., Buatier de Mongeot, F., & Valbusa, U. (2006). *Applied Physics Letters, 88*(1), 103116.
67. Kumar, M., Singh, F., Khan, S. A., Baranwal, V., Kumar, S., Agarwal, D. C., Siddqui, A. M., Tripathy, A., Gupta, A., Avasthi, D. K., & Pandey, A. C. (2005). *Journal of Physics D: Applied Physics, 38*, 637.

Chapter 7
Colloidal Nanophotonics: State-of-the-Art and Prospective

Sergey V. Gaponenko

Abstract Nanocrystalline platform based on colloidal nanophysics, nanochemistry and nanoengineering is the promising unique versatile scientific and technological basement for emerging nano-optoelectronics. The approach offers straightforward multilevel throughout bottom-up scaling including: subnanometer molecular scale interfaces, nanometer-scale semiconductor quantum dot systems, submicron photonic scale. Notably, the colloidal multilevel bottom-up approach as the technological paradigm and semiconductor quantum dots as its principal physical entity, when coupled together do offer the unprecedented road map towards versatile and affordable platform where every optoelectronic component, including light emitting diodes, LEDs, lasers, photodetectors, signal processing elements (e.g. electrooptical modulators, optical switches) and various sensors can be developed in *unified and cheap* technological processes to compete with existing *multi-base and expensive* technological approaches. Interfacing of electronic devices with biosystems is the additional essential advantageous outcome of the colloidal bottom-up approach.

7.1 Introduction

The driving technology in modern optoelectronics is epitaxial growth of monocrystalline planar structures in combination with photolithography. It is commercially used in semiconductor light-emitting diodes (LEDs), in semiconductor lasers, in solar cells and in photodetectors like charge-coupled device (CCD) matrices in video and photocameras. The same approach remains to dominate in nanostructured devices like LEDs and lasers with quantum wells, planar photonic waveguides, multiplexers, beamsplitters, and couplers [1]. At the same time many pronounced size-dependent phenomena and their optical manifestations have been demonstrated

S.V. Gaponenko (✉)
Stepanov Institute of Physics, National Academy of Sciences of Belarus, Minsk 220072, Belarus
e-mail: s.gaponenko@ifanbel.bas-net.by; nanoscience@tut.by

© Springer Science+Business Media Dordrecht 2017

B. Di Bartolo et al. (eds.), *Nano-Optics: Principles Enabling Basic Research and Applications*, NATO Science for Peace and Security Series B: Physics and Biophysics, DOI 10.1007/978-94-024-0850-8_7

and examined using colloidal nanostructures. This contribution is designed to overview the principal milestones in the fast and vast field of modern nanoscience which tends to become a new well defined technological platform, namely colloidal nanophotonics.

Actually, colloidal nanophotonics dates back to ancient times as real nanotechnology applied to color glass in cups [2], stained-glasses [3], and pottery [4]. Colloidal nanophotonics can also be treated as the dawn of nanoscience recalling first experiments by Michael Faraday on optical properties of gold colloidal solutions [5]. Notably, commercial cut-off optical filters based on semiconductor-doped glasses are known for many decades [6]. Further prerequisites can be found in first experiments on enhanced light-matter interaction with metal colloidal nanoparticles [7], in size-dependent properties of metal particle suspensions [8] and in the early predictions of inhibited spontaneous emission of light in periodic structures [9]. Nowadays there are three fields which constitute colloidal nanophotonics [10], namely

(i) quantum confinement effects on emission and absorption of light by *semiconductor* nanocrystals, quantum dots,
(ii) photon, or more accurately, light waves confinement phenomena in colloidal *dielectric* microstructures (microcavities and photonic crystals),
(iii) colloidal nanoplasmonics, i.e. a variety of phenomena related to light-matter interaction in metal-dielectric colloidal structures.

Semiconductor nanocrystals often referred to as quantum dots do form the core of colloidal nanophotonics since these offer size-controlled light emission including lasing. Dielectric structures are used to control light wave propagation, and metal nanostructures are used mainly to enhance light-matter interaction by means of enormous concentration (enhancement) of incident light intensity. Table 7.1 summarizes the main application fields of colloidal nanostructures in photonics.

The key role of quantum dots can be traced in Table 7.1 where the list of photonic devices and technologies is provided that can be implemented within the colloidal nanophotonics paradigm.

7.2 Colloidal Semiconductors: Quantum Dots

7.2.1 *Basic Properties*

In 1982–1983 research groups in Russia (Ekimov with co-workers [32, 33]) and in USA (Brus [34]) provided experimental and theoretical evidence for size-dependent optical properties of nanometer-size semiconductor crystallites (nanocrystals, quantum dots) resulting from size-dependent energy spectra of confined electros and holes (Fig. 7.1). As a reasonable reference for size-dependent absorption and

Table 7.1 Colloidal nanostructures in photonics

Photonic component/ technology	Current status	Type of colloids used			Sample refs.
		Semiconductor	Dielectric	Metal	
Filters	Commercial	+		+	[11–13]
Luminophores for display devices	First commercial applications, active research	+	+		[14, 15]
Luminophores for lighting	Active research	+	+		[15–17]
LEDs	Active research	+		+	[18–20]
Lasers	Active research	+			[21–23]
Electrooptical devices	Research	+			[24]
Photodetectors	Research	+			[25]
Sensors and biocompatibility	Active research	+		+	[26–28]
Photovoltaics	Research	+		+	[29]
Nanolithography	Research		+		[30]
Hyperbolic metamaterials	Research	+		+	[31]

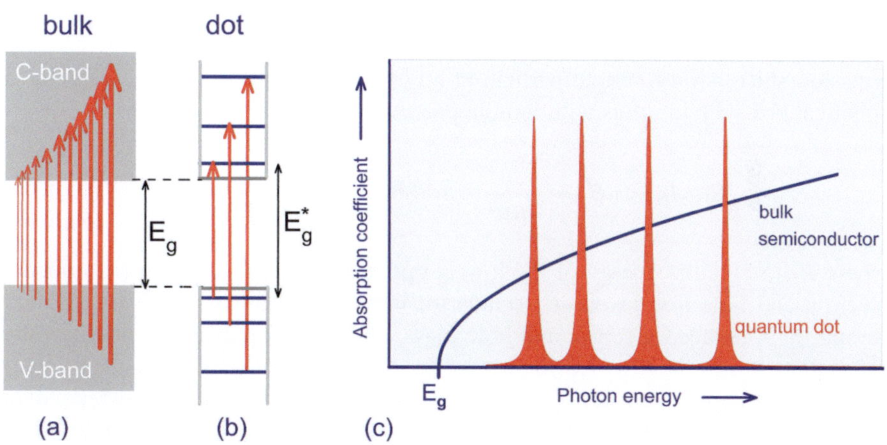

Fig. 7.1 Optical absorption by semiconductor nanocrystals in the strong confinement regime. A sketch of energy spectrum, optical transitions and absorption spectra in a simple quantum dot model versus simple bulk semiconductor model. Unlike continuous optical transitions in a bulk crystal (**a**), electron and hole energy levels obey a series of states inherent for a particle in a spherical box (**b**). Selection rules allow optical transitions coupling the hole and the electron states with the same quantum numbers. Therefore, the optical absorption spectrum of the parent bulk crystal reduces to a number of discrete bands (**c**)

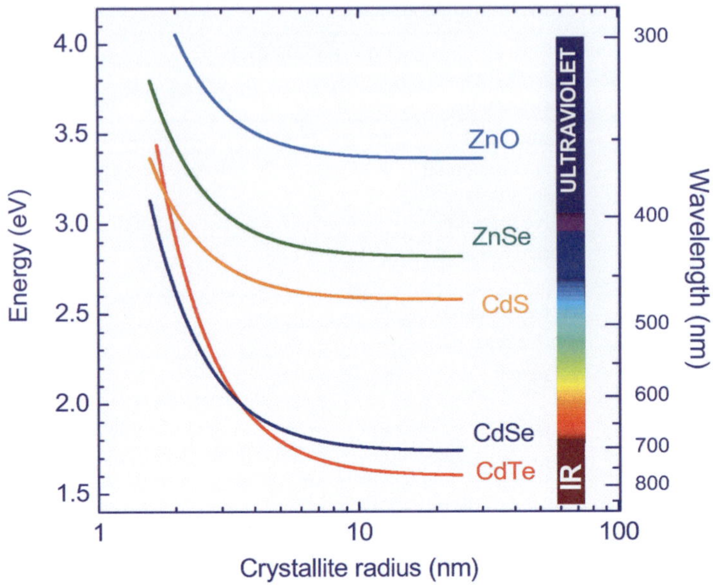

Fig. 7.2 Dependence of the photon energy for the first optical transition versus quantum dot radius for the simple model of a spherical dot with infinite potential barriers for a number of common II–VI compounds

emission shift, a very simple particle-in-a-box model is useful which gives for a spherical box with *a* radius with infinite potential barrier,

$$E_{1s1s}(a) = E_g^0 + \frac{\pi^2\hbar^2}{2\mu a^2} - 1.786\frac{e^2}{\varepsilon a}, \quad \mu = \frac{m_e^* m_h^*}{m_e^* + m_h^*}, \tag{7.1}$$

where $E_{1s1s}(a)$ is the energy of the lowest optical transition, 1 s subscripts denote electron and hole first levels, E_g^0 stands for intrinsic band gap of the parent bulk crystal, μ is reduced electron and hole mass, and ε is dielectric permittivity. The results are shown in Fig. 7.2 for a number of common II–VI compounds, their nanocrystals can be readily fabricated by colloidal techniques. Size-dependent optical properties and a number of non-trivial additional features inherent in semiconductor nanocrystals have been identified in the end of the twentieth century and summarized in the books [35–37].

Colloidal quantum dots have been synthesized for a number of II–VI compounds and their absorption and emission spectra were found to correlate pretty well with a simple formula (7.1). Figure 7.2 gives an idea how far downsizing of semiconductor nanocrystals can shift their absorption edge in terms of the first absorption peak. Apparently, quantum confinement effect on absorption is more pronounced for narrow band semiconductors. This is the consequence of smaller electron effective mass and higher dielectric permittivity inherent in narrow-band crystals. Within the same compound like CdSe or CdTe tailoring of absorption

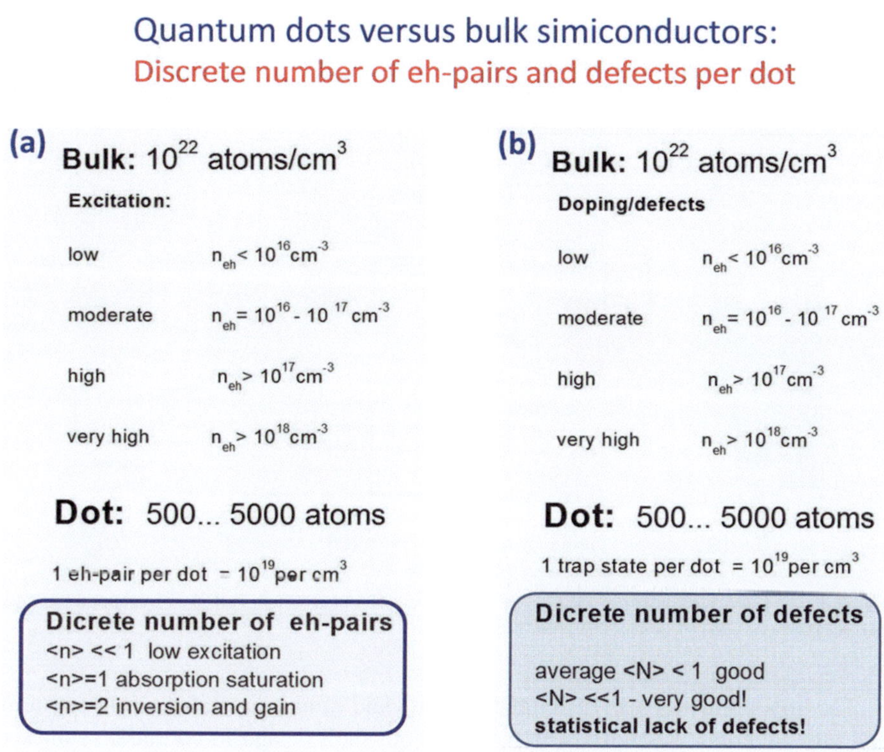

Fig. 7.3 Comparison of quantum dots versus bulk semiconductors in terms of discrete number of (**a**) electron-hole pairs and (**b**) defects per dot

onset and emission colors within the whole visible becomes possible. Absorption and emission spectra are typically inhomogeneously broadened because of size distribution. Spectral hole-burning in pump-and-probe absorption saturation, red-edge excitation technique and single dot luminescence spectroscopy were successfully used to reveal the intrinsic properties of dots smeared by size dispersion in ensembles.

Remarkable are the properties of dots resulting from the discrete number of electron-hole pairs and defects/traps (Fig. 7.3). In bulk solids, concentrations of electrons, holes, and defects are important. In a dot, the notion concentration cannot be used. Instead, the number electron-hole pairs and traps in a dot should be considered. These take discrete numbers (0, 1, 2, 3, . . .) and feature Poisson distribution over an ensemble of dots in question. E.g., if average number of electron-hole pairs per dot $<n>$ tends to 1, absorption saturation occurs, if $<n>$ tends to 2, optical gain is expected. Independent of average number of traps, there are always dots without traps and therefore luminescence yield in dots can be higher than in bulk materials. This statistical lack of defects stimulated extensive studies of light emission and excitation dynamics in dots, resulting in specific decay law for luminescence kinetics [38, 39].

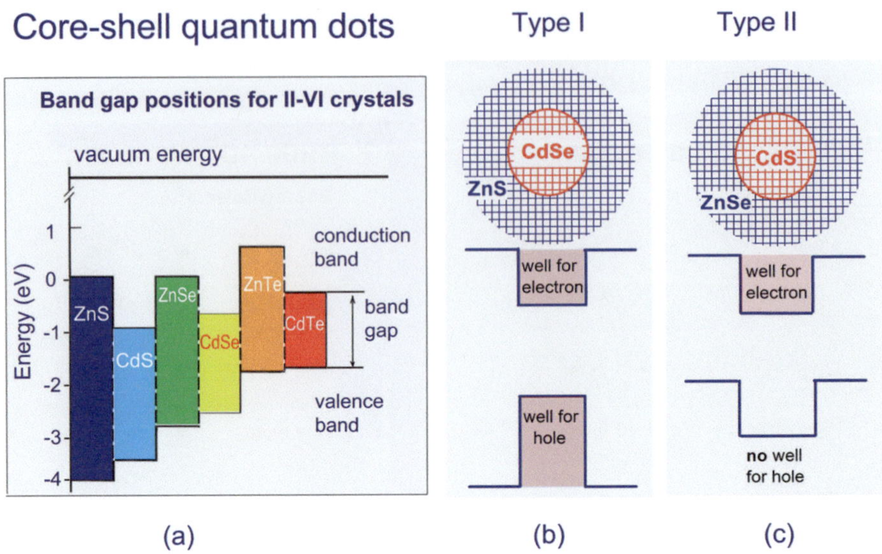

Fig. 7.4 (**a**) Energy diagram of band gaps versus vacuum of a number of II–VI compounds and (**b**), (**c**) the two types of core-shell quantum dots

The important step towards high quantum yield of nanocrystalline light emitters and luminophores was introduction of the core-shell design in colloidal synthesis of semiconductor nanocrystals (Bawendi with co-workers at MIT, middle 90-ies, see Ref.14 for detail). In this approach, the electron-hole confinement phenomena are controlled by means of the size of the core nanocrystal whereas the potential barrier and, what are most important, core-matrix interface and surface structure are defined by the crystalline shell. Depending on energy diagram shown in Fig. 7.4, different work function gives rise to different confinement condition. To develop efficient luminescent dots, Type I structures are used with simultaneous confinement of electrons and holes in the core.

In electron-hole dynamics in bulk materials, energy and momentum conservation should always be met simultaneously. E. g., at high concentrations, many-particle processes become feasible like Auger recombination (Fig. 7.5) when an electron-hole pair recombines but the energy released does not transform into radiation and is picked-up by a neighboring particle. A sketch in Fig. 7.4(left) shows pretty strict conditions for such a process to occur in terms of energy and momentum conservation in *E*-*p*-space.

Auger recombination promotes photodarkening, photoionization, luminescence blinking and quenching. It results in degradation of photoluminescent properties. Additionally, it prevents development of multi-particle states which are important for optical gain in quantum dots. Therefore to develop stable, efficient and durable light emitters and laser based on colloidal quantum dots, a way to tame Auger processes should be elaborated. Development of smooth rather than sharp potential barrier has been recently shown to inhibit Auger processes in quantum dots [40, 41].

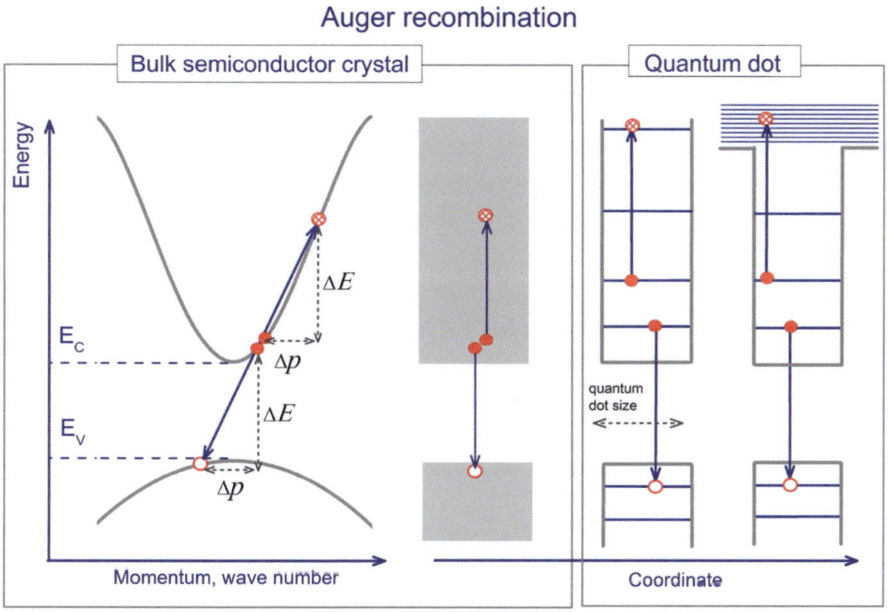

Fig. 7.5 Auger recombination. In a bulk crystal, both energy and momentum conservation are mandatory, and high electron-hole pairs concentration is necessary. In a quantum dot, two electron-hole pairs must be generated, energy conservation should be met but momentum conservation requirement vanishes

7.2.2 Colloidal Quantum Dot Luminophores

Luminescent properties of nanocrystals define important fields of their potential applications from light converters for blue or UV LEDs to fully colloidal display optoelectronic devices. In glasses, luminescence properties of nanocrystals cannot be controlled and moreover, luminescence in glasses features pronounced degradation upon continuous illumination because of photoionization phenomenon. Therefore semiconductor doped glasses cannot be considered as light-emitting materials. Fortunately, colloidal chemistry offers fine surface and interface engineering techniques which are capable to minimize trap number and to tame photoionization. In this context, the concept of the core-shell colloidal quantum dot structure coined by M. Bawendi with co-workers was extremely fruitful. These core-shell nanocrystals are called "type I" quantum dots and represent the main trend in colloidal quantum dot luminophores resulting in high quantum yield and superior photostability. To inhibit photoionization and other negative processes promoted by Auger recombination, development of smooth potential barrier has been suggested [40, 41] and implemented using a multilayer shell of different semiconductor compounds. There are first reports on commercial applications of colloidal quantum dot emitters in mobile phones and similar portable devices.

Emerging solid state lighting revolution based on white solid state light sources needs not only efficient semiconductor LEDs, but for cost-efficient devices affordable but stable and adjustable luminophores are critical as well. These should convert the blue into white meeting both energy efficiency requirements and colorimetric parameters to fit human eye sensitivity spectrum. Here colloidal quantum dots are considered as one of the key approach promising replacement of rare-earth materials from solid state lighting industry [15–17].

7.2.3 Colloidal Quantum Dot Lasers

In a traditional concept, many-particle states should be developed in a quantum dot to get optical gain by stimulated emission dominating over absorption. However, many-particle states under strong spatial confinement of electron and holes give rise to efficient Auger recombination thus bypassing stimulated emission. This issue has been the serious obstacle towards colloidal quantum dot lasers under condition of strong confinement regime. Notably, for larger semiconductor quantum dots obtained by epitaxial growth, lasing has been successfully performed and even goes close to commercial applications.

To overcome the issue of Auger recombination in colloidal quantum dot lasers, V. Klimov with co-workers suggested an alternative approach to develop optical gain [21]. Instead of generation of many-particle states, he proposed to make use of local electric field arising from spatially separated an electron and a hole within a dot. He suggested the type II dots (Fig. 7.4c) to be appropriate. Upon generation of a single electron-hole pair, an electron remains confined in a dot core whereas a hole wave function extends outside the dot core. Then local electric field disturbs electron-hole energy diagram and emission now becomes possible at lower photon energy where absorption was negligibly small. This ingenious single exciton concept of a quantum dot laser has been successfully implemented and was followed by further experiments by different groups [22].

7.2.4 Colloidal Quantum Dot LEDs

Lately, light-emitting diodes (LED) based on quantum dots have been fabricated in several laboratories. To develop a light-emitting diode (LED) based on colloidal quantum dots, a multilayer structure should be fabricated to provide electron and hole injection from conducting electrode, one of them being transparent. Core-shell colloidal quantum dots have been used in such experiments by several groups [18–20] (Fig. 7.6).

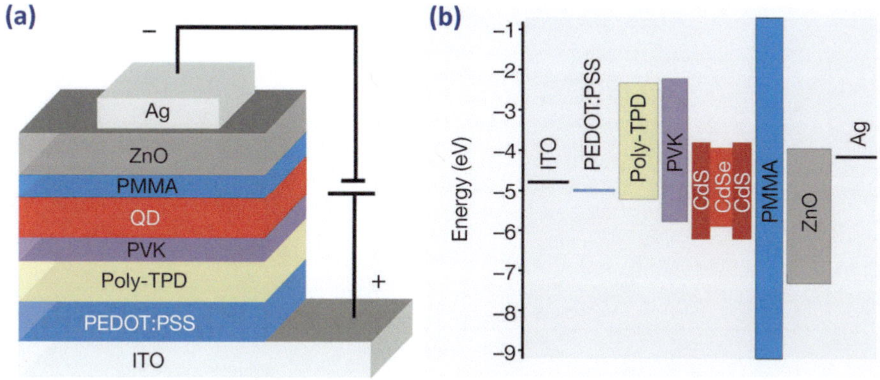

Fig. 7.6 Experimental design of efficient colloidal LED with semiconductor quantum dots as an active material. (**a**) Geometrical arrangement, (**b**) energy diagram (Reprinted with permission from Ref. [42])

Peng with co-workers suggested design (Fig. 7.5) enabling more than 10 % efficiency of a quantum dot LED in the red which is fully solution-processed [42]. The main issues towards commercial applications of colloidal LEDs are durability, efficiency and flexibility in terms of spectral properties. These LEDs forms a promising platform for cheap display devices with multiple applications. Actually, it is the recent progress in colloidal LEDs that has become a decisive step from laboratory studies on colloidal nanophotonics towards the emerging technology platform.

7.2.5 Electro-Optical Effects

Colloidal quantum dots exhibit pronounced sensitivity of optical properties to external electric fields. In thin films, a few tens of volts of external voltage can noticeably modify both absorption and emission. Absorption spectrum features mainly vanishing of the absorption peaks and can be treated as field-induced broadening of absorption bands.

Electric field typically results in luminescence quenching (Fig. 7.7) which can be viewed as mainly electron-hole wave function separation by the external field. However, the interplay of photoexcitation and external electric field in colloidal structures is more complicate and results in a number of side effects, probably related to interface phenomena in photoexcited dots promoted by strong external electric field (see, e.g. [43]). Therefore though colloidal quantum dots principally allow for modulation of both transmitted and emitted radiation, using these phenomena in optical switching needs further research towards eliminating undesirable side effects.

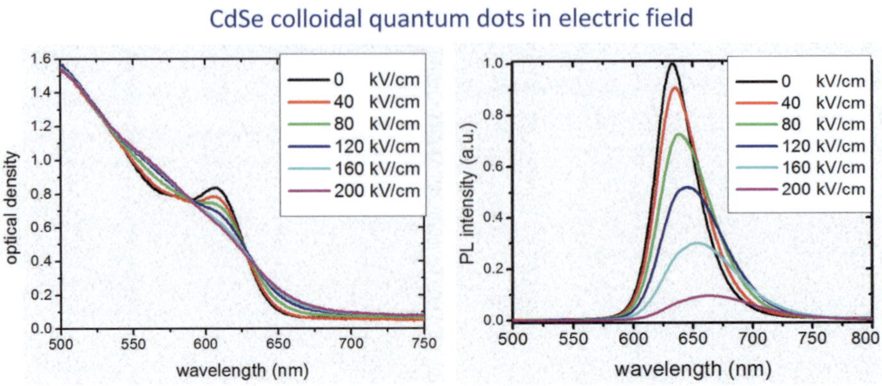

Fig. 7.7 Electric field effect on optical absorption and emission for CdSe nanocrystals in a polymer

7.2.6 *Nanoplatelets: Colloidal Quantum Well Structures*

Recently, a number of groups reported on successful synthesis of very thin monocrystalline sheets of II-VI compounds measuring a few nanometers in one direction and micrometers otherwise. These have been termed "*nanoplatelets*" and can be treated as colloidal counterparts of epitaxial quantum well structures. Similar to epitaxial quantum wells, nanoplatelets exhibit thickness-dependent multiple exciton peaks in absorption and show strong excitonic emission from the ground state of confined exciton (Fig. 7.8a). A representative example shows well pronounced exciton bands at room temperature, huge high-energy shift of band gap energy (from $E_g = 1.73$ eV inherent in bulk parent crystal to about 2.8 eV), and luminescence quantum yield about 50 %.

There is one important issue, the dielectric confinement effect, which becomes crucially important for colloidal quantum wells but is less pronounced in epitaxial ones. When an electron and a hole are within a thin layer in real space with a finite potential barrier (Fig. 7.8b), Coulomb interaction does not restricted by the well but extends outside the well. Unlike quantum well heterostructures, in case of nanoplatelets the surrounding medium (air, liquid or polymer) has much lower dielectric constant value $\varepsilon_{out} = 1 \ldots 5$ as compared to $\varepsilon_{in} > 10$ in the well region. Since interaction energy outside the well is $e^2/\varepsilon_{out} r$ whereas inside it is $e^2/\varepsilon_{in} r$, binding energy will be strongly enhanced. For the extreme thickness limit $d \ll a_B{}^*$, where $a_B{}^*$ is exciton Bohr radius for a parent bulk crystal, L. Keldysh (1979) derived the formulas for exciton binding energy E_B and exciton radius a_{ex} (SI units),

$$E_B = 2Ry^* \frac{a_B^*}{d} \left\{ \ln \left[\left(\frac{\varepsilon_{in}}{\varepsilon_{out}} \right)^2 \frac{d}{a_B^*} \right] - 0.8 \right\}, \quad a_{ex} = \frac{1}{2} \sqrt{a_B^* d}\ , \quad a_B^* = 4\pi\varepsilon_0 \frac{\varepsilon \hbar^2}{\mu_{eh} e^2}$$

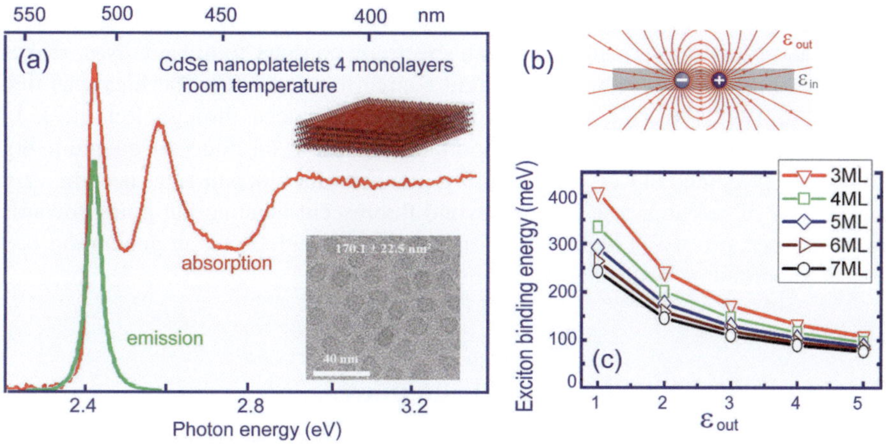

Fig. 7.8 Optical properties of nanoplatelets. (**a**) Sample absorption and emission spectra from an ensemble of 4-monolayer thick CdSe nanoplatelets (Reprinted with permission from Ref. [44]. Copyright 2015 ACS). (**b**) Coulomb interaction scheme, (**c**) Modeling of exciton binding energy for various environment with $3d$-8-band-$\boldsymbol{k}\cdot\boldsymbol{p}$-theory by [45] (Note, for bulk CdSe, band gap energy $E_g = 1.73$ eV, and exciton binding energy $Ry^* = 16$ meV)

This prediction has been principally confirmed by numerical modeling (Fig. 7.8c) and provides rationale for pronounced excitonic absorption and emission bands in nanoplatelets at room temperature.

Nanoplatelets have been recently found to exhibit superior sensitivity of optical absorption to external electric field as compared to dots and rods fabricated by the same technological route [24]. At the same time electric field effect on luminescence was found to be higher for dots but smaller for platelets. Therefore platelets can be used as efficient light emitting species in various electro-optical devices. The first successful implementation of nanoplatelets in lasers and light-emitting diodes have been already reported [23].

7.2.7 *Colloidal Quantum Dot Bioconjugates for Biosensing*

Colloidal quantum dots can be conjugated with biomolecules and owing to high quantum yield and high stability represent a new promising class of fluorescent labels in biomedical analyses based on labeled antibody detection. In these analyses antigens produced by humans under condition of certain diseases (including cancer tumors) are detected through luminescence measurements of signals from labeled antibodies which can recognize chemically desirable antigen like "key-lock" approach. Colloidal quantum dots substitute organic fluorophores in this technique. Stability of quantum dot emission is not the only advantageous feature. Additionally, narrow band emission spectrum makes detection more distinctive enabling easy multiplexing, i.e. recognizing different antibodies by using different labels, and furthermore, wide excitation spectrum, contrary to organic dyes, makes multiplexing possible with a single light source (e.g. a LED). The idea was first pushed in 1998 [46, 47], further progress can be traced through Ref. [27]. In biostudies, solubility in water is important; a review on aqueous quantum dot solution is presented in Ref. [26]. Notably, quantum dot bioconjugates become very active field of research last decade beyond fluorescent labeling but going towards photoinduced processes including energy transfer, singlet oxygen generation and others important in medical applications [28].

7.3 Colloidal Metals: Nanoplasmonics

Metal-dielectric nanostructures with characteristic surface relief of the order of 10 . . . 100 nm are known to modify spatial distribution of incident electromagnetic field owing to plasmon resonance. Local field enhancement results in enhanced absorption of photons by molecules or nanocrystals adsorbed at the surface. Additionally it results in modification of spontaneous emission and spontaneous scattering rates because of the modification the local density of photon states. A systematic application of the field enhancement in Raman scattering enhancement and in photoluminescence enhancement with respect to molecular probes is followed nowadays by application of the effect with respect to nanocrystals (quantum dots) adsorbed at metal-dielectric nanotextured surfaces. Colloidal nanoparticles of noble metals are used in solutions and in films to perform plasmonic enhancement of Raman scattering and photoluminescence. Basically, field enhancement and density of states effects should manifest themselves in the same manner in photoluminescence and scattering processes [1]. Differences in scattering and luminescence enhancement are due to quenching processes which are crucial for luminescence but not pronounced for scattering. Typically, Raman scattering is always enhanced at metal nanosurface [48] whereas luminescence is typically quenched because of enormous non-radiative decay promoted by metal proximity. When special measures are arranged to control the distance between a fluorescent object and metal surface at distance of 3–10 nm, incident field enhancement overtakes quenching and then luminescence enahncement by one order of the magnitude can be observed for fluorophore with intrinsically high quantum yield [49]. It can be used in immunoassay to detect labeled antibodies [50]. Enhancement up to 50 times for molecular luminescence has been reported [51]. For molecules with intrinsically low quantum yield much higher enhancement is possible since external quenching factors are not so valuable in this case. This property can be used in enhanced bioimaging [52]. Contrary, Raman scattering enhancement of many orders of the magnitude up to single molecule Raman detection is feasible.

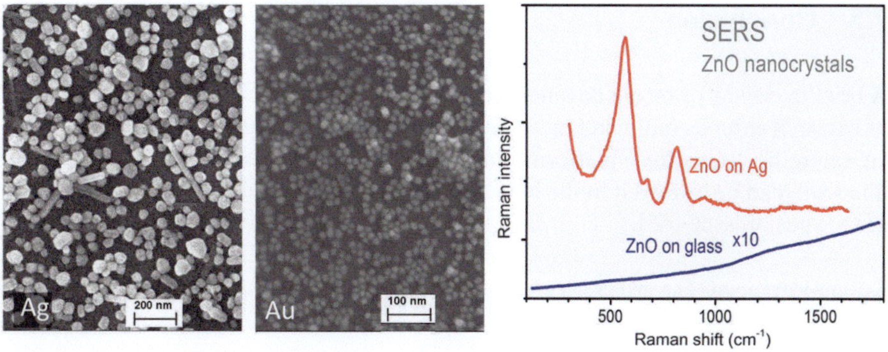

Fig. 7.9 Surface enhanced Raman scattering for ZnO quantum dots using colloidal metal nanoparticles. SEM images of silver (*left*), gold (*right*) nanoparticles and sample Raman spectrum indicating huge enhancement versus reference glass substrate (Reprinted with permission from Ref. [55]. Copyright 2013 ACS)

Plasmonic enhancement of quantum dot luminescence has been for the first time demonstrated in 2002 [53, 54] and since then studies by many groups have resulted in considerable progress.

Recently semiconductor quantum dots were shown to experience plasmonic enhancement of Raman scattering [55] and have been suggested as possible base for Raman labels when conjugated with biomolecules to complement/substitute fluorescent labels [56] (Fig. 7.9).

A multilayer stack consisting of alternating semiconductor-metal nanoparticles and dielectric spacer between layers were found to enhance not only intensity but also polarization of emission [57] which along with modified spontaneous decay provides evidence for possible hyperbolic metamaterial properties [31].

7.4 Colloidal Dielectric Structures: Microcavities and Photonic Crystals

Micrometer-size colloidal dielectric particles represent a microcavity a therefore can be used in multiple applications to control emission and Raman scattering of light for molecules or colloidal quantum dots embedded therein (see, e.g. [58, 59]). Submicrometer-size dielectric spheres are used as building blocks for photonic crystals [60]. A monolayer of monosdisperse colloidal particles can be used as a template to develop regular surfaces by means of photolithography [61].

7.5 Conclusion

A brief overview of recent advances in colloidal nanophotonics shows that this field of research extends more and more towards application issues and seems to become an emerging technology platform. Different aspects that had not been covered in this work can be retrieved in the books [62, 63], in the recent review [64] as well as in the special issue [65].

Acknowledgement The author acknowledges fruitful discussions with H.V. Demir during last years which made strong impact on the issues discussed in this work.

References

1. Gaponenko, S. (2010). *Introduction to nanophotonics*. New York: Cambridge.
2. Barber, D. J., & Freestone, I. C. (1990). An investigation of the origin of the colour of the Lycurgus Cup by analytical transmission electron microscopy. *Archaeometry, 32*(1), 33–45.
3. Colomban, P., Tournié, A., & Ricciardi, P. (2009). Raman spectroscopy of copper nanoparticle-containing glass matrices: Ancient red stained-glass windows. *Journal of Raman Spectroscopy, 40*(12), 1949–1955.
4. Colomban, P. (2009, October). The use of metal nanoparticles to produce yellow, red and iridescent colour, from bronze age to present times in lustre pottery and glass: Solid state chemistry, spectroscopy and nanostructure. *Journal of Nano Research, 8*, 109–132.
5. Faraday, M. (1857). The Bakerian lecture: Experimental relations of gold (and other metals) to light. *Philosophical Transactions of the Royal Society of London, 147*, 145–181.
6. Vargin, V. V. (1934). *Color glasses, their production and properties*. Leningrad: Goskhimtekhizdat (in Russian).
7. Baranov, A. V., & Bobovich, Y. S. (1982). Giant combinational scattering as structural analytical method in material research. *Optics and Spectroscopy, 52*, 385–387.
8. Kreibig, U., & Fragstein, C. V. (1969). The limitation of electron mean free path in small silver particles. *Zeitschrift für Physik, 224*(4), 307–323.
9. Bykov, V. P. (1972). Spontaneous emission in a periodic structure. *Soviet Journal of Experimental and Theoretical Physics, 35*, 269.
10. Gaponenko, S., Demir, H. V., Seassal, C., & Woggon, U. (2016). Colloidal nanophotonics: The emerging technology platform. *Optics Express, 24*(2), A430–A433.
11. Woggon, U. (1997). *Optical properties of semiconductor quantum dots*. Berlin/New York: Springer.
12. Gaponenko, S. V. (1998). *Optical properties of semiconductor nanocrystals*. New York: Cambridge University Press.
13. Borrelli, N. F., Hall, D. W., Holland, H. J., & Smith, D. W. (1987). Quantum confinement effects of semiconducting microcrystallites in glass. *Journal of Applied Physics, 61*, 5399–5409.
14. Shirasaki, Y., Supran, G. J., Bawendi, M. G., & Bulović, V. (2013). Emergence of colloidal quantum-dot light-emitting technologies. *Nature Photonics, 7*(1), 13–23.
15. Guzelturk, B., Martinez, P. L. H., Zhang, Q., Xiong, Q., Sun, H., Sun, X. W., Govorov, A. O., & Demir, H. V. (2014). Excitonics of semiconductor quantum dots and wires for lighting and displays. *Laser & Photonics Reviews, 8*(1), 73–93.
16. Erdem, T., & Demir, H. V. (2011). Semiconductor nanocrystals as rare-earth alternatives. *Nature Photonics, 5*(3), 126–126.

17. Erdem, T., & Demir, H. V. (2013). Color science of nanocrystal quantum dots for lighting and displays. *Nanophotonics, 2*(1), 57–81.
18. Anikeeva, P. O., Halpert, J. E., Bawendi, M. G., & Bulovic, V. (2009). Quantum dot light-emitting devices with electroluminescence tunable over the entire visible spectrum. *Nano Letters, 9*(7), 2532–2536.
19. Bae, W. K., Brovelli, S., & Klimov, V. I. (2013). Spectroscopic insights into the performance of quantum dot light-emitting diodes. *MRS Bulletin, 38*(09), 721–730.
20. Demir, H. V., Nizamoglu, S., Erdem, T., Mutlugun, E., Gaponik, N., & Eychmüller, A. (2011). Quantum dot integrated LEDs using photonic and excitonic color conversion. *Nano Today, 6*(6), 632–647.
21. Klimov, V. I., Ivanov, S. A., Nanda, J., Achermann, M., Bezel, I., McGuire, J. A., & Piryatinski, A. (2007). Single-exciton optical gain in semiconductor nanocrystals. *Nature, 447*(7143), 441–446.
22. Dang, C., Lee, J., Breen, C., Steckel, J. S., Coe-Sullivan, S., & Nurmikko, A. (2012). Red, green and blue lasing enabled by single-exciton gain in colloidal quantum dot films. *Nature Nanotechnology, 7*(5), 335–339.
23. Guzelturk, B., Kelestemur, Y., Olutas, M., Delikanli, S., & Demir, H. V. (2014). Amplified spontaneous emission and lasing in colloidal nanoplatelets. *ACS Nano, 8*(7), 6599–6605.
24. Achtstein, A. W., Prudnikau, A. V., Ermolenko, M. V., Gurinovich, L. I., Gaponenko, S. V., Woggon, U., Baranov, A. V., Leonov, M. Y., Rukhlenko, I. D., Fedorov, A. V., & Artemyev, M. V. (2014). Electroabsorption by 0D, 1D, and 2D nanocrystals: A comparative study of CdSe colloidal quantum dots, nanorods, and nanoplatelets. *ACS Nano, 8*(8), 7678–7686.
25. Talapin, D. V., Lee, J. S., Kovalenko, M. V., & Shevchenko, E. V. (2009). Prospects of colloidal nanocrystals for electronic and optoelectronic applications. *Chemical Reviews, 110*(1), 389–458.
26. Lesnyak, V., Gaponik, N., & Eychmüller, A. (2013). Colloidal semiconductor nanocrystals: The aqueous approach. *Chemical Society Reviews, 42*(7), 2905–2929.
27. Palui, G., Aldeek, F., Wang, W., & Mattoussi, H. (2015). Strategies for interfacing inorganic nanocrystals with biological systems based on polymer-coating. *Chemical Society Reviews, 44*(1), 193–227.
28. Zenkevich, E. I., Gaponenko, S. V., Sagun, E. I., & Borczyskowski, C. V. (2013). Bioconjugates based on semiconductor quantum dots and porphyrin ligands: Properties, exciton relaxation pathways and singlet oxygen generation efficiency for photodynamic therapy applications. *Reviews Nanoscience Nanotechnology, 2*, 184–207.
29. Sargent, E. H. (2012). Colloidal quantum dot solar cells. *Nature Photonics, 6*(3), 133–135.
30. Vogel, N., Weiss, C. K., & Landfester, K. (2012). From soft to hard: The generation of functional and complex colloidal monolayers for nanolithography. *Soft Matter, 8*(15), 4044–4061.
31. Zhukovsky, S. V., Ozel, T., Mutlugun, E., Gaponik, N., Eychmuller, A., Lavrinenko, A. V., Demir, H. V., & Gaponenko, S. V. (2014). Hyperbolic metamaterials based on quantum-dot plasmon-resonator nanocomposites. *Optics Express, 22*(15), 18290–18298.
32. Ekimov, A. I., & Onushchenko, A. A. (1982). Quantum size effect in the optical-spectra of semiconductor micro-crystals. *Soviet Physics Semiconductors-Ussr, 16*(7), 775–778.
33. Efros, A. L., & Efros, A. L. (1982). Interband absorption of light in a semiconductor sphere. *Soviet Physics Semiconductors-Ussr, 16*(7), 772–775.
34. Brus, L. E. (1983). A simple model for the ionization potential, electron affinity, and aqueous redox potentials of small semiconductor crystallites. *The Journal of Chemical Physics, 79*(11), 5566–5571.
35. Bányai, L., & Koch, S. W. (1993). *Semiconductor quantum dots*. Singapore/River Edge: World Scientific.
36. Woggon, U. (1997). *Optical properties of semiconductor quantum dots*. Berlin/New York: Springer.
37. Gaponenko, S. V. (1998). *Optical properties of semiconductor nanocrystals*. Cambridge/New York: Cambridge University Press.

38. Benisty, H., Sotomayor-Torres, C. M., & Weisbuch, C. (1991). Intrinsic mechanism for the poor luminescence properties of quantum-box systems. *Physical Review B, 44*(19), 10945.
39. Gaponenko, S. V., Germanenko, I. N., Petrov, E. P., Stupak, A. P., Bondarenko, V. P., & Dorofeev, A. M. (1994). Time-resolved spectroscopy of visibly emitting porous silicon. *Applied Physics Letters, 64*(1), 85–87.
40. Wang, X., Ren, X., Kahen, K., Hahn, M. A., Rajeswaran, M., Maccagnano-Zacher, S., Silcox, J., Cragg, G. E., Efros, A. L., & Krauss, T. D. (2009). Non-blinking semiconductor nanocrystals. *Nature, 459*(7247), 686–689.
41. Cragg, G. E., & Efros, A. L. (2009). Suppression of auger processes in confined structures. *Nano Letters, 10*(1), 313–317.
42. Dai, X., Zhang, Z., Jin, Y., Niu, Y., Cao, H., Liang, X., Chen, L., Wang, J., & Peng, X. (2014). Solution-processed, high-performance light-emitting diodes based on quantum dots. *Nature, 515*(7525), 96–99.
43. Gurinovich, L. I., Lyutich, A. A., Stupak, A. P., Artem'ev, M. V., & Gaponenko, S. V. (2010). Effect of an electric field on photoluminescence of cadmium selenide nanocrystals. *Journal of Applied Spectroscopy, 77*(1), 120–125.
44. Yeltik, A., Delikanli, S., Olutas, M., Kelestemur, Y., Guzelturk, B., & Demir, H. V. (2015). Experimental determination of the absorption cross-section and molar extinction coefficient of colloidal CdSe nanoplatelets. *The Journal of Physical Chemistry C, 119*(47), 26768–26775.
45. Achtstein, A. W., Schliwa, A., Prudnikau, A., Hardzei, M., Artemyev, M. V., Thomsen, C., & Woggon, U. (2012). Electronic structure and exciton–phonon interaction in two-dimensional colloidal CdSe nanosheets. *Nano Letters, 12*(6), 3151–3157.
46. Bruchez, M., Moronne, M., Gin, P., Weiss, S., & Alivisatos, A. P. (1998). Semiconductor nanocrystals as fluorescent biological labels. *Science, 281*(5385), 2013–2016.
47. Chan, W. C., & Nie, S. (1998). Quantum dot bioconjugates for ultrasensitive nonisotopic detection. *Science, 281*(5385), 2016–2018.
48. Kneipp, K., Moskovits, M., & Kneipp, H. (Eds.). (2006). *Surface-enhanced Raman scattering: Physics and applications* (Vol. 103). Berlin/New York: Springer.
49. Guzatov, D. V., Vaschenko, S. V., Stankevich, V. V., Lunevich, A. Y., Glukhov, Y. F., & Gaponenko, S. V. (2012). Plasmonic enhancement of molecular fluorescence near silver nanoparticles: Theory, modeling, and experiment. *Journal of Physical Chemistry C, 116*(19), 10723–10733.
50. Vaschenko, S. V., Ramanenka, A. A., Guzatov, D. V., Stankevich, V. V., Lunevich, A. Y., Glukhov, Y. F., Sveklo, I. F., & Gaponenko, S. V. (2012). Plasmon-enhanced fluorescence of labeled biomolecules on top of a silver sol-gel film. *Journal of Nanophotonics, 6*(1), 061710.
51. Tam, F., Goodrich, G. P., Johnson, B. R., & Halas, N. J. (2007). Plasmonic enhancement of molecular fluorescence. *Nano Letters, 7*(2), 496–501.
52. Geddes, C. D., Cao, H., Gryczynski, I., Gryczynski, Z., & Lakowicz, J. R. (2003). Metal-enhanced fluorescence (MEF) due to silver colloid on a planar surface: Potential applications of indocyanine green to in vivo imaging. *Journal of Physical Chemistry A, 107*(18), 3443–3449.
53. Kulakovich, O., Strekal, N., Yaroshevich, A., Maskevich, S., Gaponenko, S., Nabiev, I., Woggon, U., & Artemyev, M. (2002). Enhanced luminescence of CdSe quantum dots on gold colloids. *Nano Letters, 2*(12), 1449–1452.
54. Shimizu, K. T., Woo, W. K., Fisher, B. R., Eisler, H. J., & Bawendi, M. G. (2002). Surface-enhanced emission from single semiconductor nanocrystals. *Physical Review Letters, 89*(11), 117401.
55. Milekhin, A. G., Yeryukov, N. A., Sveshnikova, L. L., Duda, T. A., Zenkevich, E. I., Kosolobov, S. S., Latyshev, A. V., Himcinski, C., Surovtsev, N. V., Adichtchev, S. V., Feng, Z. C., Wu, C. C., Wuu, D. S., & Zahn, D. R. T. (2011). Surface enhanced Raman scattering of light by ZnO nanostructures. *Journal of Experimental and Theoretical Physics, 113*, 983–991.

56. Rumyantseva, A., Kostcheev, S., Adam, P. M., Gaponenko, S. V., Vaschenko, S. V., Kulakovich, O. S., Ramanenka, A. A., Guzatov, D. V., Korbutyak, D., Dzhagan, V., & Stroyuk, A. (2013). Nonresonant surface-enhanced Raman scattering of ZnO quantum dots with Au and Ag nanoparticles. *ACS Nano, 7*(4), 3420–3426.
57. Ozel, T., Nizamoglu, S., Sefunc, M. A., Samarskaya, O., Ozel, I. O., Mutlugun, E., Lesnyak, V., Gaponik, N., Eychmuller, A., Gaponenko, S. V., & Demir, H. V. (2011). Anisotropic emission from multilayered plasmon resonator nanocomposites of isotropic semiconductor quantum dots. *ACS Nano, 5*, 1328–1334.
58. Rakovich, Y. P., Donegan, J. F., Gerlach, M., Bradley, A. L., Connolly, T. M., Boland, J. J., Gaponik, N., & Rogach, A. (2004). Fine structure of coupled optical modes in photonic molecules. *Physical Review A, 70*, 051801(R).
59. Möller, B., Woggon, U., & Artemyev, M. V. (2006). Photons in coupled microsphere resonators. *Journal of Optics A: Pure and Applied Optics, 8*, 113–121.
60. Bogomolov, V. N., Gaponenko, S. V., Germanenko, I. N., Kapitonov, A. M., Petrov, E. P., Gaponenko, N. V., Prokofiev, A. V., Ponyavina, A. N., Silvanovich, N. I., & Samoilovich, S. M. (1997). Photonic band gap phenomenon and optical properties of artificial opals. *Physical Review E, 55*, 7619–7626.
61. Xiong, Z., Wei, T., Zhang, Y., Wang, J., & Li, J. (2016). Multiple-exposure colloidal lithography for enhancing light output of GaN-based light-emitting diodes by patterning Ni/Au electrodes. *Optics Express, 24*(2), A44–A51.
62. Rogach, A. L. (2008). *Semiconductor nanocrystal quantum dots*. Wien/New York: Springer.
63. Klimov, V. I. (Ed.). (2010). *Nanocrystal quantum dots*. Boca Raton: CRC Press.
64. Kim, J. Y., Voznyy, O., Zhitomirsky, D., & Sargent, E. H. (2013). 25th anniversary article: Colloidal quantum dot materials and devices: A quarter-century of advances. *Advanced Materials, 25*(36), 4986–5010.
65. Colloidal Nanophotonics. *Optics Express*, Special Issue, January 2016.

Chapter 8
Ultrafast Nano-Biophotonics

Jean-Pierre Wolf

Abstract In the quest for the next generation of imaging bio-markers, successful probes have to prove to be non toxic, bright, stable against long term excitation, and able to generate a sharp contrast against background fluorescence. Harmonics-generating Nanocrystals (HN) appeared recently as a novel labelling method with unprecedented wavelength flexibility, enabled by the non-resonant nature of the harmonics generation process. In particular, imaging using frequency doubling nanocrystals (i.e. nanodoublers), such as iron iodate, potassium niobate, barium ferrite (BFO) and KTP, has been demonstrated with laser sources in the near-infrared (800 nm) and infrared (1.55 μm) regions. The latter allows deeper penetration depth in tissues, thus especially promising for *in vivo* applications. The phase-coherent optical response of HN can also be exploited to fully characterize the excitation laser pulse in the focal spot of a high-NA objective with nanometric resolution.

HN have been used for the first time to monitor the evolution and differentiation of embryonic stem cells (ESC) by second harmonic microscopy imaging. The potential of this approach was made evident for tissue-regeneration studies, by capturing high-speed movies of ESC-derived cardiomyocytes autonomously beating within a cluster. HNs were also applied to cancer research: BFO nanoparticles were functionalized to selectively target human lung cancer cells and to serve both for cancer diagnostics and therapy (commonly referred to as "theranostics").

Although significant progress has recently been achieved in the quest for better bio-markers, such as the above mentioned HN, the ultimate goal would be the use of *no* external probe for achieving "label-free" imaging. In the last decade coherent control with optimally tailored ultrashort laser pulses revolutionized molecular spectroscopy. Here, we show that playing with quantum interference between reaction pathways, it was possible to discriminate nearly identical cellular vitamins and identify selectively proteins (antibodies) in blood serum. Immunoglobulins G and M, as well as albumin, could be directly be quantified in blood serum label-free, yielding the first "Quantum Control Based Bio-assays".

J.-P. Wolf (✉)
GAP-Biophotonics, University of Geneva, 22, chemin de Pinchat, 1211 Geneva 4, Switzerland
e-mail: jean-pierre.wolf@unige.ch

© Springer Science+Business Media Dordrecht 2017
B. Di Bartolo et al. (eds.), *Nano-Optics: Principles Enabling Basic Research and Applications*, NATO Science for Peace and Security Series B: Physics and Biophysics, DOI 10.1007/978-94-024-0850-8_8

8.1 Introduction

Nano-engineering plays an increasing role in bio-imaging and medical applications. For example, nanoparticles are now used as cargos for carrying on their surface molecular target identifiers like antibodies as well as multimodal labels, and can be simultaneously activated by light for locally releasing drugs. On the other hand plasmonic nanostructured substrates are used for identifying individual biomolecules by surface enhanced Raman spectroscopy and opto-transfecting cells with genetic material without viral vectors.

A particularly efficient way of making use of nanoparticles or nanostructured substrates is to illuminate them with ultrafast lasers, which provides high intensities at low energy and induce non-linear interactions. For instance, in the quest for the next generation of imaging bio-markers, successful probes have to prove to be non-toxic, bright, stable against long term excitation, and able to generate a sharp contrast against background fluorescence. In all these respects, Harmonics-generating Nanocrystals (HN) are receiving an increasing attention as they also open a series of alternative detection possibilities simply not accessible with the present generation of fluorescent dyes and quantum dots. The first part of this chapter will be dedicated to this novel labelling method [1–3] with unprecedented wavelength flexibility, enabled by the non-resonant nature of the harmonics generation process. In particular, imaging using frequency doubling nanocrystals (i.e. nanodoublers), such as iron iodate, potassium niobate, bismuth ferrite (BFO) and KTP, has been demonstrated with laser sources in the near-infrared (800 nm) and infrared (1.55 μm) regions. The latter allows deeper penetration depth in tissues, thus especially promising for in vivo applications [4]. The phase-coherent optical response of HN can also be exploited to fully characterize the excitation laser pulse in the focal spot of a high-NA objective with nanometric resolution. A proof-of-principle "nano-FROG" experiment [5] was performed and set the ground for further phase-sensitive self-referenced applications.

The second part of this chapter concentrates on biomedical applications of ultrafast nano-bio photonics. In particular, HN have been used for the first time to monitor the evolution and differentiation of embryonic stem cells (ESC) by second harmonic microscopy imaging [6]. The potential of this approach was made evident for tissue-regeneration studies, by capturing high-speed movies of ESC-derived cardiomyocytes autonomously beating within a cluster. Time-resolved data could be analyzed to retrieve 3D information of the contraction pattern at the cellular level. HNs were also applied to cancer research: BFO nanoparticles were functionalized to selectively target human lung cancer cells and to serve both for cancer diagnostics and therapy (commonly referred to as "theranostics") [7]. More precisely, it was shown that shining harmless IR light on the particles led to the detection of the cancer cells (via visible SHG emission), while illuminating the same nanocrystals with 540 nm (SHG at 270 nm) led to cell destruction by local DNA denaturation. Collateral damage was shown to be impressively low due to the nanometric range of action of the SHG irradiation.

Although significant progress has recently been achieved in the quest for better bio-markers, such as the above mentioned HN, the ultimate goal would be the use of no external probe for achieving "label-free" imaging. In the last two decades coherent control with optimally tailored ultrashort laser pulses revolutionized molecular spectroscopy. It allowed not only observing with unprecedented detail the atomic motion within molecules but also controlling it. In the third part of the lecture, we will show how coherent control might significantly contribute towards the goal of label-free imaging using only endogenous molecular processes.

After an introduction of the basic processes and techniques used in coherent control, such as quantum pathways interference, open and closed loop approaches, pulse shaping techniques, etc., we will concentrate on recent achievements in the field of coherent manipulation of biological objects. Particular interest will be dedicated to the identification of biological molecules that exhibit almost identical linear spectra and the wealth of related applications, such as specific serum proteins and pathogen detection. We show that by playing with quantum interference between reaction pathways, it was possible to discriminate nearly identical cellular vitamins [8, 9] and identify selectively proteins (antibodies) in blood serum. Immunoglobulins G and M, as well as albumin, could be directly be quantified in blood serum label-free, yielding the first "Quantum Control Based Bio-assays" [10].

8.2 Harmonic Nanodoublers as Inherent Nonlinear Probes for Multi-Photon Imaging

HNs, based on inorganic noncentrosymmetric crystals, exhibit a series of attractive properties as compared to fluorescent dyes, including: (i) complete absence of bleaching and blinking, (ii) narrow emission bands, (iii) excitation-wavelength tunability, (iv) orientation retrieval capability, and (v) coherent optical response.

The noncentrosymmetric crystal structure is at the origin of the second-order nonlinear properties of this family of particles. So far, a series of nanomaterials including: $LiNbO_3$, $KNbO_3$, $BaTiO_3$, $Fe(IO_3)_3$, ZnO, KTP have been studied [2]. From their common lack of crystal inversion symmetry, it follows that the second order susceptibility tensor, $\chi^{(2)}$, presents non-vanishing elements, sustaining a nonlinear polarization at frequency 2ω when excited by an electromagnetic field E^{ω} at frequency ω

$$P_i^{(2\omega)} = \varepsilon_0 \sum\nolimits_{jk} \chi_{ijk}^{(2)} E_j^{\omega} E_k^{\omega}.$$

The different materials tested belong to different symmetry point groups with characteristic $\chi^{(2)}$ tensor elements, thus they yield different responses in polarization-resolved nonlinear microscopy. Table 8.1 summarizes the measured values of hyperpolarisability β (and normalized per nanoparticle volume unit, d_{np}) for different HNs [2, 11].

Table 8.1 Hyperpolarizability, normalized hyperpolarizability, and normalized hyperpolarizability from bulk materials found in the literature of $KNbO_3$, $LiNbO_3$, $BaTiO_3$, KTP, and ZnO nanocrystals

Nanocrystal	$<\beta_{np}>$ +/− $\Delta\beta$ ($\times 10^{-24}$ esu)	$<d_{np}>$ +/− Δd, pm/V	$<d_{lit}>$, pm/V
$KNbO_3$	7.2 +/− 2.4	3.4 +/− 1.1	15.3
$LiNbO_3$	21 +/− 7.0	4.8 +/− 1.6	17.3
$BaTiO_3$	11.0 +/− 3.5	4.6 +/− 0.7	14.1
KTP	25.8 +/− 6.0	1.4 +/− 0.3	7.6
ZnO	9.6 +/− 2.8	1.9 +/− 0.6	2.8

Being function of individual nanoparticles orientation, these responses can be exploited for optical orientation retrieval and electric field probing at microscopic scale (see Refs. [1] and [12] for details).

Because harmonic generation is a nonresonant process, no absorption occurs. As a consequence, no energy is converted into heat, preventing sample photo-degradation. Moreover signal bleaching and blinking observed with fluorescent dyes and quantum dots are completely absent.

Interestingly, given that the coherence length of the nonlinear excitation process is longer than the typical dimensions of the nanoparticles (<100 nm), no phase-matching constraints apply and any wavelength can be used for excitation [1, 13]. We have demonstrated such broad excitation tunability under different experimental conditions, with laser pulses ranging from vis-UV (400 nm) to IR (1.5 um). As an example, Fig. 8.1(D) shows the responses of Polyethylene Glycol (PEG)-coated $LiNbO_3$ HNs attached to breast cells membranes stained with FM 1-43FX membrane dye. The response of the nanoparticles can be excited at any wavelength within the Ti:Sapphire spectrum (700–1100 nm). The optimal contrast obtained with respect to this standard fluorescent dye, widely used for biological assays, testifies the appeal of this approach for *in vitro* cell imaging. Longer wavelength excitation, due to the reduced tissue scattering, is indeed very promising for extending the penetration depth of imaging techniques and decreasing sample photo-damage. In Fig. 8.1(C), as a demonstration, the second and third harmonic spectra generated by a single HN excited by a 1.55 μm laser show the complete up-conversion of the original laser bandwidth (dotted line).

Finally, Fig. 8.1(E), shows the most clearly, the coherent nature of the emission of HNs as opposed to fluorescence. Two neighbouring KTP nanoparticles lying on a waveguide are excited at the second harmonic (SH) by the evanescent mode of a laser pulse propagating through the guide. The two emission patterns interfere coherently at the CCD detector [12]. The accompanying image is a numerical simulation of this observation. So far, what we believe is the immense imaging potential of this coherent response has not been fully exploited, just a series of demonstration measurements have been published on especially conceived set-ups: heterodyne detection [14] of nanoparticles for sensitivity increase, harmonic holography [15], digital phase conjugation [16, 17], and our work on spatially resolved pulse phase-retrieval for microscopy (nano-FROG) [5].

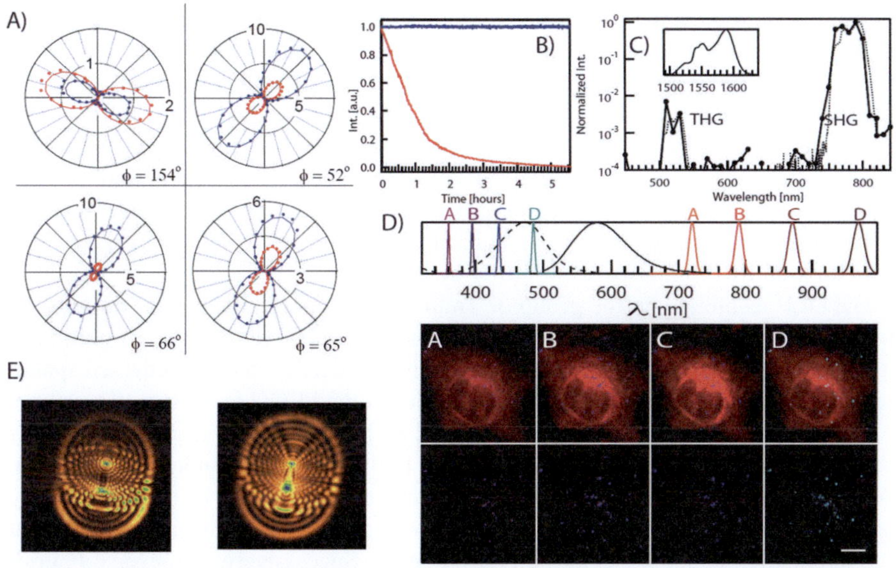

Fig. 8.1 (**A**) The polarized emissions of individual HNs bear information about their local three-dimensional orientation, in the plots Φ corresponds to the HN in-plane angle [1]. (**B**) The excellent photostability upon continuous laser irradiation for several hours of HNs is compared with the decreasing signal of a fluorescent dye imaged simultaneously. (**C**) The absence of phase-matching constraints allows complete doubling of any input pulse bandwidth. A single HN up-converts the spectrum of a Erbium ion laser at 1550 nm (*inset*) generating simultaneously SH and TH. The experimental spectral bandwidths are compared with the calculated ones (*dotted lines*). (**D**) A dye-stained cells treated with HNs is imaged at different wavelengths within the range of a Ti:Sapphire laser. *Upper plots*: all detection channels overlaid, *lower plots*: single detection channels A, B, C, and D. The frequency-doubled signal from the HNs appear each time uniquely in a narrow spectral detection region. (**E**) Left image. Coherent interference pattern between the emissions of two nanoparticles adjacently lying on a waveguide nonlinearly excited by an evanescent wave (*Right* image. Numerical simulations (in collaboration with J. Enderlein, University of Goettingen))

The motivations of the nano-FROG procedure reside on the demand for a versatile and reliable pulse phase-sensitive characterization technique, more and more critical as a number of nonlinear approaches have recently been bridged from spectroscopy to imaging. Just to mention a few examples, pulse-shaping has been applied to temporally recompress femtosecond pulses at the measuring site, to improve spatial resolution, and to increase fluorescence excitation and harmonic generation yield [18]. Intra-pulse micro-CARS [19, 20] and coherent control microscopy have demonstrated successful for obtaining higher selectivity in the imaging of structures in very diverse samples [21]. Furthermore, interest has been focused on the effects of pulse polarization, and the related capability to achieve control of optical near field [22]. To fully exploit the potential of these techniques, equally developed pulse diagnostics capabilities are necessary. In particular, considering that femtosecond pulses are prone to undergo modifications during their propagation (temporal stretching by group velocity dispersion, higher order distortions in the spectral-phase induced by microscope objectives [23], spectral amplitudes modulation by sample scattering [24] and absorption) an *in situ* measurement of the excitation pulse is customary for any advanced application. Since the early work of Muller et al. [25], several techniques, largely based on Frequency Resolved Optical Gating (FROG), have been proposed, but until present none of them incorporates the spatial resolution necessary to resolve features inside the focal region of a high numerical aperture (NA) objective [26]. The starch-based FROG approach proposed by Amat et al. [27] and the scanning SEA TADPOLE technique by the group of Trebino [28] represented a clear progress in this direction, but unfortunately both techniques are limited to a spatial resolution of a few micrometers. We have proposed a technique to achieve unprecedented spatial resolution, without renouncing at the tenability of the FROG technique.

The approach, which can be implemented on a conventional or an inverted laser scanning microscope, is based on the use of individual HNs as nonlinear medium. An exemplary FROG trace, like the one presented in Fig. 8.2(a) can be obtained by spectrally resolving the autocorrelation trace of a single HN and then by filtering out the modulation components in the Fourier space. Successively, the inverse Fourier transform can be applied to return back to the time/wavelength space. The so treated data can successively be fed into a commercial FROG-inversion program [28] to determine the actual pulse characteristics. The retrieved FROG trace is shown in Fig. 8.2(b) (lower plots series), along with pulse electric field intensity temporal profile and spectrum reported in (c) and (d). The corresponding phase-functions are also shown. The latter indicate the presence of a residual quadratic chirp on the pulse due to group velocity dispersion accumulated during propagation through the microscope optics, which is not fully corrected by the grating compressor.

8.3 3D Movie of Embryonic Stem Cell Differentiation into Cardiomyocites

A very interesting and illustrative application of the unique properties of HNs is the real time observation of the differentiation process in embryonic stem cells (ESM) [6]. More precisely, in the initial embroid bodies (EB) sample, cells underwent differentiation into cardiac cells, easily recognizable as clusters displaying rhythmic autonomous contractions. A movie of the rhythmic cluster movements could be recorded over several minutes at high speed over an area of $640 \times 640\ \mu m^2$ detecting SHG radiation from isolated HNs. It bears 3D information at sub-micrometric resolution about the contraction pattern of the cardiomyocytes. A single frame is reported in Fig. 8.3(A). The arrows indicate the direction and module of the cells'inplane movements. The absolute distances traveled by several HNs during the rhythmic EB contractions are plotted as a function of time in Fig. 8.3(B), which indicates a clear periodic pattern for all particles. To retrieve the out-of-plane

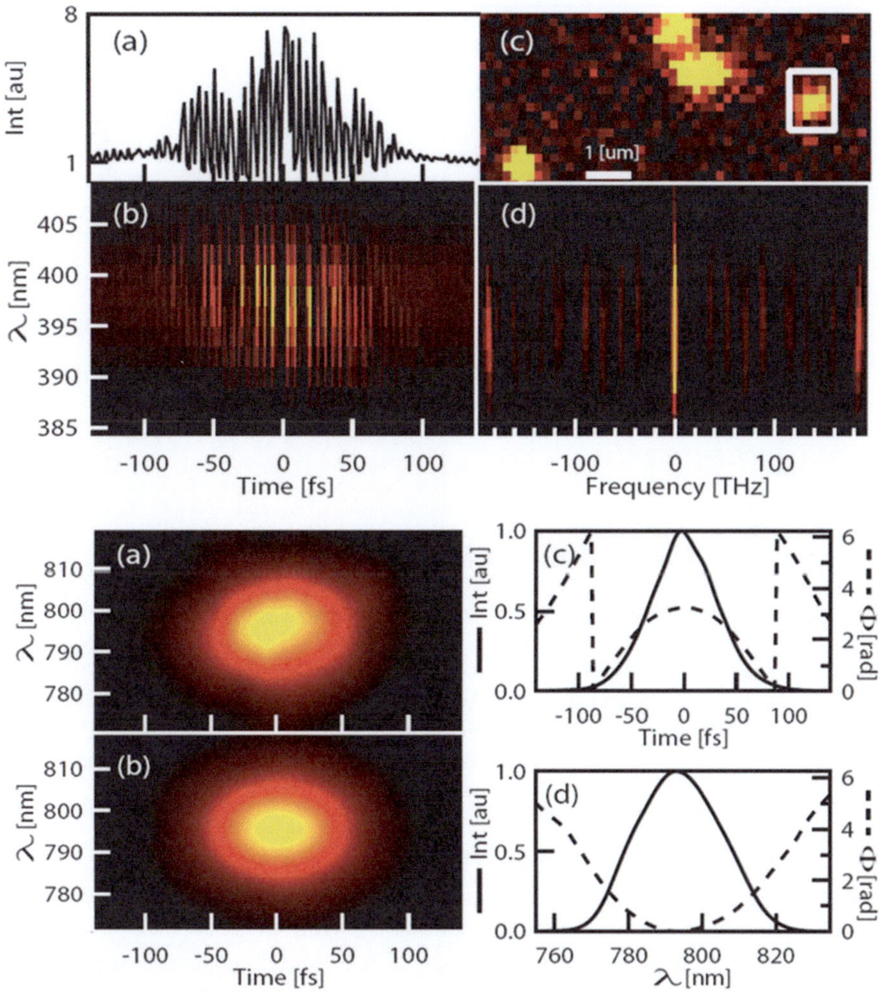

Fig. 8.2 *Upper plots*: (**a**) Interferometric autocorrelation trace. (**b**) FROG trace. (**c**) Image of the nanoparticles dispersed on a microscopic glass slide. The particle highlighted by the 1 mm square was used as nonlinear medium for acquiring the autocorrelation and the FROG trace. (**d**) Result of the Fourier transform along the time axis of the FROG trace in (**b**). *Lower plots*: (**a**) Experimental FROG trace. (**b**) Retrieved FROG trace. Electric field intensity (*solid line*) and phase (*dashed line*) as a function of time (**c**) and of wavelength (**d**)

motion, the point spread function of the microscope was determined by performing a 3D nonlinear scan of a single isolated HN on a substrate. This calibration was then applied to convert the SHG intensity modulations observed in the movie of cardiac contractions into absolute displacements along the z axis. The results of this procedure are reported in Fig. 8.3(C). The Fourier transforms of the plots in panels B and C indicate that the observed movements have a constant frequency of

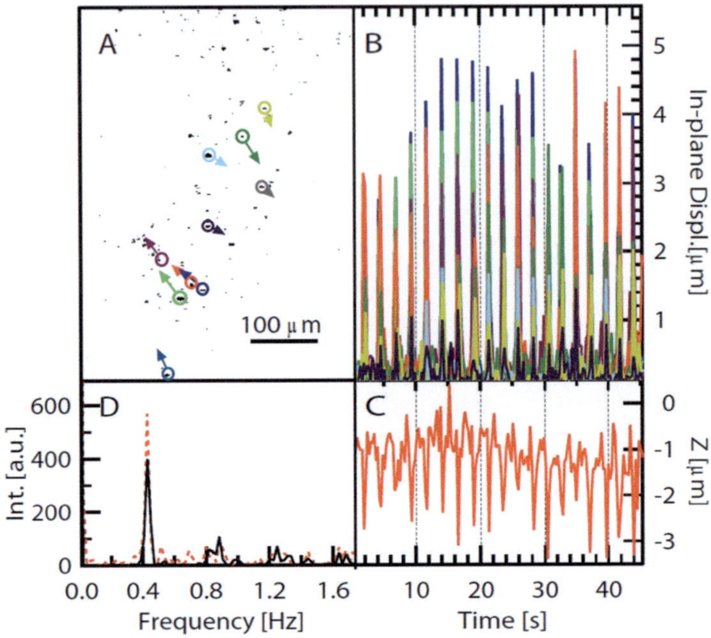

Fig. 8.3 (**A**) Individual frame of the movie of the myocytes contraction. *Arrows* indicate the directions and in-plane displacements of different HNs. (**B**) Displacements as a function of time of the HNs highlighted in A. (**C**) Same as (**B**) for movement along the optical axis. (**D**) Fourier transform of traces in (**B**) and (**C**). The beating frequency corresponds to 0.43 Hz (Ref. [6], with permission)

0.43 Hz. Of note, all contractions observed bear the same phase, within the temporal resolution of our measurement. In-plane movements range between 1 and 5 μm, out-of-plane displacements are slightly shorter (<3 μm).

Notice that the approach described above can be readily applied to monitor the evolution of various stem cell types in a transplanted tissue or their integration in a three dimensional supporting structure [29].

8.4 Nonlinear Nanocrystals Based Theranostics

The unique wavelength flexibility of HN allowed the emergence of a novel theranostics approach. More precisely, shining IR laser light on the HN generates visible radiation, that can be used for cancer diagnostics, and the very same HN leads to cell photodestruction (therapy) while illuminated with visible light (and radiating then in the UV) [7]. In order to demonstrate this innovative concept, we incubated two human lung cancer cell lines (A549 and HTB-182) with BFO

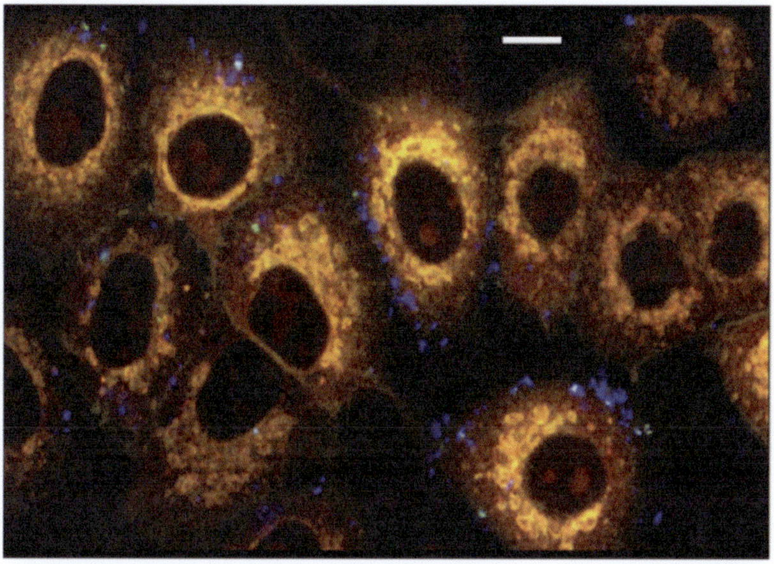

Fig. 8.4 Lung cancer diagnostics using SH emission from BFO nanocrystals labelling a Human lung cancer cell lines (From Ref. [7]) *Scale bar*: 10 um

nanocrystals and then washed the samples. As shown in Fig. 8.4, the cells were clearly labelled and identified by SH generation of a NIR illuminating laser.

Notice that the uptake, biocompability and non-toxicity of BFO nanoparticles were obviously carefully investigated and confirmed beforehand [30]. In order to demonstrate the "therapy" capability of the approach, the incoming laser wavelength was tuned to 540 nm (which is also harmless), so that the HN locally emitted 270 nm by SHG. DNA absorption is particularly efficient in the deep UV (DUV, < 300 nm), as all DNA bases possess absorption bands peaking around 260 nm with negligible intensity from 310 nm. Irradiation at this wavelength results in double-strand break and evokes a complex network of molecular responses, eventually resulting in DNA repair and/or cell apoptosis. In our work, these two different biological phenomena were monitored performing two ad hoc bioassays: gH2AX and cPARP, respectively.

The biological effects of the "therapeutic" laser irradiation at 540 nm are reported in the immunohistochemistry (IHC) images of Fig. 8.5. The left column is associated with the human malignant cell line HTB-182. Almost identical results were found for the other cancer cell line A549. The control sample (1rst row) shows no expression for both reporters, confirming that they are not present under physio-logical conditions, while the clear effect (positive cells in brown) visible in rows two and three for treated (exposed to HNs and laser) cells indicates that a strong interaction upon irradiation takes place, showing the DNA-repairing enzyme gH2AX expression well localized within the nuclei and the cPARP reporter of cell apoptosis in the cytoplasm of the damaged cells. The laser focal spot (empty white squares in Fig. 8.5 right) is superimposed to a spatially resolved pattern of 80×130

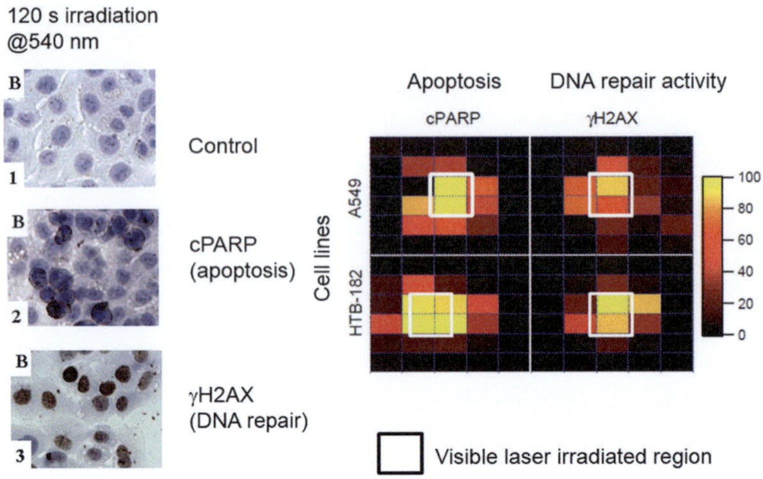

Fig. 8.5 *Left*: Immuno histochemistry. Human lung HTB-182 cancer cell line untreated (*first row*) or exposed to BFO nanoparticles and irradiated at 540 nm for 120 s (*rows 2 and 3*) demonstrating DNA photodamage
Right: spatial localization of the expression of cPARP and gH2AX reporters. Expression of cPARP (*column 1*) and g2AX (*column 2*) is expressed in false colors according to the color bar (From Ref. [7])

um rectangles and the % of positive cells to cPARP and gH2AX for both cell lines are indicated in false colors. One can appreciate how the biological effect of visible irradiation strongly co-localizes with the 170 um laser spot (>80 % positive cells) and rapidly decreases outside the focal region to negligible values.

Unlike former photodynamic approaches based on porphyrins or other organic sensitizers, the present method relies on femtosecond illumination, because the process involves nonlinear frequency conversion. This provides a significant advantage since exposition to incoherent light, like solar radiation, is unable to trigger photodamage. As a consequence, potential future patients would not be required to remain in the dark after sensitization. Moreover, direct UV photodamage to DNA in a very confined region around the nanoparticle may represent a safer way than the photo-production of oxidative radicals, which are prone to diffuse in neighboring healthy cells.

8.5 Labelling Tissues and Higher Order Harmonics

Higher order harmonics obviously require higher local intensities. However, the use of ultrashort laser pulses provides sufficiently high peak intensities to observe them, even at relatively low incoming energies (nJ), as shown for TH generation in Fig. 8.1, panel C. Odd harmonics, like the 3rd, can furthermore be generated

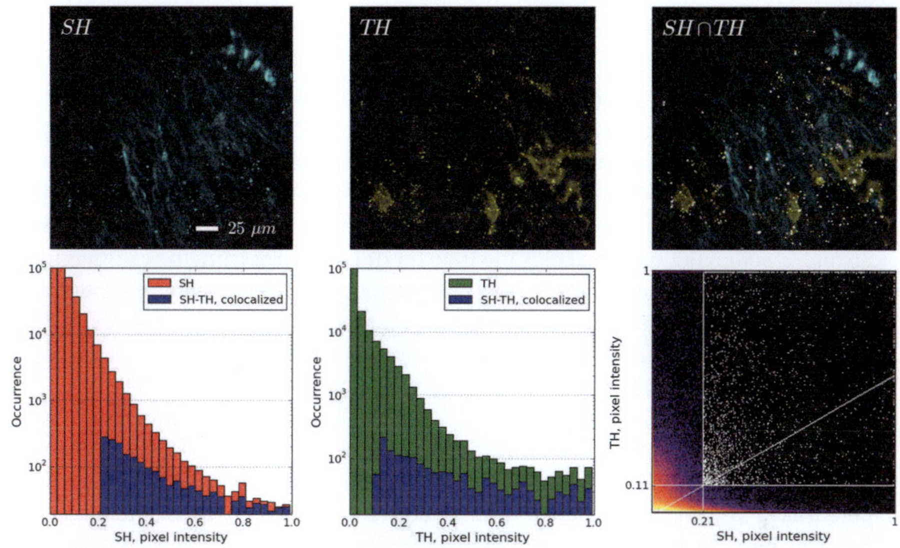

Fig. 8.6 Multi-harmonics images of breast tumor tissue labeled with BFO HNs. *First row*: SH (*cyan*), TH (*yellow*), and SH∩TH merged image with co-ocalized pixels in white. *Second row*: Semilog SH and TH intensity histograms. *Blue bars* correspond to co-localized events. *Scatter plot*: TH vs SH (From Ref. [35])

in the bulk of centrosymmetric material and are usually enhanced in some favored directions by nonlinear Mie scattering [31, 32]. The advantage of multi-harmonics imaging (SH and TH, for instance) resides in the fact that the contrast can be widely enhanced as compared to the tissue background. Collagen, with its oriented, elongated, structures, is indeed well known to produce native SH [33] while lipids is well observed as source of TH [34]. The contrast for SH and TH imaging of nanoparticles separately is therefore reduced by these endogenous emission. But co-localization in the multiplexed images uniquely identifies the positions of the nano-labels. This approach was recently demonstrated using functionalized BFO nanoparticles targeting breast cancer cells in tissues [35]. The excitation was assured by an infrared 1250 nm femtosecond laser. At the intensity applied to the sample (300–600 GW/cm^2) the SH/TH ratio emitted by the BFO nanocrystals ranges from 40 to 100.

As illustrated in Fig. 8.6, the nanocrystals localization is significantly perturbed in some regions of the tissues by endogenous emission for both SH and TH images. However, as SH and TH are not emitted by the same cell structures, the correlation between the two images (SH∩TH) clearly reveals the positions of the nanocrystals only. This approach therefore greatly reduces the risks of false positives (or false negatives if the endogenous emission is subtracted) while performing diagnostics in extended tissues.

8.6 Basics of Coherent Control

In the last decades, coherent control induced a revolution in photochemistry. Not only could the atomic motion be *observed* within molecules by snapshots of a few femtoseconds duration, it could also be *controlled*. The basic idea of this revolution was the use of quantum interference between different states that are in coherent superposition so that a target state is reached with maximum probability. A key element for achieving this control is the transfer of coherence from the exciting light field to the atomic or molecular system via the electric dipole operator. A tailored coherent laser pulse is used to excite a molecule in a superposition of states, which drives the molecule along a specific path to obtain a targeted result.

A famous example of coherent control is the Tannor, Koslov and Rice [36, 37] "pump-dump" method, in which many vibrational states of a molecule ABC are coherently excited for generating a vibrational wavepacket. This wavepacket explores, as a function of time, a large fraction of the electronically excited hypersurface. By shining a second laser at convenient times t_1 or t_2, the molecule is brought back to its ground state onto different dissociation paths, resulting in different species: AB + C or A + BC. The system can thus be driven to a specific daughter fragment by using the quantum control of the photodissociation process.

In 1992, H. Rabitz at Princeton University introduced [38, 39] the concept of "optimal control", in which a feedback loop optimizes the laser pulse characteristics in order to most efficiently reach the desired target. For this, a large number of parameters (corresponding to the amplitude and phase of each spectral component within the exciting laser spectrum) have to be controlled. This "pulse shaping" technique is usually performed by introducing a liquid crystal array in the Fourier plane between two gratings (4*f* arrangement) [40]. Excellent results in terms of efficiency have been obtained using genetic-type optimization algorithms [41, 42].

Besides the above mentioned "pump-dump" scheme, other successful control methods have been extensively applied, including stimulated Raman scattering (STIRAP), Four Wave Mixing like CARS (Coherent AntiStokes Raman Scattering), multiphoton absorption and multiphoton ionization. The controlled targets can be, for instance, specific fragments, specific isomers or specific isotopes, but also the enhancement or reduction of the fluorescence of a specific molecule (by driving it preferentially into other relaxation pathways). In this respect, a pioneering work has been performed by the group of G. Gerber [43, 44], in which the capability of distinguishing two different dyes using 2-photon absorption control was demonstrated, although the dyes had similar linear absorption and fluorescence spectra.

An attractive application is the ability of discriminating different biological systems that are usually undistinguishable with standard linear spectroscopic approaches. This is the approach that is followed in this chapter in order to reach label-free imaging of biological structures or the detection and identification of harmful bacteria.

8.7 Use of Optimal Control for Discriminating Almost Identical Bio-Systems

The identification or discrimination of molecules and quantum systems that are nearly identical (proteins, bacteria, etc..) is normally an exceedingly difficult task. The conventional means of addressing this common need is through various *static* spectroscopic techniques, which can be especially difficult in complex remote environments. However, recent theoretical work showed that two or more quantum systems, even if they differ only infinitesimally, may be drawn apart to have distinct signatures through tailored control of their *dynamics*. Ultrafast broad bandwidth lasers with adaptive pulse shaping provide a means to implement this theoretical concept through controlled quantum optimal dynamic discrimination (ODD) [45]. In collaboration with the group of H. Rabitz at Princeton University, we demonstrated the capabilities of ODD by discriminating between two almost identical molecules of riboflavin (RBF) and flavin mononucleotide (FMN) in water solution, which have virtually identical linear absorption and fluorescence spectra (see Fig. 8.7a and b) [8, 9]. The successful implementation of ODD opens up numerous applications including in fluorescence microscopy, protein identification, as well as possible remote discrimination of different bacteria. A key component of ODD is operation with shaped laser pulses, which can nonlinearly interact with the molecules to fully exploit their dynamical capabilities and create discriminating signatures. The similar optical spectra of RBF and FMN arise from the common isoalloxazine ring with the only distinguishing feature being changes at the end of the side chain tail (Fig. 8.7a).

The present application of ODD utilizes a control field consisting of a shaped ultraviolet (UV) portion at 400 nm and a near infrared (IR) component at 800 nm, which is Fourier transform limited. The unshaped IR pulse follows the shaped UV pulse by a time delay τ, and the structure of the shaped UV pulse is deduced optimally under adaptive control to achieve ODD in the fluorescence signals from RBF and FMN. The primary operation of ODD relies on the shaped UV pulse coherently transferring ground state population into the flavin S_1 or S_2 excited states with the IR component likely creating further excitation of the molecules to higher lying S_n states. The IR field disrupts the vibronic excitation in S_1 or S_2 created by the UV pulse in order to make a discriminating difference in the fluorescence spectra of RBF and FMN from S_1 recorded in the window 525–535 nm after a relaxation period of $\sim$ 5 ns. Fluorescence is therefore depleted, and this fluorescence depletion value is used as measure of the molecular response to the field. The overall mechanism also may take advantage of any beneficial intra- and inter-molecular dynamical processes, and the ODD discrimination can draw on suitable coherent and incoherent aspects of the molecular dynamics.

In the first series of experiments the RBF and FMN solutions are in separate identical flow cells exposed to the same trial shaped UV pulse and time delayed unshaped IR pulse. An acousto-optic modulator is used to shape the UV pulses.

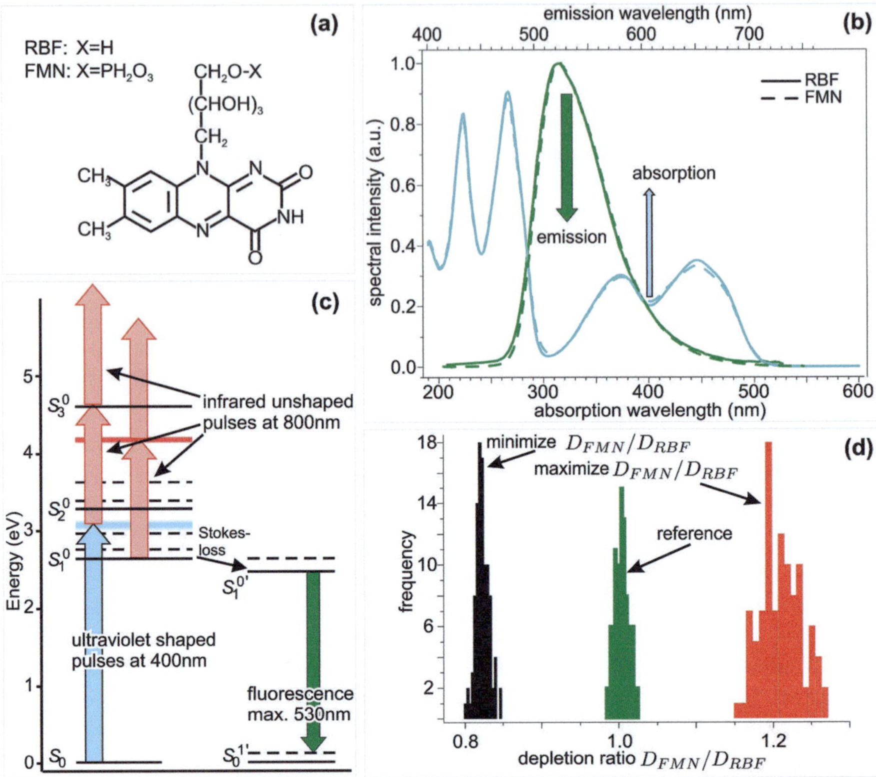

Fig. 8.7 The RBF and FMN molecules in (**a**) are very similar exhibiting nearly identical linear absorption and emission spectra in (**b**). The shaped UV control is centred at 400 nm and the flavin fluorescence signals are recorded over a window around 530 nm indicated by the respective *arrows*. The controlled optimal dynamic discrimination (ODD) of the flavins operates as indicated in (**c**) with a shaped UV pulse coordinated with a time delayed unshaped IR pulse to draw apart the vibronic dynamics of the flavins in the S_1 or S_2 excited states. The unshaped IR pulse serves to selectively disrupt the subsequent fluorescence signals from the two flavins and the red arrows indicate various possible means of action for the IR pulses. A typical outcome is shown in (**d**) where the ratio of flavin depletion signals, either *D(RBF)/D(FMN)* or *D(FMN)/D(RBF)*, is maximized. The reference case corresponds to the application of transform limited pulses (From Ref. [8])

A sequence of experiments $n = 1, 2, \ldots$ was performed to record the relative depletion signal D_n

$$D_n(RBF) = \left[F_n{}^{und}(RBF) - F_n{}^{d}(RBF)\right] / F_n{}^{und}(RBF),$$

where $F_n{}^{und}(RBF)$ is the undepleted fluorescence signal from the RBF cell (i.e., without application of the time delayed IR pulse) and $F_n{}^{d}(RBF)$ is the depleted signal in the presence of the IR pulse. An analogous expression applies to $D_n(FMN)$ for data collected from the FMN cell exposed to the same laser pulses.

The pulse shaper operated through phase modulation with 50 pixels, each on the range 0–2π, over the bandwidth of the UV pulse (only 3–4 nm). A genetic algorithm of 30 individuals was used to optimize the UV pulse phases with convergence typically occurring in $100 \sim 300$ generations.

Figure 8.7d demonstrates the ability of ODD to significantly draw apart the RBF and FMN fluorescence signatures in spite of their nearly identical linear optical spectra in Fig. 8.7b [46] For a given time delay τ, a specific optimal UV pulse is determined in the adaptive control experiment. The best discrimination was found for τ ~250–500 fs and the procedure was not able to find significant discrimination for $\tau \geq$ 1ps indicating a loss of coherence and/or insufficient overlap between the UV and IR pulses.

With a delay time τ typically around 250–500 fs, optimal pulse shapes found in different runs were very complex and showed considerable difference. They however produced exactly the same discrimination ratio when applying them, even weeks after the initial experiment. The underlying mechanisms are currently investigated, in order to identify whether this behaviour is generic or particular to this RBF/FMN system.

The assembled set of optimally discriminating laser pulses and their long-term stable performance permits exploiting ODD as a novel means for detection of nearly identical molecules when they are simultaneously present in a sample. In this work the recorded fluorescence depletion signals ${F_n}^d$ are utilized, as they are proportional to their respective flavin concentrations. Thus, the fluorescence signal $F_n(mix)$ from a mixture of flavins generated with the n-th control UV-IR pulse pair is related to its individual counterparts by

$$F_n(mix) = c(RBF){F_n}^d(RBF) + c(FMN){F_n}^d(FMN),$$

where $c(RBF)$ is the fraction of RBF present and $c(RBF) + c(FMN) = 1$. The reference signals ${F_n}^d(RBF)$ and ${F_n}^d(FMN)$ are normalized to their respective concentrations deduced in learning the n-th control field, and similarly F_n is normalized to the total sum of the two flavins present. Two distinct ODD laser pulses could successfully determine the fractional content of the two flavins (the constrained relation between the fractions was not used in order to test the capabilities of ODD). Increasing the number of interrogating optimal pulses improves the standard deviation of the extracted component concentration fractions and a typical result using six pulses was $c(RBF) = 0.35 \pm 0.04$ and $c(FMN) = 0.68 \pm 0.05$ where the exact values were 0.33 and 0.66, respectively [46].

On a fundamental point of view, the exact molecular mechanisms underlying the selectivity allowed by the application of the optimally shaped pulses could be identified by quantum chemistry calculations [9]. It was shown that the optimal laser field makes use of low-frequency vibrational modes localized on the two biochromophores side chains (terminated by a phosphate group or by hydrogen) and selects the regions of their potential energy landscapes characterized by different transition dipole moments to higher excited states leading to different ionization probabilities. The resulting modulation of the excited state populations eventually controls the emitted fluorescence intensity.

8.8 Label-Free Identification of Proteins

From the experimental standpoint, one of the major difficulties faced for the extension of the ODD scheme to proteins (or DNA) was the lack of deep UV (DUV) laser pulse shapers. This lack of ad hoc instrumentation motivated us to develop a dedicated MEMS based pulse shaper, in collaboration with CSEM-Neuchatel [47, 48]. This device proved successful for a very broad spectral range from the XUV [49] to the mid-IR.

First DUV-ODD experiments related to the fluorescing building blocks of proteins: Tryptophan and Tyrosine. The success of these experiments [50] enabled to identify tryptophan as ideal molecular target for performing ODD in proteins. In fact this bright emitting fluorophore displays a defined feature at short time-delays (<1 ps) that can be easily acted upon by an UV shaped femtosecond laser pulse and monitored by fluorescence depletion. Given tryptophan sensitivity to its local environment, it was decided to investigate the potential of ODD specifically on tryptophan-containing small peptides. We set-up a discrimination experiment aimed at the ODD fluorescence-based identification within several peptide pairs, otherwise showing identical fluorescence spectral and time-resolved response [51].

A first series of feedback driven optimizations aimed at selectivity modifying the depletion amplitude at a given time delay (600 fs) of a defined peptide against another. An exemplary plot summarizing the outcome of an optimization run (ala-trp vs. trp) is shown in Fig. 8.8. In this case, a multi-objective optimization algorithm was launched with two independent goals: (i) increasing ala-trp fluorescence and (ii) decreasing trp fluorescence. One can appreciate the result of the procedure by applying the optimal mask on the two samples and acquiring the timeresolved traces, as reported in Fig. 8.8a and b. In this example, the optimized pulse shape leads to an increase in the ala-trp time-resolved fluorescence variation by 20 %, while trp remains unaffected. Optimization runs with the same objectives were repeated several times to evaluate statistical variations on the results. Note that different pulse shapes were found to lead to similar yields, pointing out the existence of multiple solutions. The closed loop optimization procedure was then performed several times for the opposite set of goals, i.e. (i) decreasing ala-trp-fluorescence and (ii) increasing trp fluorescence. From the comparison of these traces with reference results obtained with Fourier transform pulses, we establish the histogram presented in Fig. 8.8c. The coherent manipulation of the molecular wavepacket leads to a variation of discrimination power that tears apart the trp and alatrp signatures by more than 8σ (where σ is the averaged statistical error on the results measured for a defined pulse shape). We also verified that the discrimination power DR was

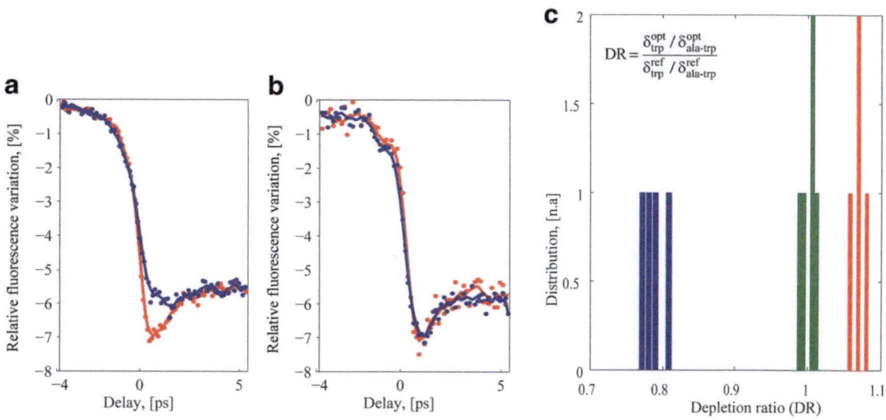

Fig. 8.8 Example of successful optimization for fluorescence depletion modulation in case of ala-trp (**a**) versus trp (**b**) with multi-objective algorithm optimization (**c**). Histogram of fluorescence depletions for ala-trp and trp. *Green* (*middle column*): reference obtained with Fourier-transform pulses, *blue* (*left*): maximization of fluorescencedepletion forala-trp, *red* (*right*): maximization of fluorescence depletion for trp

dependent on the time delay Δt. As expected for a coherent manipulation of the molecular wavepacket, DR decreases to zero for time delays larger than decoherence time (typically 1.5 ps).

The same procedure described above was applied to all pairs of dipeptides and free trp, giving rise to the matrix displayed in Fig. 8.9. In the histograms, bars indicate the ratios of the depletion values (row molecules over column molecules) obtained using phase-shaped DUV pulses retrieved by optimizations aimed at minimizing (blue) or maximizing (red) this quantity. Green bars correspond to the ratio of the depletion values measured with unshaped pulses. One can see that for several molecule pairs (trp vs. dipeptides, cyclo(-gly-trp) vs. ala-trp) the discrimination capability of the method is satisfactory, with a fluorescence difference between the optimized pulse shapes ranging from 3 s (trp vs. cyclo(-leu-trp)) to 8σ (cyclo(-gly-trp) vs. ala-trp), which is remarkably high considering the modest bandwidth available for the shaped DUV pulse. This restricted parameter space does not allow, in opposite, to discriminate other dipeptides like cyclo(-leu-trp) vs. ala-trp.

It is clear from Fig. 8.9 that even for a group of very similar molecules addressed with identical experimental resources (laser spectral bandwidth as compared to absorption bandwidth, shaping spectral resolution), large variations do occur. No obvious reason emerged from our measurements. For instance, no correlation exists between the Stokes shifts in the static spectra and discriminability. Similarly, purely dynamical considerations, based on the time-resolved measurements, would suggest that, for instance, ala-trp should be more easily distinguished from cyclo(-leu-trp) (different relaxation time constants) than from free trp, which is not the case. No

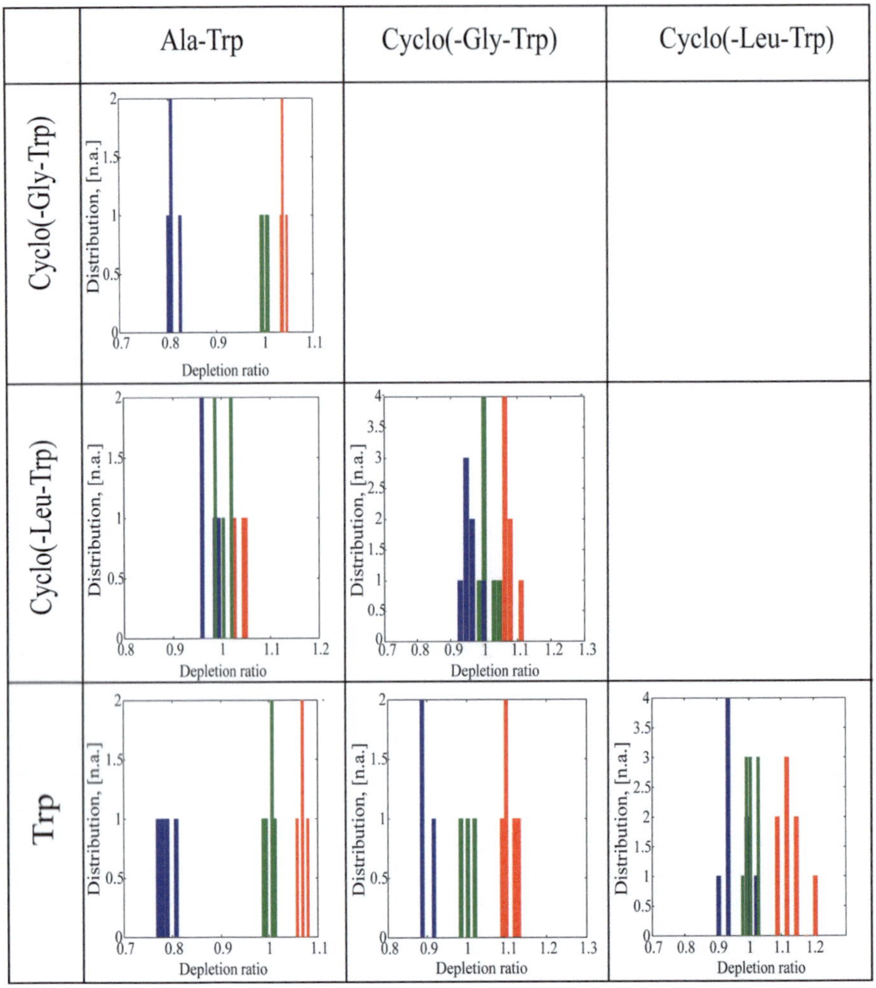

Fig. 8.9 Histograms of fluorescence depletions for various pairs of molecules. *Green*: reference obtained with Fourier-transform pulses, *blue*: maximization of fluorescence depletion for row molecule, *red*: maximization of fluorescence depletion for column molecule (From Ref. [51])

correlation with far infrared spectra was found either, stressing the importance of more complex quantum interference effects than just impulsive driving of vibrational modes.

As a natural follow up of this benchmark experiment, work is now in progress for transferring the DUV-ODD method to larger proteins with prospective applications in the field of label-free, consumable-free medical sensing. The goal of this undergoing study is currently, for instance, to quantify antibodies (such as IgG, 150 kDa) in blood serum, leading to first "quantum interference based bioassays" [10]

8.9 Conclusion

Quantum control and femtosecond spectroscopy shine a new light onto the detection and identification of biological systems, such as biochromophores but also highly complex systems like cells or bacteria. These applications are only at their beginning, and there is a reasonable hope for widely spread applications in the future, such as label-free microscopy. Major technical improvements required for reaching these goals lie in the spectral range and the spectral bandwidth accessible by future femtosecond coherent sources. On the other hand, novel labels, like HN or nitrogen vacancy (NV) nanodiamonds illustrate innovative ways to greatly improve the performance of labelling cells or tissues with organic dyes, as performed presently.

Acknowledgements The authors gratefully acknowledge the collaborators at the Universities of Geneva and Lyon, in particular L. Bonacina, F. Courvoisier, L. Guyon, V. Boutou, E. Salmon, J. Yu, G. Mejean, J. Kasparian, C. Kasparian, A. Rondi, S. Afonina, J. Extermann, P. Bejot, S. Weber, D. Kiselev, J. Gateau, S. Hermelin and M. Moret, as well as H. Rabitz and his group at Princeton, particularly M. Roth and J. Roslund, and all the collaborators of the NAMDIATREAM consortium.

We also acknowledge the financial support of the Swiss National Science Foundation (contracts No. 2000021-111688 and No 200020-124689), the Swiss NCCR MUST, and European FP7 project NAMDIATREAM (NMP-2009-4.0-3-246479).

References

1. Bonacina, L., Mugnier, Y., Courvoisier, F., Le Dantec, R., Extermann, J., Lambert, Y., Boutou, V., Galez, C., & Wolf, J. P. (2007). Polar Fe(IO3)(3) nanocrystals as local probes for nonlinear microscopy. *Applied Physics B-Lasers and Optics, 87*, 399–403.
2. Staedler, D., Magouroux, T., Joulaud, C., Extermann, J., Hadji, R., Kasparian, C., Gerber, S., Le Dantec, R., Mugnier, Y., Juillerat, L., Rytz, D., Ciepielewski, D., Bonacina, L., & Wolf, J.-P. (2012). Harmonic nanocrystals for bio-labeling: A survey of optical properties and biocompatibility. *ACS Nano, 6*(3), 2542–2549.
3. Geissbuehler, M., Bonacina, L., Shcheslavkiy, V., Bocchio, N., Geissbühler, S., Leutenegger, M., Märki, I., Wolf, J.-P., & Lasser, T. (2012). Nonlinear Correlation Spectroscopy (NLCS). *Nano Letters, 12*(3), 1668–1672.
4. Extermann, J., Bonacina, L., Cuna, E., Kasparian, C., Mugnier, Y., Feurer, T., & Wolf, J. P. (2009). Nanodoublers as deep imaging markers for multi-photon microscopy. *Optics Express, 17*, 15342–15349.
5. Extermann, J., Bonacina, L., Courvoisier, F., Kiselev, D., Mugnier, Y., Le Dantec, R., Galez, C., & Wolf, J. P. (2008). Nano-FROG: Frequency resolved optical gating by a nanometric object. *Optics Express, 16*, 10405–10411.
6. Magouroux, T., Extermann, J., Hoffmann, P., Mugnier, Y., Le Dantec, R., Jaconi, M., Kasparian, C., Ciepielewski, D., Bonacina, L., & Wolf, J. P. (2012). High-speed tracking of murine cardiac stem cells by harmonic nanodoublers. *Small, 8*(17), 2752–2756.
7. Staedler, D., Magouroux, T., Passemard, S., Schwung, S., Dubled, M., Schneiter, S., Rytz, D., Gerber-Lemaire, S., Bonacina, L., & Wolf, J.-P. (2014). Harmonic nanoparticles for cancer theranostics. *Nanoscale, 6*, 2929–2936.
8. Roth, M., Guyon, L., Roslund, J., Boutou, V., Courvoisier, F., Wolf, J. P., & Rabitz, H. (2009). Quantum control of tightly competitive product channels. *Physical Review Letters, 102*, 253001.

9. Petersen, J., Mitric, R., Bonacic-Koutecky, V., Wolf, J. P., Roslund, J., & Rabitz, H. (2010). How shaped light discriminates nearly identical biochromophores. *Physical Review Letters, 105*, 073003.
10. Afonina, S. Gateau, J. Staedler, D. Bonacina, L. Wolf, J. P. in preparation (2015)
11. Le Dantec, R., Mugnier, Y., Djanta, G., Bonacina, L., Extermann, J., Badie, L., Joulaud, C., Gerrmann, M., Rytz, D., Wolf, J. P., & Gale, C. (2011). Ensemble and individual characterization of the nonlinear optical properties of ZnO and BaTiO3 nanocrystals. *The Journal of Physical Chemistry, 115*(31), 15140–15146.
12. Baumner, R., Bonacina, L., Enderlein, J., Extermann, J., Fricke-Begemann, T., Marowsky, G., & Wolf, J. P. (2010). Evanescent-field-induced second harmonic generation by noncentrosymmetric nanoparticles. *Optics Express, 18*, 23218–23225.
13. Extermann, J., Béjot, P., Bonacina, L., Mugnier, Y., Le Dantec, R., Mazingue, T., Galez, C., & Wolf, J. P. (2009). An inexpensive nonlinear medium for intense ultrabroadband pulse characterization. *Applied Physics B, 97*, 537–540.
14. Le Xuan, L., Brasselet, S., Treussart, F., Roch, J. F., Marquier, F., Chauvat, D., Perruchas, S., Tard, C., & Gacoin, T. (2006). Balanced homodyne detection of second-harmonic generation from isolated subwavelength emitters. *Applied Physics Letters, 89*, 121118.
15. Pu, Y., Centurion, M., & Psaltis, D. (2008). Harmonic holography: A new holographic principle. *Applied Optics, 47*, A103–A110.
16. Hsieh, C. L., Pu, Y., Grange, R., Laporte, G., & Psaltis, D. (2010). Imaging through turbid layers by scanning the phase conjugated second harmonic radiation from a nanoparticle. *Optics Express, 18*, 20723–20731.
17. Hsieh, C. L., Pu, Y., Grange, R., & Psaltis, D. (2010). Digital phase conjugation of second harmonic radiation emitted by nanoparticles in turbid media. *Optics Express, 18*, 12283–12290.
18. Squier, J. A., Muller, M., Brakenhoff, G. J., & Wilson, K. R. (1998). Third harmonic generation microscopy. *Optics Express, 3*, 315–324.
19. Dudovich, N., Oron, D., & Silberberg, Y. (2002). Single-pulse coherently controlled nonlinear Raman spectroscopy and microscopy. *Nature, 418*, 512–514.
20. von Vacano, B., Wohlleben, W., & Motzkus, M. (2006). Actively shaped supercontinuum from a photonic crystal fiber for nonlinear coherent microspectroscopy. *Optics Letters, 31*, 413–415.
21. Ogilvie, J. P., Debarre, D., Solinas, X., Martin, J. L., Beaurepaire, E., & Joffre, M. (2006). Use of coherent control for selective two-photon fluorescence microscopy in live organisms. *Optics Express, 14*, 759–766.
22. Aeschlimann, M., Bauer, M., Bayer, D., Brixner, T., Cunovic, S., Dimler, F., Fischer, A., Pfeiffer, W., Rohmer, M., Schneider, C., Steeb, F., Struber, C., & Voronine, D. V. (2010). Spatiotemporal control of nanooptical excitations. *Proceedings of the National Academy of Sciences of the United States of America, 107*, 5329–5333.
23. Fuchs, U., Zeitner, U. D., & Tunnermann, A. (2005). Ultra-short pulse propagation in complex optical systems. *Optics Express, 13*, 3852–3861.
24. Tal, E., Oron, D., & Silberberg, Y. (2005). Improved depth resolution in video-rate line-scanning multiphoton microscopy using temporal focusing. *Optics Letters, 30*, 1686–1688.
25. Muller, M., Squier, J., & Brakenhoff, G. J. (1995). Measurement of femtosecond pulses in the focal point of a high-numerical-aperture lens by 2-photon absorption. *Optics Letters, 20*, 1038–1040.
26. Brixner, T., De Abajo, F. J. G., Spindler, C., & Pfeiffer, W. (2006). Adaptive ultrafast nano-optics in a tight focus. *Applied Physics B-Lasers and Optics, 84*, 89–95.
27. Amat-Roldan, I., Cormack, I. G., Loza-Alvarez, P., & Artigas, D. (2004). Starch-based second-harmonic-generated collinear frequency-resolved optical gating pulse characterization at the focal plane of a high-numerical-aperture lens. *Optics Letters, 29*, 2282–2284.
28. Bowlan, P., Gabolde, P., & Trebino, R. (2007). Directly measuring the spatio-temporal electric field of focusing ultrashort pulses. *Optics Express, 15*, 10219–10230.

29. Vunjak-Novakovic, G., Godier-Furnemont, A. F. G., Martens, T. P., Koeckert, M. S., Wan, L., Parks, J., Arai, K., Zhang, G. P., Hudson, B., & Homma, S. (2011). Composite scaffold provides a cell delivery platform for cardiovascular repair. *Proceedings of the National Academy of Sciences of the United States of America, 108*, 7974–7979.
30. Staedler, D., Passemard, S., Magouroux, T., Rogov, A., Manus Maguire, C., Mohamed, B. M., Schwung, S., Rytz, D., Justel, T., Hwu, S., Mugnier, Y., Le Dantec, R., Volkov, Y., Gerber-Lemaire, S., Prina-Mello, A., Bonacina, L., & Wolf, J.-P. (2015). Cellular uptake and biocompatibility of bismuth ferrite harmonic nanoparticles. *Nanomedicine, 11*(4), 815–824.
31. Kasparian, J., Krämer, B., Dewitz, J. P., Vajda, S., Rairoux, P., Vezin, B., Boutou, V., Leisner, T., Hübner, W., Bennemann, K., Wöste, L., & Wolf, J. P. (1997). Angular dependences of THG-generation from microdroplets. *Physical Review Letters, 78*(15), 2952.
32. Kasparian, J., Kramer, B., Leisner, T., Rairoux, P., Boutou, V., Vezin, B., & Wolf, J. P. (1998). Size dependence of Non-linear Mie scattering in microdroplets illuminated by ultrashort pulses. *Journal of the Optical Society of America B, 15*(7), 1918–1922.
33. Williams, R. M., Zipfel, W. R., & Webb, W. W. (2005). Interpreting second-harmonic generation images of collagen I fibrils. *Biophysical Journal, 88*, 1377–1386.
34. Débarre, D., Supatto, W., Pena, A.-M., Fabre, A., Tordjmann, T., Combettes, L., Schanne-Klein, M.-C., & Beaurepaire, E. (2006). Imaging lipid bodies in cells and tissues using third-harmonic generation microscopy. *Nature Methods, 3*, 47–53.
35. Rogov, A., Irondelle, M., Ramos-Gomes, F., Bode, J., Staedler, D., Passemard, S., Courvoisier, S., Yamamoto, Y., Waharte, F., Ciepielewski, D., Ph, R., Gerber-Lemaire, S., Alves, F., Salamero, J., Bonacina, L., & Wolf, J.-P. (2015). Simultaneous multi-harmonic imaging of nanoparticles in tissues for increased selectivity. *ACS Photonics, 2*(10), 1416–1422.
36. Tannor, D. J., Kosloff, R., & Rice, S. A. (1986). Coherent pulse sequence induced control of selectivity of reactions – Exact quantum-mechanical calculations. *Journal of Chemical Physics, 85*, 5805–5820.
37. Tannor, D. J., & Rice, S. A. (1985). Control of selectivity of chemical-reaction Via control of wave packet evolution. *Journal of Chemical Physics, 83*, 5013–5018.
38. Judson, R. S., & Rabitz, H. (1992). Teaching lasers to control molecules. *Physical Review Letters, 68*, 1500–1503.
39. Warren, W. S., Rabitz, H., & Dahleh, M. (1993). Coherent control of quantum dynamics – The dream is alive. *Science, 259*, 1581–1589.
40. Weiner, A. M. (2000). Femtosecond pulse shaping using spatial light modulators. *Review of Scientific Instruments, 71*, 1929–1960.
41. Bonacina, L., Extermann, J., Rondi, A., Boutou, V., & Wolf, J. P. (2007). Multiobjective genetic approach for optimal control of photoinduced processes. *Physical Review A, 76*, 023408.
42. Dantus, M., & Lozovoy, V. V. (2004). Experimental coherent laser control of physicochemical processes. *Chemical Reviews, 104*, 1813–1859.
43. Brixner, T., Damrauer, N. H., Niklaus, P., & Gerber, G. (2001). Photoselective adaptive femtosecond quantum control in the liquid phase. *Nature, 414*, 57–60.
44. Brixner, T., & Gerber, G. (2003). Quantum control of gas-phase and liquid-phase femtochemistry. *Chemphyschem, 4*, 418–438.
45. Li, B. Q., Rabitz, H., & Wolf, J. P. (2005). Optimal dynamic discrimination of similar quantum systems with time series data. *Journal of Chemical Physics, 122*, 154103.
46. Roslund, J., Roth, M., Guyon, L., Boutou, V., Courvoisier, F., Wolf, J.-P., & Rabitz, H. (2011). Resolution of strongly competitive product channels with optimal dynamic discrimination: Application to flavins. *The Journal of Chemical Physics, 134*, 034511.
47. Rondi, A., Extermann, J., Bonacina, L., Weber, S. M., & Wolf, J. P. (2009). Characterization of a MEMS-based pulse-shaping device in the deep ultraviolet. *Applied Physics B-Lasers and Optics, 96*, 757–761.
48. Weber, S., Barthelemy, M., & Chatel, B. (2010). Direct shaping of tunable UV ultra-short pulses. *Applied Physics B-Lasers and Optics, 98*, 323–326.

49. Kiselev, D., Kraus, P. M., Bonacina, L., Wörner, H. J., & Wolf, J. P. (2012). Direct amplitude shaping of high harmonics in the extreme ultraviolet. *Optics Express, 20*(23), 25843–25849.
50. Rondi, A., Bonacina, L., Trisorio, A., Hauri, C., & Wolf, J.-P. (2012). Coherent manipulation of free amino acids fluorescence. *Physical Chemistry Chemical Physics, 14*, 9317–9322.
51. Afonina, S., Nenadl, O., Rondi, A., Bonacina, L., Extermann, J., Kiselev, D., Dolamic, I., Burgi, T., & Wolf, J. P. (2013). Discriminability of tryptophan containing dipeptides using quantum control. *Applied Physics B, 111*, 541–549.

Chapter 9
Circuit Optomechanics with Diamond Integrated Optical Devices

Wolfram Pernice

Abstract Nanophotonic devices provide a powerful resource for realizing optical components with additional mechanical degrees of freedom when they are made free-standing. Using top-down fabrication many individual nanophotonic components can be reproducibly assembled into complex circuits for on-chip signal processing and sensing applications. When waveguiding devices are detached from the underlying substrate additional mechanical degrees of freedom can be harnessed for new applications in tunable optics and chipscale sensing. For optimal performance of such optomechanical elements both outstanding optical and mechanical material properties are required which makes diamond a natural choice for integrated optomechanics. Here we present an overview of chipbased photonic components made from polycrystalline diamond thin films which serve as building blocks for circuit optomechanical systems.

9.1 Introduction

Traditional integrated optical circuits are predominantly planar devices made from transparent dielectric materials. Using established fabrication routines well-known from the fabrication of electrical integrated devices, a multitude of functional elements can be realized on a waferscale and joined via waveguides into circuit-type systems. The resulting circuits contain interferometers, spectrally selective elements and microresonators for optical signal processing and sensing applications. Usually these types of devices are firmly attached to a substrate or carrier wafer from which the waveguiding devices are prepared. Further flexibility can however be achieved if portions of the waveguides or resonator structures are released from the substrate, for example by using chemical etching methods. In this case the now free-standing structures are able to mechanically move or vibrate under the influence of a suitable driving mechanism. Even though various driving forces are weak on a macroscopic level, when moving to sub-wavelength dimensions as necessary for

W. Pernice (✉)
Institute of Physics, University of Münster, Heisenbergstr. 11, 48149 Münster, Germany
e-mail: wolfram.pernice@uni-muenster.de

© Springer Science+Business Media Dordrecht 2017
B. Di Bartolo et al. (eds.), *Nano-Optics: Principles Enabling Basic Research and Applications*, NATO Science for Peace and Security Series B: Physics and Biophysics, DOI 10.1007/978-94-024-0850-8_9

integrated optical and nanophotonic devices, the picture changes and also weak forces such as radiation pressure suffice for device actuation.

Since the realization that nanoscale mechanical resonators are size-matched to nanophotonic components, merging these formerly separate research fields in a common device architecture has provided major thrust for the development of circuit-type optomechanical devices. Exploiting strong interaction between light and matter in a waveguide-based environment allows for leveraging the additional mechanical degree of freedom that optomechanics provides. This extra knob for tuning photonic components is of particular interest for material systems, where traditional methods for tunability are either not available or inefficient. Active tenability is often lacking in materials with a centro-symmetric lattice structure, such as silicon or also diamond. In these cases the advent of integrated optomechanical devices has led to a paradigm shift of how next-generation photonic devices can be used for applications in optical signal processing and sensing.

Exploiting size-matched devices leads to stable circuit-type components that are laid out in a chipscale environment analogous to integrated electrical circuits. Similar to integrated optical devices, nanophotonic waveguides are used to interconnect different optomechanical elements on chip and thus allow for realizing a bus-type architecture, where optical signals are routed throughout the chip. This approach enables fabrication of complex circuits, where optical and mechanical signal processing can be achieved in a common framework. This way many sensing elements based on nanomechanical resonators can be conveniently readout from a single optical bus. Individual optomechanical elements can be addressed using optical multiplexing, mechanical multiplexing or a mixture of both. This flexibility provides the basis of ultra-high sensitivity, commonly encountered in state-of-the-art optomechanical circuits.

By embedding nanomechanical devices in integrated optical circuits the high sensitivity of phase-sensitive optical measurement techniques can be exploited for ultra-sensitive readout of mechanical motion [1]. Typical sensitivity values are on the order of femtometer normalized to the square-root of the measurement bandwidth. Best reported values on-chip are approaching attometer resolution and provide thus an ideal platform for sensing of smallest amounts.

Using a silicon-on-insulator platform we have previously demonstrated the first use circuit architecture where optical and mechanical resonators are coupled together. Optomechanical interactions between the two technologies are mediated via gradient optical forces, which provide broadband, flexible and efficient transduction of nanoscale devices. Using substrate-coupled, doubly-clamped mechanical resonators we showed that the transduction mechanism is scalable in device frequency and material composition, enabling strong actuation that easily can drive the mechanical devices into the non-linear regime. We also showed that the actuation mechanism does not require a coherent optical driving field, which is of particular relevance for realizing integrated system for potential commercial applications [2]. In this case the optomechanical circuit can be driven through cheaper diodes, with comparable performance.

In order to enhance the driving gradient optical forces on chip several methods can be employed. Since the magnitude of the actuation force is directly proportional to the gradient in the electrical field geometries that lead to strong modal confinement offer a convenient approach to engineer light-force devices. We showed that by coupling optical waveguides through their respective near-field in slot-waveguide geometries strongly enhanced optical forces can be obtained. This geometry is also suitable for actuating stiff nanomechanical resonators, such as ultra-short doubly-clamped beam and cantilever resonators [3]. Therefore slot-geometries also provide a route towards high-frequency optomechanical resonators that operate in the fundamental mechanical mode. Reaching high resonance frequencies is of importance because air damping is strongly suppressed in the higher frequency regime. Therefore nanomechanical resonators operating in the Gigahertz frequency range can also be used under ambient conditions, which is a prerequisite for sensing applications (Fig. 9.1).

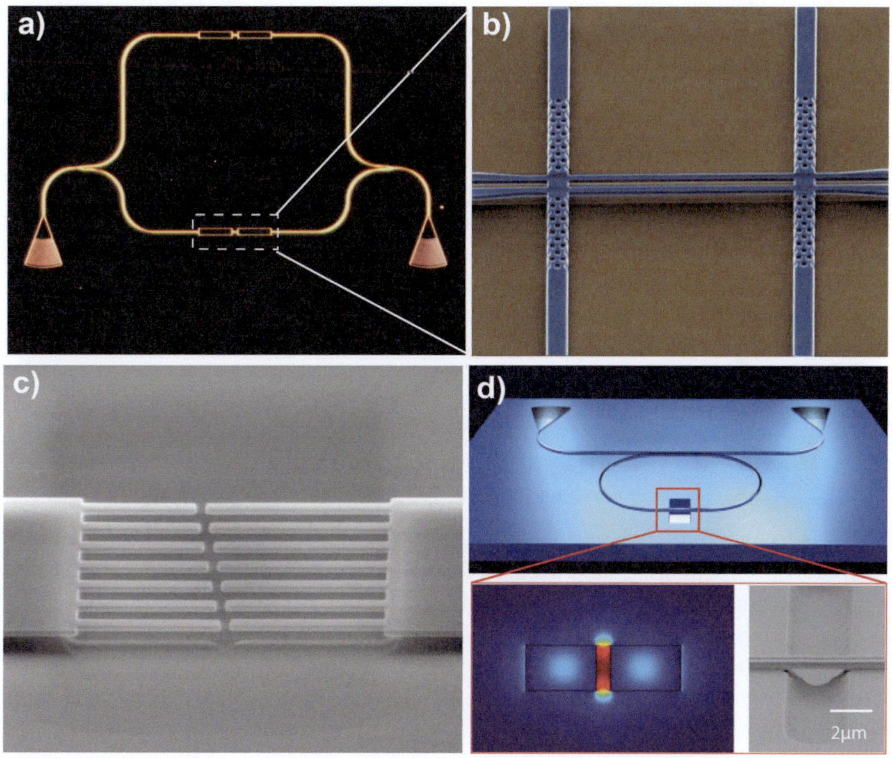

Fig. 9.1 Gradient force actuation of optomechancial circuits (**a**) Integrated Mach-Zehnder Interferometer with released nanomechanical beams. (**b**) Two side-coupled nanomechanical resonators for bipolar optical force drive. (**c**) Array of 14 nanocantilevers multiplexed through optomechanical interactions. (**d**) Slot waveguide resonantly enhanced through microring resonator. *Inset*: high-frequency actuation of ultra-short nanobeams

Furthermore, by employing in-plane coupled waveguides the distance between the nanomechanical resonators can be precisely controlled through lithography, enabling small separation and thus strong optical force drive [4]. Further enhancement of the optical forces can be achieved by employing resonant optical feedback. On chip this can be conveniently employed by selecting a suitable cavity design [5]. Depending on the target mechanical resonance frequency the cavity size can be chosen from ring resonators [6], disk resonators [7] or also photonic crystal devices [8]. Optical ring resonator cavities are also directly compatible with side-coupled optical waveguides. In this case two ring resonators are fabricated separated by a small gap, leading to a resonantly fed back slot cavity. By releasing a portion of the cavity pairs of nanomechanical resonators are realized that oscillate in-plane when driven by gradient forces.

While silicon offers superior processing capability and allows for building optomechanical circuits using high quality substrates, several limitations exist which preclude advanced applications. The relatively small electronic bandgap of silicon for instance only allows waveguiding above 1100 nm and thus excludes the important visible spectrum. In addition, strong free carrier effects lead to absorption and instabilities which are undesirable for high power optical applications. Therefore alternative material systems are explored in addition to silicon, in particular materials with a larger electronic bandgap. Popular material choices for the realization of optomechanical devices include silicon nitride [9, 10] which allows for realizing optomechanical devices with very low mechanical dissipation and thus high mechanical quality factors. Aluminum nitride provides additional attractive material parameters such as second-order optical nonlinearity and piezoelectricity and has led to the demonstration of new coupling schemes for integrated optomechanics [11]. A further interesting material option with a large bandgap is diamond which is the focus of this chapter.

9.2 Polycrystalline Diamond for Optomechanics

In order to overcome existing material limitations in optomechanical devices new templates for fabricating advanced circuits are currently being developed. In this respect diamond is a particularly promising material because it provides outstanding optical and mechanical properties simultaneously. With a bandgap of 5.5 eV diamond possesses on of the largest bandgaps among semiconducting materials. The large bandgap is the basis of ultra-broad transparency spanning ultraviolet wavelengths above 220 nm all the way into the long infrared wavelength regime. Furthermore, two-photon absorption and resulting free-carrier effects are strongly suppressed. In addition, diamond provides a relatively large refractive index of 2.4 at near infrared wavelengths, suitable for tight optical confinement and thus the realization of nanoscale photonic waveguides. On the mechanical side diamond possesses one of the largest Youngs moduli of all materials, exceeding 1100 GPa.

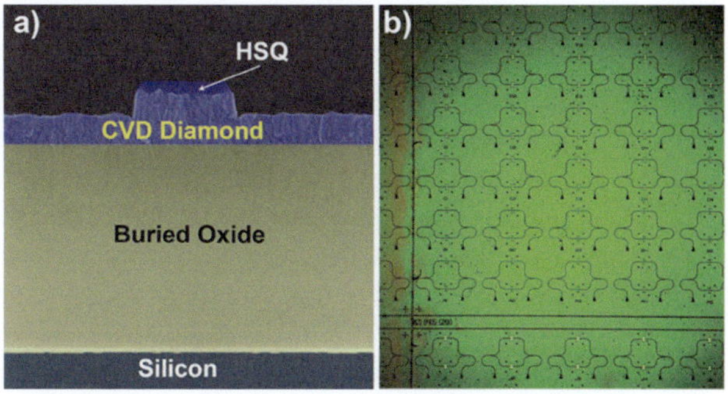

Fig. 9.2 (**a**) Cross-sectional SEM image of a diamond waveguide partially etched into a polycrystalline thin film. The diamond layer is supported by a buried oxide layer. (**b**) Optical micrograph of integrated Mach-Zehnder interferometers made from diamond. In both arms of the MZIs diamond nanomechanical beam resonators are embedded

Previously diamond mechanical and optical devices have been predominantly made from singly-crystalline substrates [12, 13]. In the case of thin-film based fabrication approaches this requires sophisticated transfer techniques, because the diamond layer needs to be surrounded by cladding layers of lower refractive index in order to enable stable waveguiding (Fig. 9.2).

Silicon dioxide is often the material of choice because oxide layers of high quality and precise thickness can be conveniently prepared by thermally oxidizing high quality silicon substrates. Onto the oxidized surface thin diamond films are then transferred. However, this approach is challenging and inherently limits the size of the template substrates. In particular for the fabrication of large-scale circuits as required for integrated optomechanical systems, the transfer approach is not a viable option. We showed however, that these difficulties can be overcome be resorting to chemical vapour deposition (CVD) diamond layers grown under controlled and high-purity conditions [14, 15]. Such films can be grown directly on oxidized silicon, leading to diamond-on-insulator templates (DOI). By nanofabricating photonic circuits from DOI we showed optical propagation over centimeter distances. Optical ring resonators fabricated from unpolished substrates provide quality factors above 10^4, limited by the surface roughness of the as-grown films. By applying polishing procedures this Q values are expected to increase further, given also that ultra-high optical Q factors in ring resonators have been previously demonstrated in other polycrystalline materials such as AlN.

Using optomechanical principles as outlined above we found superior mechanical properties by demonstrating efficient optomechanical interactions in CVD diamond [15]. Fabricated devices show mechanical non-linear behavior under strong gradient force drive with high displacement sensitivity on a fm/rtHz scale. Diamond optomechanical resonators provide the significant benefit of dramatically enhanced mechanical quality factors. In addition, mechanical quality factors exceeding 10^5 have recently been shown [16] and are thus approaching the best mechanical performance reported for silicon nitride nanostring resonators.

Diamond optomechanical devices can also efficiently be actuated using electrostatic driving, combined with on-chip optical readout [17]. The increased actuation efficiency enables the operation of such opto-electromechanical devices under ambient conditions, as a prerequisite for gas sensing. Using higher order modes we achieve resonance frequencies in excess of 100 MHz.

Diamond growth and technologies such as polishing, bonding, etching and metallization have been optimized over recent years. Especially interesting for applications in circuit optomechanics are nano-crystalline diamond films grown on SiO_2 with high Young's Modulus values approaching those of best microcrystalline and single crystal diamond [18]. As diamond grows only on diamond, special seeding techniques for the distribution of ultra-dense (10^{11} cm^{-2}) nano-diamond particle layers on the surface of Si or SiO2 have been developed. To achieve strong particle adhesion the surface charge ("zeta potential") of nano-particles was adjusted by hydrogen termination to "positive" in a way that these particles get attracted by the negative surface charge of the substrate (SiO2). For such conditions a stable diamond nano-particle layer will be formed. Growth of nanocrystalline diamond films is carried out by microwave plasma enhanced CVD (MWPECVD). In particular diamond layers grown using low methane concentration and high plasma power density exhibits outstanding mechanical properties. Thin nanocrystalline diamond films show a Young's modulus equal to that of bulk single crystal diamond. Also important is to note that the Young's modulus can be widely tuned by adjusting the growth parameters. Higher methane concentration and lower power densities give rise to increasing sp_2 incorporation. All non-diamond bands increase significantly with increasing the CH_4 admixture during the growth step which gives rise to softening of the layer.

Although diamond is ultra-hard, it can be polished either mechanically or thermo-chemical. The application of traditional mechanical polishing is not possible on thin films below 1 μm as it causes severe damages of films. We have therefore developed a new polishing technique which is a chemo-mechanical polishing (CMP) treatment allowing to polish thin diamond films on 3 inch substrates. Nanocrystalline diamond can be etched by oxygen plasma in inductive coupled (ICP) or reactive ion etching (RIE) systems which were optimized for nano-crystalline diamond (NCD) structuring for the fabrication of Micro or Nano-electro-Mechanical Systems (MEMS/NEMS).

9.3 Diamond-Based Circuit Components

DOI templates can be conveniently structured into nanophotonic components using a combination of electron beam lithography and reactive ion etching as outlined above. Because diamond consists purely of carbon, it can be efficiently etched in oxygen plasma and enables the realization of high quality waveguide based components. By relying on top-down fabrication and RIE, the available toolbox developed for silicon devices can be directly transferred onto a diamond platform.

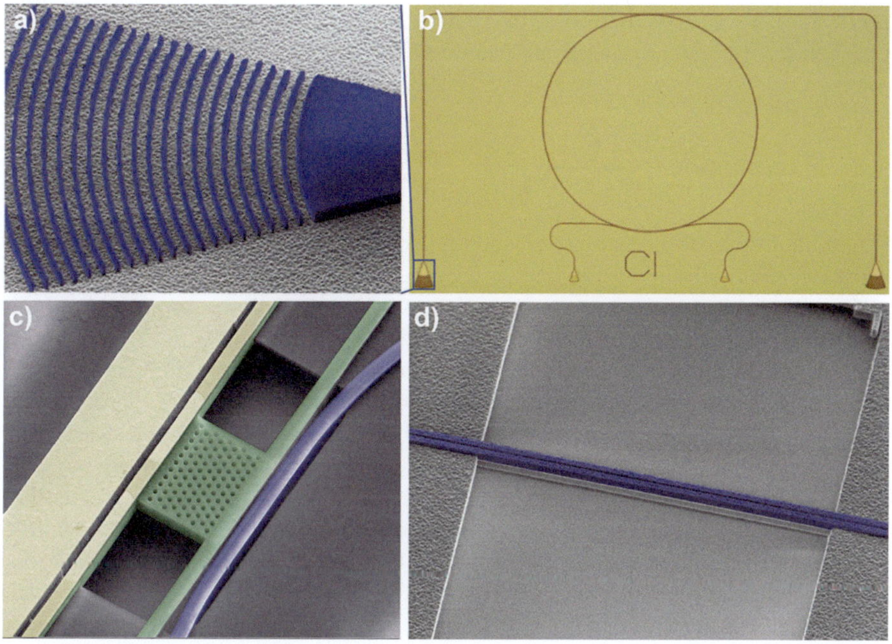

Fig. 9.3 (**a**) SEM image of a grating coupling device made from polycrystalline diamond. (**b**) Optical micrograph or a ring resonator device where grating couplers are used to inject light into the integrated optical circuit. (**c**) Diamond electro-optomechanical device where electrical actuation is combined with optical readout. (**d**) Free-standing diamond optomechanical resonators which are actuated by gradient optical forces

Several typical circuit elements that are used within diamond optomechanical circuits are collected in Fig. 9.3. These include out-of-plane optical access ports in the form of focusing grating couplers and ring resonators as discussed above.

Into diamond nanophotonic circuits suitable optomechanical resonators can then be embedded as shown in Fig. 9.3c and d. For sufficiently strong optomechanical interactions gradient forces can be used to drive diamond nanostrings as shown on the righ-hand side. Two nanostrings are coupled together via optical forces in the resulting slot waveguide and thus provide sufficiently strong actuation forces. For stiff mechanical resonators stronger actuation forces are required which can be supplied by electrostatic forces. These are provided to diamond free-standing resonators through metallic electrodes as illustrated in Fig. 9.3c. In this case capacitative forces between the thin electrodes on the diamond resonator and the thick electrode attached to the continuous diamond films can be used to pull or release the mechanical device. Optical isolation between the photonic circuitry (shown in blue as part of a MZI) and the metal electrodes (shown in yellow) is provided through a photonic crystal lattice etched into the free-standing mechanical resonator (green). Such structures are all realized through several steps of electron-beam lithography and subsequent dry etching and thus allow for scalable circuit fabrication. Therefore a multitude of optomechanical components can be made with high fabrication yield for applications in on-chip sensing.

9.4 Conclusions

Diamond offers outstanding optical and mechanical properties which makes it a natural material choice for realizing optomechanical devices. In particular for realizing mechanical devices the high Young's modulus allows for achieving high resonance frequencies which are required when optomechanical devices are to be operated under ambient conditions. This form of operation is required for practical applications in mass sensing or tunable photonic components. Here diamond also provides additional benefits because of high chemical and mechanical stability, chemical inertness and also biocompatibility for applications in nanomedicine.

While polycrystalline diamond films overcome the scalability limitations that plague current single crystal devices, the optical properties of polycrystalline films have to be improved to reduce losses because of absorption and scattering at grain boundaries. Current propagation losses in diamond circuits made from polycrystalline films are roughly a factor of ten higher than in silicon devices which limits the maximum waveguide length. With improved deposition conditions and ongoing efforts in realizing larger single-crystal diamond templates these challenges can be addressed for bring diamond photonic components to real-world applications.

References

1. Li, M., Pernice, W. H. P., Xiong, C., Baehr-Jones, T., Hochberg, M., & Tang, H. X. (2008). *Nature, 456*, 480.
2. Li, M., Pernice, W., & Tang, H. X. (2009). *Nature Nanotechnology, 4*, 377.
3. Li, M., Pernice, W. H. P., & Tang, H. X. (2010). *Applied Physics Letters, 97*, 183110.
4. Li, M., Pernice, W. H. P., & Tang, H. X. (2009). *Nature Photonics, 3*, 464.
5. Sun, X., Fong, K. Y., Xiong, C., Pernice, W. H. P., & Tang, H. X. (2011). *Optics Express, 19*, 22316.
6. Li, M., Pernice, W. H. P., & Tang, H. X. (2009). *Physical Review Letters, 103*, 223901.
7. Wiederhecker, G. S., Chen, L., Gondarenko, A., & Lipson, M. (2009). *Nature, 462*, 633.
8. Chan, J., Mayer Alegre, T. P., Safavi-Naeini, A. H., Hill, J. T., Krause, A., Gröblacher, S., Aspelmeyer, M., & Painter, O. (2011). *Nature, 478*, 89.
9. Clark, J. R., Hsu, W., Abdelmoneum, M. A., & Nguyen, C. (2005). *IEEE/ASME Journal of Microelectromechanical Systems, 14*, 1298.
10. Schliesser, A., Rivière, R., Anetsberger, G., Arcizet, O., & Kippenberg, T. J. (2008). *Nature Physics, 4*, 415.
11. Xiong, C., Pernice, W. H. P., Sun, X., Schuck, C., Fong, K. Y., & Tang, H. X. (2012). *New Journal of Physics, 14*, 095014.
12. Williams, O. A., Nesladek, M., Daenen, M., Michaelson, S., Hoffman, A., Osawa, E., Haenen, K., & Jackman, R. B. (2008). *Diamond and Related Materials, 17*, 1080.

13. Tonisch, K. L., Cimalla, V., Niebelschütz, F., Romanus, H., Eickhoff, M., & Ambacher, O. (2007). *Physica Status Solidi C, 4*, 2248.
14. Rath, P., Khasminskaya, S., Wild, C., Nebel, C., & Pernice, W. (2013). *Beilstein Journal of Nanotechnology, 4*, 300–305.
15. Rath, P., Khasminskaya, S., Wild, C., Nebel, C., & Pernice, W. (2013). *Nature Communications, 4*, 1690.
16. Ovartchaiyapong, P., Pascal, L. M. A., Myers, B. A., Lauria, P., & Bleszynski-Jayich, A. C. (2012). *Applied Physics Letters, 101*, 163505.
17. Rath, P., et al. (2014). *Applied Physics Letters, 105*, 251102.
18. Williams, O. A., Kriele, A., Hees, J., Wolfer, M., Muller-Sebert, W., & Nebel, C. E. (2010). *Chemical Physics Letters, 495*, 84–89.

Chapter 10
Terahertz Sensing at the Nanoscale

John W. Bowen

Abstract Although the terahertz frequency range covers wavelengths from around 30 μm to 3 mm, terahertz sensing can provide unique information on materials and processes occurring at much shorter length scales, even down to the nanometre range. New developments in the terahertz part of the spectrum are enabling unprecedented sub-wavelength resolution, novel devices for imaging, spectroscopy, sensing and telecommunications, new techniques for characterizing nano-materials and increased understanding of fundamental biological processes. Additionally, there are strong parallels between terahertz optics and nano-optics. After briefly describing techniques for the generation, detection and manipulation of terahertz radiation, this chapter will review recent developments in the terahertz sensing of nanoscale structures.

10.1 Introduction

The terahertz region of the electromagnetic spectrum is usually considered to extend from 100 GHz to 10 THz, corresponding to wavelengths between 3 mm and 30 μm. Nevertheless, despite the relatively large wavelengths involved, it is possible to use terahertz radiation to acquire information about materials and processes occurring at much smaller length scales, even into the nanometer range. This is due to several reasons. Firstly, the resonant modes of many molecules fall within the terahertz range, giving terahertz spectroscopy the ability to probe changes at a molecular level. Connected with this is an extreme sensitivity to the presence of water and to changes in the dynamics of water molecules. Secondly, ultrafast carrier dynamics occur on timescales that can be probed using terahertz spectroscopy, giving it an important role in the characterization of nano-materials. Thirdly, it is possible to concentrate terahertz radiation down to extreme sub-wavelength dimensions, enabling terahertz nanoscopy and enhanced sensing of small numbers of molecules.

J.W. Bowen (✉)
School of Biological Sciences, University of Reading, Whiteknights, Reading RG6 6AY, UK
e-mail: j.bowen@reading.ac.uk

© Springer Science+Business Media Dordrecht 2017
B. Di Bartolo et al. (eds.), *Nano-Optics: Principles Enabling Basic Research and Applications*, NATO Science for Peace and Security Series B: Physics and Biophysics, DOI 10.1007/978-94-024-0850-8_10

Finally, it can act as an enabling communications technology for new types of sensors, devices and systems operating at the nanoscale – the so-called internet of nano-things.

This chapter begins with a brief description of some of the main approaches to the generation, detection and manipulation of terahertz radiation. It will be noted that there are some strong parallels between terahertz optics and nano-optics. The following sections will explain how terahertz spectroscopy can be used to characterize nano-materials, the use of terahertz radiation to probe nanoscale phenomena in biological systems, terahertz nanoscopy, terahertz sensing with nanoslot antennas, and conclude with a discussion of the role terahertz radiation may play in the internet of nano-things.

10.2 The Generation, Detection and Manipulation of Terahertz Radiation

In general, the terahertz spectroscopic systems used for the applications discussed in this chapter can all be termed active systems, in that they involve the generation of terahertz radiation, allowing it to interact with the sample under test and detecting the emergent radiation that is either transmitted, reflected or scattered from the sample. The detected signal will contain information about the terahertz properties of the sample and may include both amplitude and phase information, the latter arising from delays incurred when the radiation interacts with the sample. The signal may be further processed to determine quantities such as the frequency dependence of the complex refractive index of the sample or its conductivity in the terahertz range. Some spectroscopic systems also allow the time-dependence of these quantities to be determined on ultrafast timescales.

There are a variety of ways to generate terahertz radiation, including the use of narrow band sources such as molecular gas lasers (sometimes called optically-pumped far-infrared, or FIR, lasers) and photomixers. The latter are typically based on deriving an output at the difference frequency between two slightly differently tuned distributed feedback diode lasers, often operating around 1550 nm wavelength [1]. Molecular gas lasers can be tuned to many different output frequencies in the terahertz range by changing the gas used, while photomixers allow continuous sweeping of the output across the frequency range to allow spectral information to be obtained.

An alternative is to use a broadband source, such as a high-pressure mercury arc lamp, which emits throughout the terahertz range (as well as at many other wavelengths). This can be used in conjunction with a two-beam interferometer to determine the spectral content of the resulting signal via Fourier transformation. The interferometers used are usually based on polarizing optics and known as Martin-Puplett or Michelson polarizing interferometers [2]. By placing the sample in one arm of the interferometer, both the amplitude and phase of the emergent terahertz beam can be determined, in a technique known as Dispersive Fourier

Transform Spectroscopy [3]. This allows direct determination of the refractive index and absorption coefficient of a sample, without recourse to the Kramers-Kronig relations. As the output power of the mercury arc lamp is very low, sensitive, cryogenically cooled detectors are required.

However, the majority of terahertz spectroscopy systems currently in operation are based on a technique known as terahertz Time Domain Spectroscopy (TDS). In this case, both the generation and detection of the radiation rely on the use of ultrafast (femtosecond) pulsed lasers operating at near-infrared wavelengths. The principle by which TDS operates is to produce short (several hundred femtosecond long) pulses of terahertz radiation, which, by virtue of their short pulse length in the time domain, have a broad bandwidth in the frequency domain. The bandwidth typically spans from 100 GHz to 3 THz. After interacting with the sample under test, the emergent pulse is sampled using a detector that is only activated when a gating or probe pulse derived from the ultrafast laser is incident upon it. As the probe pulse is of shorter duration (typical a few tens of femtoseconds) than the terahertz pulse, the resultant detector signal represents a sampling of the emergent terahertz pulse at a particular point in time. As the ultrafast laser produces a train of pulses, one after the other, each of which can be expected to interact with the sample under test in a similar way, the full shape of the emergent terahertz pulse can be recorded by varying the delay between the terahertz pulse and the probe pulse to sample at different points in the time domain across a sequence of terahertz pulses. The resulting time domain signal can be Fourier transformed to yield the frequency spectrum, containing both amplitude and phase information.

The terahertz pulses are produced in one of two ways. In photoconductive generation, the pump pulse from the ultrafast laser generates photocarriers in a piece of semiconductor. These photocarriers are accelerated by a bias voltage placed across the semiconductor such that they drive a current through a planar antenna, resulting in radiation of a terahertz pulse.

In optoelectronic generation, the pump pulse from the laser is incident on a non-linear crystal and generates a terahertz output pulse with a spectrum containing the difference frequencies present in the input pulse.

There are also two ways of detecting the terahertz pulses in TDS. Photoconductive detection uses a similar antenna structure to that used for photoconductive generation, again with a piece of semiconductor at its feed gap. However, rather than applying a bias voltage, the current flowing through the antenna is monitored to provide the output signal. Even though a terahertz pulse may be incident on the antenna, current can only flow when the probe beam from the laser generates photocarriers in the semiconductor, thereby gating the detector.

In electro-optic detection, the terahertz pulse is incident on a crystal that displays the Pockels effect. If the probe pulse from the near-infrared laser is incident co-linearly at the same time, it will have an elliptical output polarisation on exiting from the crystal. The degree of ellipticity will depend on the instantaneous amplitude of the terahertz pulse and can be determined by splitting the output infrared beam into two orthogonal linear polarisations, which can be detected with a balanced pair of photodiodes. The difference between the photodiode output currents is directly proportional to the terahertz field amplitude. Again the signal is gated by the laser probe beam, only being present when the probe pulse is incident at the same time as the terahertz pulse.

One thing that all of the spectroscopic systems mentioned above have in common is that they utilize the free-space propagation of terahertz radiation. However, while lenses and reflectors can be used to control the convergence and divergence of beams of terahertz radiation, they cannot be designed using geometrical optics. This is because, with wavelengths of millimeters or fractions of a millimeter, terahertz beams have to be confined to lateral dimensions of only a few wavelengths or tens of wavelengths in order to make optical systems of a manageable size. This results in the beams diffractively spreading as they propagate and so a more general treatment than geometrical optics is needed in order to take account of the diffraction. The situation is very similar to that encountered in nano-optics.

If the beams are paraxial, they can be modelled by a superposition of beam-modes, each of which maintains a characteristic amplitude profile as it propagates, although it spreads in width and has a spherical phasefront which changes its curvature [4]. As there is a slippage in phase between beam-modes of different orders, the overall profile of a beam made up of several beam-modes will change with propagation and a multi-moded beam will suffer dispersion. The beam-modes may be based on Hermite-Gaussian or Laguerre-Gaussian functions, with the fundamental beam-mode having a purely Gaussian transverse amplitude distribution.

Lenses and curved reflectors can be used to control the diffractive spreading of the beam. However it is not possible to achieve a point focus. Instead, the beam can be focused to a finite-sized beam waist, after which it will begin to diverge again under diffraction. In theory, the smallest sized spot that the beam can be focused to is around half a wavelength. However, this would require an extremely non-paraxial convergence of the beam. In practice, free-space beams can be focused down to a spot size of about one wavelength. Concentrating terahertz energy to smaller dimensions requires use of the techniques discussed in Sects. 10.5 and 10.6.

10.3 Terahertz Characterization of Nanomaterials

Terahertz spectroscopy is uniquely suited to probing the dynamic behavior of carriers in nanomaterials because typical carrier scattering rates of 10^{12}–10^{14} s^{-1} lie in the terahertz range.

The TDS technique described in Sect. 10.2 can be extended to provide time resolved measurements of nanomaterials by splitting off a further portion of the beam from the ultrafast laser that provides the pump and probe pulses to the terahertz generator and detector and using this to photo-excite the sample to be tested. By introducing a variable optical delay into this photo-exciting beam, the terahertz response of the sample can be measured at specific times after photo-excitation. Measurements over sub-picosecond to nanosecond timescales are possible. This technique is called Optical Pump – Terahertz Probe (OPTP) spectroscopy or Time-Resolved Terahertz Spectroscopy (TRTS) [5].

By holding the gating delay fixed so that the detector output signal corresponds to the peak of the detected time domain waveform, while scanning the delay in the photo-exciting beam, the relaxation behavior of photo-excited carriers within a nanomaterial can be recorded as a function of time after excitation. Thus the photoconductive lifetime can be determined. Measurements made using this technique of the photoconductive decay of semiconductor nanowires are given in [6].

Alternatively, the full time domain waveform can be recorded while holding the delay relative to the photo-exciting pulse constant. After Fourier transformation to the frequency domain, the complex refractive index of the nanomaterial sample can determined from the spectrum. In doing so, it may be necessary to make use of an effective medium approximation to take account of structure at the nanoscale [7]. The complex conductivity can then be determined from the complex refractive index. By fitting a model to the complex conductivity, such as the Drude-Smith model, parameters such as the carrier scattering time, mobility and plasma frequency can be determined. A detailed description of the whole process is given in [8]. The delay relative to the photo-exciting pulse can be adjusted to allow time-resolved measurement of the conductivity and other parameters. Examples of the characterization of a range of nanomaterials using these techniques are given in [7, 9–11].

10.4 Nanoscale Phenomena in Biological Systems

10.4.1 Solvation Dynamics

Terahertz spectroscopy is sensitive to molecular rotations and vibrations, and the collective vibrational modes of biomolecules, such as proteins, lie within the terahertz range. However, in their natural, functioning state within biological cells, biomolecules are surrounded by water molecules and it is these that dominate their spectra. The bonds between water molecules are constantly breaking and reforming on picosecond or sub-picosecond timescales, corresponding to frequencies within the terahertz range. As a consequence, water interacts strongly with terahertz radiation, of which it is a strong absorber. Because there is no long range order in the water network, liquid water does not give rise to narrow absorption lines. Instead, it produces broad bands of absorption within the terahertz range.

If a biomolecule, such as a protein, is surrounded by water molecules, it influences those closest to it. Water molecules immediately adjacent to the biomolecule can become sterically bound to it, forming a static hydration shell detectable using nuclear magnetic resonance and neutron scattering. However, the biomolecule's influence can extend to water molecules at a much greater distance from it, modifying their dynamic behavior and slowing the rate of the fluctuations in the water network [12]. This results in a dynamic hydration shell, which can extend to distances over 2 nm away from some protein molecules, and within which the water has different physical properties to those of free water. In living cells, the distance between biomolecules is typically 2–3 nm, and so almost all of the water can be in this altered state. Because the water dynamics are slowed down within the dynamic hydration shell, the resonant peak of the water absorption band is shifted to lower frequencies, resulting in a different absorption within the terahertz range to that of free water. For example, measurements between 2.3 and 2.5 THz indicate a resulting increase in the absorption over this range of frequencies [12].

The radius of the dynamic hydration shell for a given molecule in solution can be determined by measuring the terahertz absorbance as a function of molecular concentration. The molecule itself will have a lower absorbance at terahertz frequencies than the surrounding water molecules. Therefore, if the dynamically modified water were not present, the addition of greater numbers of biomolecules to the solution would be expected to decrease the absorbance. However, as each biomolecule will modify the dynamics of the surrounding water molecules such that they have an increased terahertz absorbance, the net effect will be to slow the rate of the decrease in absorbance that is seen with increasing concentration. Eventually, as the concentration is increased, a point will be reached at which the dynamic hydration shells surrounding the biomolecules begin to overlap, at which point any further increase in concentration will not result in the introduction of any additional dynamically modified water, changing the slope of a plot of absorbance against concentration. The radius of the dynamic hydration shell can be inferred from the concentration at which this change in slope occurs.

Many biological processes are dependent on the change of the conformational state of proteins. As the protein folds into a new state, not only will its collective vibrational modes change, its dynamic hydration shell will change, resulting in a change in its terahertz spectrum. Indeed, measurements show that the changes to the dynamic hydration shell occur much more rapidly than the time it takes the protein to fold into its new shape [13], highlighting the complex interaction that takes place between a protein and its surrounding water network when conformational change occurs. This process may be further moderated by the addition of other molecules. For example, it is known that plants that can survive extreme conditions, such as freezing or drought, produce sugars which have a stabilizing effect on proteins and cell membranes – a process known as anhydrobiosis – which has been harnessed for the storage of vaccines. Terahertz measurements indicate that sugars can slow down fluctuations in the water network, and thus are likely to reduce the biological activity of proteins, suggesting that this is the mechanism at play here [12]. Furthermore, site specific mutations of proteins can have a large effect on the terahertz spectrum. Thus terahertz spectroscopy can provide previously inaccessible information about the functioning of biological systems at the molecular level, occurring at nanometer and sub-nanometer length scales.

10.4.2 Terahertz DNA Sequencing

The conventional approach to DNA sequencing makes use of fluorescent chromophores for labeling. This is not without its problems, as the chromosphores can modify the DNA confirmation, are subject to degradation, variations in labelling yield, site dependent fluorescence efficiency and quenching, all lowering the precision of gene detection. Furthermore, labeling is a preparatory step that adds extra cost and complexity to the sequencing process. In contrast, terahertz DNA sequencing is a label-free technique. As a consequence, it overcomes all of these problems.

Terahertz DNA sequencing makes use of the fact that the refractive index and absorption coefficient differ for hybridized (i.e. double stranded) and denatured (i.e. single stranded) polycrystalline DNA in the terahertz range [14]. Initially, the sample of double stranded DNA in solution which is to be tested is denatured into separate strands. Short, single stranded segments of DNA a few bases long (oligos) are introduced and the DNA allowed to re-hybridize. These oligos are designed to be an exact complementary match to a sequence of bases in the long single stranded DNA in all but one base, which corresponds to the mutation being tested for. During the re-hybridization step, the oligos attach themselves to any of the long DNA strands they match – that is, those long strands which include the single base mutation. This happens more readily than two matching long DNA strands re-hybridize, and so oligos attach to all of the long single DNA strands that have the mutation, blocking pairs of mutated long strands from re-hybridizing. Oligos do not attach to any long strands that do not have the mutation, allowing complementary unmutated long strands to re-hybrized unhindered. The net result is a solution containing double stranded unmutated DNA and mutated DNA that consists of single long strands with short oligos attached. If none of the DNA sample contains the mutation that matches that being tested for by the oligo, all of the long strands will re-hybridize and will have a terahertz refractive index and absorption coefficient close to that of double stranded DNA (the oligos will have a very small effect). If all of the sample contains the mutation that matches the oligos, it will consist of long single strands with oligos attached, and thus have a terahertz refractive index and absorbtion coefficient close to that of single stranded DNA. In between these two extremes, the sample will have an intermediate refractive index and absorption coefficient which depends on the proportion of mutated DNA present in the sample.

In order to provide a simple and direct output signal, the DNA sample can be drop cast over a terahertz planar resonator to form a polycrystalline thin film of DNA [14]. The resonant frequency of the resonator will shift dependent on the refractive index of the DNA sample and can be measured using TDS. The shift in resonant frequency is dependent on the proportion of DNA that matches the mutation being tested for.

A recent improvement to this technique is based on a biochip formed from an array of split-ring resonators, each of which is functionalized with oligos having different base mutations [15]. The sample can then be applied to the whole array and each resonator tested in sequence to determine the proportion of the DNA displaying each particular mutation.

Thus, this technique allows label free single base mutation capability, doing so with femtomolar sensitivity levels.

10.5 Terahertz Nanoscopy

As noted in Sect. 10.2 above, diffraction limits the smallest possible spot size that can be formed by focusing a terahertz beam using lenses or curved reflectors to greater than half the wavelength. Nevertheless, it is possible to concentrate terahertz fields to much smaller dimensions to achieve terahertz imaging and spectroscopy with spatial resolutions of a few tens of nanometers. It is also possible to use terahertz excitation to enable ultrafast time resolved terahertz-scanning tunneling microscopy (THz-STM), with a spatial resolution of 2 nm.

One way to achieve sub-wavelength spatial resolution is to sample the terahertz evanescent field in close proximity to an object using a sub-wavelength aperture [16], or a transmission line-based probe with sub-wavelength dimensions. The latter is available commercially in a form which is easily integrated with a terahertz TDS system [17]. Imaging can be achieved by scanning the aperture or probe in the vicinity of the object. However, because of the fall off of coupling efficiency as the dimensions of the aperture or probe are reduced, these techniques have been limited to spatial resolutions greater than about 2 μm.

Concentration of the terahertz fields using tapered plasmonic waveguides can achieve a theoretical resolution of 100–250 nm [18]. However, the best that has been achieved to date is a resolution of 10 μm by 18 μm using parallel plate plasmonic waveguide [19].

Much finer spatial resolution has been achieved by using aperture-less near-field scattering techniques. By inserting an atomic force microscope tip into a terahertz beam and detecting the scattered evanescent field as the tip is scanned across an object, a resolution of 40 nm at a frequency of 2.54 THz has been demonstrated [20]. This corresponds to a spatial resolution of 1/3000 of the wavelength. A molecular gas laser was used as the terahertz source. In order to separate the scattered radiation from the background of incident radiation that would otherwise swamp it, a Michelson interferometer was used to detect the higher harmonics of the scattered field, which are present due to non-linearity in the field near the probe tip. The technique has been used to image regions of different carrier density in single nano-transisitors.

An alternative approach, with a resolution of 10 nm, is to use an atomic force microscope tip to write oxide nano-junctions around an object on the surface of a substrate composed of $LaALO_3$ and $SrTiO_3$ layers [21]. The nano-junctions can act as both terahertz sources and detectors, and can be written, erased and re-written as desired. The terahertz radiation is generated and detected through $\chi^{(3)}$ processes under illumination from an ultrafast laser.

In addition to the terahertz nanoscopy techniques described above, terahertz radiation can be used to enable simultaneous ultrafast time resolution and nanometer spatial resolution in scanning tunneling microscopy [22]. In THz-STM, an ultrafast terahertz pulse is coupled to the scanning tip of a standard STM. The terahertz pulse induces tunneling in the STM, the non-linearity of the tip-sample tunnel junction producing a rectified ultrafast tunnel current burst. Therefore, by illuminating the sample with an optical pulse, it becomes possible to carry out pump-probe techniques in conjunction with scanning tunneling microscopy. Reference [22] contains an example that demonstrates ultrafast imaging of electron trapping in an InAs nanodot. A time resolution of less than 500 fs is possible, combined with a spatial resolution of 2 nm, corresponding to 1/150,000 of the wavelength at 1 THz. Because the terahertz beam causes insignificant heating of the STM tip, THz-STM does not suffer the same thermal expansion problems as other approaches to achieving ultrafast time resolution with STMs. Furthermore, it can be accomplished without having to make any modifications to the STM tip.

10.6 Terahertz Sensing with Nanoslot Antennas

The detection of small quantities of specific molecules can be difficult in the terahertz range because their absorption cross-sections are small due to their small dimension compared to the wavelength. However, their absorption cross sections can be enhanced by orders of magnitude by placing them inside nanoslot antennas [23].

If the terahertz radiation is incident on a nanoslot antenna consisting of a narrow rectangular slot in a metal groundplane on a dielectric substrate, there is a resonant frequency for which its electric field will be enhanced by orders of magnitude compared to its magnetic field. As the wave impedance is proportional to the ratio of the electric and magnetic fields, this will be increased inside the slot. Because molecular absorption cross-section is directly proportional to wave impedance, the absorption cross-section of any molecules placed inside the slot will also be enhanced. By choosing the dimensions of the slot so that its resonant frequency matches that of a target molecule, molecular specificity can be achieved. For example, by using a 90 μm by 50 nm slot in a 50 nm thick gold film on a quartz substrate, Park et al. [23] have demonstrated detection of 40 ng of the explosive RDX drop cast over an area of 10 mm^2. This compares to around 1 mg of RDX over the same area needed to see a noticeable difference in transmission through the bare quartz substrate without the nanoslot antenna. The molecular absorption cross-section has been enhanced by a factor of 2800 times in this case, with only 22 fg of the RDX within the slot. By increasing the slot length to 150 μm, Park et al. [23] have demonstrated that the resonant frequency can be shifted so that the molecular absorption cross-section enhancement occurs for lactose rather than RDX.

10.7 The Internet of Nano-Things

The internet of nano-things is a name that has been coined for networks of distributed devices, such as sensors and actuators, each with dimensions less than a few micrometers, working together towards some common goal [24]. Networks of sensors of this sort are sometimes called wireless nano-sensor networks. At present, while no such networks have been realized in practice, they are the subject of intense theoretical investigation. There are many potential applications for networks of this sort, including: co-operative drug delivery systems, consisting of nano-devices that can be injected into the blood stream to seek out disease sites and deliver controlled quantities of drugs directly to them; small, body-warn sensors for health monitoring; sensor networks for detecting chemical and biological threats; networks of distributed nano-sensors for damage detection in materials; and networks of nano-robots for interventional surgery and damage repair.

Each element in one of these networks may contain some or all of the folowing: nano-sensors, nano- actuators, nano-memory, a nano-processor and a nano-power unit [25]. However, all elements in the network need to operate in a co-ordinated and co-operative manner, and this means that they need to be able to communicate with each other. For some applications, e.g. if the network is operating inside the body, it may be possible for the communication channel to be based upon chemical signaling. However, in many cases it is likely that the communications will be achieved using electromagnetic radiation, and so each element in the network will need to incorporate a nano-transceiver and nano-antenna. However, the scale of the elements places limitations on the frequency band that may be used.

If we consider a 1 μm long dipole antenna, it will have a resonant frequency of approximately 150 THz, corresponding to a wavelength of 2 μm. Using this frequency for communications would be impractical due to the large channel attenuation and difficulty of implementing transceivers. These factors could be overcome by operating at lower frequencies but, if classical antennas were to be used, the longer wavelengths would result in antennas that would be too large to incorporate within the elements of the network. However, a solution can be found by using graphene plasmonic nano-antennas or "graphennas".

Because the surface plasmon polariton wave velocity in graphene can be 1/100 that of the speed of light, the equivalent electrical size of a graphene antenna can be much larger than its physical dimensions. As a consequence, the resonant frequency can be two orders of magnitude lower than that of an antenna built with conventional materials. Therefore, a graphene antenna which is 1 μm long and 10 nm wide would radiate efficiently in the terahertz band.

Graphene has a number of useful properties at terahertz frequencies. Firstly, surface plasmon polaritons are much more strongly confined in the terahertz range for graphene than for gold or other noble metals. Secondly, the high electron mobility enhances the radiation efficiency. Moreover, the lateral confinement of electrons in graphene nano-ribbons enhances the conductance at terahertz frequencies, resulting in lower loss. Finally, the resistivity of graphene can be adjusted by applying a tunable bias voltage, allowing the possibility of electrically reconfigurable antenna patterns.

Reference [26] presents a graphene plasmonic nano-antenna designed for internet of nano-things use, with modeling and a theoretical analysis. It is likely that practical realization of antennas of this type will be demonstrated in the near future.

10.8 Conclusion

This Chapter has demonstrated that terahertz radiation can used to sense a wide range of phenomena at nanometer length scales, many orders of magnitude smaller than the wavelength of the radiation. Often, the information provided can only be obtained using terahertz techniques. Terahertz frequency communications are also likely to play an important enabling role in the development of the internet of nano-things. It should be remembered that the terahertz part of the electromagnetic spectrum has been the final part to be explored and exploited because of the technological challenges of operating in this range. Nevertheless, terahertz technology has been advancing extremely rapidly within recent years and it is to be expected that further possibilities for terahertz sensing at the nanoscale will be opened up in the near future.

References

1. Preu, S., Döhler, G. H., Malzer, S., Wang, L. J., & Gossard, A. C. (2011). Tunable, continuous-wave terahertz photomixer sources and applications. *Journal of Applied Physics, 109*, 06130.
2. Martin, D. H. (1982). Polarizing (Martin-Puplett) interferometric spectrometers for the near- and submillimeter spectra. In *Infrared and millimeter waves* (Systems and components, Vol. 6, pp. 66–149). New York: Academic.
3. Birch, J. R., & Parker, T. J. (1979). Dispersive fourier transform spectrometry. In K. J. Button (Ed.), *Infrared and millimeter waves* (Instrumentation, Vol. 2, pp. 137–271). Orlando: Academic.
4. Martin, D. H., & Bowen, J. W. (1993). Long-wave optics. *IEEE Transactions on Microwave Theory and Techniques, 41*(10), 1676–1690.
5. Beard, M. C., Turner, G. M., & Schmuttenmaer, C. A. (2002). Terahertz spectroscopy. *Journal of Physical Chemistry B, 106*, 7146–7159.
6. Joyce, H. J., Docherty, C. J., Gao, Q., Tan, H. H., Jagadish, C., Lloyd-Hughes, J., Herz, L. M., & Johnston, M. B. (2013). Electronic properties of GaAs, InAs and InP nanowires studied by terahertz spectroscopy. *Nanotechnology, 24*, 214006.
7. Baxter, J. B., & Schmuttenmaer, C. A. (2006). Conductivity of ZnO nanowires, nanoparticles, and thin films using time-resolved terahertz spectroscopy. *Journal of Physical Chemistry B, 110*, 25229–25239.
8. Lloyd-Hughes, J., & Jeon, T.-I. (2012). A review of the terahertz conductivity of bulk and nano-materials. *Journal of Infrared Millimeter and Terahertz Waves, 33*, 871–925.
9. Jung, G. B., Myung, Y., Cho, Y. J., Sohn, Y. J., Jang, D. M., Kim, H. S., Lee, C.-W., Park, J., Maeng, I., Son, J.-H., & Kang, C. (2010). Terahertz spectroscopy of nanocrystal-carbon nanotube and -graphene oxide hybrid nanostructures. *Journal of Physical Chemistry C, 114*, 11258–11265.

10. Parkinson, P., Joyce, H. J., Gao, Q., Tan, H. H., Zhang, X., Zou, J., Jagadish, C., Herz, L. M., & Johnston, M. B. (2009). Carrier lifetime and mobility enhancement in nearly defect-free core-shell nanowires measured using time-resolved terahertz spectroscopy. *Nano Letters, 9*, 3349–3353.
11. Strait, J. H., George, P. A., Levendorf, M., Blood-Forsythe, M., Rana, F., & Park, J. (2009). Measurements of the carrier dynamics and terahertz response of oriented germanium nanowires using optical-pump terahertz-probe spectroscopy. *Nano Letters, 9*, 2967–2972.
12. Leitner, D. M., Gruebele, M., & Havenith, M. (2008). Solvation dynamics of biomolecules: Modeling and terahertz experiments. *HFSP Journal, 2*, 314–323.
13. Ebbinghaus, S., Kim, S. J., Heyden, M., Yu, X., Gruebele, M., Leitner, D. M., & Havenith, M. (2008). Protein sequence- and pH-dependent hydration probed by terahertz spectroscopy. *Journal of the American Chemical Society, 130*, 2374–2375.
14. Haring Bolivar, P., Nagel, M., Richter, F., Brucherseifer, M., Kurz, H., Bosserhoff, A., & Büttner, R. (2004). Label-free THz sensing of genetic sequences: Towards 'THz biochips'. *Philosophical Transactions of the Royal Society of London Series A, 362*, 323–335.
15. Debus, C. (2013). A high-sensitivity THz-sensing technology for DNA detection with split-ring resonator based biochips. PhD thesis, University of Siegen, Germany.
16. Mitrofanov, O., Lee, M., Hsu, J., Brener, I., Harel, R., Federici, J., Wynn, J., Pfeiffer, L., & West, K. (2001). Collection-mode near-field imaging with 0.5-THz pulses. *IEEE Journal of Selected Topics in Quantum Electronics, 7*(4), 600–607.
17. TeraSpike, Protemics, GmbH, Otto-Blumenthal-Strasse 25, 52074 Aachen, Germany. www.protemics.com. Accessed 7 Oct 2016.
18. Rusina, A., Durach, M., Nelson, K. A., & Stockman, M. I. (2008). Nanoconcentration of terahertz radiation in plasmonic waveguides. *Optics Express, 16*, 18576.
19. Zhan, H., Mendis, R., & Mittleman, D. M. (2010). Superfocusing terahertz waves below lambda/250 using plasmonic parallel-plate waveguides. *Optics Express, 18*, 9643–9650.
20. Huber, A. J., Keilmann, F., Wittborn, J., Aizpurua, J., & Hillenbrand, R. (2008). Terahertz near-field nanoscopy of mobile carriers in single semiconductor nanodevices. *Nano Letters, 8*, 3766–3770.
21. Ma, Y., Huang, M., Ryu, S., Bark, C. W., Eom, C.-B., Irvin, P., & Levy, J. (2013). Broadband terahertz generation and detection at 10 nm scale. *Nano Letters, 13*, 2884–2888.
22. Cocker, T. L., Jelic, V., Gupta, M., Molesky, S. J., Burgess, J. A. J., De Los, R. G., Titova, L. V., Tsui, Y. Y., Freeman, M. R., & Hegmann, F. A. (2013). An ultrafast terahertz scanning tunnelling microscope. *Nature Photonics, 7*, 620–625.
23. Park, H.-R., Ahn, K. J., Han, S., Bahk, Y.-M., Park, N., & Kim, D.-S. (2013). Colossal absorption of molecules inside single terahertz nanoantennas. *Nano Letters, 13*, 1782–1786.
24. Balasubramaniam, S., & Kangasharju, J. (2013). Realizing the internet of nano things: Challenges, solutions, and applications. *Computer, 46*(2), 62–68.
25. Akyildiz, I. F., & Jornet, J. M. (2010). Electromagnetic wireless nanosensor networks. *Nano Communication Networks, 1*, 3–19.
26. Jornet, J. M., & Akyildiz, I. F. (2013). Graphene-based plasmonic nano-antenna for terahertz band communication in nanonetworks. *IEEE Journal on Selected Areas in Communications/Supplement – Part 2, 31*(12), 685–694.

Chapter 11
How Latitude Location on a Micro-World Enables Real-Time Nanoparticle Sizing

Steve Arnold, D. Keng, E. Treasurer, and M. R. Foreman

Abstract We have devised a method for using the nanoparticle induced frequency shift of whispering gallery modes (WGMs) in a microspheroid for the accurate determination of the nanoparticle size in real time. Before the introduction of this technique, size determination from the mode shift could only be obtained statistically based on the assumption that the largest perturbation occurs for binding at the equator. Determining the latitude of the binding event using two polar WGMs results in an analytic method for size determination using a single binding event. The analysis proceeds by incorporating the binding latitude into the *Reactive Sensing Principle* (*RSP*), itself containing a shape dependent form factor found using the Born approximation. By comparing this theory with experiments we find that our theoretical approach is more accurate than point dipole theory even though the optical size (circumference/wavelength) is considerably less than one.

11.1 Introduction

Our goal, since our last major advance reported at this school in 2013 [1], has been lofty: invent an accurate size spectrometer for nano-size particles that works in solution and in real-time. This is akin to a mass spectrometer that works in vacuum, and has acted as an enabler in biochemical discovery. Label-free sensors with these capabilities could identify whole viruses and exosomes not only by using bound antibodies [2, 3], but also through their size. Furthermore, a single particle technique with these capabilities could detect binding events in real-time and quantitatively test the efficacy of antibodies.

S. Arnold (✉) • D. Keng • E. Treasurer
Microparticle Photophysics Lab (MP3Lab), NYU School of Engineering, Brooklyn, NY 11201, USA
e-mail: sa1577@nyu.edu; sarnold935@aol.com

M.R. Foreman
Max Planck Institute for the Science of Light, Günther-Scharowsky-Straße 1, 91058 Erlangen, Germany

© Springer Science+Business Media Dordrecht 2017
B. Di Bartolo et al. (eds.), *Nano-Optics: Principles Enabling Basic Research and Applications*, NATO Science for Peace and Security Series B: Physics and Biophysics, DOI 10.1007/978-94-024-0850-8_11

Since Arnold et al. first proposed microcavity based single nanoparticle detection and sizing in 2003, using the perturbative frequency/wavelength shift of an optical Whispering Gallery Mode (WGM), [4] there have been a slew of papers on the subject [5]. The mechanism invoked in Ref. [4], known as the Reactive Sensing Principle (*RSP*) [6] states that "the perturbation in a resonator's photonic energy upon particle binding is equal to the energy required for the microcavity's reactive (evanescent) field to polarize the particle". In 2010 Zhu et al. established a method for determining the size of a nanoparticle in one event by using an optical micro-resonator requiring a very high Q WGM resonator ($Q \sim 10^7$–10^8) and a combination of a reactive splitting $2G$, and an increase in linewidth $\Delta\Gamma$ associated with light scattering [7]. The splitting is caused by lifting the clockwise-counterclockwise degeneracy associated with orbiting photons within WGMs. Whereas the splitting is proportional to the polarizability α of the adsorbing nanoparticle, which is in turn proportional to the particle volume, the increase in linewidth is proportional to α^2, so that the ratio $\Delta\Gamma/(2G)$ can be used to obtain the polarizability, and from that the particle size/mass. The measurable splitting vanishes however, if it is smaller than the cavity linewidth, thereby setting a lower limit for the detectable nanoparticle size [8]. The resulting "degenerate" mode still shifts, but it is only possible to obtain the polarizability through a statistical approach. In particular, the largest wavelength shift in a distribution of many events can be analytically related to the particle size, since the maximum shift of the lowest order polar mode corresponds to binding directly at the equator [4]. Unfortunately because of the statistical nature of this approach one cannot be certain that the largest event in the distribution corresponds to a particle at the equator. Consequently the size obtained is uncertain and many events are needed to generate the distribution [9]. Fortunately, we found an alternative approach that avoids these difficulties [10]. Explaining how it works and its ramifications is the subject of this chapter.

11.2 The Way Forward

As mentioned above, the need for measurements of statistical ensembles to evaluate the particle size from the frequency/wavelength shift derives from the uncertainty in the binding particle's location. Accordingly, the only distinguishing feature in the ensemble is the largest shift, which occurs for a nanoparticle binding at the equator. Our new approach avoids these difficulties by determining where the particle binds. Because of the rotational symmetry associated with a spheroid, only determination of particle's latitude is necessary. Once the latitude is determined, the *RSP* can then be invoked for size determination. The current idea is fully reactive, does not require mode splitting, and therefore can be applied in the weak coupling limit where the interaction G is considerably smaller than the linewidth, Γ, and the Q is modest ($<10^6$). Nevertheless, like mode splitting it involves analysis of the ratio of shifts from two different resonances in the same microcavity and is therefore immune to long term temperature fluctuations. Specifically, this approach involves the excitation of two resonances having the same angular momentum quantum number $\boldsymbol{l}$ but different m quantum numbers ($-l < m < l$) in a cavity for which a nanoparticle induced wavelength shift is much smaller than the linewidth (i.e. weak coupling). Although m is referred to as the magnetic quantum number in atomic physics, when considering different m modes within a microresonator for a given l we will use the term polar modes. To understand the principle of the latitude locator we must examine the shape of the WGM intensity of different polar modes. There are many such states that are characterized by $l - m + 1$ intensity peaks along the polar direction. In a sphere these m states are degenerate for a given angular momentum $\boldsymbol{l}$, but in a spheroid this degeneracy is lifted and the states separate spectrally. The first of these (i.e. shortest wavelength) is an equatorial mode for which $m = l$, thereby producing one intensity peak centered about the equator (Fig. 11.1). The next with $m = l - 1$, has two peaks, one to the North and the other to the South of the equator (Fig. 11.1). It is important to realize that the two modes depicted in Fig. 11.1 can be excited sequentially within the same slightly prolate microcavity by a fiber positioned slightly above or below the equator and directed along a line perpendicular to the symmetry axis of the spheroid. In what follows we will show that the ratio of the resonance wavelength shifts of each of these modes provides a measure of the nanoparticle's absolute latitude, from which its polarizability and size/mass may be estimated one event at a time (i.e. real time).

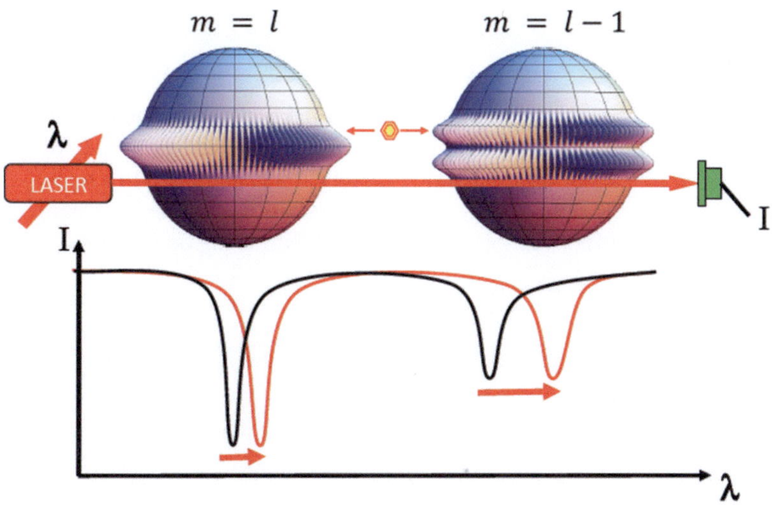

Fig. 11.1 Polar mode intensities in a single slightly prolate micro-spheroid excited sequentially during a spectral laser scan by a guided wave in a tapered fiber positioned just below the equator, and its corresponding transmission spectrum. The $m = l - 1$ mode on the *right* has a slightly longer resonance wavelength than the $m = l$ mode on the *left*. A nanoparticle adsorbed north of the equator (*orange dot*) has a larger overlap with the intensity of the $m = l - 1$ mode on the *right* and consequently this mode shifts (*red curve*) to a greater extent. The ratio of the two shifts for the same microcavity yields the nanoparticle's latitude, from which its polarizability and size/mass are calculated using the *Reactive Sensing Principle* (*RSP*)

11.3 Theoretical Approach

To start the description of the polar mode based locator we first update the wavelength shift theory to include differing polar modes. For this purpose we adopt the symbol $\Delta\lambda_{l,m}$ to describe the wavelength shift of a mode having an angular momentum number l, and polar number m. In what follows we will show that the latitude angle for particle binding is easily obtained from the ratio of two wavelength shifts, $\Delta\lambda_{l,l-1}/\Delta\lambda_{l,l}$.

To understand the importance of polar modes in nanoparticle characterization one simply has to return to the basic principle of microcavity reactive detection, the reactive sensing principle (*RSP*). To reiterate the *RSP* simply states that "the perturbation in a resonator's photonic energy upon particle binding is equal to the energy required for the microcavity's reactive (evanescent) field to polarize the particle" [4]. The principle applies to ultra-small particles such as proteins for which the radius a is much less than the characteristic evanescent intensity length L ($a<<$L), and to extended particles ($a \sim L$). On this basis, the shift in resonance wavelength $\Delta\lambda$ is the wavelength λ times the ratio of the energy required to polarize the nanoparticle to the energy in the cavity. In a dipole approximation ($a<<L$) the shift in wavelength is given by [4]

$$\Delta\lambda \simeq \frac{\alpha_{ex}\left|\boldsymbol{E}_0\left(\boldsymbol{r}_p\right)\right|^2}{2\int \varepsilon\left(r_c\right)\left|\boldsymbol{E}_0\left(\boldsymbol{r}_c\right)\right|^2 dV}\lambda, \tag{11.1}$$

where $\boldsymbol{E}_0(\boldsymbol{r}_\mathrm{p})$ is the evanescent field strength at the position of the dipole $\boldsymbol{r}_{p,}$ α_ex is the polarizability of the nanoparticle in excess of it environment (i.e. medium), and $\varepsilon(\boldsymbol{r}_\mathrm{c})$ is the permittivity of the cavity at position $\boldsymbol{r}_\mathrm{c}$. When applied to a homogeneous microsphere for $m \approx \boldsymbol{l} >> \boldsymbol{1}$ (i.e. polar modes whose mode energy is concentrated close to the equator), for a nanoparticle of radius a adsorbing on the surface, Eq. 11.1 becomes

$$\Delta\lambda_{l,m} \approx \frac{\alpha\left|Y_{l,m}\left(\xi_p\right)\right|^2 g\left(a/L\right)}{\left(n_s{}^2 - n_e{}^2\right) R_e{}^3}\lambda \tag{11.2}$$

where $\alpha = \alpha_{ex}/\varepsilon_0$ is the "geometric" polarizability that is proportional to the volume of the nanoparticle ($\alpha = D_\alpha a^3$), R_e is the microsphere equatorial radius, L is the characteristic evanescent intensity length obtained from Mie theory, $Y_{l,m}(\xi_\mathrm{p})$ is the spherical harmonic evaluated at the latitude ξ_p of the bound particle (the azimuthal dependence of the spherical harmonic is omitted since the usual $e^{im\phi}$ drops out when considering $|Y_{l,m}|$), the factor g can correct for a displacement of the point dipole from the surface (g_1), or for the more realistic case of a distributed polarization energy density (g_2), n_s, n_e and n_p are the refractive indices of the microsphere, environment and nanoparticle respectively, and finally the constant $D_\alpha = 4\pi n_e{}^2\left(n_p{}^2 - n_e{}^2\right)/\left(n_p{}^2 + 2n_e{}^2\right)$.

The simplest choice of the g factor, g_1, accounts for the displacement of the center of the dipole from the surface by a distance a. Since the evanescent intensity in the numerator of Eq. 11.1 fits well to a single exponential decay with a characteristic length L, [11] the simplest g factor is one that evaluates the evanescent intensity at the nanoparticle's center; [6]

$$g_1 = e^{-z} \tag{11.3}$$

where $z = a/L$. Alternatively we can use g_2, termed a form factor since it accounts for the shape of the nanoparticle, which can be found from making the Born approximation for scattering from a sphere [12]. Specifically, g_2 is the volume averaged surface normalized evanescent intensity within the nanoparticle. For a spherical nanoparticle, integration over its shape gives

$$g_2(z) = \frac{6}{(2z)^2}\left(1 + e^{-2z}\right) - \frac{12}{(2z)^3}\left(1 - e^{-2z}\right). \tag{11.4}$$

This form factor recognizes that the hemisphere of the nanoparticle closest to the microsphere surface will receive more polarization energy than the hemisphere furthest from the surface. Below $z \approx 1$, g_1 and g_2 are approximately equal and have the simple limiting value of $g \simeq 1$ for $a << \mathrm{L}$, therefore ensuring asymptotic consistency with the original result [4] (i.e. both describe the simple point dipole at the surface of the microcavity for $a << L$). The factor g_1 is expected to be valid so long as the ratio $X_n = 2\pi a/(\lambda/n_p) < 1$, where λ is the free-space wavelength [13]. The limits for use of the form factor g_2 have become controversial, [13] although this paper will provide a test of its use in interpreting experimental data for z values both smaller and larger than 1. Equation 11.2 is the key to our latitude locator.

Consider the ratio of wavelength shifts of the $m = l - 1$ to the $m = l$ modes at a latitude ξ_p; $\Delta\lambda_{l,l-1}(\xi_p)/\Delta\lambda_{l,l}(\xi_p)$. So long as the shifts are very small in comparison to the wavelength and $l >> 1$, it follows from Eq. 11.2 that

$$\frac{\Delta\lambda_{l,l-1}}{\Delta\lambda_{l,l}} \cong \frac{\left|Y_{l,l-1}(\xi_p)\right|^2}{\left|Y_{l,l}(\xi_p)\right|^2}. \tag{11.5}$$

As one can see the right hand side of Eq. 11.5 only depends on the latitude ξ_p of the adsorbing particle, and consequently by placing experimental wavelength shift data on the left, this equation gives the latitude of the adsorbed nanoparticle independent of its physical properties, those of the resonator, or any refractive indices. Once ξ_p is determined, Eq. 11.2 can be used to calculate the size of the nanoparticle as well as the polarizability.

Our simple latitude locator equation (Eq. 11.5) can be further simplified by utilizing a mathematical identity which connects the ratio of spherical harmonics to the tangent of the latitude angle,

$$\frac{\Delta\lambda_{l,l-1}}{\Delta\lambda_{l,l}} \cong \left|\frac{Y_{l,l-1}(\xi_p)}{Y_{l,l}(\xi_p)}\right|^2 = 2l\,\left[\tan(\xi_p)\right]^2. \tag{11.6}$$

Combining Eq. 11.4 with Eq. 11.5 gives the absolute latitude of the particle,

$$|\xi_p| \cong \tan^{-1}\sqrt{\frac{1}{2l}\frac{\Delta\lambda_{l,l-1}}{\Delta\lambda_{l,l}}} \tag{11.7}$$

Once $|\xi_p|$ is found, the size a of the nanoparticle can be obtained by re-expressing Eq. 11.2 as

$$a^3 g\,(a/L) \approx \frac{\left(n_s{}^2 - n_e{}^2\right) R^3}{\left|Y_{l,l}\left(\xi_p\right)\right|^2 D_\alpha}\frac{\Delta\lambda_{l,l}}{\lambda}. \tag{11.8}$$

The solution to Eq. 11.8 is particularly simple for $a << L$ since the form factor $g \simeq 1$ in this limit, and one gets a closed form algebraic solution. For ultra-small particles displaced from the surface, or larger particles for which the evanescent field varies significantly over the dielectric body, g can take an analytical form (e.g. g_1 or g_2), and the size can be obtained numerically from Eq. 11.8. Whether the nanoparticle is north or south of the equator is irrelevant, since the square modulus of the spherical harmonic in the denominator of Eq. 11.8 is an even function with respect to the latitude.

11.4 Experimental Approach

To test our micro-global latitude locator idea we fabricated micro-spheroids by using CO_2 laser melting at the end of a tapered silica optical fiber (inset, Fig. 11.2). Shape analysis of the images revealed that our resonators were slightly prolate [distortion $= \left(R_p - R_e\right)/\left(R_p R_e^2\right)^{1/3} < 0.03$, where R_p is the polar radius and R_e is that of the equator]. These silica micro-spheroids were then installed into our homemade microfluidic system, [2, 14, 15] where they were coupled to a tapered optical fiber. In the inset the fiber is below the equator of the spheroid.

A typical under-coupled spectrum taken through the coupling fiber is shown in Fig. 11.2. All of the resonances were excited with a 1063 nm tunable DFB laser polarized along a meridian [Transverse Electric (TE) polarization] by using an integrated system provided by MP3Laser.com. The laser was current tuned with a saw tooth wave that accounts for the rising backbone of the spectrum; increasing the drive current tunes the laser to greater wavelength but also increases its output power. It should be noted that the resonance dip on the left has no neighbor at shorter wavelength. This is the signature of the $m = l$ equatorial mode of a prolate spheroid; the $m = l$ mode has the shortest wavelength, since the equator has the smallest circumference [16]. To the right of this mode (longer wavelength) is the $m = l - 1$ mode which is narrower with a smaller dip. Note that the $m = l - 2$ mode is of similar depth to the $m = l$ mode, while the $m = l - 3$ mode (at yet longer wavelength) looks similar in depth to the $m = l - 1$ mode. This sequence of deep-

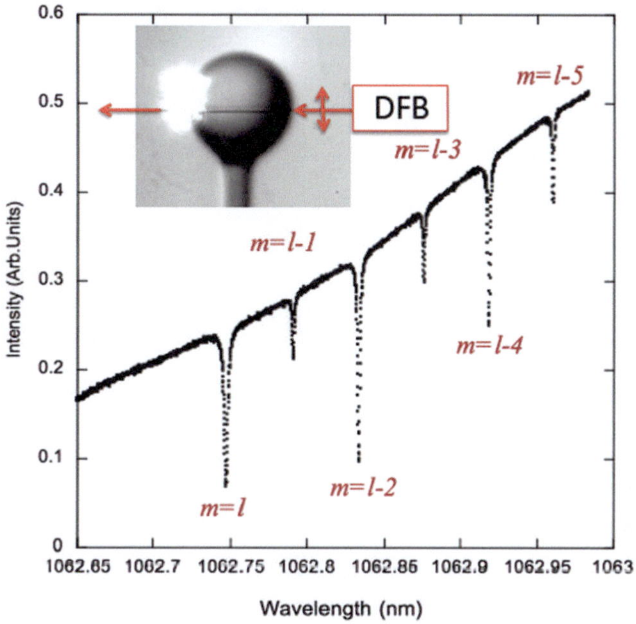

Fig. 11.2 Spectrum of a slightly prolate microcavity immersed in 30 mM NaCl solution and excited by tapered fiber coupling just below the equator as seen in the *inset*. The equatorial radius was 41 microns. From Mie theory $l = 340$ with all modes having TE polarization. The DFB laser was scanned with a saw tooth drive having a period of 100 msec

shallow-deep-shallow dips in Fig. 11.2 is a consequence of the overlap between the fiber field and the polar symmetries of the WGMs; the coupling constant requires performing volume integration over the product of the optical fiber field with the WGM field [17]. Whereas the $m = l$ mode is symmetric in latitude about the equator as is the fiber field, the $m = l - 1$ mode is antisymmetric. Exciting the antisymmetric WGM mode requires that the centerline of the exciting fiber be slightly above or below the equator due to the greater modal overlap this gives. Before performing nanoparticle binding experiments the fiber contacted the cavity which reduced noise due to fiber vibration, but caused the resonances to be both red shifted and broadened due to over-coupling. All the modes were identified as being of first radial order.

The latitude locator idea was tested by using nanoparticles for which $X_n < 1$. For convenience polystyrene beads were chosen with a manufactured size of $\langle a_m \rangle \pm \sigma_m = 96.7 \pm 4.5$ nm [Polysciences], for which $X_n = 0.90$ at 1063 nm. These particles were injected into our microfluidic system at a 175 fM concentration in the presence of a resonator similar to that depicted in Fig. 11.2, but with an equatorial radius of 40.5 μm. The solution had a 100 mM NaCl concentration to promote binding to the silica surface by decreasing the Debye length associated with ionized silanol groups [6]. The fiber is placed in contact with the resonator to reduce mechanical vibration

Table 11.1 Data (*2nd and 3rd column*) taken for polystyrene particles with a size of $\langle a_m \rangle \pm \sigma_m = 96.7 \pm 4.5$ nm, as determined by the manufacturer. The mean size found using our latitude locator scheme differs from $\langle a_m \rangle$ by 1.3 % when using g_1, and by 0.4 % when using g_2. The common parameters used to calculate columns 5 and 6 are $R = 40.5$ μm, $l = 336$, $L = 194.4$, $n_p = 1.5718$, $n_s = 1.449$ and $n_e = 1.326$

Event		$\Delta\lambda_{336,336}$ (*fm*)	$\Delta\lambda_{336,335}$ (*fm*)	$\lvert\xi_p\rvert$ (deg) Eq. 11.7	a_{rsp} (*nm*) Eq. 11.8, g_1	a_{rsp} (*nm*) Eq. 11.8, g_2
1.	90 s	100	58	1.69	103.9	102.8
2.	130 s	120	18	0.86	102.6	101.4
3.	485 s	37	74	3.13	92.7	91.9
4.	1358 s	25	85	4.08	104.7	103.6
5.	1612 s	38	66	2.92	88.9	88.2
6.	2601 s	63	68	2.30	95.4	94.5
7.	4038 s	45	75	2.86	93.8	92.9
8.	4412 s	110	51	1.51	105.5	104.3
9.	4917 s	64	83	2.52	100.3	99.2
10.	5225 s	64	55	2.05	91.9	91.1
11.	6863 s	62	79	2.50	98.6	97.6
12.	7182 s	58	73	2.48	95.7	94.8
13.	7310 s	73	77	2.27	100.7	99.7
Mean					**98.0**	**97.1**
$< a_{rsp} > - < a_m >$					**1.3**	**0.4**

noise. This results in a red shift and broadening of the resonances. Upon coupling the $m = l$ and $m = l - 1$ resonances had Qs of 5×10^5 and 2×10^5, respectively.

The second and third columns of Table 11.1 show the experimental shifts for 13 detected events, measured over a period in excess of 2 h, while the fourth column provides the latitude angle found from Eq. 11.7. The fifth and sixth columns provide the nanoparticle size determined from Eq. 11.8 using form factors g_1 or g_2, respectively. As one can see the mean radius determined using the *RSP* for both g_1 (98.0 nm) and g_2 (97.1 nm) are near the manufacturer's specified size, although the mean radius using g_2 is considerably closer to the manufacturer's value of 96.7 nm.

The standard deviation of all the particle results determined using Eq. 11.8 (5.0 nm) is very close to the manufacturer's standard deviation of 4.5 nm, implying that our standard deviation is the result of particle size variation, and not the computational approach.

11.5 Concluding Remarks

For the data in Table 11.1 $X_n = 0.9$, which easily qualifies for testing our sizing technique based on g_1, [13] however it appears that g_2 provides better agreement with our data. The mean size arrived at using our global positioning scheme differs from $\langle a_m \rangle$ by deviations of about +1.3 % using g_1, and by + 0.4 % using g_2. It is not difficult to understand the superiority of g_2 over g_1.

Reference [13] suggests that the intensity in the numerator of Eq. 11.1 should be evaluated at the center of the nanoparticle so long as $X_n < 1$. The upper limit in size consistent with this inequality is $a_{max} = \lambda/(2\pi n_p)$, which for polystyrene particles at 1063 nm is about 108 nm. For the microcavity used in Table 11.1, $L = 194.4$ nm, so the intensity at the furthest part of the bound nanoparticle is only 33 % of what it is at the microcavity surface. This is expected to lead to a decidedly inhomogeneous energy density within the particle. In other words the centroid of energy density within the particle cannot be at the center as it would be for a uniform intensity; **considering the current high precision of our measurements, a point dipole model is less reasonable than the shape dependent Born approximation even for $X_n < 1$.**

To further support our conclusion a COMSOL simulation was performed for microcavity and nanoparticle parameters similar to those in Table 11.1. Figure 11.3 shows a typical calculation. The intensity within the particle falls off by a factor of 2.87 from the closest point to the microcavity to the furthest point even though X_n <1. This image allows us to calculate the total polarization energy of the particle. The results are consistent with the *RSP* theory for a non-uniform polarization density, [18] and with the use of the form factor g_2 [19].

Whereas the factor g_1 requires that the particle be viewed as a displaced dipole having $X_n < 1$, [13] the limitation on the Born approximation as applied to polarization by an transverse electric (TE) evanescent field is not as apparent. To test this we increased the particle size so that X_n reached 2.1, by performing experiments on polystyrene particles having mean radii of 178 nm and 228 nm [19]. Surprisingly, when using g_2 the results were nearly as good as those for the smaller particles in Table 11.1; with the measured mean radii deviating from the manufacturer specified means of 178 nm and 228 nm by +0.5 % and +0.8 % respectively. By comparison the displaced point dipole factor g_1 does not compare well as expected; the mean radii calculated with g_1 deviate away from the manufacturer specified means of 178 nm and 228 nm by +3.6 % and +8.9 % respectively; ~10× the deviation encountered by using g_2.

We have just begun our work on constructing high precision nanoparticle size measurements in real time. One thing which is quite clear from our measurements and simulations is that the dipole theory even in a "first principled" form [13] is not sufficient, and that perturbation theory can prove more accurate because it allows for variation in the internal field.

At its inception we called the technique to locate the nanoparticle's binding site, a micro-global positioning system (WGM-gps) [10]. Ordinarily a global positioning system needs to report latitude, longitude and altitude, however it would seem from Table 11.1 that we only located the latitude. In point of fact by identifying the particle's size we have determined the "altitude" to its center at its landing site. Consequently a pre-binding record of the frequency shift by a single particle can determine its "altitude" history before and at the time of landing. In that sense we can have both the latitude and altitude. Although not necessary for particle

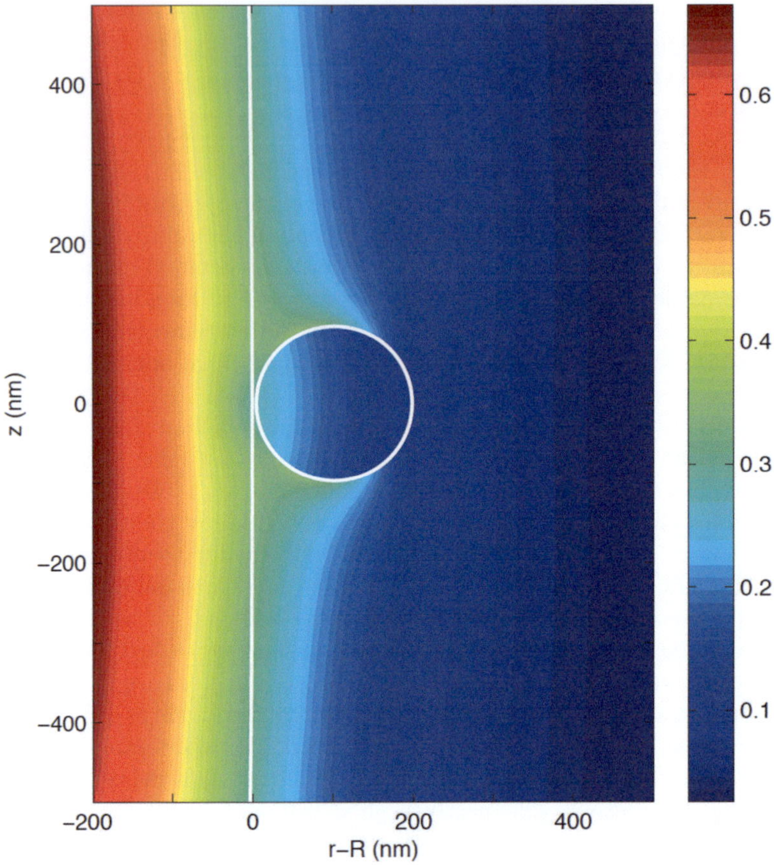

Fig. 11.3 COMSOL calculation showing the internal field within a 96.7 nm radius polystyrene particle located at the equator of a spheroidal resonator and bathed in a TE field of a 1st order WGM with $l = 340$. The position of the *vertical white line* closest to the nanoparticle represents the surface of the microcavity. The intensity within the particle falls off by a factor of 2.87 from the closest point to the microcavity to the furthest point even though $X_n < 1$

sizing, the determination of longitude would complete our WGM-gps sensor leading to massive multiplexing by comparison to the manner in which WGM sensors are currently used (i.e. one assay per resonator). This should revolutionize sensor technology on our micro-world, just as it revolutionized navigation on the Earth centuries ago [19].

Acknowledgements The research described herein was supported by the National Science Foundation grant EECS 1303499.

References

1. Arnold, S., Holler, S., & Fan, X. (2015). Taking microcavity label-free single molecule detection deep into the protein realm: Cancer marker detection at the ultimate sensitivity. In *Nano-structures for optics and photonics* (pp. 309–322). Dordrecht: Springer.
2. Arnold, S., Ramjit, R., Keng, D., Kolchenko, V., & Teraoka, I. (2008). Microparticle photophysics illuminates viral bio-sensing. *Faraday Discussions, 137*, 65–83.
3. McClellan, M. S., Domier, L. L., & Bailey, R. C. (2012). Label-free virus detection using silicon photonic microring resonators. *Biosensors and Bioelectronics, 31*, 388–392.
4. Arnold, S., Khoshsima, M., Teraoka, I., Holler, S., & Vollmer, F. (2003). Shift of whispering-gallery modes in microspheres by protein adsorption. *Optics Letters, 28*, 272–274.
5. Foreman, M. R., Swaim, J. D., & Vollmer, F. (2015). Whispering gallery mode sensors. *Advances in Optics and Photonics, 7*, 168–240.
6. Arnold, S., Keng, D., Shopova, S. I., Holler, S., Zurawsky, W., & Vollmer, F. (2009). Whispering gallery mode carousel. *Optics Express, 17*, 6230–6238.
7. Zhu, J., Ozdemir, S. K., Xiao, Y. F., Li, L., He, L., Chen, D. R., & Yang, L. (2010). On-chip single nanoparticle detection and sizing by mode splitting in an ultrahigh-Q microresonator. *Nature Photonics, 4*, 46–49.
8. Kim, W., Özdemir, Ş. K., Zhu, J., & Yang, L. (2011). Observation and characterization of mode splitting in microsphere resonators in aquatic environment. *Applied Physics Letters, 98*, 141106.
9. Lu, T., Lee, H., Chen, T., Herchak, S., Kim, J. H., Fraser, S. E., Flagan, R., & Vahala, K. (2011). High sensitivity nanoparticle detection using optical microcavities. *Proceedings of the National Academy of Sciences, 108*, 5976–5979.
10. Keng, D., Tan, X., & Arnold, S. (2014). Whispering gallery micro-global positioning system for nanoparticle sizing in real time. *Applied Physics Letters, 105*, 071105.
11. Khoshsima, M. (2004). *Perturbation of whispering gallery modes in microspheres by protein adsorption: Theory and experiment*. Doctoral dissertation, Polytechnic University.
12. Vollmer, F. (2004). *Resonant detection of nano to microscopic objects using whispering gallery modes*. Doctoral dissertation, The Rockefeller University.
13. Deych, L., & Shuvayev, V. (2015). Spectral modification of whispering-gallery-mode resonances in spheroidal resonators due to interaction with ultra-small particles. *Optics Letters, 40*, 4536–4539.
14. Keng, T. K. D. (2009). *Whispering gallery mode bioparticle sensing and transport*. Doctoral dissertation, Polytechnic Institute of New York University.
15. Keng, D., McAnanama, S. R., Teraoka, I., & Arnold, S. (2007). Resonance fluctuations of a whispering gallery mode biosensor by particles undergoing Brownian motion. *Applied Physics Letters, 91*, 103902.
16. Lin, G., Qian, B., Oručević, F., Candela, Y., Jager, J. B., Cai, Z., Lefevre-Segun, V., & Hare, J. (2010). Excitation mapping of whispering gallery modes in silica microcavities. *Optics Letters, 35*, 583–585.
17. Little, B. E., Laine, J. P., & Haus, H. A. (1999). Analytic theory of coupling from tapered fibers and half-blocks into microsphere resonators. *Journal of Lightwave Technology, 17*, 704–715.
18. Teraoka, I., & Arnold, S. (2006). Theory of resonance shifts in TE and TM whispering gallery modes by nonradial perturbations for sensing applications. *Journal of the Optical Society of America B, 23*, 1381–1389.
19. Sobel, D. (2007). *Longitude: The true story of a lone genius who solved the greatest scientific problem of his time*. New York: Bloomsbury Publishing.

Chapter 12
Nanoplasmonic and Microfluidic Devices for Biological Sensing

G. Perozziello, A. Giugni, M. Allione, B. Torre, G. Das, M. L. Coluccio, M. Marini, L. Tirinato, M. Moretti, T. Limongi, P. Candeloro, and Enzo Di Fabrizio

Abstract In this chapter we report about recent advances on the development and application of 2D and 3D plasmonic nanostructures used for sensing of biological samples by Raman spectroscopy at unprecedented resolution of analysis. Besides, we explain how the integration of these nanodevices in a microfluidic apparatus can simplify the analysis of biological samples. In the first part we introduce and motivate the convenience of using nanoplasmonic enhancers and Raman spectroscopy for biological sensing, describing the phenomena and the current approaches to fabricate nanoplasmonic structures. In the second part, we explain how specific multi-element devices produce the optimal enhancement of the Raman scattering. We report cases where biological sensing of DNA was performed at few molecules level with nanometer spatial resolutions. Finally, we show an example of microfluidic device integrating plasmonic nanodevices to sort and drive biological samples, like living cells, towards the optical probe in order to obtain optimal conditions of analysis.

12.1 Introduction

Plasmonic properties of metal nanoparticles have been exploited since long ago. For instance, artists have been using optical properties of metal nanodrops to generate brilliant colors, when shined by the direct sunlight, in glass artefacts and artwork for centuries.

G. Perozziello • M.L. Coluccio • P. Candeloro
Bionem Lab, Departement of Experimental Clinics, Campus "Salvatore Venuta", University Magna Graecia, Viale Europa, 88100 Germaneto-Catanzaro, Italy

A. Giugni • M. Allione • B. Torre • G. Das • M. Marini • L. Tirinato • M. Moretti • T. Limongi • E. Di Fabrizio (✉)
Physical Science and Engineering Division (PSE), King Abdullah University of Science and Technology (KAUST), Thuwal 23955-6900, Saudi Arabia
e-mail: Enzo.DiFabrizio@kaust.edu.sa

© Springer Science+Business Media Dordrecht 2017

B. Di Bartolo et al. (eds.), *Nano-Optics: Principles Enabling Basic Research and Applications*, NATO Science for Peace and Security Series B: Physics and Biophysics, DOI 10.1007/978-94-024-0850-8_12

When an electromagnetic field interacts with a metallic particle, among different possible processes, it can also excite a plasma oscillation, an electric-field-induced oscillation of the free electrons of the particle that strongly enhances the local electromagnetic field. The study and the control of the interaction between electromagnetic field and free charges is the topic of Plasmonics research. Plasmonics also investigates the mutual induction between conductive nanostructures through their electromagnetic fields in a classical and quantum framework, commonly applied to 2D and 3D metal nanostructures.

These electronic excited states can be classified, for example, as surface or bulk plasmons identified by their property to be confined or not at the interface between the metal and the dielectric. Each of them has a specific dispersion relation, very different the one of light. Energy and momentum conservation, in general, hinders their excitation by direct photons irradiation on the metal except for specific conditions and shape adopted, that define the optimal coupling geometries. That is why we commonly use nanostructured grating on a flat surface, or specific geometrical shapes and orientations of nanoparticles in respect to the incoming electromagnetic field. For scientific application, surface plasmons polaritons, SPPs, are the most fruitful ones; they are characterized by an evanescent electric field, fact that confines their actions tightly to the metal surface on the range of few tens of nanometers. For a single metal nanoparticle immersed in the electromagnetic field, the excitation of SPPs gives rise to a major spatial reshaping of the electromagnetic energy, generating intense subwavelength-localized spots at specific resonant frequencies. In clustered nanostructures, the enhancement of the electromagnetic field depends also by their interaction with each other. In general, the resonance also shift in response of a change in the dielectric constant of the medium around the nanoparticles, as it happens when metal nanoparticles merge with substances like water solutions or organic media.

In particular, for nanoparticles sensibly smaller than the incident wavelength, an efficient localized surface plasmon resonance (LSPR) occurs.

In this case, the external field $\boldsymbol{E_0}$ can be approximated as uniform across each particle, and the electric dipole induced within the nanoparticle (Figs. 12.1 and 12.2a) generates an additional dipolar field $\boldsymbol{E_{dip}}$, so that:

$$E_{tot} = E_\mathbf{0} + E_{dip} \tag{12.1}$$

By solving the electrostatic problem in the outer space of a spherical nanoparticle, as schematically represented in Fig. 12.2a, the electric potential comes out as the sum of a term due to $\boldsymbol{E_0}$ and another one generated by a dipole moment $\boldsymbol{p}$:

$$\mathbf{p} = \pi\varepsilon_0\varepsilon_d a^3 \left(\frac{\varepsilon_m - \varepsilon_d}{\varepsilon_m + 2\varepsilon_d}\right)\mathbf{E_0} \tag{12.2}$$

where ε_m and ε_d are respectively the dielectric constant of the metal and the dielectric surrounding medium, and a is the radius of the nanoparticle.

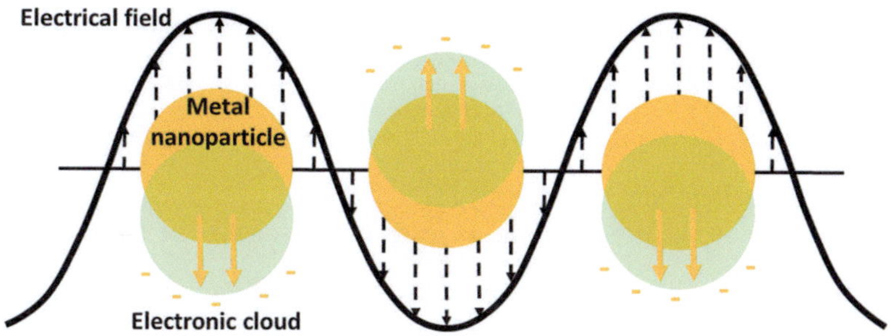

Fig. 12.1 Schematic representation of sinusoidal electric field acting on a metal nanoparticle

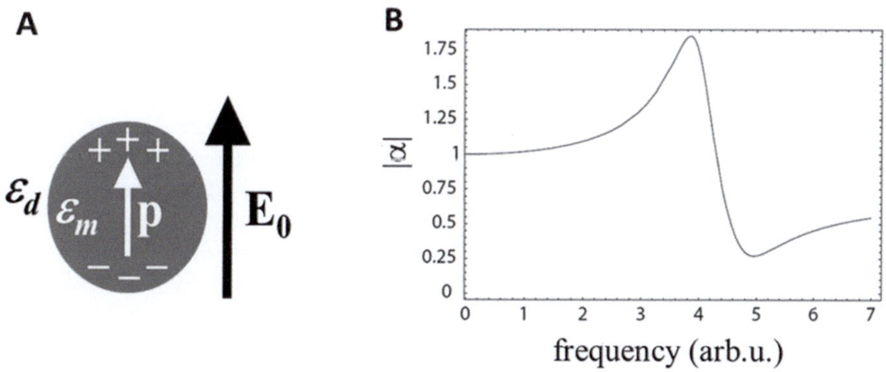

Fig. 12.2 (**a**) Schematic representation of a metal nanoparticle polarization under the effect of a uniform electrical field; (**b**) Graphical representation of the polarizability of that nanoparticle as a function of the electric field frequency

Introducing the electrical polarizability α and considering explicitly its frequency dependence, we can rewrite $\boldsymbol{p}$ as:

$$\mathbf{p}(\omega) = \varepsilon_0 \varepsilon_d \alpha(\omega) \mathbf{E_0}(\omega) \tag{12.3}$$

with:

$$\alpha = 4\pi a^3 \frac{\varepsilon_m(\omega) - \varepsilon_d}{\varepsilon_m(\omega) + 2\varepsilon_d} \tag{12.4}$$

As shown in Fig. 12.2b, the polarizability α has a maximum when:

$$\varepsilon_m(\omega) = -2\varepsilon_d \tag{12.5}$$

Without demonstrating it Ref. [1], the resonance condition occurs when:

$$\omega = \frac{\omega_p}{\sqrt{1 + 2\varepsilon_d}} \tag{12.6}$$

where ω_p is the bulk plasmon frequency [1].

The resonance in the polarizability α leads to a resonance of the local electric field:

$$\mathbf{E}_{dip} = \frac{1}{4\pi\varepsilon_0\varepsilon_d}\left[\frac{3(\mathbf{p}\cdot\mathbf{r})\,\mathbf{r}}{a^5} - \frac{\mathbf{p}}{a^3}\right] \tag{12.7}$$

An important application of LSPRs is in surface enhance Raman scattering (SERS) spectroscopy that takes place when a molecule is located on a metallic nanostructure or a roughened metal surface. At the base of the enhancement of Raman scattering is an increased electromagnetic near field due to excitations of LSPRs. Other contributions comes from more complex effects like a local increasing of optical density of states and a modified molecular Raman cross-section due to the changed environment, that contribute to increase the Raman excitation rate. Chemical effects also contribute to the final result. The highest SERS enhancement recorded to date is about 10^{14}, achieved on roughened silver surfaces [1].

Raman spectroscopy is an optical technique that allows obtaining information on the vibrational spectrum of a sample. The combination of a Raman spectrometer with an optical microscope allows obtaining a chemical bonds map of the sample at a spatial resolution typically around 1 μm (Fig. 12.3a). The main advantages of this technique are: (i) the capability to perform chemical analysis in a label-free environment, for example in the case of a biological sample this reduces the steps necessary to pretreat the sample and avoids any spurious contribution due

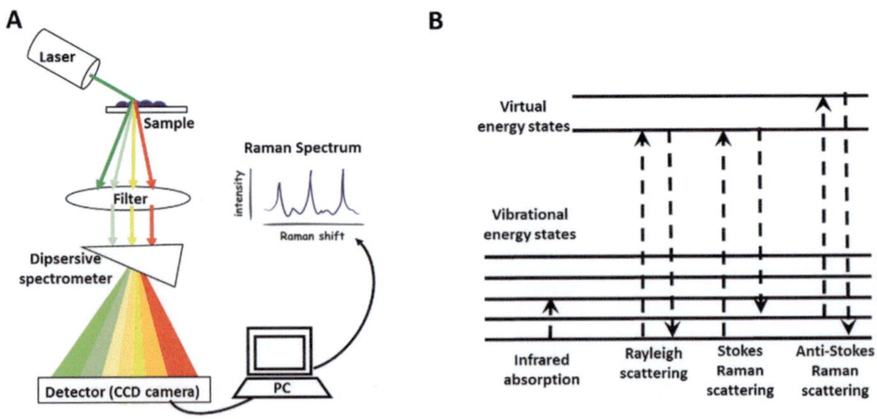

Fig. 12.3 (**a**) Schematic representation of the working principle of a Raman Spectroscopy Instrument (**b**) Scheme of energy levels of scattered light

to introduction of fluorescent agents; (ii) the specificity of this analysis due to the unique fingerprint that each chemical has; (iii) the possibility to analyze biological substances in water/buffer solutions that support life conditions.

The vibrational spectrum is determined by the chemical composition of the sample. In particular, the Raman effect describes the inelastic scattering between a photon and a molecule mediated by its thermally populated vibrational or rotational modes. Due to the interaction, the energy of the incoming photon is shifted by the energy of probed mode. When the photon delivers its energy to the molecular vibration, the phenomenon is called Stokes scattering, while it is called Anti-Stokes scattering when it gains energy by de-excitation of a vibrational mode. This is schematically shown in the Fig. 12.3b and more information can be found in Ref. [2].

Raman transitions can have very sharp lines enabling a detailed analysis of the molecule under study. Typically the Raman cross sections are more than ten orders of magnitude smaller than those of fluorescence processes ($\sigma_{Raman} \approx 10^{-30}$ cm^2/molecule). This is the reason why it is fundamental to enhance the Raman signal through local surface plasmon resonance effects to detect few molecules. For molecules in proximity of metal nanoparticles (typically Au or Ag are the most commonly used metals), the Raman signal is strongly amplified by the coupled LSPR.

Plasmonics undergo a steadily increasing interest in Science for its countless applications in almost any research field in technology and life science. These studies are further motivated by the recent development of nano-fabrication technologies (electron beam lithography or ion beam milling) and of modern characterization techniques (as laser microscopy and scanning near field and electron microscopies).

Nowadays, advances in medicine mainly relies on new technical solutions and more sensitive methods to develop devices with very high sensitivity and selectivity.

When a biological sample, like the body fluids (blood, saliva, plasma, etc.) needs to be analyzed or screened, several challenges and issues need to be faced. These include the severe dilution of the biomolecules of interest, that are often well below the resolution limit of the most used screening procedures or the complexity of the same sample in terms of number of substances contained into it, which require a sample pretreatment allowing to resolve the analyses of a single particular substance. Nanoplasmonic devices offer realistic tools to face these problems. Artificial nano-devices can interact with biological objects (from cells, as large as ten micrometers, to biomolecules sizing few nanometers or less), and arc, therefore, realistic candidates for solving these detection issues. The transition from the micron to the nanoscale dimensional control on surface features translates into increases in information quality, quantity and density, whereby the molecular detection, identification and diagnostics are significantly improved. In particular, nanoscale biophotonic devices, and especially SERS substrates, are capable to detect biological moieties in the limits of very low abundance ranges [3]. A large variety of plasmonic nano structures has been realized for applications in photonics, spectroscopic characterizations, biological sensing and detection of single molecules. These devices feature different designs, sometimes somehow exotic, including electrochemically modified electrodes, colloids, island films [4, 5], adiabatic cones [6], or regular arrays of metallic nano-dots [7].

In the following sections we will show some example of plasmonic nanostructures used for biosensing. In particular we will show how nanospheres, aggregates of nanoparticles and nanostars can achieve SERS effect and could be used to analyze DNA at very low concentrations.

The last section of the chapter will explain how integrating nanoplasmonic structures in a microfluidic device it is possible to simplify biological analysis of single living cells. This sensor allows trapping single cells in proximity of the optical probe. This analysis strategy sorts single cell among a large population thus reducing acquisition time consistently.

12.2 Novel Approaches for Optical Nanostructures Implementation

12.2.1 Plasmonic Self Similar Chains

An electromagnetic field interacting with metallic nanoparticles induces a local field,, as a consequence of the plasma oscillation, as explained in the previous section. For example a single silver nanosphere can provide a field enhancement $g = E/E_0$ around 30, a value that can be enormously increased in the case of several close nanoparticles, for the collective plasmonic oscillation in each nanoparticle. It was found that the highest values of field enhancement occur in particular points, called "hot spots", which fall in the gaps between metal nanoparticles [8]. On the basis of these observations the group of Mark I. Stockman had the idea to theoretically study self-similar systems composed of silver nanospheres, with the aim to create efficient nanolenses displaying hot spots with giant field enhancements [9]. These systems, designed and measured for application in Raman spectroscopy, offer specific active sites that can trigger giant Raman scattering from trapped molecules.

A plasmonic self-similar system may be composed of a chain of metal nanospheres. In this case, we indicate the radius of the ith nanosphere by R_i and its surface-to-surface separation, or gap, from the $(i+1)$th nanosphere by $d_{i;i+1}$. To obtain the self-similarity, the radiuses and gaps must obey to the relations:

$$\begin{aligned} R_{i+1} &= kR_i \\ d_{i+1,i+2} &= kd_{i,i+1} \end{aligned} \tag{12.8}$$

k is a constant.

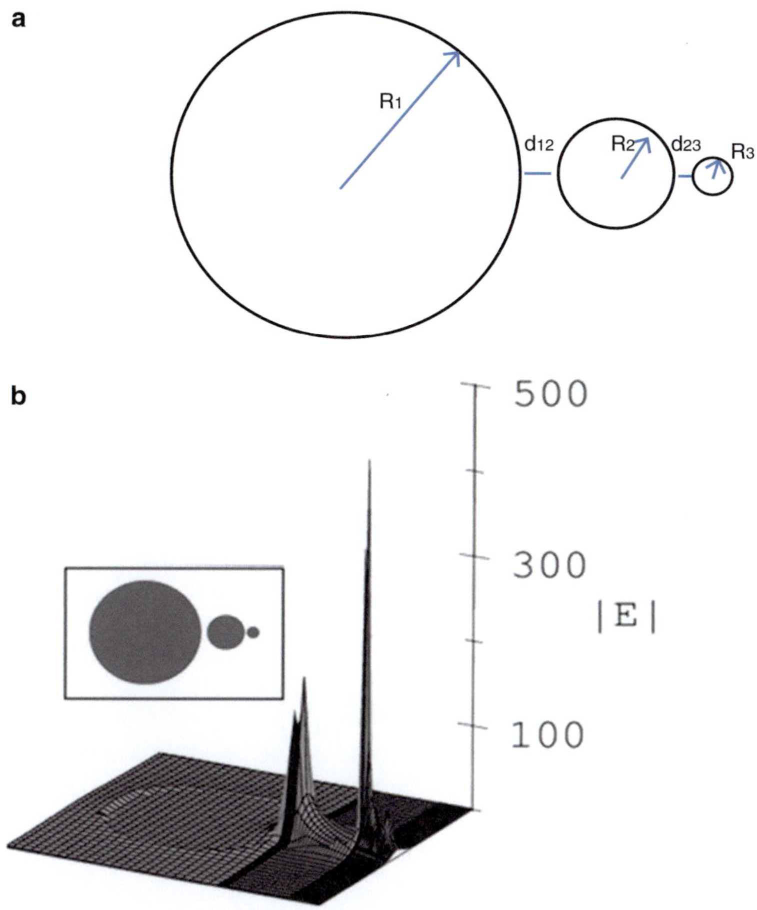

Fig. 12.4 (**a**) Schematic representation of a self-similar chain of three nanospheres. (**b**) Theoretical amplification of the electrical field in proximity of nanospheres in resonance conditions. The inset shows the cross section of the spherical nanoparticles used in the simulation, where the radius of the smallest nanosphere is 10 nm. The values of the modulus of the electric field are normalized on the value of the electric field of the incident beam

In Fig. 12.4, it is represented a schematic drawing of a self-similar chain (SSC) consisting of three metallic nanospheres and a numerical simulation of the total electric field enhancement.

The efficacy of this system relies on the smallest gap, where a very large local field enhancement can be obtained. In fact, if Q is the resonance quality factor

$$Q \approx \frac{\mathrm{Re}\ \varepsilon(\omega)}{\mathrm{Im}\ \varepsilon(\omega)} \tag{12.9}$$

$\varepsilon(\omega)$ is the relative dielectric function of the metal and ω is the photon frequency, the local field E near the biggest nanosphere is enhanced by a factor Q with respect to the exciting field E_0. It is noticeable that for a noble metal $Q \gg 1$. The local field of this sphere excites the next nanosphere. For this second sphere, the excitation contribution from the exciting field E_0 is negligible if compared to the excitation induced by the first bigger one because its polarizability, proportional to its volume, is much smaller (in this case $k \ll 1$). The same phenomenon happens to the following couple of nanospheres, whose gap shows a field enhancement of Q^2, and so on for the following nanospheres. Consequently the *(n-1)th* gap in this cascade produces a field enhancement of Q^n.

The great potentiality of a plasmonic SSC is clearly evident in Raman spectroscopy, where the cascade enhancement finds an optimal application in SERS. The SERS amplification factor g^R goes approximately as the forth power of the local field enhancement g:

$$g^R \cong g^4 = \left(\frac{|E|}{|E_0|}\right)^4 \tag{12.10}$$

where E is the local optical electric field, and E_0 is the electric field of the excitation wave [Ref]. In the case of n nanospheres, g can be expressed as:

$$g(n) = \frac{E_n}{E_0} \approx Q\left(\frac{R_n}{R_1}\right)^{\frac{\ln Q}{\ln k}} \tag{12.11}$$

A SSC of three nanospheres is already an optimal plasmonic nanolens. For example, in the specific case of nanospheres of silver for which Q 10, the smallest gap has an enhancement factor $Q^3 = 10^3$, corresponding to a Raman scattering enhancement of 10^{12} [10, 11].

Experimentally a similar nanolens can be realized using a two steps process (Fig. 12.5). In the first step nanoholes which reproduce a SSC of spheres are drawn on a layer of electronic-resist spun onto a silicon wafer, by means of electron beam lithography. In the last step the nanoholes are filled of silver or gold nanoparticles by electroless deposition (ED).

ED is an autocatalytic redox reaction in which electrons are exchanged between metal ions in solution (here in particular silver or gold) and the substrate, which in this case is a silicon wafer. This specific substrate is particularly suitable for the ED technique due to the presence on its surface of dangling bonds, which act as deep-center defects, with a bound state that lies well below the conduction band [12]. The deposited metal nanoparticles size and density are controlled by parameters such as the concentration of the metal salt in solution (typically $AgNO_3$ or $AuCl_3$), the temperature and the time of the process. According to recent studies, the characteristic of lithographic pattern, in terms of absolute size and density of nanostructures in the pattern affect the process [13].

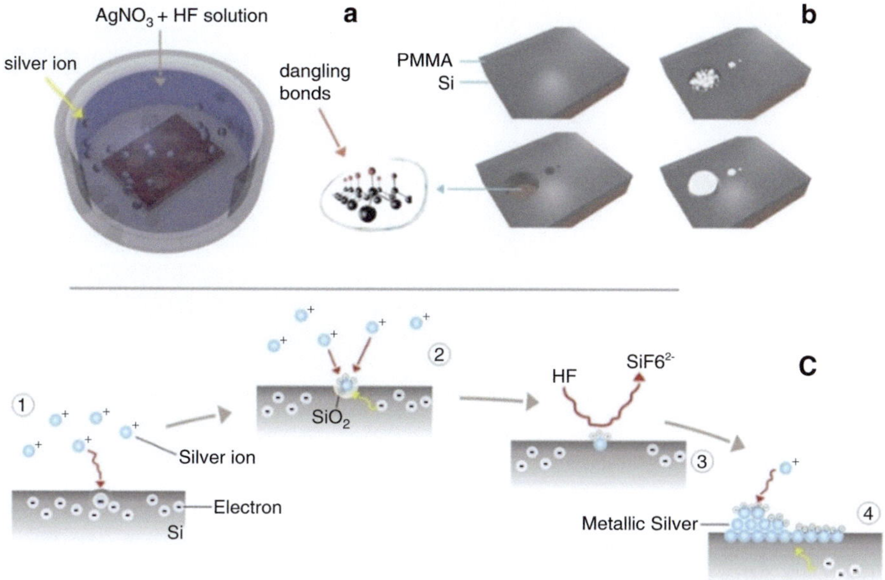

Fig. 12.5 Fabrication process of silver SSC. (**a**) After electron beam lithography and surface treatment with 2 M HF, the sample is immersed in HF/$AgNO_3$ aqueous solution, where Ag + is reduced to silver metal through a redox reaction chain. (**b**) In nanowells (reduction surface), silver growth follows a spherical symmetry and generates three spheres of appropriate diameter and interdistance. (**c**) Redox reactions inside a nanowell starting from the silicon surface

Taking into account the previous considerations, silver SSC can be fabricated by immersion of the patterned silicon wafer in a solution of HF 0.15 M and $AgNO_3$ 0.05 mM for 20 s, at a temperature of 50 °C [14]. Into the patterned nanoholes the clusters of metal nanoparticles take a spherical shape, which in turn is advantageous for the plasmonic applications (Fig. 12.6). The method to create such devices ensures the reproducibility of the fabrication, the high controllability of position, size, and shape of the hot spots and the homogeneity in the field enhancement.

12.2.2 3-D Nanostar Dimers

Couples of 3D star-shaped gold nanostructures on silicon pillars (NOSP) represent another interesting kind of optical system. They can exploit the very high field enhancement that can be achieved when plasmonic nanostructures are tightly coupled.

Electron beam lithography (EBL) on a thick layer of poly(methyl methacrylate) (PMMA) spin-coated on a p-type Silicon wafer was used to define the shape of

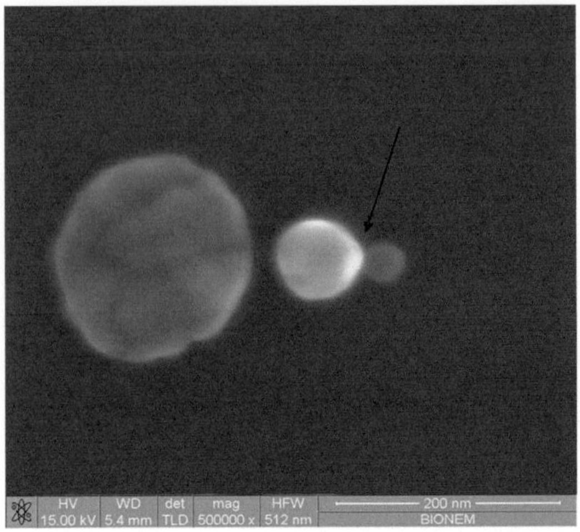

Fig. 12.6 SEM images of a silver SSC. The *arrow* indicates the smallest gap that is about 3 nm

nanostars. This was followed by a deep reactive ion etching process (DRIE) to obtain the pillars, as explained in details in ref [Chirumamilla et al. *Adv. Mater.* 2014, *26*, 2353–2358].

The star shape has the advantage with respect to other nanostructures of having multiple fine tips in the dimer and at the same time large inter-branches spaces important for the success of the DRIE process. Moreover, their complex design is reproducible with an inter-particle spacing (IPS) below 10 nm.

The plasmonic behavior of these nanostructures depends from some parameters: the height of the dimer, the substrate material, and the IPS. In the cited paper, numerical simulations, before experimental tests, calculated electric field enhancement (E/E_0, where E and E_0 are the actual and initial electric fields) versus the dimer height, with the IPS constant. When dimers are directly in contact with Si bulk ($h = 0$ nm), low electric field enhancement is observed, due to the strong spread of the local electric field within the substrate; with the increasing of dimer height, the electric field enhancement increases too until the induced electric field in the dimers is completely decoupled from the substrate.

At small IPS (<10 nm), a strong mutual interaction of the surface plasmons of each nanostar causes a further localization and enhancement of the electromagnetic field in proximity of the IPS region (hot-spots).

These structures enable the detection of biomolecules at very low concentrations (1 pM), i.e. when there is only few or a single molecule per hot-spot. In Fig. 12.7a a sketch of the SERS working principle of a nanostars dimer is reported. Figure 12.7b, c show representative SEM images of the device and in Fig. 12.7d shows the SERS signal from a single or a few adenine molecules deposited on a NOSP device by chemisorption from a 1 pM solution.

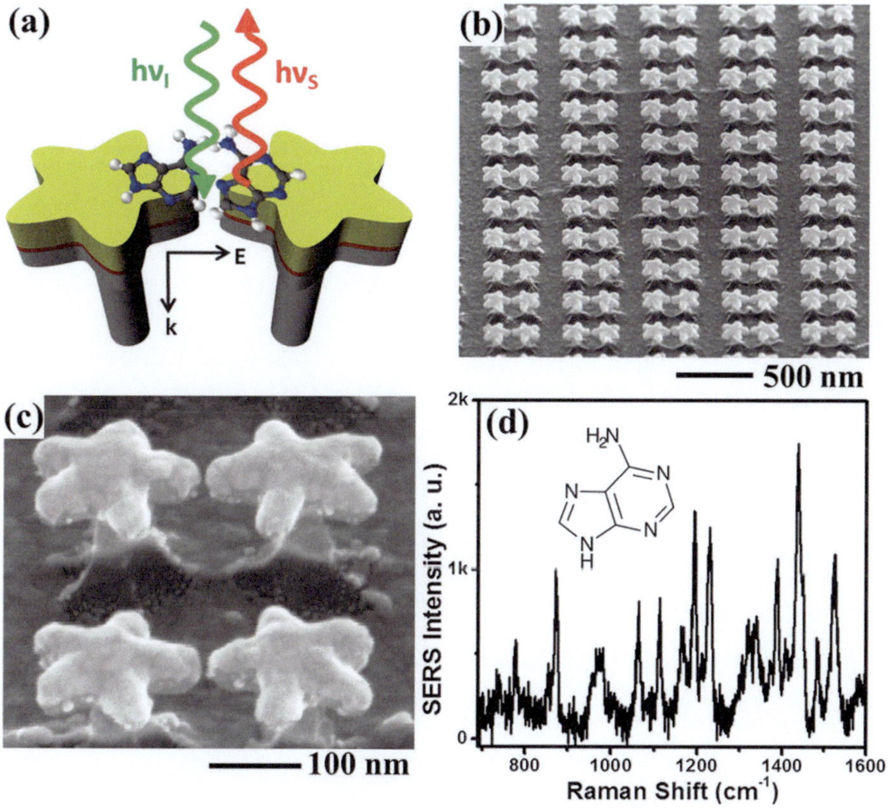

Fig. 12.7 (**a**) Illustration of the SERS working principle for a free-standing 3D nanostar dimers. *E* and *k* are the electric field and wave vector, respectively. *H*νi and *h*νs represent the incoming and scattered photon energy, respectively. (**b**, **c**) SEM images at a 52° tilted view of the NOSP dimers with a 6-nm IPS and 150-nm Si pillar height. (**d**) Raman spectrum recorded from adenine that was chemisorbed from a solution at 1-p M concentration. The k in the intensity scale denotes 1000. *Inset*: chemical structure of adenine. [14]

12.2.3 Silver Decorated Nanotips

Metallic cones can be used as plasmonic concentrators of electromagnetic energy. Their ability to focus and to amplify the optical radiation was studied by theoretical calculations and also experimentally demonstrated. In particular, silver and gold nano-cones can sustain adiabatic propagation and concentration at their apex of SPPs, with efficiency depending on their shape. Small apex angles provide optimal focusing and signal enhancement. This can be understood considering that SPPs coupled at the base of the cone from an incident far-field laser propagate

Fig. 12.8 SEM micrographs of cones. Cone before (*left*) and after silver electroless deposition (*right*), displayed here for two geometries: $\varphi = 1.7°$ (*top*) and 25° (*bottom*), realized with a fixed silver salt concentration of 1 mM [18]

towards the tip along the taper with decreasing wavelength and speed. As a result, SPPs accumulate at the tip of the cone providing a strong amplification of the electromagnetic fields.

The nano-cones shown in this section in Fig. 12.8, are realized by Electron Beam-Induced Deposition (EBID), injecting into the chamber of a dual-beam system an organometallic platinum-carbon gas precursor $(CH_3)_3Pt(CpCH_3)$ [15–17]. A pattern generator was used to adjust the parameters and consequently obtain different sizes of the cones.

Bare nanocones can be used for SERS applications, exploiting the field localization at the apex. By deposition on the cone surface of gold or silver nanoparticles, e.g. by electroless deposition, Raman signal emitted from molecules on the cones can be increased [18].

The peculiarity of this integrated system is the possibility to have a tridimensional probe in which the SERS efficiency may be adjusted simply varying the size and density of the silver nanoparticles from which the plasmonic behavior depends. At this purpose, it is possible to act on one hand on the concentration of the silver salt solution and on the other on the apex angle and height of the tip, since metal nanoparticles deposition is guided from the tip geometry itself. Silver nanoparticles in each case would preferentially distribute along the cone, with higher concentration at its apex (Fig. 12.8). For example, silver decorated tips has been used as SERS substrates for analysis of Rhodamine 6G. That result suggests their implementation in microfluidic channels for analysis of complex mixtures of biological elements.

12.3 SERS for Ultra-sensitive Detection of Biological Samples

As described above, plasmonic nanostructures, in particular SSCs of silver nanospheres or dimers of metal nanostructures, allow the analyses of complex mixture or of molecules also in highly diluted conditions by Raman spectroscopy.

The strong plasmon localization, which is the peculiar characteristic of these nanostructures, produces a field enhancement ideal for SERS applications, so that their employment in Raman spectroscopy constitutes a powerful means to solve analytical problems with great sensitivity and specificity.

12.3.1 Single Amino Acid Mutation Detection

The micro-Raman spectrum of a biological sample is a complex mixture of signals coming from different macromolecules. The Raman bands originated from each component or molecular portion overlap each other, consequently the method is limited and not suitable for single-molecule or few-molecules detection so important in molecular medicine.

The advantage of silver SSCs with respect to another SERS device relies on the predetermined geometrical localization of the enhancement effect: a SERS device without this property amplifies Raman signal coming from all the substances deposited on it, without any spatial discrimination capability.

SSCs of silver nanospheres have been designed, fabricated and utilized as SERS substrate for early disease detection of breast cancer [19].

Coluccio et al. [19] adopted silver SSCs to detect biomarkers implicated in breast cancer development in human samples. Specifically they searched the single amino-acid mutation promoting cancer in a mixture of peptides. The BRCA1 protein is a tumor suppressor gene whose germline mutations predispose to breast and ovarian

Fig. 12.9 SEM image at low magnification of a m × n matrix of nanolenses (SSCs). Each brigh spot in the figure is due to a single nanolens

cancers [20]. The protein is functionally organized in domains located at the N terminus and C terminus. The C terminus has two amino-acid repeats, named BRCT, having important roles in the stability of the BRCA1 protein conformation and in DNA transcription. This is the reason for which a mutation in the BRCT domain is an index of increased risk of developing breast and ovarian cancers. M1775R peptide is one of these missense mutations, where a methionine is substituted by an arginine at the fixed sequential position 1775 [21].

The direct detection of these single amino-acid mutations in the BRCA1 protein is impossible through Raman or other spectroscopic techniques because the signal of the mutation is overwhelmed by that of the whole protein.

m × n matrices of SSCs, with their high sensitivity, represent a good solution for this problem (Fig. 12.9).

The steps necessary for this kind of analysis by SSCs are: (i) preparation of the biological sample by breaking into mixtures of peptides the BRCA1 protein after the extraction from blood; (ii) deposition of each mixture on a SSCs array by drop-cast; (iii) drying of the samples in air; (iv) SERS measurement on each single nanolens of the matrix; (v) multivariate analysis.

At this purpose, Coluccio et al. utilized a matrix of 10 × 10 SSCs, each made of three silver nanospheres, prepared by electron beam lithography and silver electroless deposition, as described before. Each single nanolens of the matrix is characterized by an intense hotspot localized in the smallest gap, whose enhancement ability depends mainly from the control of the dimension of the smallest nanosphere (10 nm) and the smallest gap itself (<10 nm). For obtaining good results with this detection system, the concentration of the biological sample should be accurately calibrated and the SSCs must be built with the proper dimensions. In fact, the concentration of the mixture determines the number of molecules that can fit into the volume of each hotspot: the average mutual distance of the molecules after deposition should be comparable with the linear size of the hotspot, in order that only few molecules can be found in the smallest gap after the solution has dried.

Regarding the choice of the matrix dimensions ($m \times n$), a number of elements high enough to have all types of molecules statistically represented in the entire device is a fundamental condition to discriminate the components inside the mixture.

Exploiting the 10×10 matrix of SSCs, two fractions of around 12 peptides were analyzed by Raman spectroscopy in properly diluted solutions. One solution contained the M1775 peptide and the other the M1775R peptide.

At this point, detection of mutated peptides in BRCA1 becomes an analytical problem of finding the elemental composition of the multicomponent mixture. The full matrix is then measured, point by point, by a Raman spectrometer and the multicomponent spectrum at each position is fitted with a linear combination of the base elements, previously collected. A Raman spectrum, s_M, containing different peptids, can be fitted by the function:

$$Y(f) = \sum_{i,j}^{n,m} x_{i,j} s_{i,j}(f) \tag{12.12}$$

the index i runs over the number of peptides ($i = 1, \ldots, m$), whereas j runs through the number of intervals ($j = 1, \ldots, n$) in which spectra are divided. m is the wave vector range in the Raman spectra. The coefficient $x_{i,j}$ Are calculated by a least squares fit of Y with the experimental values s_M. By repeating the fitting procedure for all $m \times n$ matrix elements, all peptides extracted from BRCA1 are recognized [19].

This procedure ensures the possibility of detection of the mutated species in the mixtures. In Fig. 12.10b the Raman spectra of the peptide M1775 from pure samples, wild type and mutated, highlight the differences in the spectra between the two samples.

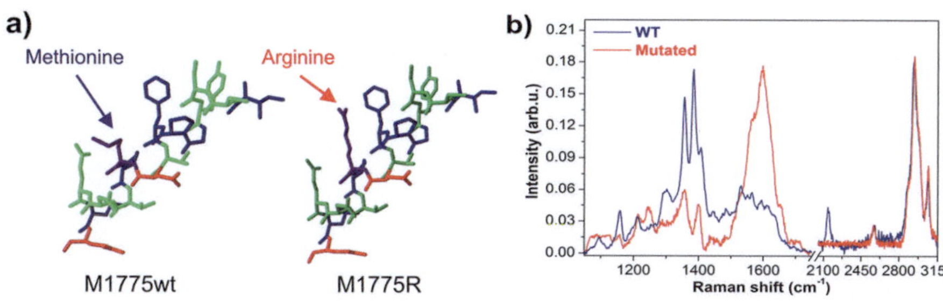

Fig. 12.10 Structure of wild-type and mutated peptides (**a**) and their Raman spectra (**b**) showing a clear difference between the spectra of the two peptides despite they are different by only the exchange of one amino acid (a methionine replaced by an arginine)

12.3.2 From Nucleotides to DNA Analysis

Another interesting application of SSCs is the analysis of macromolecules and, in particular, the DNA sequencing. The working principle is the same as for the peptide mixture analysis.

The use of the SERS effect for oligonucleotides and DNA analysis could be attractive thanks to the efficiency of the technique and to its label-free character. Moreover, SERS can be combined with a microfluidic device to analyze oligonucleotides, label-free RNA or DNA in diluted solutions [22]. The advances in nanotechnology have allowed the development of such kind of devices, potentially very useful in medicine as instruments of diagnosis.

Recently an electrochemical surface-enhanced Raman spectroscopy aptasensor based on a functionalized silver nanoparticle was implemented for direct and specific detection of DNA hybridization for diagnosis in tuberculosis [23].

On the same line, the paper of Coluccio et al. [13] proposes the matrices of SSCs combined with data analysis to achieve label-free detection of DNA portions. Improving conventional SERS systems, it demonstrates the possibility to detect a single mutation or a very small fragment of nucleic acid by measuring changes in the spectrum that would be undetectable using conventional Raman.

The implemented device is still a matrix of $m \times n$ SSCs. The measurement was performed collecting the signal from any individual SSC generated by the fraction of nucleotides drop-cast on them.

The authors report the analysis of 6 and 9 bases oligonucleotides, which differ only for few bases. For each one, the most relevant spectra were compared to each other and to the spectra of isolated DNA bases and then fitted using a linear combination of the nucleobases. An example of the obtained spectra and of the array used is reported in figure (Fig. 12.11).

The SERS spectra of the oligonucleotide reported in figure exhibit few common bands at which other vibrational bands are superimposed, with differences that are attributed to the different orientations of the adsorbed molecules on the surface. The spectra were fitted to the expression in Eq. 12.12.

This led, in the case of a nucleotide with 9 bases (CCC GCC GCT), to a calculated relative content of the bases that resulted in very good agreement with the known real values, as can be seen by the following comparison

$A = 0.044$
$C = 0.565$
$G = 0.2256$
$T = 0.161$

These values are comparable to the theoretical, which are:

$A = 0$
$C = 0.6667$
$G = 0.2223$
$T = 0.1111.$

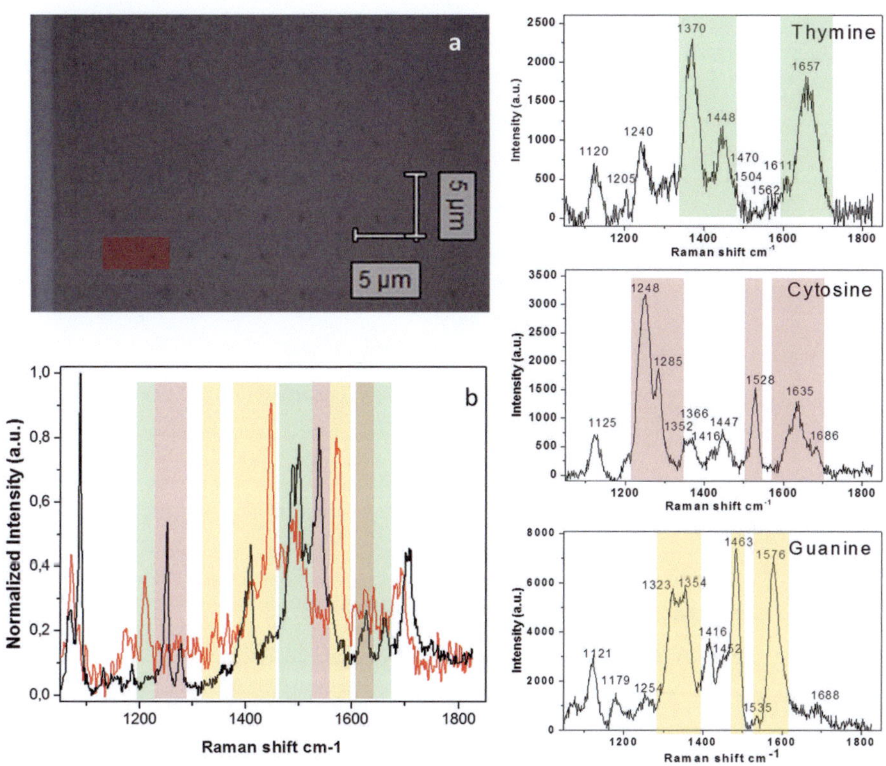

Fig. 12.11 Oligonucleotide p9 (CCCGTT): (**a**) optical image of the device. The *red* rectangle highlights the Raman mapped area around two nanolenses; (**b**) SERS spectra obtained in two hot spots of the Raman mapping compared with SERS spectra of the three DNA bases. The colored bands evidence some of the principal peaks of the three nucleotides [13]

The method just described can be applied in all cases in which the dimension of the nucleotide is less than 3 nm and thus two or three molecules at most fall into the hot spot of each nanolens. Figure 12.12 sketches how the Raman signal only comes from the hot spot of a nanolens discriminating the molecules, or the portions of a molecule, held in the area where the strong localization of the electromagnetic field is confined.

In the case of DNA analysis, the molecule is much longer than oligonucleotides. Since the hot spot for field enhancement in one of our SSCs has a dimension that is limited by the value of the last gap to 10 nm, which also poses a limit on the maximum extension of the hot spot in all directions, it turns out that the maximum number of basis that can be probed at the same time is about 30, being single nucleotide 0.33 nm in length. As a consequence, DNA or oligonucleotides with a larger number of bases will give very different spectra depending on their specific portion that falls into the hot spot (Fig. 12.13).

This SSCs, coupled with a device for realizing strand and scrolling of a DNA filament, could become a relevant technique for DNA sequencing.

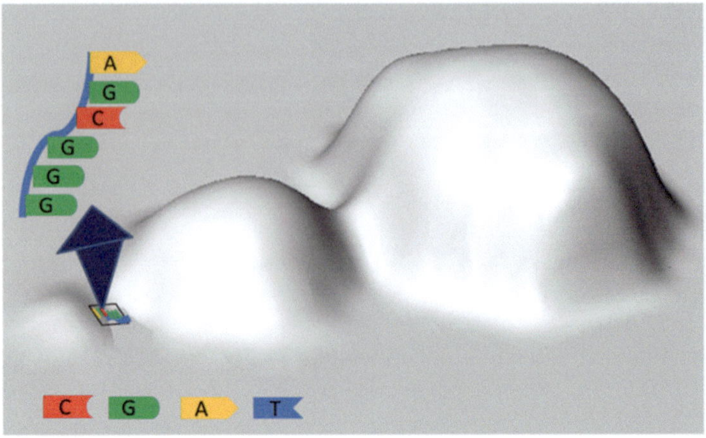

Fig. 12.12 Schematic representation of a Raman analysis on a layer of biomolecule in the smallest gap of a SSC: the Raman signal from the smallest gap dominates the intensity spectra

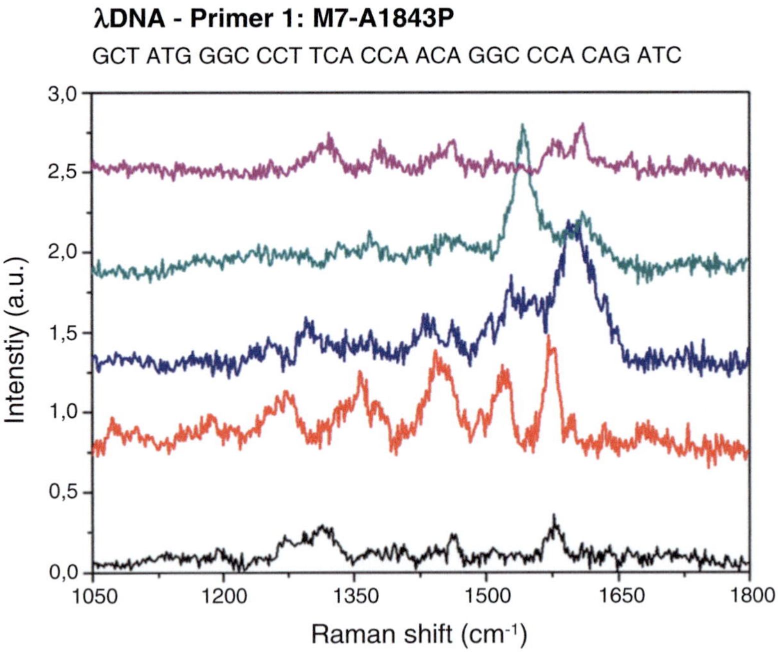

Fig. 12.13 Spectra of 33 bases portions of λDNA measured on various nanolenses

12.4 Microfluidic Devices Integrating Plasmonic Nanostructures for Biological Sensing

12.4.1 Raman Spectroscopy Compatible Microfluidic Devices for Cell Analysis

Microfluidics focuses on the transport, manipulation and analysis of small amount of liquids, cells and particles. Microfluidic devices have been developed and exploited from several research groups as bioanalytical tools to give several advantages over conventional bioanalytical technologies. In fact, these devices guarantee high portability, simplified pretreatment protocols, accurate handling and low consumption of samples and reagents, as well as high resolution analysis and low production costs [24].

There are many examples in literature of microfluidic devices used for analytical purposes [25, 26]. These have been developed for DNA detection [27–29], protein analysis [30–32] and biomarkers [33, 34], discriminating healthy from tumor cells [35, 36], phenotypes screening [37, 38], and cells mechanical stress and deformability investigations [39–41].

Among these applications, a microfluidic device capable of analyzing the composition of a single cell at a subcellular level would push a step forward the conventional analysis on cells.

It is well known that individual cells, even those identical in appearance, differ in many characteristics. Due to this heterogeneity, traditional biochemical assays, which analyze cells in bulk, do not allow to get specific information at single cell level allowing investigating their activities or testing possible new biomarkers. Microfluidics allows a sequential handling and manipulation of cells which is significant for scientific research and clinical diagnostic applications.

Commonly, monitoring of cells over a long period of time is most frequently performed by fluorescence microscopy. However, it is quite common that the fluorescent dyes used for specific subcellular staining interfere with the development of cells, changing their phenotype and metabolism. Another limitation is that fluorescent dyes bleach quite fast. Finally, this technique only allows investigating molecules already known. These aspects make fluorescence staining strategies less desirable with respect to label free sensing.

This can be done using Raman Spectroscopy combined with plasmonic nanostructures integrated into microfluidic devices.

Raman spectroscopy has been largely used to determine cell conditions, such as living cells [42], dead cells [43], apoptotic cells [44], proliferating cells [45], and differentiating cells [46]. Raman spectroscopy can be integrated in microfluidic devices to achieve an accurate monitoring of single cells, allowing them to stay over long period in physiological or controlled environments [47–49].

Finally, the integration of plasmonic nanodevices in the microfluidic tool allow increasing significantly the resolution of analysis.

Raman spectra are only minimally affected by the presence of water. This peculiarity is crucial when dealing with measurements on living cells that are always immersed in an aqueous buffer.

On the other hand, when microfluidic devices are combined with Raman Spectroscopy it is necessary to select only materials that give negligible contribution to the Raman spectra. This constrains leads to exclude all the polymers which give a strong Raman signal in the characteristic spectral range of cells, being polymers made by carbon chains, similarly to biomaterials. The concurrent strategies are: (i) use materials which give different Raman bands from the biological species or negligible signal amplitude; (ii) design a proper geometrical configuration to reduce the background, like for instance using metal shields and working in reflection [39];

In the following section, we show an example of a microfluidic device used to take measurements by Raman Spectroscopy on living cells [50]. The device is based on the use of a microfluidic trap integrated with a plasmonic Raman enhancer. It is composed by two slides of calcium fluoride, giving a negligible Raman signal, between which a photosensitive resist is placed. Five through-channels, integrating a microfluidic trap, are fabricated in the resist by photolithography, connected to external tubes by means of fluidic gaskets and a mechanical holder. A picture of the final device is shown in Fig. 12.14.

The trap is realized by means of a simple volumetric constriction of the channels (Fig. 12.15a).

Cells flowing in the microchannel can be blocked at each trap where a Raman spectrum can be collected with a minimum acquisition time of few seconds. The

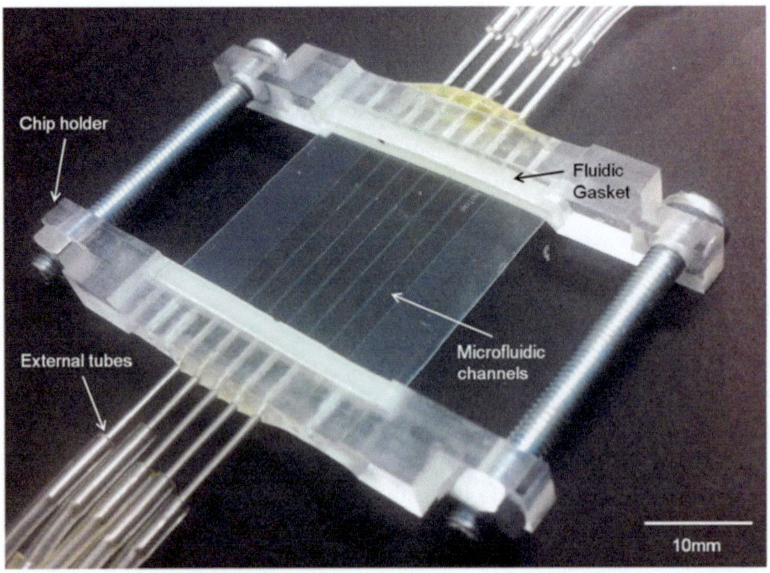

Fig. 12.14 Microfluidic device for Raman Spectroscopy on living cells

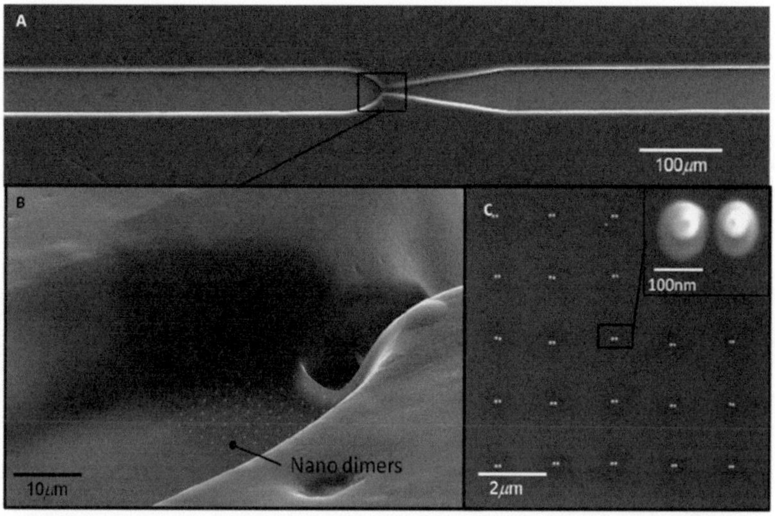

Fig. 12.15 Scanning electron image of the microfluidic trap and the integrated array of nanodimers. (**a**): Top view; (**b**): isometric view; (**c**): zoom in the area where the nanodimers are integrated

traps integrate plasmonic nanodimers (Fig. 12.15b, c). These plasmonic nanodevices enhancing Raman scattering increase the sensitivity and enrich the information about the cellular membrane and the cytoplasm of the trapped cell.

In principle, this geometry can be tailored for every kind of cell simply adjusting the size of the channels and of the traps.

These devices are designed to control a temporarily trapping by dynamically changing the flow conditions. When a constant flow rate is applied, a cell is moved toward the trap where it stops. Here the fluidic resistance and the pressure drop across the cell increase locally. The resulting force acting on the cell causes its deformation and pushes it through the trap (Fig. 12.16). After that, another cell can approach the trap. The controlled flow determines the cell transit time, approximately of 3 s, enough to record a Raman spectrum. In such a way, it is possible to perform a continuous analysis of cells by Raman Spectroscopy in a passive way.

Such a device can be used in different ways: (i) the same cell can be moved forth and back in the trap by changing direction of the flow so that measurements on different portion of the same cell can be taken [Perozziello et al. microelectronic Engineering]; (ii) one single measurement on a cell sitting on the nanodimers in the field of view can be taken, getting information from one single cell membrane. The measurements from single cells of the same type can be also integrated together; (iii) micro-Raman by using objectives with higher NA can be taken, in this case the measurement can be used to search for specific information about the cell membrane and cytoplasm.

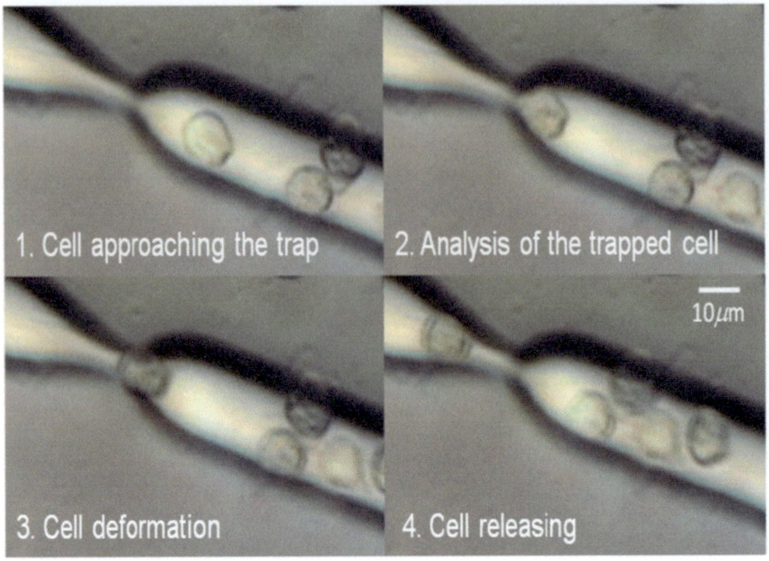

Fig. 12.16 Sequence of the cell analysis mechanism: (*1*) the cell approaches the trap followed by other cells (t = 0 s); (*2*) the cell is trapped and the flow is stopped (t = 2 s), in this phase Raman measurements can be performed on the cell (t = 4 s); (*3*) the increasing pressure due to clogged channel deform the cell forcing it through the trap (t = 6 s); (*4*) the cell is released continuing to move over the trap while other cells are approaching the trap aligning themselves to it (t = 8 s)

The microfluidic devices were fabricated on calcium fluoride (CaF_2) slides on which a laminated dry photoresist was attached. The microchannels were defined by means of UV-photolithography. Subsequently, the plasmonic nanodevices were fabricated on top of the CaF_2 substrate (Fig. 12.15b, c). These consist in a matrix of dimers of 100 nm diameter gold nanodots, separated by a gap of 10 nm. These were fabricated using a FEI Nova Nanolab 600 dual beam focused ion beam system introducing an organometallic gas precursor (CH3)3Pt(CpCH3) to produce the PtC alloy. The electron acceleration was set at 30 keV and the current at 0.15 nA. The resulting matrix, covering a square area of 20 μm in size with a 2 μm pitch, was then sputtered with 20 nm of gold and ion beam milled [50]. Subsequently, another slide of calcium fluoride was bonded using a hot press on top, to seal the micro channels.

12.4.2 *Single Living Cells Study*

The instrument used to perform Raman spectroscopy on the cells was a Renishaw inVia Raman Microscope mounting a 50X objective. Syringe pumps were connected to the microfluidic devices by Teflon tubes (0.4 mm ID -Internal Diameter), loaded with the biological samples. Such set-up was used to inject 1 ml of biological

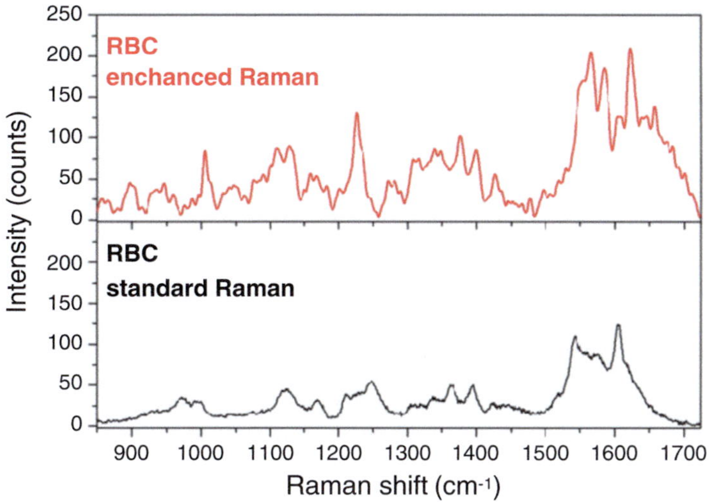

Fig. 12.17 Raman spectra collected from *red* blood cells (RBCs) on nanodimers (used power for the enhanced Raman measurements is 10 % of the total) and on a flat substrate (used power for the standard Raman measurements is 100 % of the total). The spectrum recorder on nanodimers exhibits a stronger signal and with higher detail spatial mapping

sample containing cells diluted at a concentration of about 5000 cells/μl in PBS. The flow rate was set at 5 μl/min so that the cells could proceed very slowly and be stopped in correspondence of the trap. For the measurements, a laser with a near-infrared wavelength (832 nm) was used. The power used to analyze the sample was 10 mW and the signal collection time was 3 s. During Raman measurements the objective was focused at the level of the dimer gap in order to fully exploit the signal enhancement.

The large enhancement of the electromagnetic field at the gap of the dimers produces an enhancement of the Raman signal of biomolecules, as proved comparing the measurements taken on red blood cells trapped on top of the nanodimers and on a flat surface (Fig. 12.17). While the measurements taken on a flat substrate needed a laser power of a100 mW to record a spectrum, only a power of 10 mW was needed to obtain a comparable signal on nanodimers. This is a very important point to consider when dealing with living biological samples since a high laser power can affect the viability of the cells.

The SERS spectrum showed not only a higher signal, compared to standard Raman, but above all the curve exhibits a richer peak profile due to plasmon-enhancement of biomolecular vibrations occurring at the dimers locations.

Figure 12.18 shows Raman spectra collected over red blood cells (RBCs), peripheral blood lymphocytes (PBLs) and K562 tumor cells from leukemia in correspondence of the trap. Each spectrum is the average of the signals recorded from ten cells of the same type to provide a typical spectrum for each specific type. We stress that each SERS spectrum brings information from a very localized part

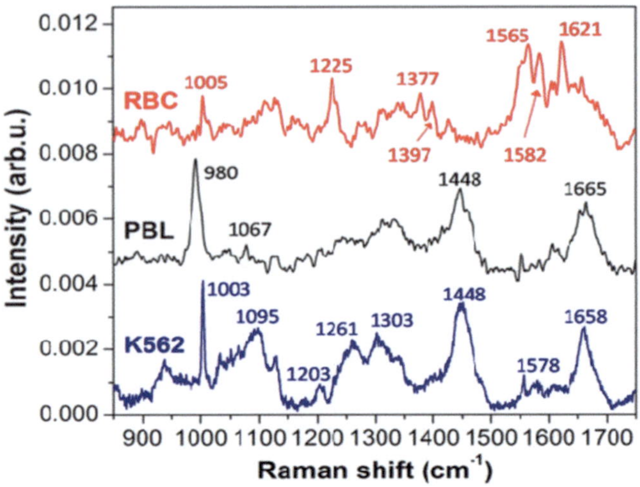

Fig. 12.18 Raman spectra collected in three different experiments with different cell lines, in the proximity of the microfluidic traps. From *top* to *bottom* average spectra of red blood cells (RBC), peripheral blood lymphocytes (PBL), and K562 tumor cells from leukemia are reported

of the cell and in particular from its membrane. In standard Raman spectroscopy, even using a high NA objective, the measurements on cells is an average of this information with that from the cell cytoplasm.

Concerning the red blood cells, it is possible to observe the typical Raman peaks of phenylalanine at 1005 cm^{-1}, the deformation of the C_m-H bond at 1225 cm^{-1}, the half- and quarter-ring stretching of pyrimidine respectively at 1377 and 1397 cm^{-1}, and finally several peaks of the multi-structured band between 1540 and 1630 cm^{-1}. The oxygenated state of RBCs can be deduced by the relative intensities of the 1565 cm^{-1} peak of the CβCβ stretching, the one at 1582 cm^{-1} of CαCm asymmetric stretching, and the 1621 cm^{-1} peak of Ca = Cb stretching, as already reported in literature [51],

The signature of PBLs is evidenced by the deoxyribose vibrations, at 980 and 1448 cm^{-1} [52], the phospholipid C-C stretching at 1067 cm^{-1}, and the Amide I vibration from proteins secondary structure at 1665 cm^{-1}. The broadening of 1448 cm^{-1} peak is due to the overlap of deoxyribose signature with C-H deformation from protein.

Finally, the last curve shows the Raman signal from K562 cells. Comparison with literature shows a very good agreement: the phenylalanine ring mode is at 1003 cm^{-1} [53, 54]; the PO2- backbone vibration arises at 1095 cm^{-1}; Amide III vibrations from proteins appear at 1203 and 1303 cm^{-1}, where the former one overlaps with DNA bases signal; C-H bending and deformation are found at 1261 and 1448 cm^{-1}, respectively. A small DNA signal is also observed at 1578 cm^{-1}, and finally the Amide I strong vibration from proteins is located at 1658 cm^{-1}. Compared to PBL, the Amide I shift towards lower frequencies (from 1665 to

1658 cm^{-1}) is likely due to a larger presence of alpha helix structure of proteins. This is also supported by the small peak observed at 935 cm^{-1} (alpha helix C-C skeletal mode) for the K562 curve.

12.5 Conclusions

In this paper we collected part of our activity on nanoplasmonics for sensing and showed the application in biosensing. It is clearly appearing that the combination of plasmonics, microfluidics and spectroscopy is an interesting strategy for improving the detection sensitivity, specificity and complexity of mixtures.

The combination with microfluidics opens the possibility to extend the detection to single molecule but increasing the throughput analysis. It is our belief that this approach will become of commercial interest and medicine with accurate and personalized diagnosis will benefit from this approach.

Acknowledgments We acknowledge all contributors that contributed in this work from other institution, in particular the researchers from the previous Nanostructure Division at IIT (Italian Institute of Technology) in Genoa.

We thank for the financial support the King Abdullah University of Science and Technology start-up fund and the Italian Minister of Health (projects nos. GR-2010-2320665 and GR-2010-2311677)

References

1. Maier, S. A. (2007). *Plasmonics: Fundamentals and applications*. New York: Springer. ISBN 0-387-33150-6.
2. Giugni, A., Torre, B., Allione, M., Gentile, F., Candeloro, P., Coluccio, M. L., Perozziello, G., Limongi, T., Marini, M., Raimondo, R., Tirinato, L., Francardi, M., Das, G., Proietti Zaccaria, R., Falqui, A., & Di Fabrizio, E. (2015). Novel plasmonic probes and smart superhydrophobic devices, New tools for forthcoming spectroscopies at the nanoscale. *NATO Science for Peace and Security Series B: Physics and Biophysics, 68*, 209–235.
3. De Angelis, F., Gentile, F., Mecarini, F., Das, G., Moretti, M., Candeloro, P., et al. (2011). Breaking the diffusion limit with super-hydrophobic delivery of molecules to plasmonic nanofocusing SERS structures. *Nat Photonics, 5*, 682–7.
4. Le Ru, E. C., Etchegoin, P. G., Grand, J., Felidj, N., Aubardì, J., & Lévi, G. (2008). Surface enhanced Raman spectroscopy on nanolithography-prepared substrates. *Current Applied Physics, 8*, 467–70.
5. Qiu, T., & Chu, P. K. (2008). Self-selective electroless plating: An approach for fabrication of functional 1D nanomaterials. *Materials Science and Engineering, 61*, 59–77.
6. De Angelis, F., Das, G., Candeloro, P., Patrini, M., Galli, M., Bek, A., et al. (2010). Nanoscale chemical mapping using three-dimensional adiabatic compression of surface plasmon polariton. *Nature Nanotechnology, 5*, 67–72.
7. Das, G., Mecarini, F., Gentile, F., De Angelis, F., Kumar, M. H. G., Candeloro, P., et al. (2009). Nano-patterned SERS substrate: Application for protein analysis vs. temperature. *Biosensors and Bioelectronics, 24*, 1693–9.

8. Michaels, A. M., Jiang, J., & Brus, L. (2000). Ag nanocrystal junctions as the site for surface-enhanced Raman scattering of single rhodamine 6G molecules. *The Journal of Physical Chemistry B, 104*, 11965.
9. Li, K., Stockman, M. I., & Bergman, D. J. (2003). Self-similar chain of metal nanospheres as an efficient nanolens. *Physical Review Letters, 91*, 227402.
10. Dai, J., Čajko, F., Tsukerman, I., & Stockman, M. (2008). Electrodynamic effects in plasmonic nanolenses. *Physical Review Letters, 77*, 115419.
11. Schofield, S. R., Studer, P., Hirjibehedin, C. F., Curson, N. J., Aeppli, G., & Bowler, D. R. (2013). Quantum engineering at the silicon surface using dangling bonds. *Nature Communications, 4*, 1–7.
12. Coluccio, M. L., Gentile, F., Francardi, M., Perozziello, G., Malara, N., Candeloro, P., Di Fabrizio, E., et al. (2014). Electroless deposition and nanolithography can control the formation of materials at the nano-scale for plasmonic applications. *Sensors, 14*, 6056–6083.
13. Coluccio, M. L., Gentile, F., Das, G., Perozziello, G., Malara, N., Alrasheed, S., Candeloro, P., & Di Fabrizio, E. (2015). From nucleotides to DNA analysis by a SERS substrate of a self similar chain of silver nanospheres. *Journal of Optics, 17*, 114021.
14. Chirumamilla, M., Toma, A., Gopalakrishnan, A., Das, G., Zaccaria, R. P., Krahne, R., Rondanina, E., Leoncini, M., Liberale, C., De Angelis, F., & Di Fabrizioe, E. (2014). 3D nanostar dimers with a sub-10-nm gap for single-/few-molecule surface-enhanced Raman scattering. *Advanced Materials, 26*, 2353–2358.
15. De Angelis, F., Proietti, R. Z., Francardi, M., Liberale, C., & Di Fabrizio, E. (2011). Multi-scheme approach for efficient surface plasmon polariton generation in metallic conical tips on AFM-based cantilevers. *Optics Express, 19*, 22268–79.
16. Proietti, R. Z., Alabastri, A., De Angelis, F., Das, G., Liberale, C., & Toma, A. (2012). Fully analytical description of adiabatic compression in dissipative polaritonic structures. *Physical Review B, 86*, 035410.
17. Proietti, R. Z., De Angelis, F., Toma, A., Razzari, L., Alabastri, A., & Das, G. (2012). Surface plasmon polariton compression through radially and linearly polarized source. *Optics Letters, 37*, 545–7.
18. Coluccio, M. L., Francardi, M., Gentile, F., Candeloro, P., Ferrara, L., Perozziello, G., & Di Fabrizio, E. (2016). Plasmonic 3D-structures based on silver decorated nanotips for biological sensing. *Optics and Lasers in Engineering, 76*(9), 45–51.
19. Coluccio, M. L., Gentile, F., Das, G., Nicastri, A., Perri, A. M., Candeloro, P., Perozziello, G., Proietti, R., Totero-Gongora, J. S., Alrasheed, S., Fratalocchi, A., Limongi, T., Cuda, G., & Di Fabrizio, E. (2015). Detection of single amino acid mutation in human breast cancer by disordered plasmonic self-similar chain. *Science Advances, 1*(8), e1500487.
20. Kiyotsugu, Y., & Yoshio, M. (2004). Role of BRCA1 and BRCA2 as regulators of DNA repair, transcription, and cell cycle in response to DNA damage. *Cancer Science, 95*, 866–871.
21. Scully, R., Ganesan, S., Vlasakova, K., Chen, J., Socolovsky, M., & Livingston, D. M. (1999). Genetic analysis of BRCA1 function in a defined tumor cell line. *Molecular Cell, 4*, 1093–1099.
22. Prado, E., Colin, A., Servant, L., Lecomte, S., & Phys, J. (2014). SERS spectra of oligonucleotides as fingerprints to detect label-free RNA in microfluidic devices. *Journal Physical Chemistry C, 118*, 13965–71.
23. Wu, Y. C., Lo, W. Y., Chiu, C. R., & Yang, T. S. (2006). Surface enhanced Raman spectra of oligonucleotides induced by spermine. *Journal of Raman Spectroscopy, 3*, 799–807.
24. Perozziello, G., Møllenbach, J., Laursen, S., di Fabrizio, E., Gernaey, K., & Krühne, U. (2012). Lab on a chip automates in vitro cell culturing. *Microelectronic Engineering, 98*, 655–658.
25. Whitesides, G. M. (2006). The origins and the future of microfluidics. *Nature, 442*(7101), 368–373.
26. Simone, G., Perozziello, G., & Nanosci, J. (2011). UV/Vis visible optical waveguides fabricated using organic-inorganic nanocomposite layers. *Nanotechnology, 11*(3), 2057–2063.

27. Yu, A., Savas, T., Cabrini, S., Difabrizio, E., Smith, H. I., Stellacci, F., & Am, J. (2005). High resolution printing of DNA feature on poly (methyl methacrylate) substrates using supramolecular nano-stamping. *Chemical Society, 127*(48), 16774–16775.
28. Zhang, Y., Park, S., Yang, S., & Wang, T. H. (2010). An all-in-one microfluidic device for parallel DNA extraction and gene analysis. *Biomedical Microdevices, 12*(6), 1043–1049.
29. Keramas, G., Perozziello, G., Geschke, O., & Christensen, C. B. V. (2004). Development of a multiplex microarray microsystem. *Lab on a Chip, 4*(2), 152–158.
30. Liu, K., & Fan, Z. H. (1011). Thermoplastic microfluidic devices and their applications in protein and DNA analysis. *Analyst (London), 136*(7), 1288–1297.
31. Wu, J., Wu, X. Z., Huang, T., & Pawliszyn, J. (2004). *Analysis of proteins by CE, CIEF, and microfluidic devices with whole-column-imaging detection.* In M. A. Strege, & A. L. Lagu (Eds.), *Methods in molecular biology, vol. 276: capillary electrophoresis of proteins and peptides* (pp. 229–252). Totowa: Humana Press.
32. Perozziello, G., Candeloro, P., Gentile, F., Coluccio, M. L., Tallerico, M., De Grazia, A., Nicastri, A., Perri, A. M., Parrotta, E., Pardeo, F., Catalano, R., Cuda, G., & Di Fabrizio, E. (2015). A microfluidic dialysis device for complex biological mixture SERS analysis. *Microelectronic Engineering, 144*, 37–41.
33. Tekin, H. C., Scherz, C., & Gijs, M. A. M. (2011). Microfluidic device for analysis of protein biomarkers using magnetic bead surface coverage detection. In *Proceedings of the 15th international conference on miniaturized systems for chemistry and life sciences* (MicroTAS, 2011).
34. Perozziello, G., Candeloro, P., Gentile, F., Nicastri, A., Perri, A., Coluccio, M. L., Adamo, A., Pardeo, F., Catalano, R., Parrotta, E., Espinosa, H. D., Cuda, G., & Di Fabrizio, E. (2014). Microfluidics & nanotechnology: towards fully integrated analytical devices for the detection of cancer biomarkers. *RSC Advances, 4*(98), 55590–55598.
35. Perozziello, G., La Rocca, R., Cojoc, G., Liberale, C., Malara, N., Simone, G., Candeloro, P., Anichini, A., Tirinato, L., Gentile, F., Coluccio, M. L., Carbone, E., & Di Fabrizio, E. (2012). Microfluidic devices modulate tumor cell line susceptibility to NK cell recognition, Small. *Small, 8*(18), 2886–2894.
36. Simone, G., & Perozziello, G. (2010). Ca^{2+} mediates the adhesion of breast cancer cells in self-assembled multifunctional microfluidic chip prepared with carbohydrate beads. *Micro and Nanosystems, 2*(4), 261–268.
37. Simone, G., Perozziello, G., Battista, E., De Angelis, F., Candeloro, P., Gentile, F., Malara, N., Manz, A., Carbone, E., Netti, P., & Di Fabrizio, E. (2012). Cell rolling and adhesion on surfaces in shear flow. A model for an antibody-based microfluidic screening system. *Microelectronic Engineering, 98*, 668–671.
38. Perozziello, G., Simone, G., Malara, N., La Rocca, R., Tallerico, R., Catalano, R., Pardeo, F., Candeloro, P., Cuda, G., Carbone, E., & Di Fabrizio, E. (2013). Microfluidic biofunctionalisation protocols to form multi-valent interactions for cell rolling and phenotype modification investigations. *Electrophoresis, 34*(13), 1845–1851.
39. Perozziello, G., Catalano, R., Francardi, M., Rondanina, E., Pardeo, F., De Angelis, F., Malara, N., Candeloro, P., Morrone, G., & Di Fabrizio, E. (2013). A microfluidic device integrating plasmonic nanodevices for Raman spectroscopy analysis on trapped single living cells. *Microelectronic Engineering, 111*, 314–319.
40. Kwan, J. M., Guo, Q., Kyluik-Price, D. L., Ma, H., & Scott, M. D. (2013). Microfluidic analysis of cellular deformability of normal and oxidatively damaged red blood cells. *American Journal of Hematology, 88*(8), 682–689.
41. Faustino, V., Pinho, D., Yaginuma, T., Calhelha, R. C., Ferreira, I. C., & Lima, R. (2014). Extensional flow-based microfluidic device: deformability assessment of red blood cells in contact with tumor cells. *BioChip Journal, 8*(1), 42–47.
42. Notingher, I., Verrier, S., Romanska, H., Bishop, A. E., Polak, J. M., & Hench, L. L. (2002). In situ characterisation of living cells by Raman spectroscopy. *Spectroscopy, 16*, 43–51.
43. Verrier, S., Notingher, I., Polak, J. M., & Hench, L. L. (2004). In situ monitoring of cell death using Raman microspectroscopy. *Biopolymers, 74*, 157–162.

44. Uzunbajakava, N., Lenferink, A., Kraan, Y., Volokhina, E., Vrensen, G., Greve, J., & Otto, C. (2003). Nonresonant confocal Raman imaging of DNA and protein distribution in apoptotic cells. *Biophysical Journal, 84*, 3968–3981.
45. Short, K. W., Carpenter, S., Freyer, J. P., & Mourant, J. R. (2005). Raman spectroscopy detects biochemical changes due to proliferation in mammalian cell cultures. *Biophysical Journal, 88*, 4274–4288.
46. Notingher, I., Jell, G., Lohbauer, U., Salih, V., & Hench, L. L. (2004). In situ non-invasive spectral discrimination between bone cell phenotypes used in tissue engineering. *Journal of Cellular Biochemistry, 92*, 1180–1192.
47. Cheng, I. F., Chang, H.-C., Hou, D., & Chang, H.-C. (2007). An integrated dielectrophoretic chip for continuous bioparticle filtering, focusing, sorting, trapping, and detecting. *Biomicrofluidics, 1*, 21503.
48. Huh, Y. S., Chung, A. J., & Erickson, D. (2009). Surface enhanced Raman spectroscopy and its application to molecular and cellular analysis. *Microfluidics and Nanofluidics, 6*, 285–297.
49. Ramser, K., Enger, J., Goksoer, M., Hanstorp, D., Logg, K., & Kaell, M. (2005). A microfluidic system enabling Raman measurements of the oxygenation cycle in single optically trapped red blood cells. *Lab on a Chip, 5*, 431–436.
50. Perozziello, G., Candeloro, P., De Grazia, A., Esposito, F., Allione, M., Coluccio, M. L., Tallerico, R., Valpapuram, I., Tirinato, L., Das, G., Giugni, A., Torre, B., Veltri, P., Kruhne, U., Della Valle, G., & Di Fabrizio, E. (2016). Microfluidic device for continuous single cells analysis via Raman spectroscopy enhanced by integrated plasmonic nanodimers. *Optics Express, 24*(2), A180–A190.
51. Wood, B. R., Caspers, P., Puppels, G. J., Pandiancherri, S., & McNaughton, D. (2007). Resonance Raman spectroscopy of red blood cells using near-infrared laser excitation. *Analytical and Bioanalytical Chemistry, 387*(5), 1691–1703.
52. Bankapur, A., Zachariah, E., Chidangil, S., Valiathan, M., & Mathur, D. (2010). Raman tweezers spectroscopy of live, single red and white blood cells. *PLoS One, 5*(4), e10427.
53. Tallerico, R., Todaro, M., Di Franco, S., Maccalli, C., Garofalo, C., Sottile, R., Palmieri, C., Tirinato, L., Pangigadde, P. N., La Rocca, R., Mandelboim, O., Stassi, G., Di Fabrizio, E., Parmiani, G., Moretta, A., Dieli, F., Kärre, K., & Carbone, E. (2013). Human NK cells selective targeting of colon cancer-initiating cells: A role for natural cytotoxicity receptors and MHC class I molecules. *The Journal of Immunology, 190*(5), 2381–2390.
54. Ong, Y. H., Lim, M., & Liu, Q. (2012). Comparison of principal component analysis and biochemical component analysis in Raman spectroscopy for the discrimination of apoptosis and necrosis in K562 leukemia cells. *Optics Express, 20*(20), 22158–22171.

Chapter 13
High-Throughput and Ultra-Sensitive Biosensing and Spectroscopy by Plasmonics

Hatice Altug

Abstract New healthcare initiatives including personalized medicine, global health, point-of-care and early disease diagnostics are demanding breakthrough developments in biosensing technologies. Unfortunately, current biosensors are time consuming, costly, bulky, require a relatively advanced infrastructure and trained laboratory professional, making them unsuitable for disease control and patient care in the field. To address these challenges Altug Lab is developing cutting edge optical biosensor and spectroscopy systems by exploiting plasmonics. In a recent work, Altug's lab introduced a high-throughput and label-free protein microarray technology with nearly one million sensor elements that enable reliable and quantitative detection of bio-chemicals. By coupling wide-field plasmonic arrays with lens-free computational on-chip imaging they demonstrated handheld, lightweight and low-cost diagnostic tools suitable for point-of-care applications. The sensor is based on metallic nanohole arrays and exploits highly sensitive surface plasmons and associated extra-ordinary optical transmission phenomenon. Her lab showed that such optical sensors can detect live and intact viruses in biological media at medically relevant concentrations. Furthermore, by uniquely integrating nanofluidics and optics on sub-wavelength nanohole arrays, they innovated a way to overcome mass-transport problem severely limiting sensor performance at low-analyte concentrations. Altug's lab also working on novel ultrasensitive infrared vibrational spectroscopy techniques. Infrared absorption spectroscopy is a powerful biochemical analysis tool that can fingerprint molecules in a label-free fashion. Its molecular specificity renders the technique to be sensitive to the subtle conformational changes exhibited by biochemical such as proteins. Yet, its low sensitivity and strong water absorption severely limit infrared spectroscopy to perform sensitive and real-time measurements of biomolecules and molecular interactions in aqueous/water environment. By engineering on-chip optical nano-antennas, her lab demonstrated for the first time that fundamental Beer-Lambert law can be overcome and the intrinsic signals of proteins can be enhanced by more than 10,000 times. Significantly using extreme field concentration at interfaces,

H. Altug (✉)
Bioengineering Department, BioNanoPhotonics System Laboratory EPFL, Klosters, Switzerland
e-mail: hatice.altug@epfl.ch

© Springer Science+Business Media Dordrecht 2017
B. Di Bartolo et al. (eds.), *Nano-Optics: Principles Enabling Basic Research and Applications*, NATO Science for Peace and Security Series B: Physics and Biophysics, DOI 10.1007/978-94-024-0850-8_13

they showed real-time and in-situ monitoring of biomolecular interactions from low quantities of molecules. The method, exploiting dramatically strong light-matter interactions enabled by sculptured nano-scale metallic particles, opens up new paradigms in ultra-sensitive vibrational spectroscopy.

13.1

Devastating diseases such as Alzheimer's, Parkinson's and cancer are still without cures and continue to undermine the quality of life for millions of people worldwide. Biosensors that can allow study of building blocks of life such as proteins, lipids and decipher biological mechanisms responsible for these diseases are needed. Likewise, next generation healthcare initiatives such as personalized medicine, global healthcare, point-of-care and early disease diagnostics demand breakthrough developments in biosensors. The recent Ebola epidemic, for example, clearly illustrates how the rapid detection and surveillance of infectious pathogens remains a significant challenge in resource-limited settings. In such cases, it should be possible to screen suspected patients rapidly using a minimal amount of biofluid, preferably outside of existing hospital support. Unfortunately, today's biosensors are unsuitable for decentralized and personalized patient control in the field because they are costly, bulky and require several days for a trained laboratory professional, working in a relatively advanced infrastructure, to obtain a result. To meet these needs, in our lab at EPFL we are working to develop new nanophotonic devices enabling ultra-sensitive mid-infrared spectroscopy and biosensing for label-free, multiplexed, high-throughput and real-time detection. We employ physics and engineering toolboxes, such as nanophotonics, metamaterials, nanofabrication, micro/nanofluidics, together with biology and chemistry. We also exploit novel nano- optics phenomena on nano-engineered surfaces and recently discovered low dimensional materials such as graphene to achieve new biosensing methods.

Among various different nanophotonic phenomena plasmonics is particularly suitable for biosensing. In fact, surface plasmon resonance sensors are already considered to be the gold standard for label-free biodetection. Plasmons are quasiparticles associated with the collective oscillations of the free electrons, such as in noble metals and graphene, at optical frequencies. Metallic nanostructures with defined shapes and arrangements whose dimensions are smaller than the wavelength of light are fueling significant excitement in the field. This is because of their unique ability to scatter and absorb light with efficiencies many orders of magnitude larger than dielectric nanoparticles and dye molecules, as well as their ability to confine and enhance electromagnetic radiation at the nanometer scale. Plasmons could also concentrate electromagnetic fields at interfaces with unique propagation characteristics. These features are attractive for fundamental research as well as breakthrough technologies. Metallic nanoparticles are, for example, already being used in low-cost pregnancy tests and photothermal ablation therapy for cancer treatment.

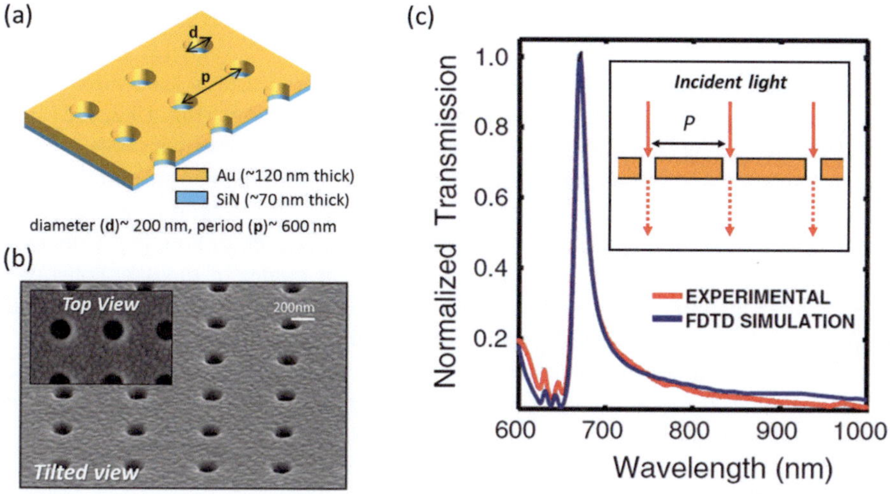

Fig. 13.1 (**a**) Schematic of plasmonic nanohole arrays fabricated in an optically *thick gold film*. The nanohole structures mentioned in this article are on a thin Si_3N_4 film. (**b**) SEM images of plasmonic nanohole arrays (*top & side view*). (**c**) Transmission spectrum showing EOT resonance peak around 680 nm

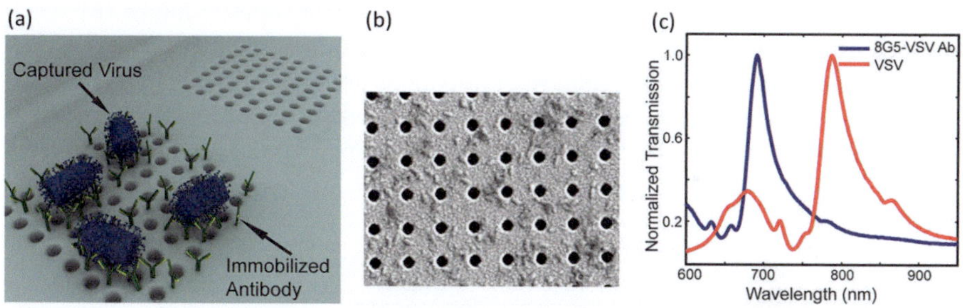

Fig. 13.2 (**a**) Schematic of two plasmonic nanohole array sensors side by side. The left sensor is functionalized with immobilized antibodies. They enable to specifically capture only the targeted viruses. (**b**) SEM image of a functionalized nanohole array sensor with captured viruses (**c**) EOT transmission spectrum of antibody (8G5-VSV Ab) functionalized nanohole array sensor (*blue curve*). After binding of live and intact VSV-Ebolo viruses the EOT peak rea-shifts (*red curve*)

In our lab we have been utilizing nanohole-perforated metallic films for biosensing (Fig. 13.1a, b)[1]. These structures support extraordinary optical transmission (EOT) at a specific resonance wavelength determine by the periodicity of the hole array (Fig. 13.1c). The resonance condition of EOT is very sensitive to the optical refractive index of the environment near the surface of the device which is the basis for using them as a label-free biosensor. We functionalize their surface with biomolecules to provide biospecificity (Fig. 13.2a). When target analytes (such as viruses as shown in Fig. 13.2a, b) bind on the nanostructured gold surface, the effective refractive index of the medium increases. This results in a redshift in the

EOT spectrum (Fig. 13.2c). The magnitude of this spectral shift is proportional to the amount of molecules bound to the sensor surface, thus enables quantitative refractometric biosensing. We used plasmonic nanohole arrays sensor to detect live pseudo-Ebola viruses at medically relevant concentrations (Fig. 13.2c) [2]. By uniquely integrating nanofluidics and optics on sub-wavelength nanohole arrays, we also showed that mass-transport problem severely limiting sensor performance at low-analyte concentrations could be overcome [3]. This approached allowed us to detect virus like particles rapidly even at low concentrations [4]. Coherent interference between the propagating and evanescent modes of the nanoholes can be used to excite ultrasharp asymmetric Fano resonances. We used this sharp optical response to detect single monolayers of biomolecules with the naked eye [5].

Compared with conventional surface plasmon resonance sensors, the nanohole array eliminates the need for a prism to excite plasmons, thus enabling the excitation and detection of the EOT signal in the same optical path [6]. In many cases, the use of gratings to couple the far-field light source to the surface plasmon modes significantly simplifies the optical requirements of the system and reduces the sensor footprint, thereby increasing its ability to detect multiple analytes simultaneously. Recently, we integrated a plasmonic protein microarray with a lens-free on-chip microscope to create a compact, handheld and high-throughput biosensor that uses low-cost LED sources and a CMOS camera for operation (Fig. 13.3a, b) [7]. The system also integrated a microfluidic chamber to control solution intake [8]. Such a new sensor could be particularly desirable for screening large numbers of disease biomarkers for the early identification of infections and point-of-care applications.

Another topic that has been covered in the talk is our research lines with mid-infrared nanophotonics [9]. Within the Mid-IR spectrum of $\tilde{3}$——10 μm (1000–

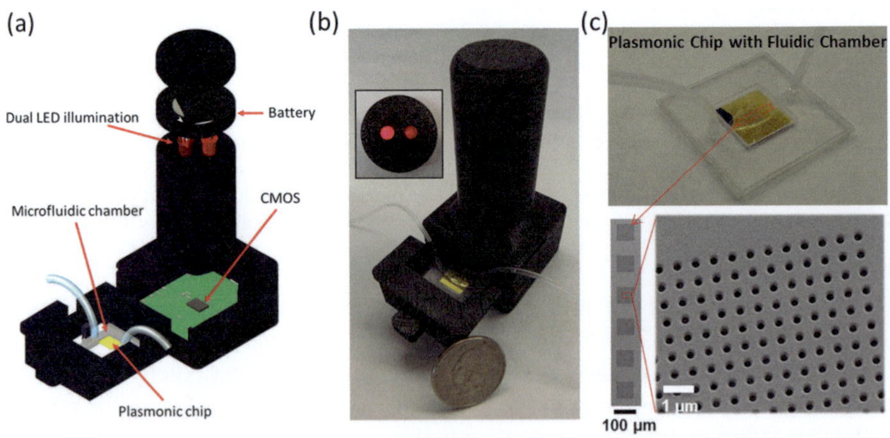

Fig. 13.3 (**a**) Schematic of the hand-held plasmonic sensor system. (**b**) Actual photo of the manufactured sensor. (**c**) *Top image* show photo of the plasmonic sensor chip integrated with microfluidic chips with inlet and outlet. *Bottom SEM images* show the lay-out the plasmonic sensor arrays on the chip

3500 cm^{-1}), biomolecules such as proteins, lipids and nucleic acids have distinct vibrational motions [10]. For proteins the strong chemically specific absorption bands are amide I to VII and amide A to B. Among these, amide-I is highly useful for the analysis of their structure and conformational changes. Mid-IR spectroscopy even allows simultaneous study of proteins and lipids in intact biological membranes without external labels as their absorption bands do not overlap. However, despite its significant advantages, Mid-IR spectroscopy has been so far limited for biological studies due to several shortcomings, including low sensitivity and difficulties with measuring in water. Sensitivity is limited as a result of Beer's law, which implies that in thin samples such as protein monolayers and lipid bilayers, absorption signals are prohibitively weak. Water is essential for biological processes, yet it causes a large absorption background that often obscures protein signal. A nanophotonic based approach called surface enhanced IR absorption (SEIRA) can fundamentally overcome these limitations [11].

SEIRA exploits tiny metallic nanoparticles, called plasmonic antennas [12] to dramatically enhance the effective absorption cross section of molecules and hence the resulting Mid-IR absorption signal. Early SEIRA studies have relied on random metal island films i.e. prepared by chemical means. Instead, our recent studies show that properly designed and nano-fabricated plasmonic nanoantennas (Fig. 13.4a) can provide at least three unique functionalities [13–17]. (1) strong near-field intensity enhancements, (2) sub-wavelength photon confinement, (3) spectroscopic and spatial control over far-field light scattering.

Regarding point (1): More than 3 orders of magnitude intensity enhancements can be obtained with antennas at their resonant frequencies (Fig. 13.4d, y-axis). By overlapping such resonance frequencies with the absorption bands of the targeted molecules, specific molecular bonds can be selectively excited and large signals even from ultra-small sample volumes can be achieved. Using tailored plasmonic antennas, we achieved more than 10,000 fold enhancements of backbone signatures of silk proteins and demonstrated detection of amide-I and II absorption signals with sensitivity levels approaching zepto-moles [13]. Regarding point (2): The field intensity decays exponentially away from the nanoantenna surface with a penetration length of less than 100 nm (Fig. 13.4d, x-axis). Thus one can selectively

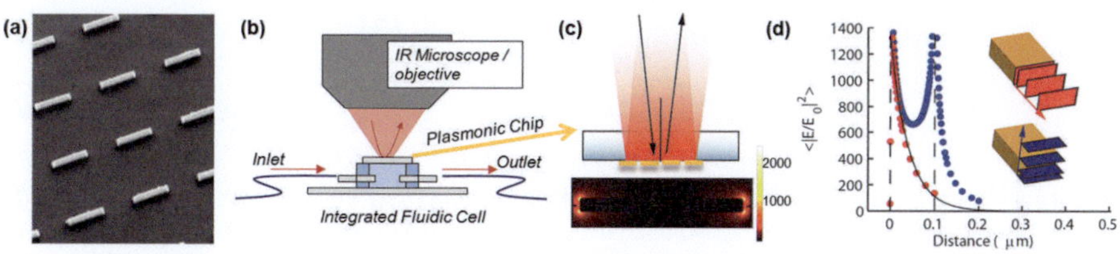

Fig. 13.4 (**a**) SEM images of engineered plasmonic gold nano-rod antennas array. (**b-c**) Set-up scheme of the surface enhanced infrared spectroscopy using plasmonic chip (on an IR transparent CaF_2 window) integrated with fluidics. (**d**) Calculated spatial near-field intensity distribution shows strong confinement (x-axis) and high intensities (y-axis)

probe the molecules in the nanoantenna proximity and minimize interference from the rest of the sample (i.e. water molecules in solution). This makes SEIRA ideal for studying thin protein monolayers and lipid membranes in solution. Regarding point (3): Nanoparticles can be engineered to efficiently reflect only the resonant photons carrying chemical information without requiring highly specialized, expensive and difficult to use optical setups.

In our 2013 Nature Communications paper [14], by exploiting these properties, we integrated engineered Mid-IR plasmonics with fluidics for the first time. Schematics is shown in Fig. 13.4b, c. Using this platform we demonstrated ultra-sensitive in-situ absorption spectroscopy enabling real-time kinetic measurements of protein-protein interactions in water (Fig. 13.5b) [14]. This protein immunoassay demonstration also confirmed that SEIRA is quantitative: the absorption signal increases with increased protein accumulation on the sensor, for example due to higher bioaffinity (Fig. 13.5d, time >200 min). Our results showed that plasmonic field enhancements are still high even after surface functionalization and multiple protein layers (>20 nm thick).

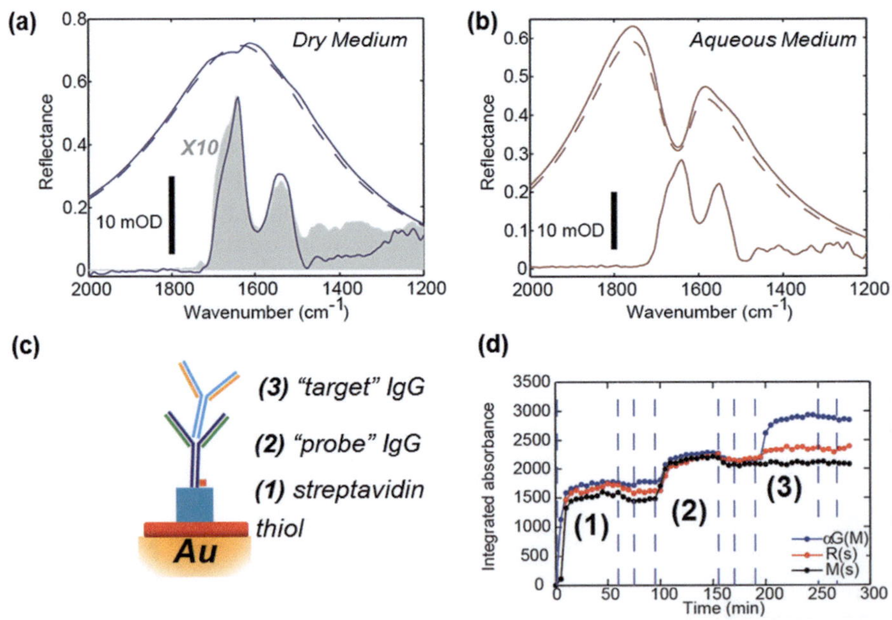

Fig. 13.5 (**a**) Experimental IR reflection of nanoantennas before (*dashed*) and after (*solid*) <10 nm thick protein layer. The absorbance (obtained from difference of before/after spectrum) at the bottom shows clear Amide I-II bands. (**b**) Sensor performance in solution is comparable to dry condition. (**c–d**) Immunoassay scheme. Real-time protein-protein interactions are quantitatively monitored using SEIRA. Different proteins and washing solutions (*dashed vertical lines* and colored curves) are introduced using fluidics

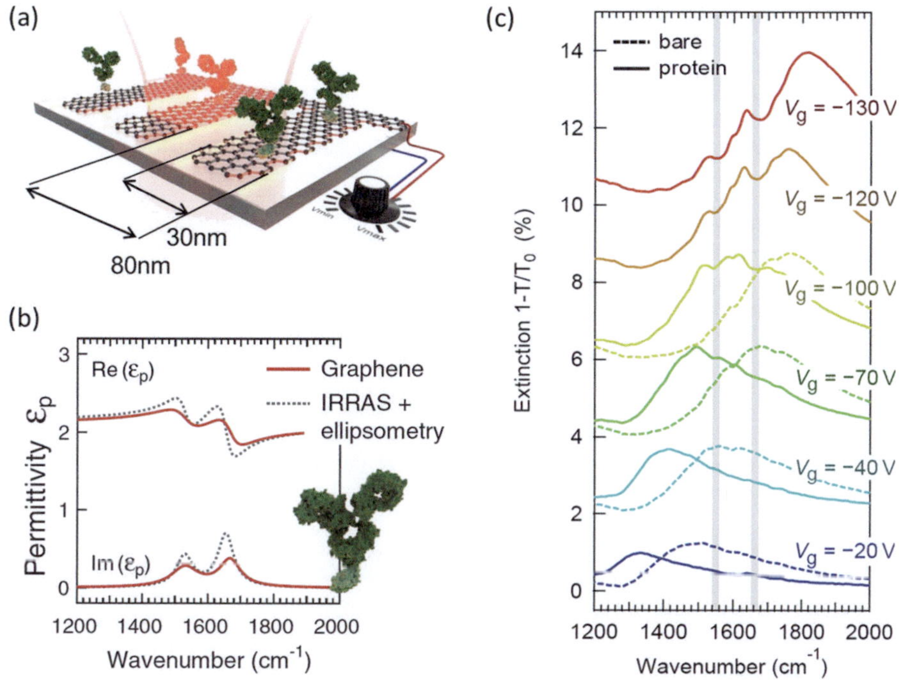

Fig. 13.6 (**a**) Schematic of tunable MidIR graphene biosensor. (**b**) Extracted complex refractive index with tunable graphene sensor (*red curves*). The fit is in good agreement with the values determined with the combination of two other methods, infra-red reflection absorption spectroscopy (IRRAS) and ellipsometry (*dotted black curves*). (**c**) Measured extinction spectrum from graphene sensor at different voltage settings before and after coating with a protein layer, *colored dashed* and *solid curves*, respectively

Most recently we showed that graphene could revolutionize biosensing due to its exceptional opto-electronic properties [18]. Graphene plasmons can be tuned by electrostatic gating, in contrast to conventional plasmonic materials such as noble metals (Fig. 13.6a). By exploiting this unique feature we demonstrated a dynamically tunable plasmonic mid-infrared biosensor that can extract complete optical properties of proteins over a broad spectrum (Fig. 13.6b, c). In addition, the extreme light confinement in graphene – up to two orders of magnitude higher than in metals – produces an unprecedentedly strong overlap with nanometric biomolecules, enabling superior sensitivity. The combination of tunable spectral selectivity and enhanced sensitivity of graphene opens exciting prospects for sensing, not only proteins but also a wide range of chemicals and thin films.

In conclusion, plasmonic biosensors have evolved quickly in recent years. Further significant innovations are expected in the future, given the advances in low-cost and large-area nanofabrication methods, the use of alternative materials, and the integration of plasmonics with nano/microscale fluidic and electromechanical systems.

References

1. Altug, H. (2015). Nano-optics gets practical. *Nature Nanotechnology, 10*, 11–15.
2. Yanik, A. A., Huang, M., Kamohara, O., Artar, A., Geisbert, T. W., Connor, J. H., & Altug, H. (2010). An optofluidic nanoplasmonic biosensor for direct detection of live viruses from biological media. *Nano Letters, 10*, 4962–4969.
3. Yanik, A. A., Huang, M., Artar, A., Chang, T.-Y., & Altug, H. (2010). Integrated nanoplasmonic-nanofluidic biosensors with targeted delivery of analytes. *Applied Physics Letters, 96*, 021101
4. Chang, T.-Y., Huang, M., Yanik, A. A., Tsai, H.-Y., Shi, P., Aksu, S., Yanik, M. F., & Altug, H. (2011). Large-scale plasmonic microarrays for label-free high-throughput screening. *Lab on a Chip, 11*, 3596–3602.
5. Yanik, A. A., Cetin, A. E., Huang, M., Artar, A., Mousavi, S. H., Khanikaev, A., Connor, J. H., Shvets, G., & Altug, H. (2011). Seeing protein monolayers with naked eye through plasmonic Fano resonances. *Proceedings of the National Academy of Sciences of the United States of America, 108*, 11784–11789.
6. Huang, M., Galarreta, B. C., Cetin, A. E., & Altug, H. (2013). Actively transporting virus like analytes with optofluidics for rapid and ultrasensitive biodetection. *Lab on a Chip, 13*, 4841–4847.
7. Cetin, A. E., Coskun, A. F., Galarreta, B. C., Huang, M., Herman, D., Ozcan, A., & Altug, H. (2014). Handheld high-throughput plasmonic biosensor using computational on-chip imaging. *Light: Science & Applications, 3*, e122.
8. Coskun, A. F., Cetin, A. E., Galarreta, B. C., Alvarez, D. A., Altug, H., & Ozcan, A. (2014). Lensfree optofluidic plasmonic sensor for real-time and label-free monitoring of molecular binding events over a wide field-of-view. *Scientific Reports, 4*, 6789.
9. Adato, R., Aksu, S., & Altug, H. (2015). Engineering mid-infrared nanoantennas for surface enhanced infrared absorption spectroscopy. *Materials Today, 18*, 436–446.
10. Koch, C., Posch, A. E., Goicoechea, H. C., Herwig, C., & Lendl, B. (2014). Multi-analyte quantification in bioprocesses by Fourier-transform-infrared spectroscopy by partial least squares regression and multivariate curve resolution. *Analytica Chimica Acta, 807*, 103–110.
11. Jiang, X., Zaitseva, E., Schmidt, M., Siebert, F., Engelhard, M., Schlesinger, R., Ataka, K., Vogel, R., & Heberle, J. (2008). Resolving voltage-dependent structural changes of a membrane photoreceptor by surface-enhanced IR difference spectroscopy. *Proceedings of the National Academy of Sciences of the United States of America, 105*, 12113–12117.
12. Novotny, L., & van Hulst, N. (2011). Antennas for light. *Nature Photonics, 5*, 83–90.
13. Adato, R., Yanik, A. A., Amsden, J. J., Kaplan, D. L., Omenetto, F. G., Hong, M. K., Erramilli, S., & Altug, H. (2009). Ultra-sensitive vibrational spectroscopy of protein monolayers with plasmonic nanoantenna arrays. *Proceedings of the National Academy of Sciences of the United States of America, 106*, 19227–19232.
14. Adato, R., & Altug, H. (2013). In-situ ultra-sensitive infrared absorption spectroscopy of biomolecule interactions in real time with plasmonic nanoantennas. *Nature Communications, 4*, 2154.
15. Limaj, O., Etezadi, D., Wittenberg, N. J., Rodrigo, D., Yoo, D., Oh, S.-H., & Altug, H. (2016). Infrared plasmonic biosensor for real-time and label-free monitoring of lipid membranes. *Nano Letters, 16*, 1502–1508.
16. Chen, K., Adato, R., Altug, H. (2012). Dual-band perfect absorber for multispectral plasmon-enhanced infrared spectroscopy. *ACS Nano, 6*, 7998–8006.
17. Wu, C., Khanikaev, A. B., Adato, R., Arju, N., Yanik, A. A., Altug, H., & Shvets, G. (2012). Fano-resonant asymmetric metamaterials for ultrasensitive spectroscopy and identification of molecular monolayers. *Nature Materials, 11*, 69–75.
18. Rodrigo, D., Limaj, O., Janner, D., Etezadi, D., García de Abajo, F. J., Pruneri, V., & Altug, H. (2015). Applied physics. Mid-infrared plasmonic biosensing with graphene. *Science (New York, N.Y.), 349*, 165–168.

Chapter 14
Photoemission from Nanomaterials in Strong Few-Cycle Laser Fields

Qingcao Liu, Philipp Rupp, Benjamin Förg, Johannes Schötz, Frederik Süßmann, William Okell, Johannes Passig, Josef Tiggesbäumker, Karl-Heinz Meiwes-Broer, Lennart Seiffert, Thomas Fennel, Eckart Rühl, Michael Förster, Peter Hommelhoff, Sergey Zherebtsov, and Matthias F. Kling

Abstract The application of ultra-short waveform-controlled laser fields to nanostructured materials enables the generation of localized near-fields with well-defined spatiotemporal field evolution. The optical fields that can be tailored on sub-wavelength spatial and attosecond temporal scales have a high potential for the control of ultrafast processes at the nanoscale, with important implications for laser-driven electron acceleration, extreme ultraviolet (XUV) light generation, and nanoscale electronics operating at optical frequencies.

Q. Liu (✉) • P. Rupp • B. Förg • J. Schötz • F. Süßmann • S. Zherebtsov • M.F. Kling (✉)
Department of Physics, LMU Munich, Am Coulombwall 1, D-85748 Garching, Germany

Max Planck Institute of Quantum Optics, Hans-Kopfermann-Str. 1, D-85748 Garching, Germany
e-mail: liuqingcao@gmail.com; matthias.kling@mpq.mpg.de

W. Okell
Max Planck Institute of Quantum Optics, Hans-Kopfermann-Str. 1, D-85748 Garching, Germany

J. Passig • J. Tiggesbäumker • K.-H. Meiwes-Broer • L. Seiffert • T. Fennel
Institut für Physik, Universität Rostock, D-18051 Rostock, Germany

E. Rühl
Physical Chemistry, Freie Universität Berlin, Takustr. 3, D-14195 Berlin, Germany

M. Förster • P. Hommelhoff
Max Planck Institute of Quantum Optics, Hans-Kopfermann-Str. 1, D-85748 Garching, Germany

Department of Physics, Friedrich-Alexander University Erlangen-Nürnberg (FAU), Staudtstrasse 1, D-91058 Erlangen, Germany

© Springer Science+Business Media Dordrecht 2017

B. Di Bartolo et al. (eds.), *Nano-Optics: Principles Enabling Basic Research and Applications*, NATO Science for Peace and Security Series B: Physics and Biophysics, DOI 10.1007/978-94-024-0850-8_14

14.1 Waveform-Control of Strong Field Dynamics in the Near-Field of Isolated Nanoparticles

14.1.1 Introduction

Field localization by non-resonantly excited nanostructures opens new avenues in strong-field physics with optical fields that can be tailored on sub-wavelength spatial and attosecond temporal scales. Enhanced strong-field electron acceleration in nanolocalized fields has been observed in isolated dielectric nanospheres, metal nanotips, and surface based nanostructures. Here, we present our results on strongfield-induced waveform-controlled dynamics in isolated nanoparticles.

14.1.2 Imaging of Laser-Induced Electron Emission from Isolated Nanoparticles

In experiments exploring the waveform control of electron photoemission from isolated nanoparticles, few-cycle laser fields with controlled carrier-envelope phase (CEP) were employed [1, 2]. Few-cycle laser pulses of ~4.5 fs duration centered at 720 nm were generated from the output of an amplified laser system (25 fs pulse duration, 790 nm central wavelength) [3] that was spectrally broadened in a capillary filled with 2.8 bar Ne gas, and compressed by a chirped mirror compressor. For each laser pulse the CEP was measured with a stereo-ATI phasemeter [4, 5] using a small fraction of the laser beam (~15 %). The main part of the beam was focused into the center of the electrostatic optics of a velocity-map imaging (VMI) detector, where it intersected a nanoparticle beam (see Fig. 14.1a). The electron emission distribution was projected onto a microchannel plate (MCP) and phosphor screen assembly, and the image on the phosphor screen was recorded by a high-speed CMOS camera at the full repetition rate of the laser (1 kHz) [6]. Acquisition of single-shot images at these high rates was achieved by only storing pixels whose brightness exceeded a threshold level. The SiO_2 nanospheres were prepared by using wet chemistry methods [7, 8] for producing spheres with diameters in the range 50–550 nm and polydispersity of less than 10 % [1, 9] (e.g. Fig. 14.1b). The isolated nanospheres were delivered into the interaction region by injection of the nanospheres suspension into a carrier gas, evaporation of the solvent (ethanol), focusing of the nanospheres stream with an aerodynamic lens, and differential pumping of the carrier gas.

With the current efficiency of the nanoparticle source the particle density in the interaction region is limited and typically only one out of a few (5–10) frames contains nanoparticle signal. As the number of electrons emitted from the nanoparticles is significantly larger than the electron emission from the residual gas, single-shot detection allows significant improvement of the signal-to-noise ratio by selecting only those frames that contain nanoparticle signal. Figure 14.1c shows histograms

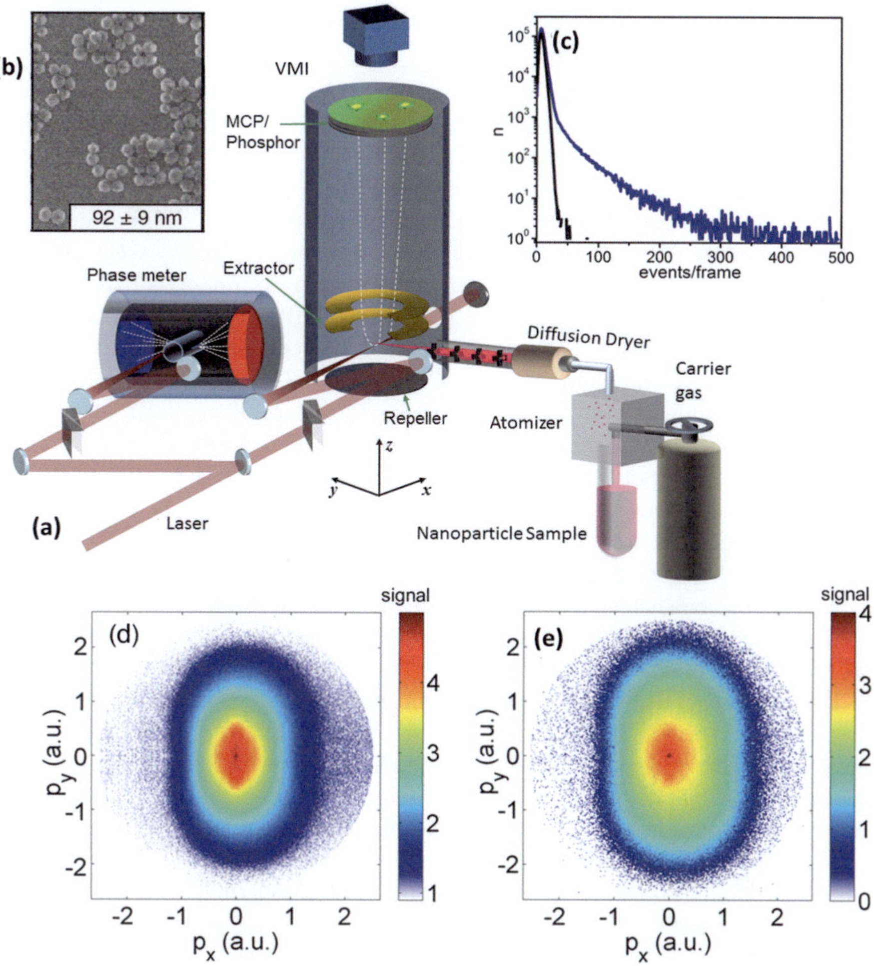

Fig. 14.1 (**a**) Schematic of the VMI setup with an aerodynamic nanoparticle source and single-shot phasemeter. The polarization of the laser was in the plane of the detector. (**b**) TEM image of 92 nm diameter SiO_2 nanospheres. (**c**) Histograms of the number of events per frame obtained from the solvent ethanol only (*black line*) and SiO_2 nanospheres in ethanol (*blue line*) measurements. (**d, e**) CEP-averaged VMI images (log color scale) obtained from measurements with a suspension of SiO_2 nanospheres in ethanol. The image in (**d**) is the average over all the frames, and the image in (**e**) is the average over only the frames containing more than 30 events

of the number of events per frame for the ethanol-only (black) and suspension of nanoparticles in ethanol (blue) measurements. While the ethanol-only measurement signal is concentrated at low event numbers, in nanoparticle measurements many frames contain a much higher number of events. Figure 14.1d, e illustrate efficient suppression of the background signal by neglecting the frames with low event numbers.

14.1.3 Waveform-Controlled Electron Near-Field Acceleration

Figure 14.2a, b show typical results for VMI of waveform-controlled electron photoemission from 95 nm diameter SiO_2 nanoparticles illuminated with CEP-controlled few-cycle laser fields. The electron momentum distribution has an elliptical shape and is elongated along the polarization direction (p_y axis) (Fig. 14.2a). To quantify the CEP dependence of the directional emission an asymmetry parameter $A(p_y, \varphi) = (P_{up}(p_y, \varphi) - P_{down}(p_y, \varphi))/(P_{up}(p_y, \varphi) + P_{down}(p_y, \varphi))$ was plotted as a function of the momentum along polarization axis (p_y) and CEP (φ). Here $P_{up}(p_y, \varphi)$ and $P_{down}(p_y, \varphi)$ $P_{down}(p_y, \varphi)$ are the angle-integrated electron yields (within [$-25°$, $+25°$] angular range) in the up (positive p_y momentum) and down (negative p_y momentum) directions (Fig. 14.2b). The cutoff of the momentum map is in agreement with the cutoff of the CEP-dependent electron emission and is at about 50 U_p, where U_p is the ponderomotive potential of an electron in the driving laser field. For the studied intensity range (1–4.5) $\times$ 10^{13} W/cm^2 the measurements show a nearly linear intensity dependence of the cutoff energy with an average scaled cutoff of about 53.0 U_p (Fig. 14.2e). The measured cutoff energy is much higher than what is expected for the classical atomic emission in a dielectrically enhanced field near a SiO_2 nanosphere (cutoff of $\sim$24 U_p). The maximum asymmetry phase $\varphi_{\max}$ increases with the laser intensity (except the lowest intensity point) (Fig. 14.2f).

The mechanism of the enhanced electron acceleration was analyzed with quasi-classical trajectory-based simulations using the Meanfield Monte-Carlo model developed in the group of one of the authors (Th. Fennel) [2]. Results of these calculations performed for the same parameters as in the experiment are presented in Fig. 14.2c–f. The simulations reproduce the main features of the experiment such as overall shape of the momentum and asymmetry maps as well as the cutoff value. The simulation model explained these effects through many-particle charge interaction of ionized electrons and positively charged ions in the nanoparticle. The charge interaction can be decomposed into a short-range nonlinear surface potential due to self-consistent electron motion in the local field and a long-range potential due to ionization. While the long-range potential traps the low-energy direct electrons, both components contribute to the backscattered electron acceleration. Persistence of the observed CEP controlled electron photoemission over a wide intensity range indicates robustness of the near-field induced electron recollision process even at laser intensities when the multi-particle nonlinear response of the system becomes significant. The simulation shows a good agreement with the experiment except for the lowest intensity point (Fig. 14.2e, f). The discrepancy at the lowest intensity can be ascribed to the deviation of the initial ionization mechanism from the instantaneous tunneling assumed in the model.

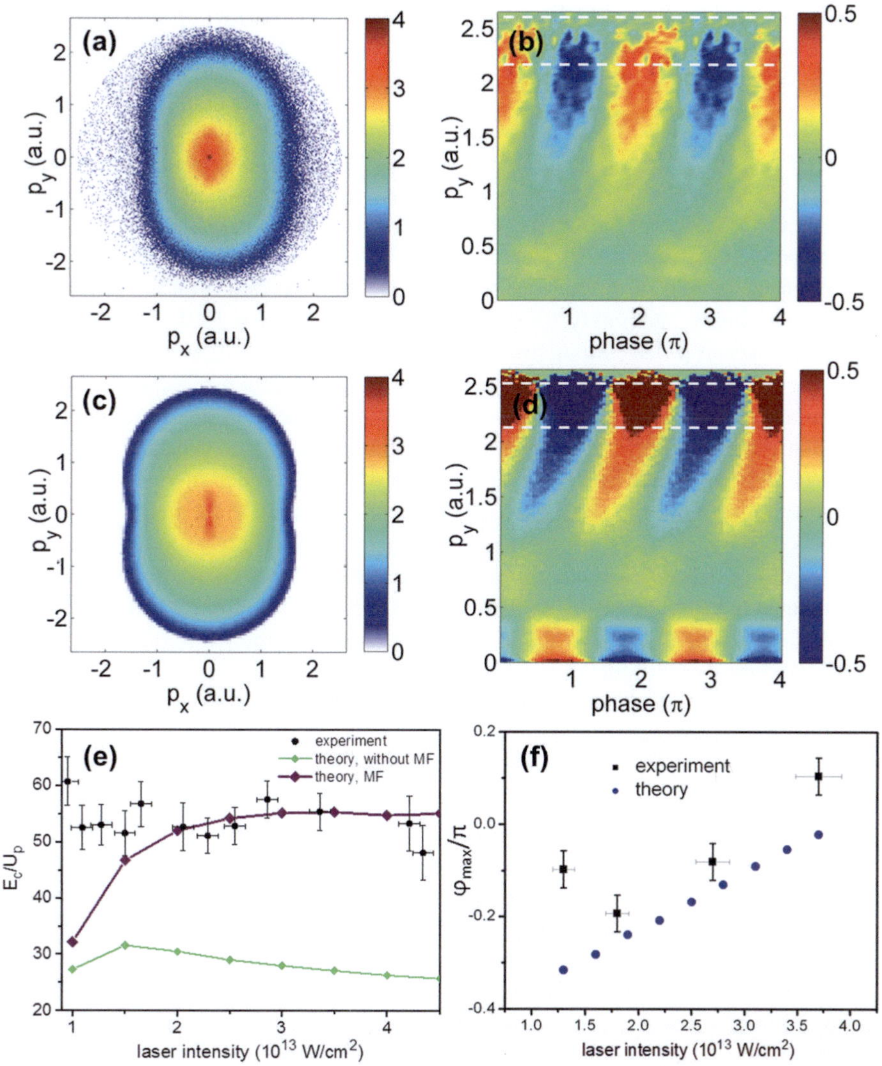

Fig. 14.2 (**a**) Photoelectron momentum map (projected along p_z) averaged over the CEP (log color scale) and (**b**) asymmetry of the electron emission as a function of the electron momentum and the CEP measured for 95 nm SiO_2 nanoparticles at 3.7×10^{13}W/cm^2. (**c**, **d**) Photoelectron momentum and asymmetry map calculated for the same parameters as in (**a**, **b**). (**e**) Intensity dependence of the electron emission cutoff in terms of U_p measured in SiO_2 nanoparticles (*filled black circles*) and results of the model simulations without mean field (*green line with filled circles*) and including mean field contribution (*purple line with filled circles*). (**f**) Dependence of the CEPs at the maximum asymmetry φ_{max} of the electron emission from SiO_2 nanoparticles of 95 nm diameter on the laser intensity measured (*black boxes*) and calculated (*blue filled circles*). To obtain φ_{max} the asymmetry maps were integrated over p_y in the cutoff region (indicated by *white dashed lines* in (**b**, **d**)) and fitted with a function $f(\varphi_{\max}) = A * \cos(\varphi_{CEP} - \varphi_{\max})$.

14.1.4 Effect of Field-Propagation Deformation on Electron Photoemission from Isolated Nanospheres

Local field sensitivity of the near-field-driven electron photoemission prompted further work on tailoring near-fields for control of strong-field-induced electron dynamics. In the first proof of principle experiment we employed isolated nanospheres of different diameters for the generation of near-fields with adjustable polarization and spatial characteristics. Isolated nanospheres have the advantage that their linear near-field properties can be analytically described. The linear response of a sphere can be accurately described by the Mie solution [10] and for a given refractive index, it is determined by the sphere diameter d and the central wavelength of the excitation field λ through the dimensionless parameter $\rho = \pi d/\lambda$. For spheres much smaller than the excitation wavelength ($\rho \ll 1$) the field distribution exhibits quasi-static dipole character with the poles aligned along the polarization vector. For large sphere diameters ($\rho \gg 1$) excitation of higher order modes results in a significant shift of the region of maximal field enhancement in the field propagation direction. Figure 14.3a shows the near-field distribution of the radial electric field as predicted by Mie theory for 100 and 550 nm diameter SiO_2 particles. For large nanospheres the field propagation also induces a nontrivial elliptical local field due to a phase difference between the radial and tangential components (Fig. 14.3a). The CEP-modulation amplitude maps of SiO_2 nanoparticles for diameters of 95 nm and 400 nm are shown in Fig. 14.3b and c, respectively. The shape of the amplitude maps is as expected dipolar without a noticeable asymmetry around $p_x = 0$. A bending of the momentum distribution towards the laser propagation axis can be observed for $d = 400$ nm.

We utilized isolated SiO_2 nanospheres of different diameters ($d = 50$–550 nm) to study the size-dependence of the near-field deformation on the strong field induced photoemission. The effect of field propagation on the phase controlled electron photoemission is illustrated in Fig. 14.4a, b. For this energetic electron emission near the cutoff, the yield is presented as a function of emission angle and CEP. Comparison of the electron emission for nanospheres with diameters between 95 and 550 nm (Fig. 14.4e, f) shows concentration of the energetic emission within a narrow angular range, and a pronounced CEP dependence that is in phase for small and large particles. While for the small nanospheres the signal is concentrated along the polarization vector, for the large nanospheres the emission is tilted towards the propagation direction, reflecting a field-propagation-induced change in the near-field distribution. For modeling the experiment the quasi-classical Monte-Carlo trajectory-based model was extended to take into account the field deformations by means of the Mie solution (Mean-Field Mie Monte-Carlo (M^3C)) [10, 11]. The model simulations (Fig. 14.4c, d) reproduce the main experimental results indicating that all major effects contributing to the electron acceleration are included in the model. The size dependence of the critical emission parameters is illustrated in Fig. 14.4e, f. The systematic tunability of the photoemission direction and CEP

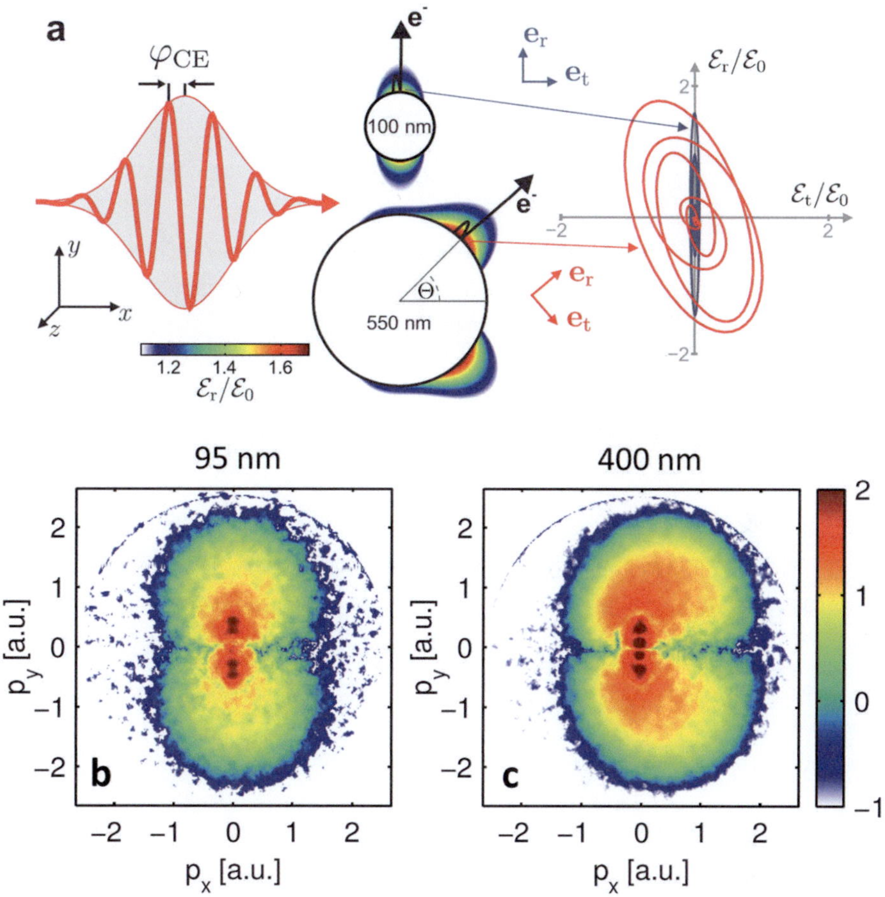

Fig. 14.3 (**a**) Peak radial field enhancement in the x-y plane at z = 0 obtained by Mie solution for SiO_2 spheres illuminated with a 4 fs linear polarized laser pulse centered at 720 nm and a CEP of $\varphi_{CEP} = 0$ (*left*). Field evolution in the local reference frame at the points of maximum field enhancement (*right*). (**b**, **c**) Amplitude of the CEP-dependent part of the electron emission obtained from measurements of SiO_2 nanoparticles of 95 nm (**b**) and 400 nm (**c**) diameters (logarithmic scale, arbitrary units)

dependence of the electron emission indicates a robust attosecond control over the recollision process in sculpted near fields. Quantitative analysis of different many-particle contributions to the acceleration process (Fig. 14.4g) shows a weak size-dependence of the short-range nonlinear surface potential that can be explained by the local character of this potential. On the other hand the contribution from the long-range space-charge repulsion increases strongly with the particle size, indicating its sensitivity to the full electron distribution.

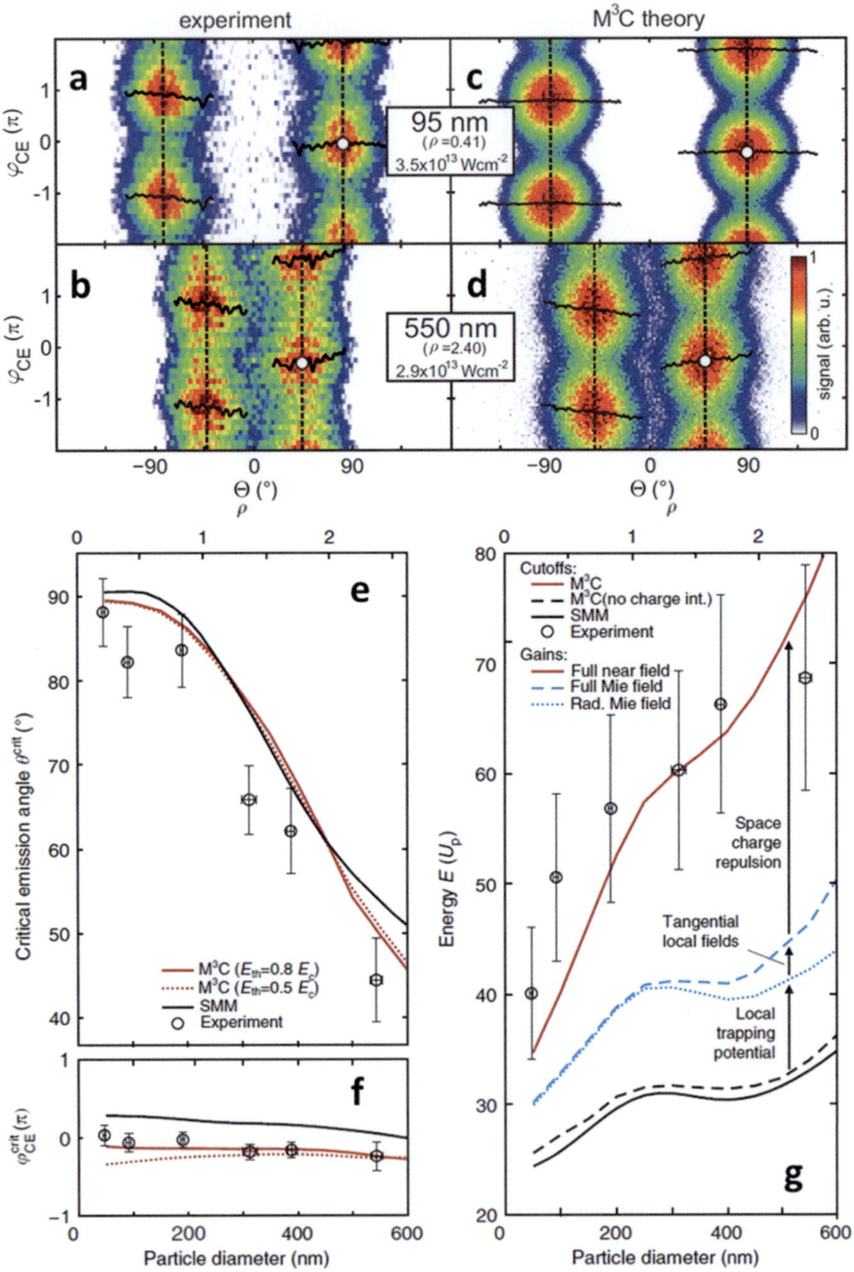

Fig. 14.4 (**a–d**) Measured (**a**, **b**) and simulated (**c**, **d**) angle and CEP-resolved electron yields of energetic electrons near cutoff. The white dots indicate CEP values φ_{CEP}^{crit} and emission angles θ^{crit} of maximum upward emission. (**e**, **f**, **g**) Particle size dependence of the critical emission angle (**e**), critical phase (**f**), and cutoff energies (**g**). The symbols and lines indicate measured and calculated parameters. The Simple Man's Model (SMM) is described in details in Refs. [10, 11]

14.1.5 Nonlinear Regime Near Optical Breakdown

The work presented in the previous section is focused on the response of nanoparticles excited by relatively moderate laser fields in the regime where the optical response of the material can still be assumed linear. The response of solid materials at laser intensities close to the damage threshold has been a subject of both experimental and theoretical studies [12, 13]. These studies are motivated by interest in the fundamental laser-matter interaction, and by prospective applications in material processing, and ultrafast computation. As typical time scales for energy transfer in solids range from fs for electron-electron interaction to ps for electron-phonon and tens of ps for phonon-phonon interaction, the laser energy deposition process strongly depends on the laser pulse duration. For pulse durations of tens of ps the melting process occurs during the interaction of the laser pulse with the solid. For sub-picosecond laser pulses the ultrafast deposition of energy into the electron system leads to a local Coulomb explosion of ions (laser ablation) [14]. For pulses longer than 100 fs the creation of free electrons is attributed mainly to avalanche ionization [14, 15]. For pulses of 5 fs duration the ionization process is dominated by multi-photon and tunnel ionization, and peak intensities exceeding 10^{14} W/cm^2 are necessary to induce optical breakdown [16].

Because the detection range of the electron kinetic energies of the VMI spectrometer used in the previous studies is limited to about 100 eV, we used a stereo-TOF spectrometer (Fig. 14.5a). Isolated nanoparticles are especially well suited for these

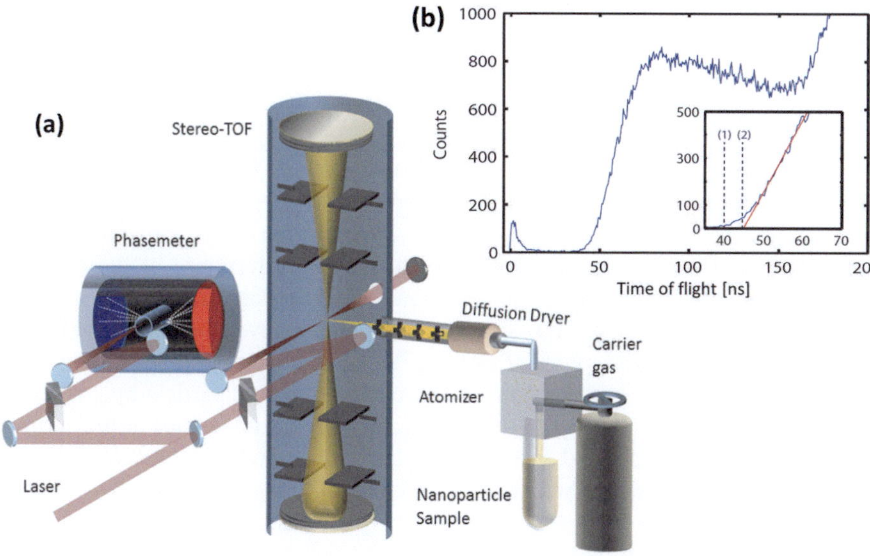

Fig. 14.5 (**a**) Schematic of the single-shot stereo-TOF setup for high intensity waveform controlled photoemission from isolated nanoparticles. (**b**) Typical TOF spectrum obtained from SiO_2 nanoparticles excited by a few-cycle laser field. The inset shows the two methods for determining the cutoff

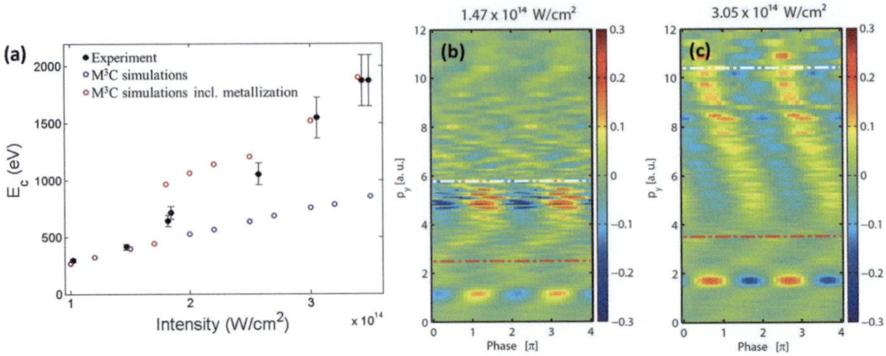

Fig. 14.6 (**a**) Intensity dependence of the electron emission cutoff measured in SiO_2 nanoparticles (*filled black circles*) and results of the M^3C model simulations for SiO_2 without metallization (*open blue circles*) and including metallization (*open red circles*). (**b**, **c**) Asymmetry maps obtained from the measurements of SiO_2 nanoparticles at the laser intensities indicated above the corresponding panels. The cutoff momentum obtained from the TOF spectrum is indicated for the nanoparticle (*white dashed line*) as well as for the reference scan (*red dashed line*)

studies as the target is replaced for each laser shot. A typical TOF spectrum is presented in Fig. 14.5b. The sharp peak at small delays is induced by light scattering (and possibly light emission) from the nanoparticles and was used to calibrate the time-zero of the TOF trace.

The results of intensity-dependent measurements are presented in Fig. 14.6. For intensities up to $\sim 1.5 \times 10^{14}$ W/cm^2 the cutoff energies are nearly linear with the intensity of the incident laser field and are in agreement with the low intensity limit of about 50 U_p obtained from VMI measurements. At higher laser intensities the emission cutoff in terms of U_p monotonically increases and reaches 100 U_p at an intensity of $\sim 3.5 \times 10^{14}$ W/cm^2 (Fig. 14.6a). The M^3C simulations reproduce the experiment at intensities below $\sim 1.5 \times 10^{14}$ W/cm^2 while they significantly underestimate the cutoff energies at higher laser intensities. This observation together with a change of the asymmetry map shape above the critical intensity (Fig. 14.6b, c) indicates an onset of nonlinear response of the material that is not included in the model. It was shown recently that electric polarizability and electronic structure of SiO_2 can be manipulated reversibly with few-cycle laser fields at similar laser intensities [17]. To account for the nonlinear process in SiO_2 we simulated the high-intensity response of the nanoparticle by switching the permittivity of the material to that of Au, as the field intensities at the surface of the nanoparticle reach $\sim 1.5 \times 10^{14}$ W/cm^2. The modified (and still crude) model is in better agreement with the experiment at the highest intensities, while it still requires improvement to correctly describe the transition process at intensities around 1.5×10^{14} W/cm^2.

14.2 Attosecond Streaking in Nanolocalized Plasmonic Fields

In the waveform-controlled electron emission experiments presented above, reconstruction of the time evolution of the nanoscale near-fields relies to a large extent on model calculations. In contrast, a pump-probe approach provides direct access to the near-field dynamics. So far, ultrafast plasmonic near-fields surrounding nanowires, nanoantennas, and nanotips have been studied using femtosecond pulse characterization techniques [18–22]. By virtue of an enhanced temporal resolution, attosecond streaking promises to yield an even deeper understanding of the collective electron dynamics governing plasmon formation and decay, where transport and interaction effects on sub-optical-cycle timescales are expected to be important. In nano-plasmonic streaking measurements, the plasmonic field is excited by a few-cycle laser pulse incident on the nanostructures and probed by photoemission induced by an attosecond extreme ultraviolet (XUV) pulse with variable delay. The XUV-induced photoelectrons are accelerated by the plasmonic field [23]. From the delay-dependent spectrogram information on the temporal structures of the streaking field and the XUV pulse, as well as the attosecond electron dynamics taking place in and at the nanostructure can be extracted. For streaking in nanolocalized fields, the situation is more complex than in a standard streaking measurement in the gas phase [24] or at a plain macroscopic surface [25–27]. This is because the electron can, under certain conditions, experience the spatial inhomogeneity of the nanolocalized fields [23]. For few-cycle streaking pulses three different regimes of streaking can be identified, distinguished by the ratio of the time it takes an electron to leave the inhomogeneous field t_0 to the streaking field period T and pulse duration t_p [23, 28, 29]. In the "ponderomotive" limit the streaking field pulse duration t_p is much shorter than the time it takes the electron to leave the near-field t_0 ($t_p << t_0$) and the electron does not experience any spatial variation of the near-field. This is the same situation as in a standard streaking measurement. In the "instantaneous" limit the electron leaves the localized field within a fraction of the optical cycle T ($t_0 << T$). In contrast to conventional, ponderomotive streaking, in the instantaneous regime the electron streaking curve follows the electric field evolution instead of its vector potential. In the intermediate regime the electron traverses the localized field in about one optical oscillation $t_0 \approx T$ and the streaking trace shows a behavior, which lies in-between the two limits.

14.2.1 *Attosecond Streaking in the Near-Field of Isolated Nanospheres*

Isolated nanospheres can serve as a reference system for tracing plasmonic excitations because their simple shape permits an analytical description of the near-fields. Experimentally, isolated nanoparticles of well-defined size and shape can be introduced into the interaction region by employing aerodynamic lens focusing [1, 2] or optical trapping [30]. These methods allow isolated nanoparticles to be studied without the influence of a substrate or particle-particle interactions.

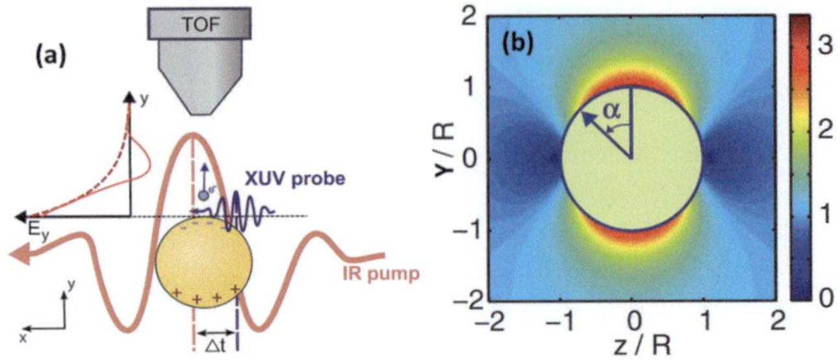

Fig. 14.7 (**a**) Schematic of an attosecond streaking measurement with an isolated gold nanosphere. (**b**) Simulation of the amplitude of the field (E_y) distribution at an Au sphere of 100 nm diameter illuminated with a 5 fs pulse centered at 720 nm. The field is normalized to the incident field

Attosecond streaking on isolated gold nanospheres, shown schematically in Fig. 14.7a, was theoretically analyzed to determine what information could be extracted from experiments [31]. The simulations considered plasmonic oscillations excited by a few-cycle NIR laser pulse and probed by photoemission induced by a delayed attosecond XUV pulse. Electron detection was assumed to be performed using a time-of-flight (TOF) spectrometer along the polarization direction of the NIR pulse. To achieve an ultrafast broadband response, an off-resonant 5 fs laser pulse centered at 720 nm was used as the excitation field [31]. The fields around the nanoparticle were calculated by finding the Mie solution at the central laser wavelength. The simulated near-field in Fig. 14.7b exhibits symmetry relative to the polarization vector of the incident field, with the maximum field enhancement at the poles along the polarization vector. The photoemission was initiated by a 250 as XUV pulse (7 eV bandwidth centered at 105 eV). Photoemission from the whole surface facing the TOF was simulated, taking into account XUV attenuation by the nanoparticle. The photoemitted electrons were assumed to have initial velocity vectors parallel to the y-axis (along the TOF axis). The electron initial position is represented by the angle α with respect to the polarization vector of the laser field (Fig. 14.7b). To achieve good statistics approximately 1.5×10^5 trajectories were initialized from the surface at each delay step.

Figure 14.8a, b show streaking curves simulated for different electron initial positions for gold nanoparticles of 10 nm (a) and 100 nm (b) diameter. For the 10 nm diameter nanoparticle the phase shift of the streaking curves relative to the streaking field suggests a gradual change from the ponderomotive regime for electrons emitted at the poles of the particle ($\alpha = 0^\circ$), to a nearly instantaneous regime for electrons emitted at $\alpha = 90^\circ$. This emission position dependence can be attributed to the dipolar character of the near-field, where the spatial extent of the near-field is much larger at the poles of the sphere than at the equator ($\alpha = 90^\circ$). For a much larger 100 nm diameter particle most of the photoemitted electrons experience ponderomotive streaking.

The corresponding streaking spectrograms consisting of contributions from trajectories originating from different parts of the surface are shown in Fig. 14.8c,

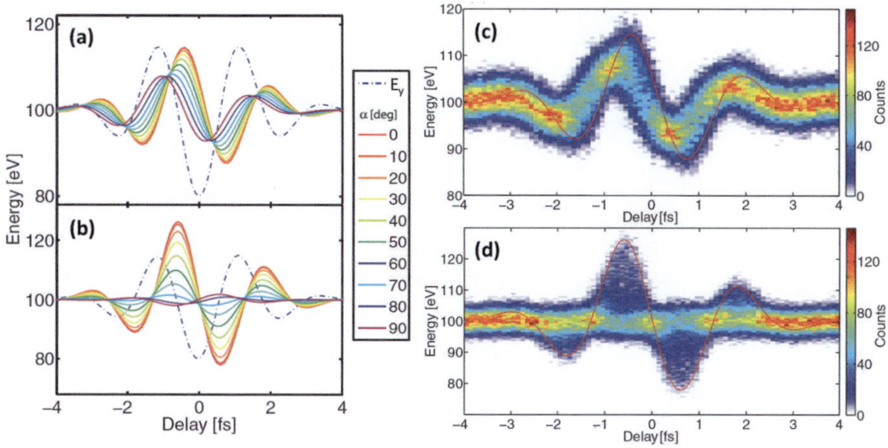

Fig. 14.8 (**a**, **b**) Simulated streaking waveform for electrons emitted at different positions on a sphere of 10 nm (**a**) and 100 nm (**b**) diameter. The *blue dashed lines* show the plasmonic near-field at the surface (After [31]). (**c**, **d**) Simulated streaking spectrograms for spheres of 10 nm (**c**) and 100 nm (**d**) diameter. The *red line* indicates the streaking curve for electrons emitted at the particle pole

d. In comparison to typical streaking measurements in an atomic gas, streaking on a nanoparticle results in a blurred spectrogram due to different acceleration conditions for electrons emitted from different parts of the nanoparticle. Trajectory analysis shows that the electrons emitted from the poles contribute to the largest energy shifts of the photoelectron spectra (red line in Fig. 14.8c, d). As an analytical description of streaking is generally not available for the complex inhomogeneous fields around nanostructures, numerical simulations combined with appropriate feedback may be employed for the retrieval of the spatiotemporal evolution of the near-fields.

14.2.2 Attosecond Nanoscale Sampling of the Near-Field at Gold Nanotips

Attosecond-resolved experiments with isolated nanoparticles typically require a high XUV photon flux to obtain a practical XUV photoelectron yield due to the limited target density. Nanotips, however, feature relaxed flux requirements due to a higher target density while still maintaining the advantages of studying a single, well-controlled nanostructure. The near-fields at the nanotaper's apex and shank play, exemplarily, a crucial role in the study of strong-field photoemission at nanotapers [32–36] and have recently been examined with femtosecond lasers [37–40]. Here we present our recent results on attosecond near-field retrieval with a tapered gold nanowire using streaking spectroscopy [41].

Figure 14.9a shows a schematic of attosecond nanoplasmonic near-field sampling around a nanotaper. Phase-stabilized few-cycle laser pulses of 4.5 fs duration

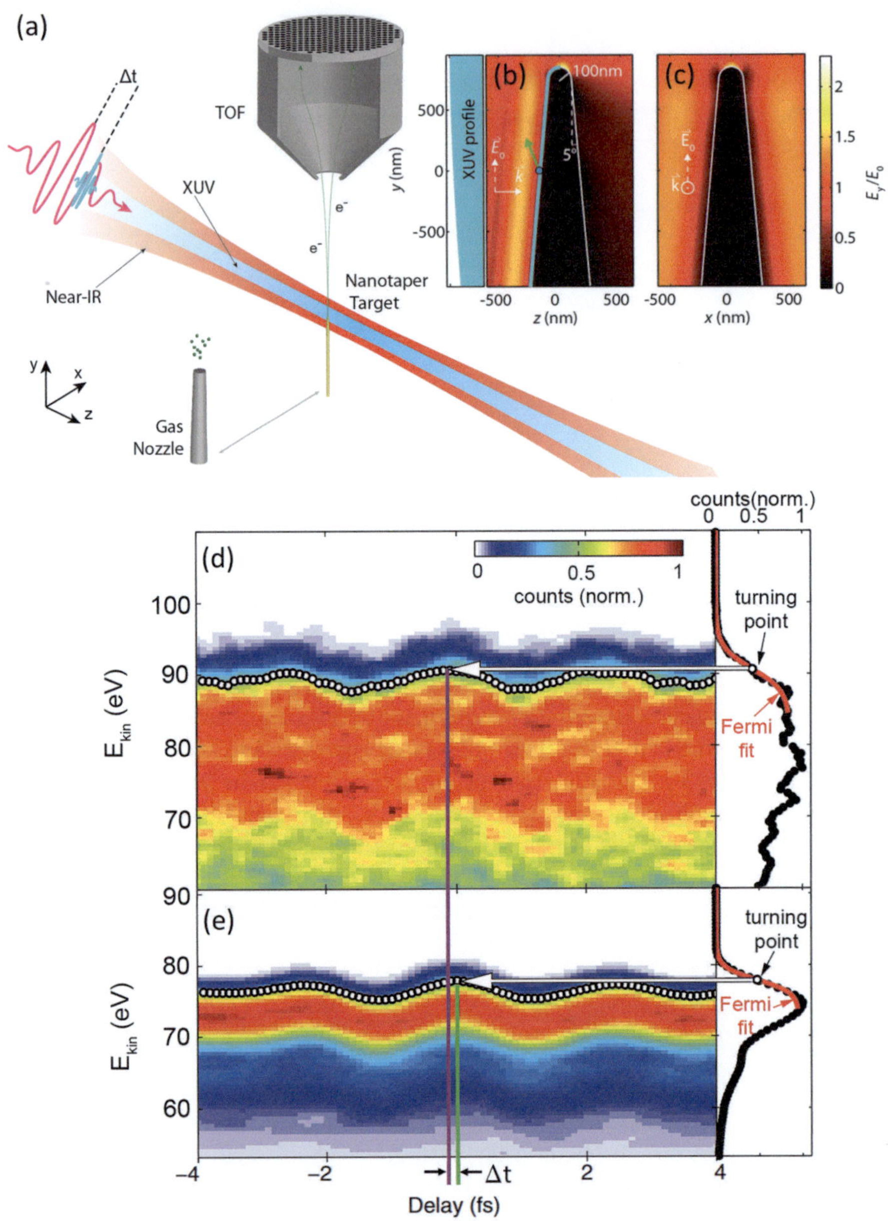

Fig. 14.9 (**a**) Schematic of experimental setup for attosecond nanoplasmonic near-field sampling. Isolated attosecond XUV and few-cycle NIR pulses are focused onto an Au nanotaper. For gas-phase reference measurements in Ne the taper is replaced by a gas nozzle. The electron emission is

centered at 720 nm excite collective electron dynamics in the nanotaper, resulting in spatially-varying near-fields (Fig. 14.9b, c). Delayed attosecond XUV pulses (220 as duration at 95 eV central photon energy) release electrons from the sample, which are subsequently accelerated by the near-fields. The tips were fabricated by electrochemical etching [42]. The delay-dependent momentum distribution of the emitted electrons is recorded with a TOF spectrometer aligned parallel to the laser polarization. For independent characterization of the incident NIR field and the XUV pulse by means of standard attosecond streaking, the taper can be replaced by a gas target (Ne). Neglecting the absolute photoemission time delay from Ne atoms, the gas phase streaking gives access to the vector potential of the incident NIR field.

Typical streaking spectrograms obtained from the Au nanotaper and Ne gas for the same laser parameters are presented in Fig. 14.9d and e, respectively. The panel to the right depicts the spectrum at a fixed delay. The high-energy edge of the Au spectrum is assigned to the photoemission of 5d electrons. For each time delay the high-energy part of the spectra was fitted with a Fermi function and from these fits the streaking curve was obtained (open circles in Fig. 14.9d, e). A relative shift between the streaking traces for tip and Ne is clearly visible, and is evaluated by fitting the streaking traces as $\Delta t = (200 \pm 50)$ as. In contrast to the previous attosecond studies on metal surfaces [25–27], where the streaking field component perpendicular to the surface was probed, in the presented experiment the field component parallel to the surface is probed. The decay of this field component inside the material is given by the skin depth ($\sim$30 nm) and is much larger than the XUV photoemission depth ($\sim$1 nm). It is therefore safe to assume that the delays due to the transport of the electrons to the surface are negligible in our experiments, and the shift is thus mostly attributed to the collective free-electron polarization response of the gold nanowire. The comparison with trajectory simulations in near-fields obtained with the macroscopic Maxwell's equations places the experiment in the ponderomotive streaking regime and provides access to the temporal evolution of the electric near fields around the nanotaper.

Fig. 14.9 (continued) detected by a TOF spectrometer at different relative delays of the XUV and NIR pulses. (**b**, **c**) Maximal normalized local field strengths of the electrical field component parallel to the taper axis in the plane of the laser propagation direction (**b**) and in the plane perpendicular to the laser propagation direction (**c**) as obtained from FDTD-simulations. The *green arrow* shows the maximum electron detection angle. The *blue* lineout illustrates the XUV photoemission area. (**d**, **e**) Experimental spectrograms obtained from measurements of the Au nanotaper (**d**) and Ne gas (**e**). The *right* hand panels of the spectrograms show electron spectra for a fixed delay of −0.2 fs (nanotaper) and 0 fs (gas) illustrating extraction of the streaking curves. A Fermi function (*red*) is fitted to the cut-off edge of the spectrum. The turning points of the Fermi functions for different delay times provide the curves depicted by open circles in (**d**) and (**e**)

In summary, with a streaking approach our experiment allowed the complete temporal characterization of near-fields at nanostructures for the first time. Given the universality of the technique, we anticipate that it will help to understand and tailor collective electron dynamics on the nanoscale.

Acknowledgements We are grateful for support from our collaborators and colleagues that contributed to the presented work. We acknowledge funding by the DFG via SPP1840 (QUTIF), SPP1391, LMUexcellent and the center of excellence "Munich Centre for Advanced Photonics" and by the EU via the ERC grants "ATTOCO" (no. 307203) and "Near Field Atto" (no. 616823).

References

1. Zherebtsov, S., et al. (2011). Controlled near-field enhanced electron acceleration from dielectric nanospheres with intense few-cycle laser fields. *Nature Physics, 7*, 656.
2. Zherebtsov, S., et al. (2012). Carrier-envelope phase-tagged imaging of the controlled electron acceleration from SiO_2 nanospheres in intense few-cycle laser fields. *New Journal of Physics, 14*, 075010.
3. Ahmad, I., et al. (2009). Frontend light source for short-pulse pumped OPCPA system. *Applied Physics B: Lasers and Optics, 97*(3), 529–536.
4. Wittmann, T., et al. (2009). Single-shot carrier-envelope phase measurement of few-cycle laser pulses. *Nature Physics, 5*, 357–362.
5. Rathje, T., et al. (2012). Review of attosecond resolved measurement and control via carrier-envelope phase tagging with above-threshold ionization. *Journal of Physics B: Atomic, Molecular and Optical Physics, 45*, 074003.
6. Süßmann, F., et al. (2011). Single-shot velocity-map imaging of attosecond light-field control at kilohertz rate. *The Review of Scientific Instruments, 82*, 093109.
7. Stöber, W., Fink, A., & Bohn, E. (1968). Controlled growth of monodisperse silica spheres in the micron size range. *Journal of Colloid and Interface Science, 26*, 62.
8. Sau, T. K., & Murphy, C. J. (2004). Room temperature, high-yield synthesis of multiple shapes of gold nanoparticles in aqueous solution. *Journal of the American Chemical Society, 126*(28), 8648–8649.
9. Süßmann, F. (2013). *Dissertation LMU Munich*.
10. Süßmann, F., et al. (2015). Field propagation-induced directionality of carrier-envelope phase-controlled photoemission from nanospheres. *Nature Communications, 6*.
11. Seiffert, L., et al. (2015). Competition of single and double rescattering in the strong-field photoemission from dielectric nanospheres. *Applied Physics B: Lasers and Optics, 122*, 101.
12. Otobe, T., Yabana, K., & Iwata, J. I. (2009). First-principles calculation of the electron dynamics in crystalline SiO 2. *Journal of Physics. Condensed Matter, 21*(6), 064224.
13. Durach, M., et al. (2010). Metallization of nanofilms in strong adiabatic electric fields. *Physical Review Letters, 105*(8), 086803.
14. Stuart, B. C., et al. (1996). Nanosecond-to-femtosecond laser-induced breakdown in dielectrics. *Physical Review B, 53*(4), 1749–1761.
15. Gamaly, E. (2011). *Femtosecond laser-matter interactions: Theory, experiments and applications*. Singapore: Pan Stanford Publishing.
16. Lenzner, M., et al. (1998). Femtosecond optical breakdown in dielectrics. *Physical Review Letters, 80*(18), 4076–4079.
17. Schiffrin, A., et al. (2013). Optical-field-induced current in dielectrics. *Nature, 493*(7430), 70–74.
18. Dombi, P., et al. (2010). Observation of few-cycle, strong-field phenomena in surface plasmon fields. *Optics Express, 18*(23), 24206–24212.

19. Vogelsang, J., et al. (2015). Ultrafast electron emission from a sharp metal nanotaper driven by adiabatic nanofocusing of surface plasmons. *Nano Letters, 15*(7), 4685–4691.
20. Hanke, T., et al. (2009). Efficient nonlinear light emission of single gold optical antennas driven by few-cycle near-infrared pulses. *Physical Review Letters, 103*(25), 257404.
21. Anderson, A., et al. (2010). Few-femtosecond plasmon dephasing of a single metallic nanostructure from optical response function reconstruction by interferometric frequency resolved optical gating. *Nano Letters, 10*(7), 2519–2524.
22. Rewitz, C., et al. (2012). Ultrafast plasmon propagation in nanowires characterized by far-field spectral interferometry. *Nano Letters, 12*(1), 45–49.
23. Stockman, M. I., et al. (2007). Attosecond nanoplasmonic-field microscope. *Nature Photonics, 1*(9), 539–544.
24. Kienberger, R., et al. (2004). Atomic transient recorder. *Nature, 427*(6977), 817–821.
25. Okell, W. A., et al. (2015). Temporal broadening of attosecond photoelectron wavepackets from solid surfaces. *Optica, 2*(4), 383–387.
26. Cavalieri, A. L., et al. (2007). Attosecond spectroscopy in condensed matter. *Nature, 449*(7165), 1029–1032.
27. Neppl, S., et al. (2012). Attosecond time-resolved photoemission from core and valence states of magnesium. *Physical Review Letters, 109*(8), 087401.
28. Skopalova, E., et al. (2011). Numerical simulation of attosecond nanoplasmonic streaking. *New Journal of Physics, 13*, 083003.
29. Kelkensberg, F., Koenderink, A. F., & Vrakking, M. J. J. (2012). Attosecond streaking in a nano-plasmonic field. *New Journal of Physics, 14*, 093034.
30. Hansen, P. M., et al. (2005). Expanding the optical trapping range of gold nanoparticles. *Nano Letters, 5*(10), 1937–1942.
31. Süßmann, F., & Kling, M. F. (2011). Attosecond nanoplasmonic streaking of localized fields near metal nanospheres. *Physical Review B, 84*, 121406(R).
32. Schenk, M., Kruger, M., & Hommelhoff, P. (2010). Strong-field above-threshold photoemission from sharp metal tips. *Physical Review Letters, 105*(25), 257601.
33. Bormann, R., et al. (2010). Tip-enhanced strong-field photoemission. *Physical Review Letters, 105*(14), 147601.
34. Krüger, M., Schenk, M., & Hommelhoff, P. (2011). Attosecond control of electrons emitted from a nanoscale metal tip. *Nature, 475*(7354), 78–81.
35. Herink, G., et al. (2012). Field-driven photoemission from nanostructures quenches the quiver motion. *Nature, 483*(7388), 190–193.
36. Piglosiewicz, B., et al. (2014). Carrier-envelope phase effects on the strong-field photoemission of electrons from metallic nanostructures. *Nature Photonics, 8*(1), 37–42.
37. Yanagisawa, H., et al. (2009). Optical control of field-emission sites by femtosecond laser pulses. *Physical Review Letters, 103*(25), 257603.
38. Thomas, S., et al. (2013). Probing of optical near-fields by electron rescattering on the 1 nm scale. *Nano Letters, 13*(10), 4790–4794.
39. Park, D. J., et al. (2013). Characterizing the optical near-field in the vicinity of a sharp metallic nanoprobe by angle-resolved electron kinetic energy spectroscopy. *Annals of Physics (Berlin), 525*(1-2), 135–142.
40. Lienau, C., Raschke, M. B., & Ropers, C. (2015). *Ultrafast nano-focusing for imaging and spectroscopy with electrons and Light, in attosecond nanophysics: From basic science to applications* (pp. 281–324). Weinheim: Wiley-VCH.
41. Förg, B., et al. (2016). Attosecond nanoscale near-field sampling. *Nature Communications, 7*, 11717.
42. Eisele, M., et al. (2011). Note: Production of sharp gold tips with high surface quality. *The Review of Scientific Instruments, 82*(2), 026101.

Chapter 15
Luminescence of Er^{3+} Ions in Nanocrystalline Glass-Ceramics

Rolindes Balda, R. Morea, J. Gonzalo, and J. Fernandez

Abstract This work analyzes the influence of the nanocrystals on the luminescence properties of Er^{3+} ions in transparent oxyfluoride tellurite glass-ceramics obtained by heat treatment of the precursor Er-doped glass. The comparison between the fluorescence properties of Er-doped glass and glass-ceramic suggests that a fraction of Er^{3+} ions are forming nanocrystals while the rest remain in a glass environment. The presence of ErF_3 nanocrystals has a strong effect in the upconverted red emission from $^4F_{9/2}$ level. The time evolution of the upconverted red emission shows that energy transfer upconversion processes are responsible for the increase of this emission in the glass-ceramic sample.

15.1 Introduction

Rare earth-doped transparent oxyfluoride glass-ceramics have been the subject of several recent investigations since they possess optical properties of interest for photonic applications such as lasers, optical amplifiers, sensors, light converters and so on. In particular, these materials containing one or more crystalline phases embedded in the glass matrix, combine the low phonon energy, optical transparency and rare-earth (RE) ions solubility of fluoride crystals with the good mechanical, thermal, and chemical properties of oxide glasses [1].

The first transparent glass-ceramic (TGC) based on aluminosilicate glass doped with Er^{3+} and Yb^{3+} ions was developed by Wang & Ohwaki in 1993 [2]. In this glass, cadmiun lead fluoride nanocrystals were nucleated by suitable thermal treatments. It was observed that, exciting Yb^{3+} ions at 970 nm, the green-red upconversion emission from Er^{3+} ions was 100 times more intense than observed

R. Balda (✉) • J. Fernandez
Departamento de Física Aplicada I, Escuela Superior de Ingeniería, Universidad del País Vasco UPV/EHU, Alda. Urquijo s/n 48013 Bilbao, Spain and Materials Physics Center CSIC-UPV/EHU, 20080 San Sebastián, Spain
e-mail: rolindes.balda@ehu.es; joaquin.fernandez@ehu.es

R. Morea • J. Gonzalo
Laser Processing Group, Instituto de Optica, CSIC, Serrano 121, 28006 Madrid, Spain

© Springer Science+Business Media Dordrecht 2017
B. Di Bartolo et al. (eds.), *Nano-Optics: Principles Enabling Basic Research and Applications*, NATO Science for Peace and Security Series B: Physics and Biophysics, DOI 10.1007/978-94-024-0850-8_15

from fluoride glass, which suggested the segregation of Yb^{3+} and Er^{3+} ions in the nanocrystals. Since then, different rare earth doped oxyfluoride glass ceramics mainly based on silicate glass matrices and other fluoride crystal phases, such as LaF_3 [3–5], CaF_2 [6, 7], BaF_2 [8, 9], etc. have been proposed as active media for solid state lasers, optical amplifiers, phosphors or to enhance the efficiency of photovoltaic cells [10, 11]. However, the limited RE solubility of silica-based compositions and small nonlinear behavior as well as the low chemical durability of pure fluoride compositions has motivated the search for other glass precursors for TGC.

Among different oxide glasses, tellurite glasses combine good mechanical stability, chemical durability, and high linear and nonlinear refractive indices, with a wide transmission window (typically 0.4–6 μm), which make them promising materials for photonic applications [12]. In fact, broadband Er-doped fiber amplifiers have been achieved by using tellurite-based fibers with a gain of 25.3 dB and a noise figure of less than 6 dB from 1561 to 1611 nm [13] and more recently, efficient laser emission around 2 μm has been demonstrated in a tellurite fiber doped with Tm^{3+} ions [14].

Mixed fluorotellurite glasses which combine the low phonon energies of fluorides with the high chemical durability and thermal stability of tellurites can reduce the OH content which has a great influence in the quenching processes of the radiative emission of excited levels of rare earth ions, which increases fluorescence lifetimes [15, 16]. The transformation, by the adequated heat treatment, of flurotellurite glasses into glass-ceramics in which the RE ions are incorporated in the crystalline phase can increase the maximum cross-sections of the RE ions and the emission efficiency. However, up to now, there are only a few works dealing with Er^{3+}-doped fluorotellurite glass ceramics [17, 18].

In this work, we report the effect of the heat treatment on absorption spectra, near infrared luminescence, and upconversion emission of Er^{3+}-doped transparent oxifluoride tellurite glass-ceramics obtained by the heat treatment of the Er^{3+}-doped 73.6 TeO_2–17.6 ZnO–8.8 ZnF_2 glass. The comparison of the fluorescence properties of Er^{3+}-doped precursor glass and glass-ceramic confirms the successful incorporation of the rare-earth into the nanocrystals.

15.2 Theoretical Background

The rare-earth ions, referred to as the lanthanides, comprise the series of elements in the sixth row of the periodic table after lanthanum from cerium to ytterbium. These atoms are usually incorporated in crystalline or amorphous hosts as trivalent ions and occasionally as divalent ions. The ground state configuration of the trivalent rare-earth ions is $[Xe]4f^n$ where n varies from 1 (Ce^{3+}) to 13 (Yb^{3+}) and indicates the number of electrons in the unfilled 4f shell.

The 4f electrons are shielded by external full outer shells which limits the interaction between the ion and the crystalline field to a small perturbation giving the RE ions their characteristic sharp and well defined spectral features.

The optical activity of the RE ions in solids occurs mainly between electronic states within the 4f configuration. These transitions are parity forbidden and are made partially allowed by crystal field interactions mixing states of different parity. The result is that the transitions between 4f states are weak, with oscillator strengths of the order of 10^{-6}, and radiative lifetimes in the micro and miliseconds range. The long lifetime plays an important role to increase the probability of sequential excitations in the excited states of a single ion as well as in permitting ion-ion interactions in the excited states to allow energy transfer.

The lifetime of an emitting level is governed by a sum of probabilities for several competing processes: radiative decay, nonradiative decay by multiphonon (MPH) relaxation and by energy transfer to other ions. MPH relaxation processes in which the energy difference between the emitting level and the next lower level is converted into many lattice phonons are strongly dependent on the host matrix. Generally, the larger the number of phonons needed to convert the excitation energy, the lower is the efficiency of the non radiative processes. Therefore to enhance the emission efficiency hosts with low phonon energy are required.

Energy levels in rare-earth ions are labeled according to their angular momentum and spin quantum numbers using terms symbols such as $^4I_{15/2}$, or $^4S_{3/2}$. Here the letter refers to the total orbital angular momentum of the ion, the superscript is the number of possible orientation of the total spin of the ions, given as $2S + 1$, where S is the total spin angular momentum. The subscript refers to the total angular momentum of the ion and is determined using the Russell-Saunders coupling scheme [19]. These levels represent the energy levels of a free ion. However, when the ion is in a host, electron-host interactions further split these levels into Stark sublevels, due to the effect of the electric field of the matrix (crystal field effect). This effect is quite small since the 4f orbitals are shielded from the environment by the filled 5s and 5p sub-shells. Nevertheless, the crystal field (CF) induced by the host is what determines the shape of the absorption and emission spectra of RE ions. When comparing the spectra of RE ions in crystals and glasses, the absorption and emission bands are narrow with well resolved Stark components, even at room temperature, in the case of crystals when RE ions occupy a well defined crystal field site. On the contrary, the spectra of RE ions in an amorphous matrix, like a glass, present an inhomogeneous broadening due to the different sites occupied by the ions in the host and thus, different crystal fields. As a consequence the Stark structure of RE spectra in glasses is only partially resolved even at low temperature.

On the other hand, the symmetry and strength of the CF also affect the splitting of the levels. Moreover, the probability of radiative and nonradiative transitions is strongly affected by the host matrix.

The radiative transition probabilities between energy levels in rare-earth ions can be calculated by using the Judd-Ofelt theory [20, 21]. In the framework of this theory, and considering the electric-dipole and magnetic-dipole contributions, the radiative transitions within the $4f^n$ configuration from the initial state |a J>, to the final state |b J′ > can be described in terms of the oscillator strength by using the expression,

$$f_{cal}\left(aJ;bJ'\right)=\frac{8\pi^2 m\nu}{3h\left(2J+1\right)e^2n^2}\left[\chi_{ed}S_{ed}\left(aJ;bJ'\right)+\chi_{md}S_{md}\left(aJ;bJ'\right)\right] \quad (15.1)$$

where m is the mass of the electron, h the Planck's constant, ν is the frequency of the transition, J is the total angular momentum of the initial state, e the electron charge, and n is the refractive index of the host. $\chi_{ed}=\frac{n\left(n^2+2\right)^2}{9}$ and $\chi_{md}=n^3$ are the effective field correction at a well-localized center in a medium of refractive index n. S_{ed} and S_{md} are the line strengths for electric-dipole and magnetic-dipole transitions respectively, expressed by,

$$S_{ed}=e^2\sum_{t=2,4,6}\Omega_t\left|\left\langle aJ\left\|U^{(t)}\right\|bJ'\right\rangle\right|^2 \quad (15.2)$$

$$S_{md}=\frac{e^2\hbar^2}{4m^2c^2}\left|\left\langle aJ\left\|\vec{L}+2\vec{S}\right\|bJ'\right\rangle\right|^2 \quad (15.3)$$

$\langle\|U^{(t)}\|\rangle$ are the double-reduced matrix elements of the unit tensor operators which are considered to be independent of the host matrix and Ω_t are the JO parameters.

The calculated oscillator strengths can be compared with the experimental ones obtained from the absorption spectra by,

$$f_{\exp}=\frac{mc}{\pi e^2N}\int_{band}\alpha\left(\nu\right)d\nu \quad (15.4)$$

where N is the number of ions per unit volume and $\alpha(\nu)$ is the absorption coefficient. Since the experimental oscillator strength contains electric-dipole and magnetic-dipole contributions, the latter has to be subtracted from the experimental one to obtain the electric-dipole contribution. The magnetic-dipole contribution, f_{md}, can be obtained from the equation $f_{md}=nf'$ [22], where n is the refractive index of the studied sample and f' is a quantity calculated based on the energy-level parameters for lanthanide aquo ions [23].

The JO parameters are derived from the electric-dipole experimental oscillator strengths by using a least squared fitting between experimental and calculated oscillator strengths. The magnetic dipole contribution is only present in transitions with $\Delta S=\Delta L=0$ and $\Delta J=0, \pm 1$ [23].

The radiative transitions probability for electric- and magnetic-dipole transitions are given by,

$$A\left(aJ;bJ'\right)=A_{ed}+A_{md}=\frac{64\pi^4\nu^3}{3h\left(2J+1\right)c^3}\left[n\frac{\left(n^2+2\right)^2}{9}S_{ed}+n^3S_{md}\right] \quad (15.5)$$

The magnetic-dipole contributions can be calculated using the values given by Weber for LaF_3 [24] and correcting for the refractive index,

$$A_{md} = \frac{n}{n'} A'_{md} \tag{15.6}$$

where n and A_{md} are the values reported for LaF_3.

The radiative lifetime of an emitting level is related to the total spontaneous emission probability from this level by,

$$\tau_R = \left\{ \sum_{bJ'} A\left(aJ; bJ'\right) \right\}^{-1} \tag{15.7}$$

The emission branching ratio of a transition is defined by,

$$\beta = \frac{A\left(aJ; bJ'\right)}{\sum_{bJ'} A\left(aJ; bJ'\right)} \tag{15.8}$$

The experimental lifetimes and the radiative lifetimes can be related by,

$$\frac{1}{\tau_{\exp}} = \frac{1}{\tau_R} + W_{NR} = \frac{1}{\tau_R} + W_{MPH} + W_{ET} \tag{15.9}$$

where W_{NR} is the nonradiative probability. This nonradiative probability can be due to multiphonon relaxation, which depends on the host matrix, and/or energy transfer which is dependent on ion concentration.

Finally, one important parameter for laser applications and optical amplification is the emission cross-section. The stimulated emission cross-section can be obtained from the absorption spectrum by using the McCumber approach [25], which relates the absorption and emission cross-sections by,

$$\sigma_{em}\left(\nu\right) = \sigma_a\left(\nu\right) \exp\left[\frac{\left(\varepsilon - h\nu\right)}{KT}\right] \tag{15.10}$$

where σ_a and σ_{em} are the absorption and emission cross-sections respectively, ν is the photon frequency, ε is the net free energy required to excite one Er^{3+} ion from the $^4I_{15/2}$ to the $^4I_{13/2}$ state at temperature T, h is the Planck's constant, and K is the Boltzmann constant. The absorption cross-section is the absorption coefficient divided by the concentration of active ions and ε can be determined by using the simplified procedure provided in reference 26. The effective emission cross-section can also be calculated from the experimental emission spectrum and radiative transition probability by using the following expression [27],

$$\sigma_{eff} = \frac{\lambda_p^4}{8\pi n^2 c} \frac{1}{\Delta\lambda_{eff}} A\left(aJ; bJ'\right) \tag{15.11}$$

where c is the velocity of light, λ_p is the peak fluorescence wavelength, $\Delta\lambda_{eff}$ is the effective linewidth, A is the radiative transition probability of the emitting level, and n the refractive index. The effective linewidth of the emission is defined by,

$$\Delta\lambda_{eff} = \int \frac{I(\lambda)\, d\lambda}{I_{\max}} \tag{15.12}$$

15.3 Structural Characterization

Precursor oxyfluoride tellurite glass with a composition of 73.6 TeO_2–17.6 ZnO–8.8 ZnF_2 mol% doped with 1 wt% of ErF_3, which corresponds to $1.4 \cdot 10^{20}$ ions/cm^3, was prepared by the conventional melt-quenching technique. To nucleate a glass-ceramic (GC) phase a two step treatment was performed according to the values of the glass transition and crystallization temperatures ($T_g \approx 293$ °C and $T_c \approx 390$ °C respectively) [18]. The first treatment was done at 310 °C for 10 h, followed by a 3 h treatment at 340 °C, which is below that of the onset of crystallization.

The microstructure of the samples and chemical composition of the nanocrystals have been analyzed by Transmission Electron Microscopy (TEM) and Energy Dispersive X-ray spectroscopy (EDX). TEM analysis clearly demonstrates the formation of nanocrystals in the heat treated sample as it is shown in Fig. 15.1a. They present a dendritic structure with typical sizes of 45 ± 10 nm and clear lattice fringes. The Fast Fourier Transform (FFT) of this image in Fig. 15.1b shows different brilliant spots in the reciprocal space which can be indexed according to the planes of ErF_3 crystals obtained in the [101] direction. This analysis was performed based on the compositional profile obtained from EDX spectroscopy. The EDX analysis of the nanocrystals (Fig. 15.2) shows intense peaks corresponding to F and

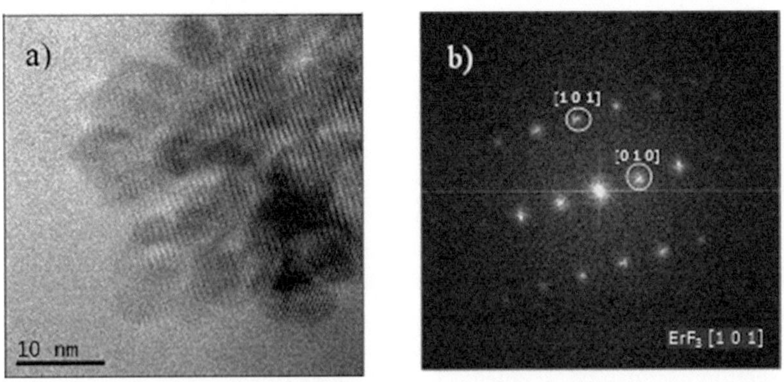

Fig. 15.1 (**a**) HRTEM image of the GC sample. (**b**) Fast Fourier Transform (FFT) image indexed according to the ErF_3 structure

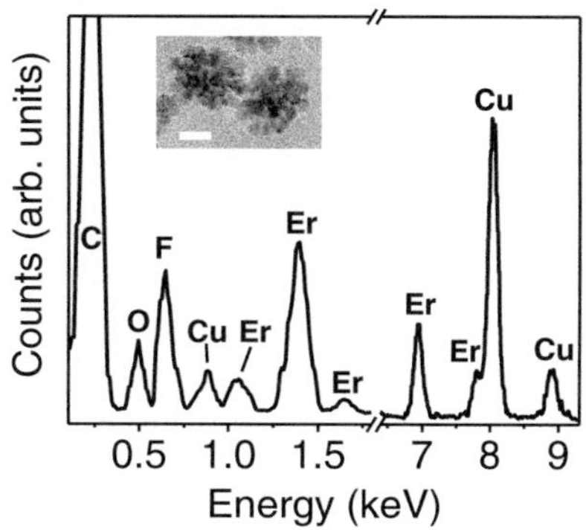

Fig. 15.2 EDX spectrum in the energy ranges 0–1.8 and 6.5–9.3 keV of the nanocrystals observed in the GC sample. The elements responsible for the peaks observed are indicated. The inset shows a TEM image of two of these nanocrystals, where the horizontal line corresponds to 20 nm

Er in addition to those associated to the C layer of the replica and the Cu grid (C, O, and Cu). The absence of peaks related to Zn or Te suggests that the nanocrystals are formed by ErF_3.

15.4 Influence of the Nanocrystals on the Luminescence Properties of Er^{3+} Ions

15.4.1 Absorption Properties

The influence of the nanocrystals on the spectroscopic properties of Er^{3+} ions in glass-ceramic samples based on fluorotellurite glasses has been studied by using absorption and steady-state and time-resolved laser spectroscopy. The absorption spectra of the glass (GS) and glass-ceramic (GC) samples are displayed in Fig. 15.3. The absorption bands are assigned to the transitions from the $^4I_{15/2}$ ground state to the excited states of Er^{3+} ions. Due to the absorption of the host in the UV, some transitions are not observed. As can be seen, there are some difference between both spectra due to the presence of the nanocrystals (NCs). The absorption edge is shifted to longer wavelengths in the GC sample due to the scattering of short wavelength light by the NCs present in this sample. Moreover, the intensity of the hypersensitive transition $^4I_{15/2} \rightarrow {}^2H_{11/2}$ is strongly reduced in the glass-ceramic sample which indicates that the ligand field around Er^{3+} ions has changed as a consequence of the heat treatment. The strength of the hypersensitive transitions is greatly affected by the environment of rare earth ions. Ligand polarizability as well as an asymmetric surrounding environment increases the intensity of hypersensitive transitions of RE ions [28–30]. Another evidence of the incorporation of Er^{3+} ions in the nanocrystals is the splitting of the inhomogeneously broadened absorption bands in the absorption spectrum of the GC sample. As an example, the inset in Fig. 15.3 shows a detail of the $^4I_{15/2} \rightarrow {}^4I_{13/2}$ absorption band for the glass and GC samples.

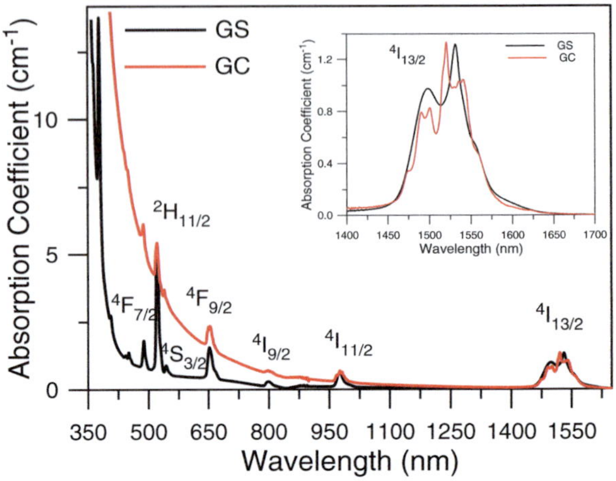

Fig. 15.3 Room temperature absorption spectra for precursor glass and glass-ceramic samples. The inset shows the detail of the $^4I_{15/2} \rightarrow ^4I_{13/2}$ absorption bands

Table 15.1 Judd-Ofelt parameters and r.m.s. deviation for precursor glass (GS) and glass ceramic (GC) samples

Sample	Ω_2 ($\times 10^{-20}$)	Ω_4 ($\times 10^{-20}$)	Ω_6 ($\times 10^{-20}$)	r.m.s.
GS	4.71	1.57	1.13	$3.24 \cdot 10^{-7}$
GC	1.28	1.04	1.00	$3.47 \cdot 10^{-7}$

The absorption spectra have been used to calculate the radiative transition probabilities, the radiative lifetimes, and branching ratios of the emissions by using the JO intensity parameters. From these spectra the experimental oscillator strengths have been calculated and following the procedure described in Sect. 2, the theoretical oscillator strengths and the JO parameters of Er^{3+} ions in the glass and glass-ceramic samples. The values are displayed in Table 15.1. The matrix elements given by Carnall et al [23] have been used in the calculation. The low value of the root mean square (rms) indicates a good agreement between the experimental and calculated oscillator strengths.

The value of Ω_2 in the GC sample is lower than in the GS, which means that the heat treatment induces a reduction in the degree of covalency in the rare-earth site. It is well known that Ω_2 is most sensitive to local structure and its value is indicative of the amount of covalent bonding between RE ions and ligand anions. The decrease of Ω_2 further confirms the incorporation of Er^{3+} ions into the fluoride NCs. Moreover, the sum of the JO parameters decreases for the GC sample due to the decrease of covalency of the chemical bond between the Er^{3+} ion and the ligand anions. This is in accordance with reported results that predict that the sum decrease in the order oxide > oxyfluoride > fluoride. In addition, the Ω_2 parameter is closely related to the hypersensitive transitions. The more intense the hypersensitive transition is, the larger the value of Ω_2 is. The lower value of this parameter in the heat treated sample is in agreement with the decrease of the strength of the hypersensitive transition $^4I_{15/2} \rightarrow ^2H_{11/2}$ which indicates the variation of the local structure around the Er^{3+} ions and further confirms their incorporation into the nanocrystals.

The JO parameters have been used to calculate the radiative transition probabilities and radiative lifetimes. The radiative lifetimes of level $^4I_{13/2}$ are 3.23 and 3.73 ms for the GS and GC sample respectively.

15.4.2 Near Infrared Emission

The room temperature near infrared emission spectra have been obtained by exciting at 801 nm in the $^4I_{9/2}$ level with a Ti-sapphire ring laser. The fluorescence was analyzed with a monochromator and the signal detected with a photomultiplier and finally amplified by a standard lock-in technique. After excitation of level $^4I_{9/2}$, the next lower levels are populated by multiphonon relaxation. Two emission bands were observed at around 980 and 1530 nm, which correspond to transitions from $^4I_{11/2}$ and $^4I_{13/2}$ levels to the ground state respectively. Figure 15.4 shows the emission spectra corresponding to the $^4I_{13/2} \rightarrow {}^4I_{15/2}$ transition. The spectrum of the GC sample shows a more resolved structure with seven Stark components. Moreover, the effective bandwidth is reduced from 73.4 nm in the Gs to 64 nm in the GC sample probably due to the reduction of inhomogeneous broadening.

Figure 15.5 shows the absorption and emission cross-sections obtained for the GC sample by using the McCumber approach together with the normalized measured emission spectrum. The stimulated emission cross-section has a value of $8.9 \cdot 10^{-21}$ cm^2, similar to the one found in the glass sample ($8.8 \cdot 10^{-21}$ cm^2) [16] and higher than those found in other fluorotellurite glass-ceramics [31].

The comparison of the excitation spectra of the $^4I_{15/2} \rightarrow {}^4I_{9/2}$ transition obtained by collecting the luminescence at 1540 nm for the glass and glass-ceramic samples also confirms the incorporation of Er^{3+} ions in the nanocrystals. As can be seen in

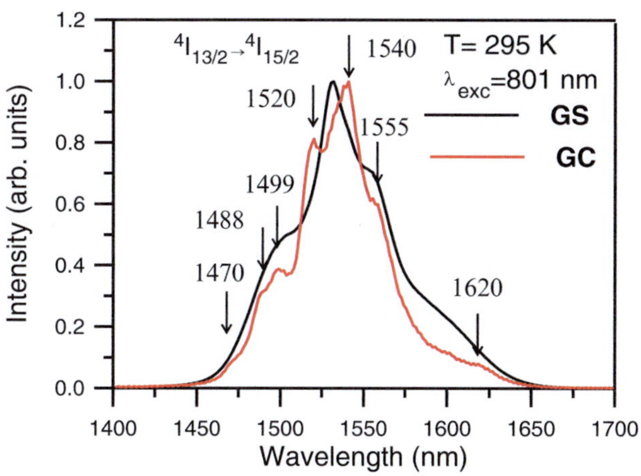

Fig. 15.4 Room temperature emission spectra obtained under excitation at 801 nm in the glass and glass-ceramic samples

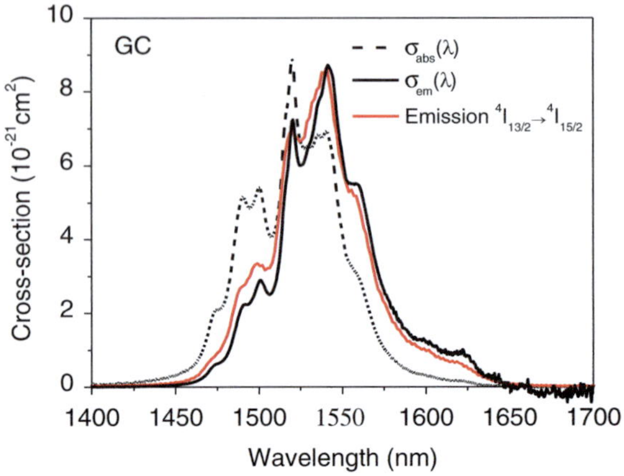

Fig. 15.5 Absorption and emission cross-sections of the $^4I_{13/2}\leftrightarrow{}^4I_{15/2}$ transitions for the GC sample together with the normalized emission spectrum

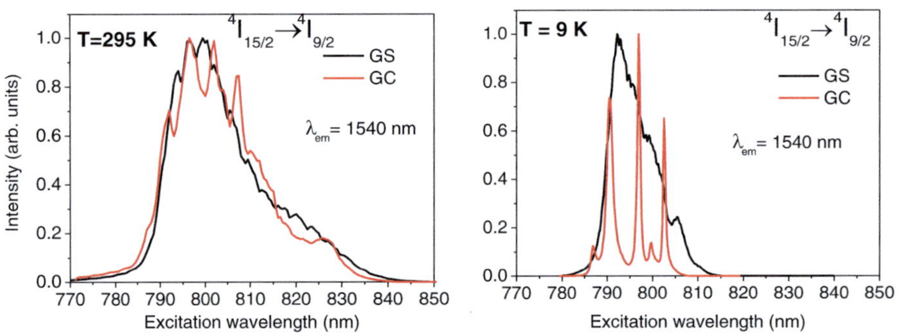

Fig. 15.6 Excitation spectra of the $^4I_{15/2}\rightarrow{}^4I_{9/2}$ transition obtained by collecting the luminescence at 1540 nm in the glass and glass-ceramic samples at 295 K and 9 K

Fig. 15.6, the spectrum corresponding to the GC sample shows well defined peaks not observed in the excitation spectrum of the glass sample. This is more evident from the low temperature excitation spectrum of the GC sample which shows narrow peaks according with the presence of Er^{3+} ions in a crystalline environment.

The experimental decays of luminescence from levels $^4I_{11/2}$ and $^4I_{13/2}$ were obtained at room temperature for both samples by exciting at 801 nm in level $^4I_{9/2}$ by using a Ti-sapphire laser pumped by a pulsed frequency doubled Nd:YAG laser (9 ns pulse width). The decay from level $^4I_{13/2}$ shows, in both samples, an initial rise time due to the population from the higher $^4I_{11/2}$ level. Figure 15.7 shows the experimental decay curves for GC and GS samples. The decay of the glass sample can be described by a single exponential function with a lifetime of 3.6 ms, however the experimental decay of the GC sample is well described by a double exponential function with lifetimes of 1.2 and 3.1 ms which suggest contributions from different environments around Er^{3+} ions. In the GC, a fraction of Er^{3+} ions remains in a glass environment while the rest is forming NCs. The reduction in lifetimes after heat treatment has been previously reported [32] and attributed to the segregation

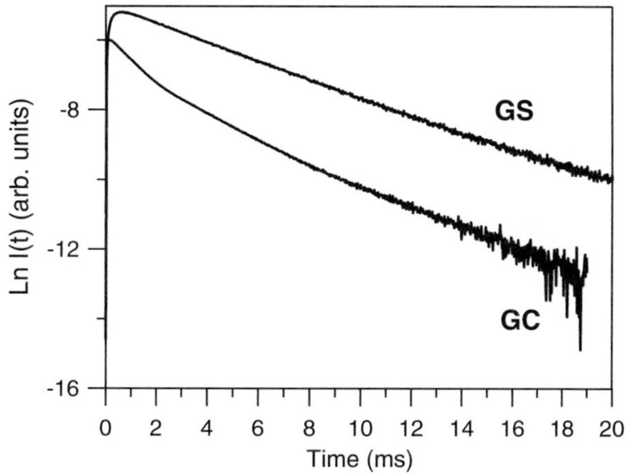

Fig. 15.7 Semilogarithmic plot of the fluorescence decays of the $^4I_{13/2}$ level for precursor glass and glass-ceramic samples obtained after excitation at 801 nm

of the rare-earth ions into the crystalline phase which reduces Er^{3+}-Er^{3+} distances. Shorter distances between Er^{3+} ions favour dipole-dipole energy transfer between them, which increases non-radiative rates and reduces the lifetime of level $^4I_{13/2}$. A similar behavior is observed for the lifetime value of the $^4I_{11/2}$ level which is reduced from 420 μs in the glass sample to 220 μs in GC sample.

15.4.3 Upconversion Emission

Up-conversion (UC) is an anti-Stokes process that converts the absorbed lower energy photons, usually in the NIR range, into higher energy photons in UV, visible, or shorter NIR ranges. UC can occur mainly by two fundamental mechanisms: (i) excited state absorption (ESA); (ii) energy transfer upconversion (ETU). In the case of ESA, excitation involves sequential absorptions of pump photons by a single ion. This mechanism requires a nearly equal separation for the first and second steps, as well as the long lifetime of the intermediate level. In an ETU process, two ions interact, one of then loses energy and goes to the ground state whereas the other one gains energy and goes to the upper level. The UC efficiency of an ETU process is strongly dependent on the active ions concentration. In contrast to ETU, the efficiency of an ESA process is independent of the dopant concentration, because ESA involves sequential excitation in the same ion.

Visible upconversion emission has been observed for both glass and GC samples at room temperature under continuous wave (cw) and pulsed laser excitation in resonance with the $^4I_{9/2}$ level. The upconverted emission spectra obtained under cw excitation were measured by using a Ti-sapphire ring laser. Cut-off filters were used to remove the pumping radiation. Figure 15.8a shows the upconversion emission spectra corresponding to the glass and glass-ceramic samples. The upconversion emission of the glass sample shows the characteristic green emission attributed

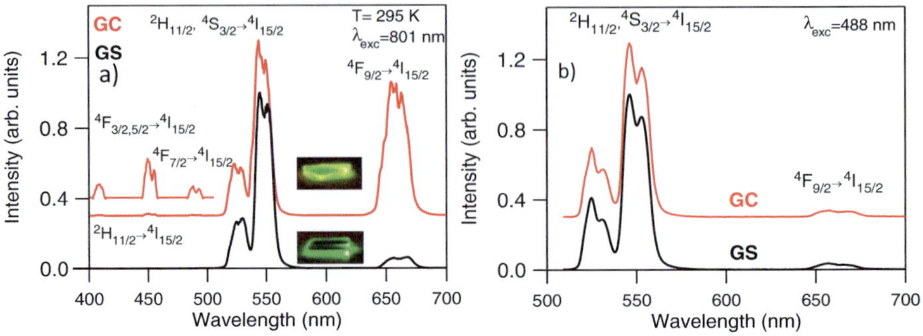

Fig. 15.8 Visible emission obtained (**a**) under 801 nm and (**b**) under 488 nm excitation for the glass and glass-ceramic samples

to the transitions from the two thermallized levels ($^2H_{11/2}$, $^4S_{3/2}$) together with a weak red emission from the $^4F_{9/2}$ level. However, the spectrum of the GC sample shows in addition to the green and red emissions very weak blue emissions from higher levels which do not appear in the upconversion emission of the glass sample. As can be seen, the emission bands are better resolved in the GC spectrum in agreement with the incorporation of Er^{3+} ions in the crystalline phase. Moreover, the intensity of the red emission increases significantly in the GC sample. This increase of the red emission, not observed in the spectra obtained by exciting at 488 nm (Fig. 15.8b), indicates that after NIR excitation, level $^4F_{9/2}$ is populated by energy transfer processes in addition to multiphonon relaxation from level $^4S_{3/2}$. This behavior could be associated to a higher concentration of Er^{3+} ions in the nanocrystals, which reduces the Er^{3+}-Er^{3+} distances and increases the probability of energy transfer processes [33, 34].

To investigate the excitation mechanisms for populating the $^4S_{3/2}$ and $^4F_{9/2}$ levels after IR excitation in the GC sample, we have measured the upconverted emission intensities for different pumping powers. The upconversion emission intensity (I_{em}) depends on the incident pump power (P_{pump}) according to the relation $I_{em} \propto (P_{pump})^n$, where n is the number of photons involved in the pumping mechanism. Figure 15.9 shows a logarithmic plot of the integrated emission intensity of the upconverted green and red fluorescence as a function of the pump laser intensity. The dependence in both cases is nearly quadratic which indicates a two photon (TP) upconversion process to populate the $^4S_{3/2}$ and $^4F_{9/2}$ levels. This process may be associated with excited state absorption and/or energy transfer upconversion.

As it is well known, the time evolution of the upconversion luminescence after an excitation pulse provides an useful tool in discerning which the operative mechanism is. The radiative ESA process occurs during the excitation pulse and leads to an immediate decay of the upconversion luminescence after excitation. Upconversion by energy transfer leads to a time-dependent emission that shows a rise of the upconverted population after the laser pulse, followed by a decay of

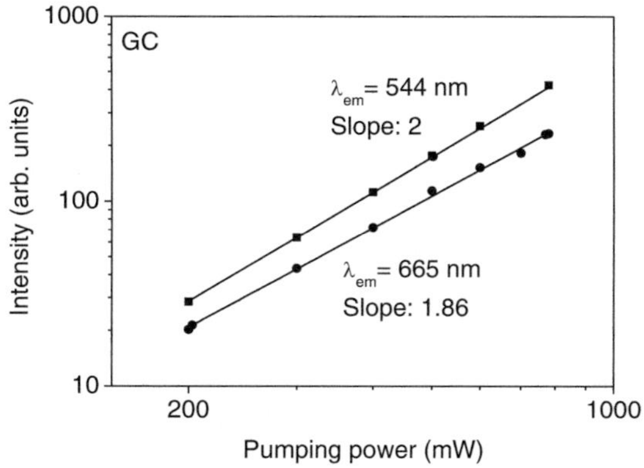

Fig. 15.9 Logarithmic plot of the integrated intensities of the upconverted emission from $^4S_{3/2}$ (544 nm) and $^4F_{9/2}$ (665 nm) levels obtained under excitation at 801 nm

the population, with a lifetime longer than the one after direct excitation. The rise and decay times are determined by both the intermediate and the upper excited state lifetimes. This distinction is possible when the pump pulse width is much shorter than the time constant of the relevant energy transfer step.

According to the energy level diagram of Er^{3+} ion in the GC sample, the $^4S_{3/2}$ level can be populated by ESA and/or ETU in the case of the green emission (Fig. 15.10). In the first case, the absorption of an IR photon excites the electrons to the $^4I_{9/2}$ level, then multiphonon relaxation populates the next lower level, and the absorption of a second IR photon promotes the electrons to the $^4F_{3/2,5/2}$ levels from where $^4S_{3/2}$ level is reached by nonradiative relaxation. Part of the excitation energy in the $^4I_{11/2}$ level further relaxes, radiatively and nonradiatively to level $^4I_{13/2}$ and ESA from $^4I_{13/2}$ to $^2H_{11/2}$ can occur. This last process is nearly resonant. Another possibility is an energy transfer process which involves the interaction between two Er^{3+} ions in the $^4I_{11/2}$ level. In this process one ion gains energy and reaches level $^4F_{7/2}$, whereas the other one loses energy and goes to the ground state.

The lifetime behavior of the upconverted green emission suggests that both processes ESA and ETU are present. Figure 15.11a shows the decays of the green emission obtained by exciting at 488 and 801 nm respectively for the GC sample. In both cases, the decays can be described to a good approximation by two components. The fast one has a lifetime of 36 and 41 μs for the 488 and 801 nm excitation respectively but the lifetime of the long component (129 μs), is longer than the one obtained by direct excitation (107 μs). The observation of the instantaneous rise after the 9 ns pulse excitation indicates the presence of an ESA process; however, the longer lifetime of the slow component after IR excitation indicates that an additional ETU process is also necessary to describe the lifetime of the decay. The lifetime of the long component (129 μs) is about half the lifetime of the $^4I_{11/2}$ level (220 μs), which suggests that the upconversion green emission from level $^4S_{3/2}$ can be produced by the interaction between two Er^{3+} ions in the $^4I_{11/2}$ level via the transitions ($^4I_{11/2} \rightarrow {}^4I_{15/2}$) and ($^4I_{11/2} \rightarrow {}^4F_{7/2}$).

Instead, the time evolution of the upconverted red emission confirms that the $^4F_{9/2}$ level is populated by ETU processes. As an example, the time evolution of

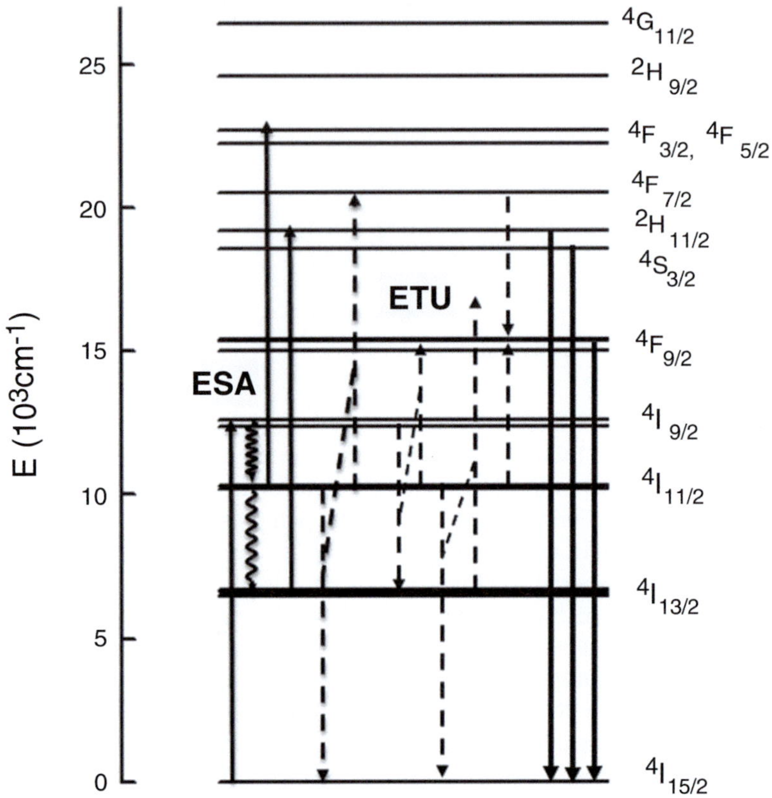

Fig. 15.10 Energy level diagram of Er^{3+} ions in the GC sample and possible upconversion mechanisms

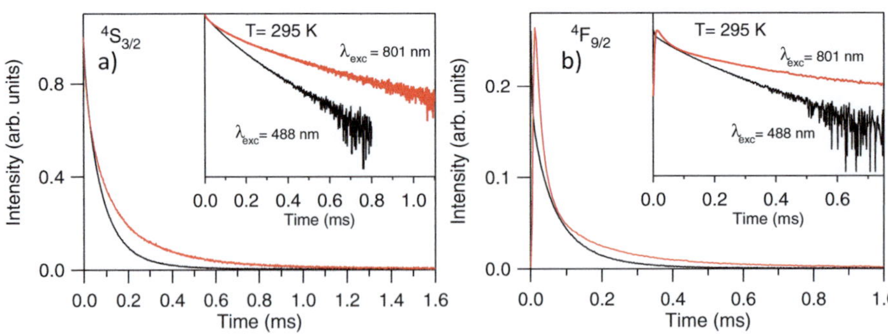

Fig. 15.11 Semilogarithmic plot of the experimental decay curves of the (**a**) *green* ($^4S_{3/2}$) and (**b**) *red* ($^4F_{9/2}$) emissions under excitation at 488 nm (*black line*) and 801 nm (*red line*) for the GC sample

the red emission obtained under excitation at 801 nm and 488 nm for the GC sample is shown in Fig. 15.11b. In both cases, the decays can be described by two components. The fast one (~21 μs) is similar for both excitation wavelengths whereas the lifetime of the slow component increases from 89 μs to 183 μs when the excitation wavelength changes from 488 nm to 801 nm. Moreover, there is an initial rise time of about 15 μs. The presence of a rise time together with the lengthening of the slow component of the decay supports the hypothesis that ETU processes are populating level $^4F_{9/2}$. According to the energy level diagram of Er^{3+} ions in the GC sample (Fig. 15.10), there are different ETU processes to populate the $^4F_{9/2}$ level. This level can be populated via ($^4I_{9/2} \rightarrow {}^4I_{13/2}$) and ($^4I_{11/2} \rightarrow {}^4F_{9/2}$) transitions and/or ($^4I_{11/2} \rightarrow {}^4I_{15/2}$) and ($^4I_{13/2} \rightarrow {}^4F_{9/2}$). There exists another possible process for populating level $^4F_{9/2}$ in which two Er^{3+} ions interact, one of them in level $^4I_{11/2}$ and the other one in level $^4F_{7/2}$, going both to level $^4F_{9/2}$. By considering that the lifetime of a higher-energy level excited by ETU reflects those of the intermediate levels from which upward excitation occurs, and that the long component of the decay of the red emission after excitation at 801 nm is very close to the lifetime of level $^4I_{11/2}$, the ($^4I_{9/2} \rightarrow {}^4I_{13/2}$);($^4I_{11/2} \rightarrow {}^4F_{9/2}$) and ($^4F_{7/2} \rightarrow {}^4F_{9/2}$);($^4I_{11/2} \rightarrow {}^4F_{9/2}$) processes seem to be the likeliest ones to explain the population of the $^4F_{9/2}$ level by ETU.

15.5 Conclusions

Transparent glass-ceramics containing ErF_3 nanocrystals have been prepared by suitable thermal treatment of an Er^{3+}- doped 73.6 TeO_2–17.6 ZnO–8.8 ZnF_2 mol% fluorotellurite glass.

The comparison of the fluorescence properties of Er^{3+}-doped precursor glass and glass-ceramic confirms the successful incorporation of the rare-earth into the nanocrystals. The strong reduction of the Ω_2 JO parameter and the intensity of the hypersensitive transition $^4I_{15/2} \rightarrow {}^2H_{11/2}$ in the heat treated sample indicate that Er^{3+} ions are in a crystalline environment. The $^4I_{13/2}$ decay in the glass-ceramic sample can be described by two single exponential functions which suggests that a fraction of Er^{3+} ions are forming nanocrystals while the rest remains in a glass environment. Moreover, the emission lifetime is shorter than in the glass sample probably due to the segregation of the Er^{3+} ions into the crystalline phase which reduces Er^{3+}-Er^{3+} distances and increases energy transfer processes.

The presence of ErF_3 nanocrystals has a strong effect in the upconverted red emission from $^4F_{9/2}$ level. This emission increases significantly as compared with the green one in the GC sample. The time evolution of the upconverted red emission suggests that ETU processes are responsible for the increase of this emission in the glass-ceramic sample.

Acknowledgments We deeply acknowledge A. Tomas from the National Centre for Metallurgical Research (CENIM, CSIC) and A. Ruiz-Caridad from the Laboratory of Electron Nanoscopies (LENS, University of Barcelona) for TEM sample preparation and HRTEM analysis, respectively.

This work was supported by the Spanish Government MEC under Projects No. TEC2012-38901-C02-01 and MAT2013-48246-C2-2-P and Basque Country Government IT-659-13 and IT-943-16. R. Morea acknowledge a FPI grant from the Spanish Government (BES-2010-038084).

References

1. Dejneka, M. J. (1998). Transparent oxyfluoride glass-ceramics. *MRS Bulletin, 11*, 57.
2. Wang, Y., & Ohwaki, J. (1993). New transparent vitroceramics codoped with Er^{3+} and Yb^{3+} for efficient frequency upconversion. *Applied Physics Letters, 63*, 3268.
3. Chen, D., Yu, Y., Huang, P., Lin, H., Shan, Z., & Wang, Y. (2010). Color-tunable luminescence of Eu^{3+} in LaF_3 embedded nanocomposite for light emitting diode. *Acta Materialia, 58*, 3035.
4. Luo, Q., Qiao, X., Fan, X., Yang, H., Zhang, X., Cui, S., et al. (2009). Luminescence behavior of Ce^{3+} and Dy^{3+} codoped oxyfluoride glasses and glass ceramics containing LaF_3 nanocrystals. *Journal of Applied Physics, 105*, 043506.
5. Velazquez, J. J., Rodriguez, V. D., Yanes, A. C., del-Castillo, J., & Mendez-Ramos, J. (2010). Increase in the Tb^{3+} green emission in SiO_2-LaF_3 nano-glass-ceramics by codoping with Dy^{3+} ions. *Journal of Applied Physics, 108*, 113530.
6. Secu, M., Secu, C. E., Polosan, S., Aldica, G., & Chica, C. (2009). Crystallization and spectroscopic properties of Eu-doped CaF_2 nanocrystals in transparent oxyfluoride glass-ceramics. *Journal of Non-Crystalline Solids, 355*, 1869.
7. Sun, X., Gu, M., Huang, S., Jin, X., Lui, X., Liu, B., et al. (2009). Luminescence behavior of Tb^{3+} ions in transparent glass and glass-ceramics containing CaF_2 nanocrystals. *Journal of Luminescence, 129*, 773.
8. Qiao, X., Fan, X., & Wang, M. (2006). Luminescence behavior of Er^{3+} in glass ceramics containing BaF_2 nanocrystals. *Scripta Materialia, 55*, 211.
9. Luo, Q., Fan, X., Qiao, X., Yang, H., Wang, M., & Zhang, X. (2009). Eu^{2+}-Doped glass ceramics containing BaF_2 nanocrystals as a potential blue phosphor for UV-LED. *Journal of the American Ceramic Society, 92*, 942.
10. Goncalves, M. C., Almeida, R. M., & Santos, L. F. (2002). Rare-earth-doped transparent glass ceramics. *Comptes Rendus Chimie, 5*, 845.
11. de Pablos-Martín, A., Durán, A., & Pascual, M. J. (2012). Nanocrystallisation in oxyfluoride systems: mechanisms of crystallisation and photonic properties. *International Materials Reviews, 57*, 165.
12. El-Mallawany, R. A. H. (2001). *Tellurite glasses handbook-physical properties and data*. Boca Raton: CRC Boca Raton.
13. Mori, A. (2002). 1.58-μm Broad-band erbium-doped tellurite fiber amplifier. *IEEE Journal of Lightwave Technology, LT-20*, 822.
14. Richards, B., Tsang, Y., Binks, D., Lousteau, J., & Jha, A. (2008). Efficient 2 μm Tm^{3+}-doped tellurite fiber laser. *Optics Letters, 33*, 402.
15. Sidebottom, D. L., Hruschka, M. A., Potter, B. G., & Brow, R. K. (1997). Increased radiative lifetime of rare earth-doped zinc oxyhalide tellurite glasses. *Applied Physics Letters, 71*, 1963.
16. Miguel, A., Morea, R., Gonzalo, J., Arriandiaga, M. A., Fernandez, J., & Balda, R. (2013). Near-infrared emission and upconversion in Er^{3+}-doped TeO_2-ZnO-ZnF_2 glasses. *Journal of Luminescence, 140*, 38.
17. Yu, C., Zhang, J., Wen, L., & Jiang, Z. (2007). New transparent Er^{3+}-doped oxyfluoride tellurite glass ceramic with improved near infrared and up-conversion fluorescence properties. *Materials Letters, 61*, 3644.
18. Miguel, A., Morea, R., Arriandiaga, M. A., Hernández, M., Ferrer, F. J., Domingo, C., et al. (2014). Structural, optical, and spectroscopic properties of Er^{3+}-doped TeO_2-ZnO-ZnF_2 glass-ceramics. *Journal of the European Ceramic Society, 34*, 3433.

19. Russell, H. N., & Saunders, F. A. (1925). New regularities in the spectra of the alkaline earths. *Astrophysical Journal, 61*, 38.
20. Judd, B. R. (1962). Optical absorption intensities of rare-earth ions. *Physics Review, 127*, 750.
21. Ofelt, G. S. (1962). Intensities of crystal spectra of rare-earth ions. *Journal of Chemical Physics, 37*, 511.
22. Carnall, W. T., Fields, P. R., & Rajnak, K. (1968a). Spectral intensities of the trivalent lanthanides and actinides in solution. II. Pm^{3+}, Sm^{3+}, Eu^{3+}, Gd^{3+}, Tb^{3+}, Dy^{3+}, and Ho^{3+}. *Journal of Chemical Physics, 49*, 4412.
23. Carnall, W. T., Fields, P. R., & Rajnak, K. (1968b). Electronic energy levels in the trivalent lanthanide aquo ions. I. Pr^{3+}, Nd^{3+}, Pm^{3+}, Sm^{3+}, Dy^{3+}, Ho^{3+}, Er^{3+}, and Tm^{3+}. *Journal of Chemical Physics, 49*, 4424.
24. Weber, M. J. (1967). Probabilities for radiative and nonradiative decay of Er^{3+} in LaF_3. *Physics Review, 157*, 262.
25. McCumber, D. E. (1964). Einstein relations connecting broadband emission and absorption spectra. *Physics Review, 136*, A954.
26. Miniscalco, W. J., & Quimby, R. S. (1991). General procedure for the analysis of Er^{3+} cross sections. *Optics Letters, 16*, 258.
27. Weber, M. J., Ziegler, D. C., & Angell, C. A. (1982). Tailoring stimulated emission cross sections of Nd^{3+} laser glass: Observation of large cross sections for $BiCl_3$ glasses. *Journal of Applied Physics, 53*, 4344.
28. Ebendorff-Heidepriem, H., Ehrt, D., Bettinelli, M., & Speghini, A. (1998). Effect of glass composition on Judd-Ofelt parameters and radiative decay rates of Er^{3+} in fluoride phosphate and phosphate glasses. *Journal of Non-Crystalline Solids, 240*, 66.
29. Jorgensen, C. K., & Reisfeld, R. (1983). Judd-Ofelt parameters and chemical bonding. *Journal of Less-Common Metals, 93*, 107.
30. Tanabe, S., Ohyagi, T., & Soga, N. (1992). Compositional dependence of Judd-Ofelt parameters of Er^{3+} ions in alkali-metal borate glasses. *Physical Review B, 46*, 3305–3310.
31. Jlassi, I., Elhouichet, H., Hraiech, S., & Ferid, M. (2012). Effect of heat treatment on the structural and optical properties of tellurite glasses doped erbium. *Journal of Luminescence, 132*, 832.
32. Dantelle, G., Mortier, M., Patriarche, G., & Vivien, D. (2006). Er^{3+} doped PbF_2: Comparison between nanocrystals in glass-ceramics and bulk single crystals. *Journal of Solid State Chemistry, 179*, 1995.
33. Balda, R., García-Revilla, S., Fernández, J., Seznec, V., Nazabal, V., Zhang, X. H., et al. (2009). Upconversion luminescence of transparent Er^{3+}-doped chalcohalide glass-ceramics. *Optical Materials, 31*, 760.
34. Balda, R., García-Adeva, A. J., Fernández, J., & Fdez-Navarro, J. M. (2004). Infrared-to-visible upconversion of Er^{3+} ions in GeO_2-PbO-Nb_2O_5 glasses. *Journal of the Optical Society of America B, 21*, 744.

Chapter 16
Localization of Yb^{3+}, Er^{3+} and Co^{2+} Dopants in an Optical Glass Ceramics of $MgAl_2O_4$ Spinel Nano-crystals Embedded in SiO_2 Glass

G. Boulon, Y. Guyot, G. Alombert-Goget, M. Guzik, T. Epicier, L. Chen, L. Hu, and W. Chen

Abstract The main goal of this research is the localization of Yb^{3+} and Er^{3+} rare earth ions and Co^{2+} transition metal ion as dopants in a glass ceramics composed of $MgAl_2O_4$ spinel nano-crystals of 10–20 nm size embedded in SiO_2 glass. This is the first step of a compact self-Q-switched microchip laser. We conjugate, that is rather rare, both TEM-EDX and optical spectroscopy techniques. The use of TEM-EDX technique associated with both the elemental mapping of each dopant and the direct visualization of the heavier rare earth ions is unique way in luminescent materials which has led to the result that Er^{3+} and Yb^{3+} rare earth ions are preferentially located in the spinel nano-crystals. Regarding the Co^{2+} low concentration, this technique was not enough accurate and finally absorption spectroscopy technique have probed the main presence of Co^{2+} ions in the spinel nano-crystals. The use of site selective spectroscopy technique applied to Yb^{3+} structural probes allows to identify the 0-phonon broad line at 975 nm with both that of the disordered glass and that of the spinel inverted phases. A new Yb^{3+} radiationless center has been pointed out by the presence of a strong absorption line at 970 nm which has been assigned to the strongly perturbed area of the spinel nano-crystallite surface. This dopant

G. Boulon (✉) • Y. Guyot • G. Alombert-Goget
Univ Lyon, Université Claude Bernard Lyon 1, CNRS, Institut Lumière Matière, Lyon F-69622, Villeurbanne, France
e-mail: georges.boulon@univ-lyon1.fr

M. Guzik
Faculty of Chemistry, University of Wrocław, 14 F. Joliot-Curie, PL-50-383, Wrocław, Poland

T. Epicier
MATEIS, UMR 5510 CNRS, INSA of Lyon, University of Lyon, Bât. B. Pascal, 69621 Villeurbanne, France

L. Chen • L. Hu • W. Chen
High Power Laser Components R&D Center, Shanghai Institute of Optics and Fine Mechanics, Chinese Academy of Sciences, Jiading, Shanghai 201800, China

© Springer Science+Business Media Dordrecht 2017
B. Di Bartolo et al. (eds.), *Nano-Optics: Principles Enabling Basic Research and Applications*, NATO Science for Peace and Security Series B: Physics and Biophysics, DOI 10.1007/978-94-024-0850-8_16

characterization worthwhile to be shown in this School by applying both TEM-EDX and optical spectroscopy techniques as a pedagogical case in the research of rare earth ions and transition metal ions localization.

16.1 Introduction

16.1.1 Evolution of Solid State Lasers from Nd^{3+}/Yb^{3+}-Doped Single Crystals to Nd^{3+}/Yb^{3+}-Doped Transparent Ceramics

Inorganic solid-state lasers based on light-pumped insulating host materials containing optically active dopants as metal transition ions (3d configuration) and rare earth ions (4f configuration) continue to occupy a prominent place among the different kinds of laser devices. Indeed, solid-state lasers are mainly used as oscillators or amplifiers in industrial, medical, military and scientific fields. The 1st laser has been demonstrated by Maiman at Hughes Lab with Cr^{3+} ions of $3d^3$ configuration-doped Al_2O_3 in 1960 [1], emitting the sharp line $^2E \rightarrow {}^4A_2$ at 694.3 nm. However, this ruby laser is an inefficient three-level scheme and it was necessary to look on more efficient four-level scheme. An important breakthrough was the discovery of Nd^{3+}-doped $Y_3Al_5O_{12}$ (YAG) garnet crystal by Geusic as four-level scheme with right parameters and its extension to Nd^{3+}-doped glasses [2]. The improvement of laser efficiency of Nd^{3+}-doped YAG laser crystal is a slow process: it won a factor of only two, 20 years after the first growth in 1964 [3]. During this period of time, the focus of the respective fields, namely phosphors for lighting application and solid-state lasers, has converged toward common targets, in the sense that phosphor research and development has been heavily concerned with the luminescence properties of trivalent rare earth ions, while in the research for tunable, coherent sources, the interest in laser materials has veered toward the type of band emitters traditionally investigated in phosphor research. Then scientific know-how has somehow percolated between the two fields. The information exchange from the laser field to the phosphor field has not been restricted to theoretical models but has been important in the field of materials (apatites, garnets, sesquioxides) [4]. The most important driving forces and noteworthy developments of laser materials since 1960 were the following ones [3]:

- Understanding of optical spectroscopy of luminescent ions used as laser ions,
- Laser fundamentals understandings,
- Choice and improvement of crystal growth techniques,
- Laser-diode-pumped solid state lasers as new sources,
- Polycrystalline ceramic lasers as a new generation of laser materials

16.1.2 Nd^{3+}/Yb^{3+}-Doped Transparent Ceramics

The recent development of ceramic laser materials is probably the most important innovation in the field of laser material fabrication technology. The large-scale laser materials are necessary for high-power laser systems. Unfortunately, the size of single-crystal is limited and, the boule cross-section shows facets giving rise to difficulties during the growth. Single crystals are generally grown from the melt and they suffer from drawbacks such as segregation of the dopant from the host, optical inhomogeneity caused by stress during crystal growth, high cost and low productivity due to high temperature processing. Transparent polycristalline are advantageous over single crystals in many ways. The process is simple, cost effective and typically carried out at lower temperature. Most important, much higher doping concentrations in ceramics can be obtained without phase segregation as we often observe in single crystals. As a result, laser ceramics are quite attractive since 1995 and it has been possible to fabricate the large-scale laser materials by using ceramic technology applied on isotropic hosts like $Y_3Al_5O_{12}$ cubic garnets, (Y_2O_3) sesquioxides, $MgAl_2O_4$ spinels, CaF_2 fluorides [5–10]. Producers are challenging to make various large size ceramics with high optical quality under rods, disks and slabs shapes and by playing with engineer profiles and structures and also composites like un-doped and doped ceramics [11].

However, although high efficiencies have been demonstrated, one limitation can be associated with Nd^{3+} segregation at grain boundaries in transparent Nd^{3+}-doped YAG ceramics for lasers [12], Ce^{3+}-doped (Y,Gd)Al_3O1_2 [13], Ce^{3+}-doped YAG [14] for scintillators and also phosphors for white LED sources. Segregation was not observed in Yb^{3+}-doped YAG [15, 16] or Yb^{3+}-doped Y_2O_3 [16]. We have interpreted such data by the variation of segregation coefficients of rare earth ions in oxide optical ceramics like in growth of crystals from the melt and/or flux. The lower is the segregation coefficient of the luminescent ions in the melt growth, the greater amount of luminescent ions will be collected on the grain boundaries of the ceramic material [17]. It means that such segregation phenomenon can perturb kinetics of the fluorescence in transparent ceramics and then the best laser sources should be obtained with the best uniformity of Yb^{3+} and Nd^{3+} ions in optical ceramics.

16.1.3 Essays to Use Glass Ceramics

Another field of optical materials consists to develop rare earth/ transition metal ions-doped glass ceramics for laser sources or scintillators or phosphors. Until today, glass ceramics have been applied for thermal and mechanical properties. In spite of numerous experimental essays, optical properties of glass ceramics are not yet really exploited.

An original idea was proposed recently with transparent glass ceramics [18, 19] containing Yb^{3+}, Er^{3+} ions in glass matrix and Co^{2+}-doped $MgAl_2O_4$

nano-crystals of sizes of 10–20 nm, which can be used as compact self-Q-switched microchip laser. First of all, we have to show that the main conditions are fulfilled in glass ceramics for obtaining self-Q-switched microchip laser. Passively Q-switched, diode-pumped solid state lasers are currently being used as miniature or micro-lasers capable of delivering high peak output power at high repetition rates and short nanosecond (ns) temporal pulse width. These lasers are of great interest due to their potential applications in micro-machining, remote sensing, target, ranging, and microsurgery. Cr^{4+} and Co^{2+} dopants have been proposed as saturable absorbers for passive Q-switching of solid-state lasers operating in the near-infrared region. Most commonly, systems presently used are based on Nd^{3+}: YAG or Nd^{3+}:YVO_4 lasers, passively Q-switched by Cr^{4+}: YAG, where the unique well-known characteristics of Cr^{4+} garnets as saturable absorbers are applied [20–23]. Another crystal like Cr^{4+}:Mg_2SiO_4 was also proposed as saturable absorber [24]. The most important application of the oxide crystals doped with tetrahedrally coordinated Co^{2+} ions is passive Q-switching of solid-state lasers operating in the near-infrared region. A number of crystals such as YAG, $LaMgAl_{11}O_{19}$ (LMA), ZnSe was doped with Co^{2+} ions [25, 26]. As an example, passively Q-switched 1.34 μm Nd^{3+}:$Y_xGd_{1-x}VO_4$ laser with Co^{2+}: $LaMgAl_{11}O_{19}$ saturable absorber was demonstrated [27].

Indeed, glasses containing nano-crystals embedded in a glass matrix can enhance existing or offer completely new properties compared to those of the precursor glass. These transparent glass ceramics possess the spectral requirements of both eye safe lasers by Er^{3+} $^4I_{13/2} \rightarrow ^4I_{15/2}$ emission in glass matrix between 1.5 and 1.6 μm under Yb^{3+} pumping and saturable absorbers by $^4A_2$ (F)$\rightarrow ^4T_1$ (F) transition of Co^{2+}-doped $MgAl_2O_4$ nano-crystals so that, together, might be developed as self-Q-switched microchip laser. Obviously, the morphology of this new class of optical materials requires to solve the specific characterisations of the location of dopants in the two glassy and nano-spinel phases in order to check the feasibility of this system.

Consequently, the main goal of this lecture presented at the School is to conjugate TEM-EDX and optical spectroscopy tools to localize Yb^{3+}, Er^{3+} and Co^{2+} dopants in a glass ceramics composed of $MgAl_2O_4$ spinel nano-crystals of 10–20 nm size embedded in the SiO_2 glass [28].

16.2 Experimental Techniques

16.2.1 Sample Preparation

Glass with a composition (in wt. %) of $1.8Yb_2O_3$-$1.7Er_2O_3$-11MgO-$31.5Al_2O_3$-$47SiO_2$-$4.5TiO_2$-$2.5ZrO_2$-0.03CoO (CYE glass) was prepared, where TiO_2 and ZrO_2 were used as mixed nucleating agents [19]. Concentrations of Co^{2+}, Yb^{3+} and Er^{3+} ions are 5.77×1018 cm^{-3}, 1.5×1020 cm^{-3}, and 1.46×1020 cm^{-3}, respectively. For comparison, CoO-un-doped glass (YE glass) with the same composition was also prepared. For two glasses, homogeneous mixtures (300 g) of analytical reagent grade raw materials, obtained by ball milling, were melted in a corundum crucible in a laboratory electric furnace at 1580 °C for 4 h in air atmosphere with stirring. According to differential scanning calorimetry (DSC) analysis, CYE and YE glass samples were heat treated at 760 °C for 12 h, and then at 930 °C for 4 h. As a result, transparent CYE GC and YE GC were synthesized. The purity of Yb_2O_3 and Er_2O_3 is 4 N, while other reagents are analytical grade.

For comparative absorption analysis of samples containing different concentrations of Yb^{3+} and one Er^{3+} concentration we used the compositions (in wt.%) of 2Yb2Er, 4Yb2Er, and 6Yb2Er, it means xYb_2O_3-1.1Er_2O_3-(3.3-x)La_2O_3-11MgO-30.6Al_2O_3-47SiO_2-4.5TiO_2-2.5ZrO_2-0.03CoO ($x = 1.1, 2.2, 3.3$), respectively.

16.2.2 *Experimental Section*

16.2.2.1 TEM-EDX Experiments

Experiments were conducted on the Cs-corrected TITAN Environmental Transmission Electron Microscope (ETEM) from FEI installed at CLYM (Centre Lyonnais de Microscopie, University of Lyon, www.clym.fr) and operated at 300 kV. Thin foils were prepared by ion bombardment (PIPS system from GATAN). According to this preparation method, some Ar^+ has been detected during EDX analysis with a SDD analyzer from Oxford Instruments (classical implantation of Ar^+ ions in glassy phases). Cu is also detected as spurious elements due to the sample environment in the microscope. During TEM work, irradiation effects were encountered due to the high electron flux at 300 kV. However, all reported observations were obtained in reasonably short times and prior to any significant degradation (e.g. drift, deformation or detectable structural chemical variation), unless otherwise specified. Especially, chemical analysis in EDX were limited to reduced doses and short acquisition times, thus limited the signal-to-noise ratio and the sensitivity of the measurements to detect the very small concentrations of Yb^{3+}, Er^{3+}, Co^{2+} dopants. Under these conditions, both the detection limit and the accuracy of the present results are estimated to be about 0.2 at.% for all species of special interest, i.e. the Yb^{3+}, Er^{3+} and Co^{2+} doping ions. Nano-probe analysis and EDX elementary maps in the Scanning TEM (STEM) mode was performed using a probe of about 2–4 nm (Full-Width at Half-Maximum). Micrographs were acquired in both Cs-corrected High Resolution, and High Angle, Annular dark Field (HAADF) STEM modes.

16.2.2.2 Absorption Spectroscopy

Absorption spectra of glass and glass ceramic samples were measured both, at SIOM (Shanghai), at 293 K, using a Perkin Elmer UV/Visible/NIR Lambda 900 spectrophotometer in 300–2000 nm, and at the Faculty of Chemistry of the

University of Wroclaw (Poland), at 293 K and 4 K, in the same spectral range with a Cary-Varian 5000 Scan spectrometer equipped with an Oxford CF 1204 helium flow cryostat. In addition, a special thin layer technique was used to detect the 4f → 4f transitions of well-grounded nano-powder of Yb^{3+}-doped spinel samples. These homogenous samples were compressed in the form of pellets under 20 MPa pressure. Transparent samples prepared in this way were used for the absorption measurements only of the Yb^{3+}-doped spinel nano-powder.

16.3 Results and Discussion

16.3.1 Morphology of Samples Observed by TEM Measurements

A typical STEM-HAADF image is presented in Fig. 16.1a where spinel nanoparticles appear with a brighter contrast with respect to the vitreous matrix. Their size ranges typically between 4 to 20 nm with an average at around 10 nm. Lattice imaging clearly indicates the crystalline nature of the spinel phase as shown in Fig. 16.2. Cystallographic indexing was performed either in the conventional diffraction mode or by using Fourier transforms, or diffractograms, of the HREM images as reported as an insert in Fig. 16.2: all acquired data are fully consistent with the usual spinel, Fd-3 m structure of $MgAl_2O_4$ with $a = 0.808$ nm. A confirmation of this finding is presented in Fig. 16.3 where a spinel crystallite is observed at High Resolution in the [001] viewing direction.

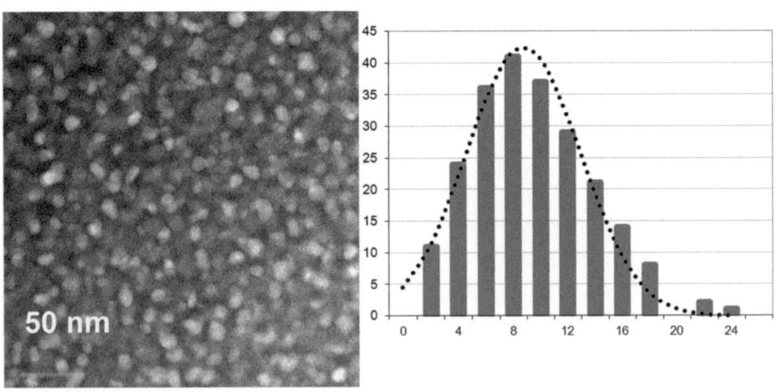

Fig. 16.1 (**a**) Typical STEM-HAADF image showing the distribution of spinel nanocrystallites embedded in the silicate glassy phase (spinel nanoparticles appear with a brighter contrast with respect to the vitreous matrix). (**b**) Size histogram of micrograph shown in a); a mean equivalent (Ferret's) diameter of 10 ± 0.3 nm can be deduced from a Gaussian fit (*dotted curve*)

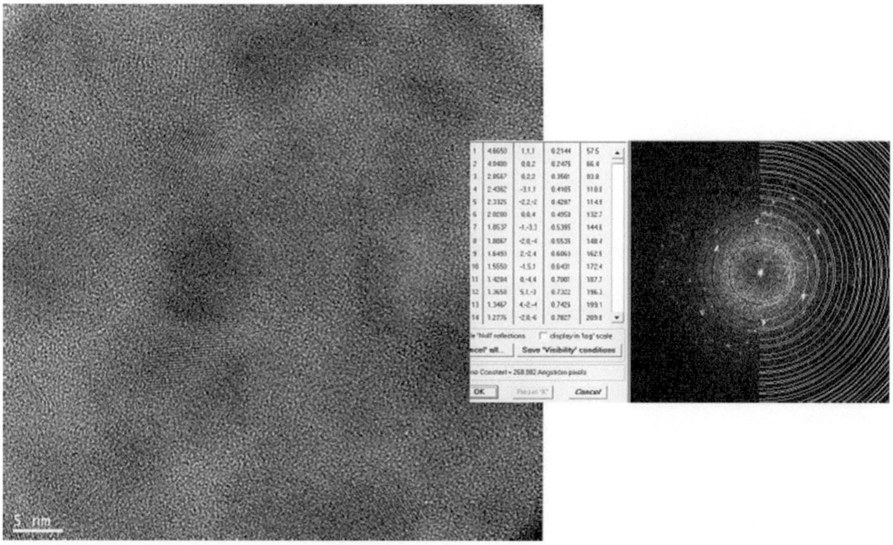

Fig. 16.2 Low magnification HRTEM micrograph and associated Fourier transform (diffractogram): diffraction rings produced by the nano-crystallite lattice planes can be consistently indexed as the prominent reflections from the expected spinel phase $MgAl_2O_4$ (Fd-3 m with $a = 0.808$ nm)

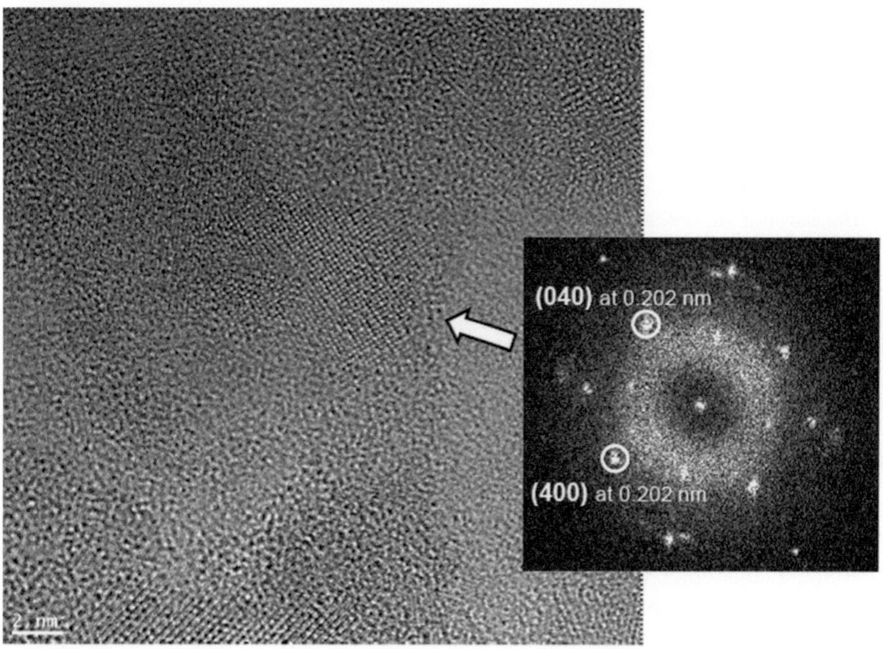

Fig. 16.3 Typical HRTEM image and indexed diffractogram of a nano-crystallite (*arrowed*). Lattice distances correspond to the usual spinel structure within a 2 % accuracy

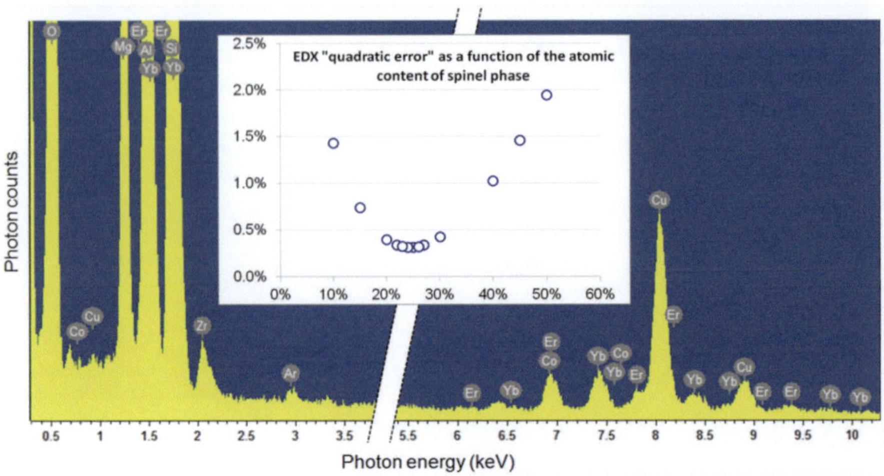

Fig. 16.4 Quantification of the spinel vs. vitreous matrix ratio: the chemical composition deduced form an EDX analysis of a large area is fitted by the following linear combination (LC): x $MgAl_2O_4$ + (1-x) SiO_2 where x represents the atomic fraction of the spinel phase; the quadratic error define d as: $R^2 = ([O]_{LC} - [O]EDX)^2 + ([Mg]_{LC} - [Mg]_{EDX})^2 + ([Al]_{LC} - [Al]_{EDX})^2 + ([Si]_{LC} - [Si]_{EDX})^2$presents a minimum for abound 25 at% of spinel phase. The detection of Zr and Ti (no shown) arises from the presence of ZrO_2 and TiO_2 constituents added in the staring compound to make the glass ceramics (they were not considered in the quantification)

This is important to evaluate the ratio between the $MgAl_2O_4$ nano-spinel crystallites and the SiO_2 glassy phase. To do so, EDX results obtained on large areas of about 1 μm^2 were fitted by a linear combination of the expected chemical formulae of both phases: x $MgAl_2O_4$ (1-x) SiO_2, where x represents the atomic fraction of the spinel phase. According to the usual consideration of absorption effects, linked to the sample thickness (estimated here to about 50 nm, with little incidence on the final spinel fraction in the interval 0–100 nm), the elemental contents for the major constituents O, Mg, Al, Si were deduced from the experimental and the spinel atomic fraction x was refined from a simple mean-least square fitting procedure (see Fig. 16.4): using this approach, an optimum value of x = 25 % was found.

16.3.2 Distribution of Yb^{3+}, Er^{3+}, Co^{2+} Dopants as Revealed by Nano-probe EDX

One main challenge of the present structural characterization is to know not only the distribution of Yb^{3+}, Er^{3+}, Co^{2+} dopants in both glassy and crystalline phases, but also at the interface between the nano-spinel crystals and the glass. The greatest difficulty when applying TEM-EDX technique is to find out adequate thin regions where no or very little overlap occurs between nano-cystals, due to high concentration of nano-crystals. EDX nanoprobe analysis was performed using the classical line scan procedure as employed previously to study the distribution of Yb^{3+} ions in grains and grain boundaries in Ce^{3+}-doped YAG ceramic bulky samples [13, 14], Yb^{3+}-doped YAG and Yb^{3+}-doped Y_2O_3 ceramic bulky samples [15, 16] and Yb^{3+}-doped YAG nano-ceramics [29].

16.3.2.1 TEM-EDX Well Adapted for Rare Earth Dopants But Not for Cobalt Dopant

A typical analysis is shown in Fig. 16.5. As can be seen from the HREM micrograph, we have chosen here a region where two spinel crystallites are surrounded by the glassy matrix, where no lattice fringes were resolved. 15 measurements were performed using a probe of about 3 nm, as indicated by the circle marks. The diagram in Fig. 16.5a, b shows the evolution of the Yb^{3+}, Er^{3+} and Co^{2+} contents: enrichment in rare earth ions is clearly evidenced in the spinel phase whereas slight enrichment of the vitreous matrix is suggested by these data. We have to mention that the very low content measured for Co^{2+} and the strong overlapping between Co^{2+} and Er^{3+} high intensity lines as seen in Fig. 16.6, do not permit any definitive conclusion about any possible Co^{2+} distribution heterogeneity.

16.3.2.2 Interest of EDX Elementary Maps

To give a better global view of the dopant distribution, EDX elementary maps were recorded. Because of irradiation effects severely limiting the acceptable dose on the

Fig. 16.5a Typical HRTEM: Energy-Dispersive X-ray spectroscopy line scan applied to the same Yb-Er-Co-doped glass ceramic sample analyzed in Fig. 16.2b (probe size is about 3 nm). The two *yellow* zones are analyzed as spinel 1 and spinel 2 in Fig. 16.5b

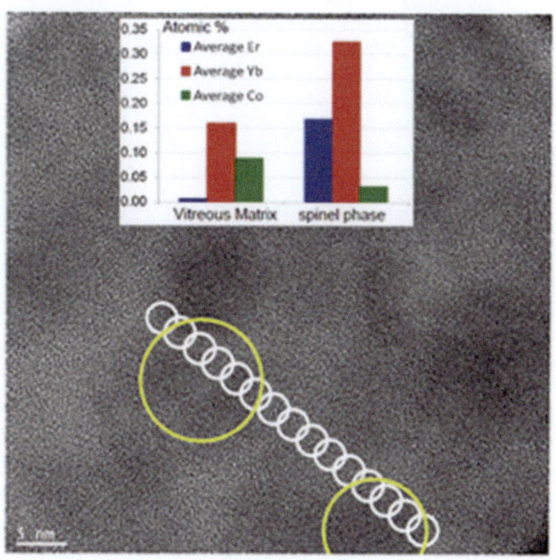

a

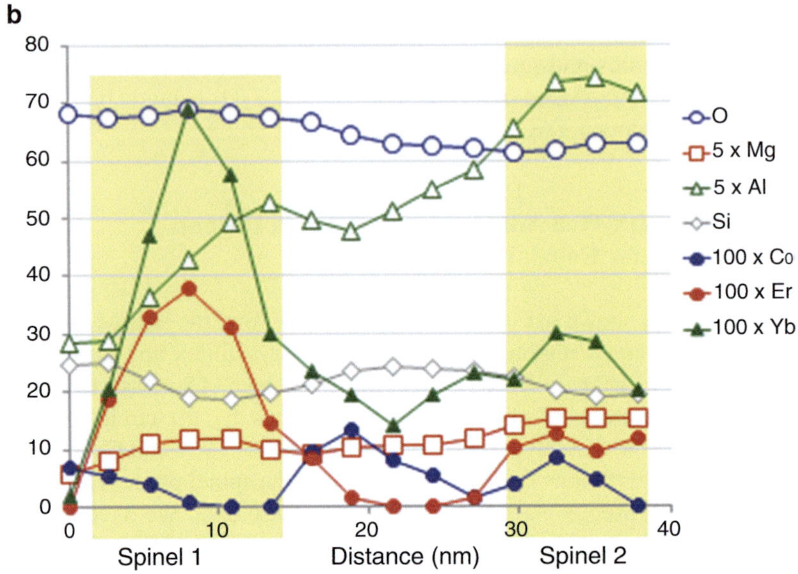

Fig. 16.5b Single spot analysis of the line scan of Fig. 16.5a. Two nano-crystallites so-called spinel 1 and spinel 2 are emphasized by the maxima of Mg, Al, Yb, Er, Co elements. Amplified intensities are applied (X 5) for Mg and Al constituents of $MgAl_2O_4$ nano-crystals and (X 100) for Yb, Er, Co dopants

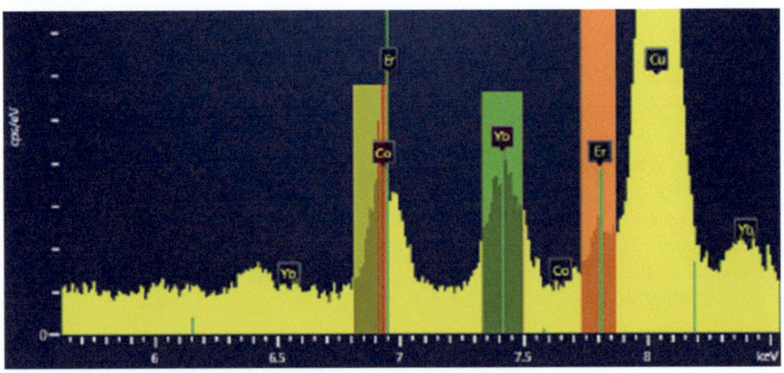

Fig. 16.6 Illustration of the major difficulty met with Co, which do not exhibit 'isolated' X-ray lines. The overlap between Er and Co high intensity lines is clearly seen

sample, it could not be possible to acquire maps with sufficient X-rays counting for an accurate quantitative treatment. However, semi-quantitative analysis gave reasonable results in agreement with previous nano-probe data, as shown by the representative montage reported in Fig. 16.7.

The elemental maps for the major species (Al, Si) allow easily the $MgAl_2O_4$ spinel and the SiO_2 vitreous matrix to be separated; as already shown in Fig. 16.1,

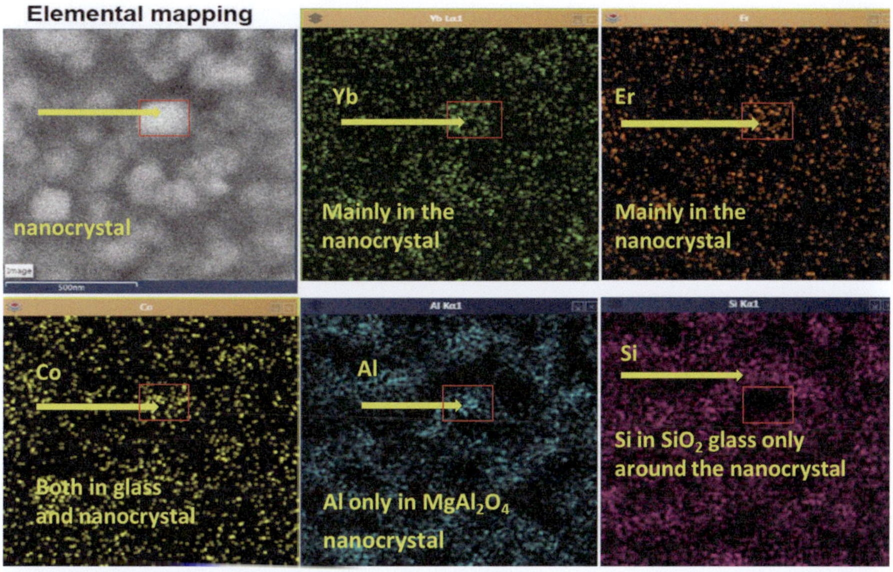

Fig. 16.7 Typical semi-quantitative elemental mapping of the Yb^{3+}, Er^{3+}, Co^{2+} dopants and the Al^{3+} component of the nano-crystalline phase and Si^{4+} component of the vitreous phase. The scale of 500 nm for all diagrams is shown in the *left* side of the first image showing the position of the analyzed nano-crystal

darker areas in STEM-HAADF images corresponds to the vitreous SiO_2-based phase. These maps are treated in a quantitative way by the Aztec commercial software provided by the manufacturer of the EDX system, which means that overlap effects between X-ray lines is in principle correctly taken into account. According to this, the Yb^{3+} and Er^{3+} maps clearly confirm the greatest concentrations of Yb^{3+}, Er^{3+} dopants inside the nano-crystalline phase and much lower traces inside the SiO_2 glassy phase. Again, the Co^{2+} content appears too low in average and its strongest line overlaps too strongly with Er^{3+} line one, to rightly conclude about Co^{2+} ion distribution.

16.3.2.3 Direct Observation of Heavy Yb^{3+} and Er^{3+} Rare Earth Individual Ions

Another original worthwhile observation should be mentioned here concerning the HRTEM imaging of individual dopants. According to (i) the resolution permitted by the spherical aberration corrector, (ii) the large 'weight' (atomic number) difference between rare earth ions (Er^{3+}, Yb^{3+}) compared to the spinel and glassy phase elements (O^{2-}, Al^{3+}, Mg^{2+}, Si^{4+}), it can be postulated that in the thinnest regions, an 'absorption-type' contrast could allow the direct visualization of heavier rare earth individual ions. In convenient so-called 'Scherzer images', atoms are expected

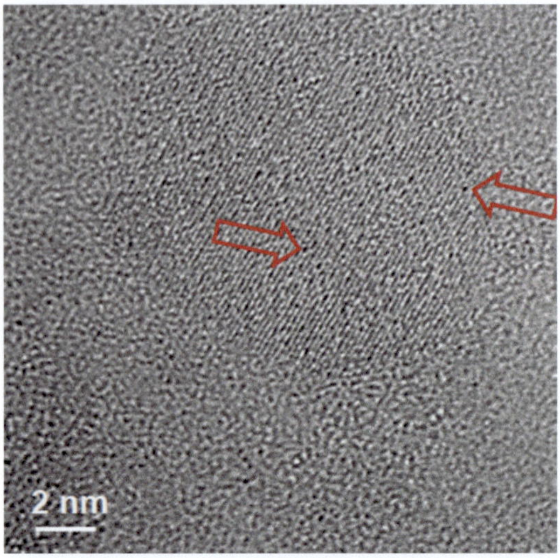

Fig. 16.8a Typical Yb^{3+} and Er^{3+} ions as observed by *black* points in spinel nanoparticle as seen with a set of lattice planes resolved and also in glass.

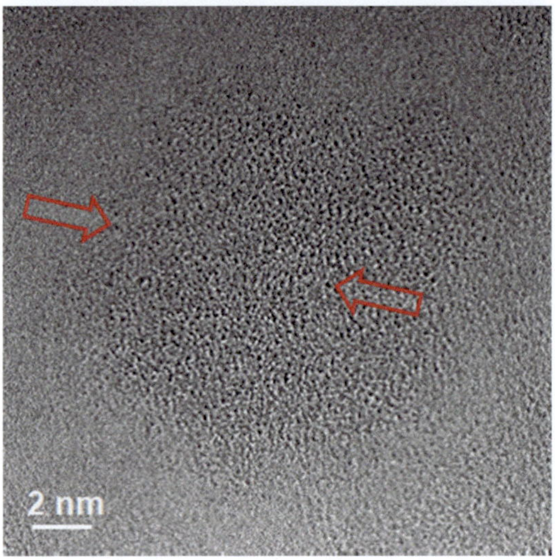

Fig. 16.8b Typical Yb^{3+} and Er^{3+} ions as observed by *black* points in spinel nanoparticle as seen "of axis"

to appear as dark dots on a brighter background. Figures 16.8a and 16.8b can be examples of such an experimental evidence in which this is even possible to distinguish rare earth pairs. We note that rare earth ions are mainly detected in the spinel nano-crystal phase. An interesting data is the possibility to observe Yb^{3+}-Yb^{3+} pairs also detected by cooperative luminescence in many Yb^{3+}-doped crystal or ceramic materials or here in glass ceramics which should be only doped by Yb^{3+}.

As a general conclusion of this TEM-EDX analysis, elemental maps of Fig. 16.7 (and, to some extent, HREM imaging in Fig. 16.8) confirm the results obtained from simple line scans in Figs. 16.5 and 16.6, less representative from the point of view of spatial distribution but more representative from a quantitative point of view: Er^{3+}

and Yb^{3+} ions are preferentially located in the spinel nano-crystallites. This is "a priori" the same conclusion for Co^{2+} ions but again the low concentration of this ion and the overlapping with Er^{3+} line cannot permit any unambiguous location so that we need some useful help from the spectroscopic data as related in the next section.

In parallel, we did not measure any significant gradient or segregation phenomenon, especially for Yb^{3+} and Er^{3+} ions, in the area of the interface between the spinel nano-crystals and the vitreous phase, although the induced damages from irradiation did not make possible to perform long acquisitions which would have been needed to gain the desirable sensitivity. This segregation was never seen for Yb^{3+} heavy ions in garnets and sesquioxides, but was encountered for example in the grain boundaries of the YAG ceramics doped with the first Nd^{3+} and Ce^{3+} rare earth ions [14–17]. As we will discuss it later, Yb^{3+} spectroscopic characterizations give some information on Yb^{3+} location on the disturbed surface of the nano-crystals.

16.3.3 Spectroscopic Properties of the Glass and the Glass Ceramic Samples

16.3.3.1 Absorption Spectra of Co^{2+}, Yb^{3+} and Er^{3+} Ions in the Visible and the IR Spectral Ranges

The absorption spectra of a glass ceramic sample shown in Fig. 16.9a, b are composed of both sharp lines from Er^{3+} and Yb^{3+} rare earth ions and broader bands from Co^{2+} transition metal ions.

We can easily see the usual Er^{3+} ion transitions from the $^4I_{15/2}$ ground state: $^4I_{15/2} \rightarrow {}^2H_{9/2}$ (~407 nm), $^4I_{15/2} \rightarrow {}^4F_{3/2}$ (442 nm), $^4I_{15/2} \rightarrow$ 4 F5/2 (451 nm), $^4I_{15/2} \rightarrow {}^4F_{7/2}$ (~488 nm), $^4I_{15/2} \rightarrow {}^2H_{11/2}$ (~521 nm), $^4I_{15/2} \rightarrow {}^4S_{3/2}$ (~543 nm), $^4I_{15/2} \rightarrow {}^4F_{9/2}$ (~653 nm), $^4I_{15/2} \rightarrow {}^4I_{9/2}$ (~797 nm) (not presented here), $^4I_{15/2} \rightarrow {}^4I_{11/2}$ is hidden by $^2F_{7/2} \rightarrow {}^2F_{5/2}$ of Yb^{3+} ions which have much higher absorption strength (965–985 nm) and $^4I_{15/2} \rightarrow {}^4I_{13/2}$ (1510–1540 nm), respectively. In order to better understand the assignments of the electronic transitions we remind the energy level diagram of Er^{3+} and Yb^{3+} ions in Fig. 16.10.

As for Co^{2+} ions, we detect the presence within two spectral ranges by the transition of both $^4T_{1g} \rightarrow {}^4T_{1g}$ (^4P) (expected approximatively O_h symmetry of glass) and $^2A_1I^2T_2(^2G) + {}^4T_1(^4P) + {}^2E^2T_1(^2G)$ (T_d Co^{2+}) (expected T_d symmetry of Mg^{2+} site in $MgAl_2O_4$ spinel nano-crystals) between 500 and 680 nm and, in the near IR, between 1100 and 1700 nm, mainly by the $^4A_2 \rightarrow {}^4T_1(^4F)$ (T_d Co^{2+}) of the T_d symmetry in nano-spinels. This superposition of the two types of transitions justify the potentiality of this glass ceramic composition as saturable absorbers since the Co^{2+} broad absorption overlaps widely the expected laser emission from the Er^{3+} emission near 1550 nm. It has been already shown that this sample possesses excellent transparency up to nearly 87 % at the near-infrared band [19].

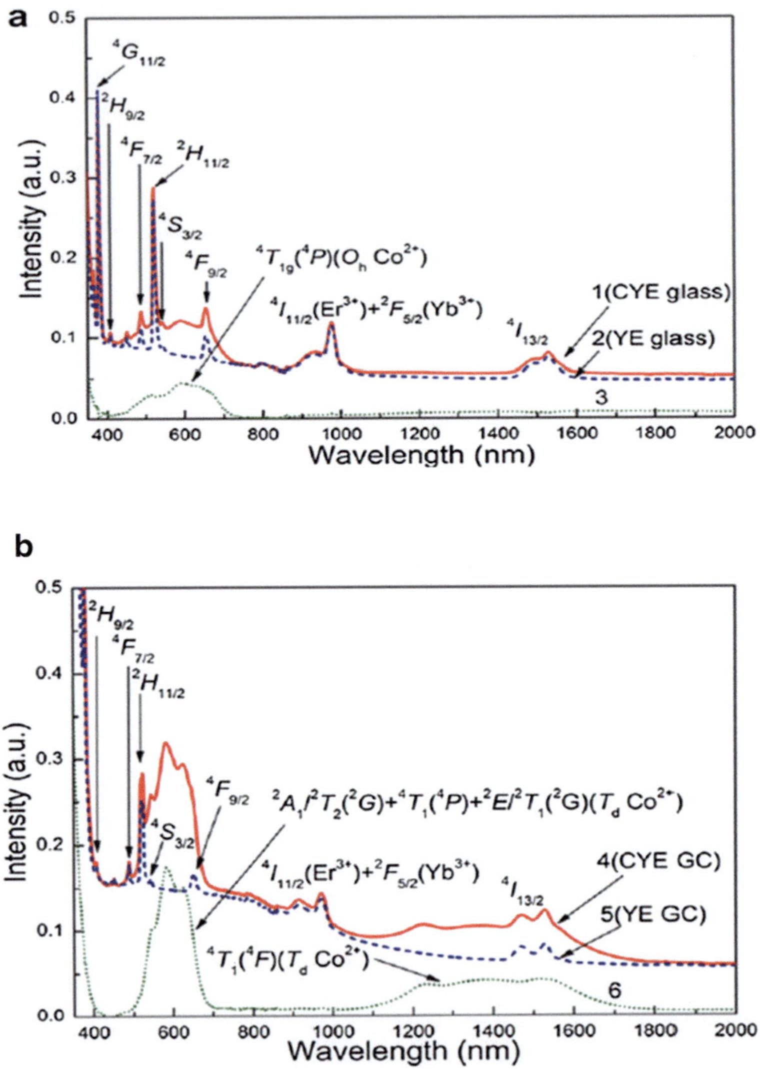

Fig. 16.9 (**a**) Comparison of the absorption of Co-Yb-Er-doped silicate glass (CYE glass) sample with respect of Yb-Er-doped silicate glass (YE glass) at RT. Co^{2+} ion absorption in glass has been deduced [19]. (**b**) Comparison of the absorption of Co-Yb-Er-doped glass ceramic (CYE-GC) sample with respect of Yb-Er-doped silicate glass (YE glass) at RT. Co^{2+} ion absorption mainly into Mg^{2+} sites of $MgAl_2O_4$ nano-crystals glass ceramics has been deduced [19]

16.3.3.2 Discussion on the Location of Co^{2+} Ions in Glass Ceramics

In the previous paper [19], the question on the location of Yb^{3+} and Er^{3+} rare earth ions as dopants in the glass ceramics from spectroscopic data has not been

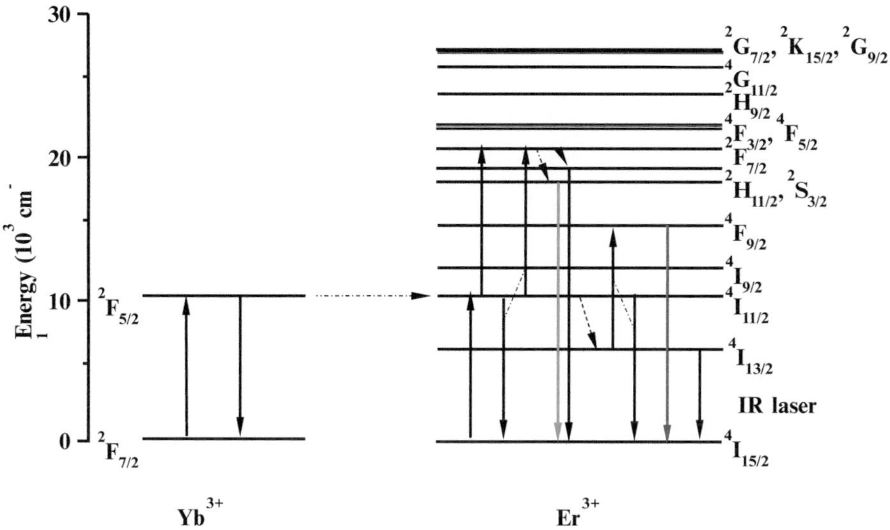

Fig. 16.10 Energy level diagram of Yb^{3+} and Er^{3+} ions in materials. The main *down* and *up* transitions are mentioned

unambiguously solved, the assignment of Co^{2+} ions was only clearly established between glass and $MgAl_2O_4$ spinel nano-crystals of the glass ceramics. Let's remind that Co^{2+} ions optical transitions belong to transition metal ions with high oscillator strength (f $\sim 10^{-2}$) and high absorption cross-section in crystals much easier to detect than rare earth ions which are characterized by much smaller f $\sim 10^{-6}$ one. Consequently, Co^{2+} ions can be seen at low concentration on absorption spectra as it has been described here and can allow to give complementary results which were not obtained by TEM-EDX due to the overlapping of Co^{2+} and Er^{3+} X-Ray lines.

We show in Fig. 16.9a the weak Co^{2+} near infrared absorption in glass as mentioned by the spectrum 3, intensity difference of the spectrum 1 of Co^{2+} ions in the glassy phase of Co^{2+}-Yb^{3+}-Er^{3+}-doped silicate glass (CYE glass) and the spectrum 2 of Yb^{3+}-Er^{3+}-doped silicate glass (YE glass). The strongest absorption of Co^{2+} ions in glass ceramics as observed in Fig. 16.9b by the spectrum 6, intensity difference between 4 and 5 spectra, is assigned to Co^{2+} into Mg^{2+} sites of T_d symmetry in Co^{2+}-doped $MgAl_2O_4$ nano-crystals of the (CYE glass ceramics). As a result the absorption results are in favor of the strongest population of Co^{2+} ions rather in glass ceramics than in glass.

16.3.3.3 Discussion on the Location of Yb^{3+} Ions in Glass Ceramics

The previous analysis of TEM-EDX experiments indicates we should expect the optical transitions of the three Co^{2+}, Yb^{3+} and Er^{3+} dopants in both the SiO_2

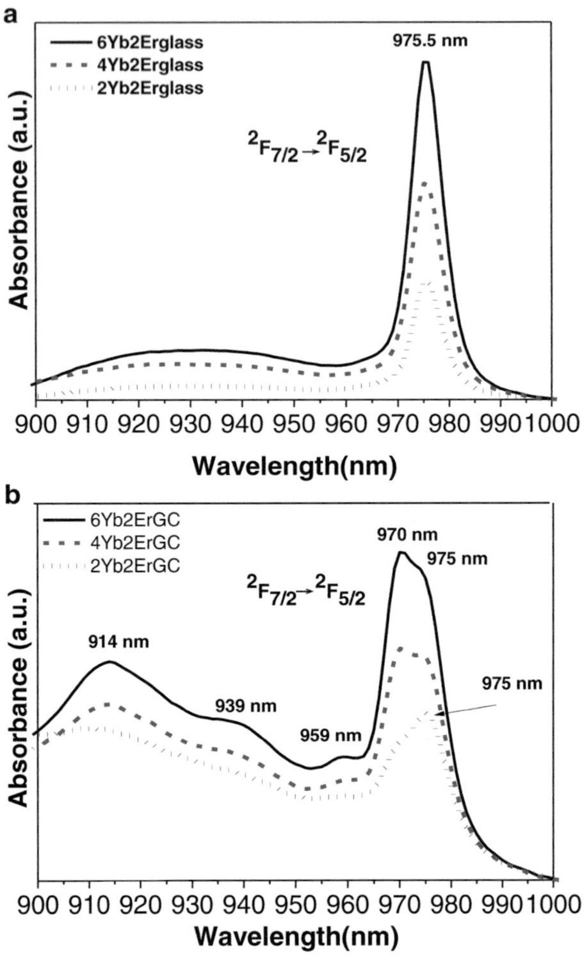

Fig. 16.11 Concentration dependence of x% Yb^{3+} ion absorption in both glass (**a**) and glass ceramics (**b**) samples at RT. X = 2, 4, 6

glassy phase and the $MgAl_2O_4$ nano-crystallite phase. We will report now that spectroscopic data brings new results for these glass ceramic samples.

It was found out in the absorption spectra the appearance of a new absorption line between 4 K and room temperature near 970 nm, so that it was necessary to compare the absorption of different samples containing different concentrations of Yb^{3+} and Er^{3+} ions in the two hosts of glasses and glass ceramics. First, we have pointed out the double structure of the Yb^{3+} ions between 965 and 985 nm by two broad lines of practically similar intensity, so-called 0-phonon lines, with maximum at 970 and 975 nm respectively as seen in Fig. 16.11b. Previously only one peak at 977 nm was observed [19]. These two broad lines characterize the $^2F_{7/2} \rightarrow ^2F_{5/2}$ transitions of Yb^{3+} ions between the lowest Stark energy level of each $^2F_{7/2}$ and $^2F_{5/2}$ manifolds rather than those of Er^{3+} ions, due to the much larger absorption coefficient of Yb^{3+} ions.

Concentration dependences of the Yb^{3+} absorption spectra of both glasses and glass ceramic samples containing the same amounts of Yb^{3+} and Er^{3+} rare earth ions are shown in Fig. 16.11a, b. Absorption spectra intensities presented for both glasses and glass ceramics are dependent on the Yb^{3+} concentration increasing. One can see the $^2F_{7/2}$ (1) $\rightarrow$ $^2F_{5/2}$ (5) double broad lines of Yb^{3+} 0-phonon lines with maxima at 970–975 nm. Absorption spectra of Yb^{3+}-doped glass ceramics exhibit a higher degree of structuration than that glass spectra. By comparing these absorption spectra one might assume that the line at 975 nm in glass ceramics has the same glassy origin as that in glass samples with maximum at 975.5 nm. It might be due to the presence of a double phase, like for example, both the glassy one and the nano-spinel crystalline phase with different line intensities, but this hypothesis has to be discussed as the following.

This is interesting to compare the spectra of Yb^{3+}-doped silica glass with those we obtained recently for Yb^{3+}-doped calcium alumino-silicate glasses (CAS and LSCAS) for excitation and emission spectra [30]. The spectra in these glasses are much more resolved than that of Yb^{3+}-doped silica glass, showing distinctive three $^2F_{7/2} \rightarrow {}^2F_{5/2}$ components of the absorption spectra and four $^2F_{5/2} \rightarrow {}^2F_{7/2}$ components of the emission spectra, as expected. The main reason of this difference between the two types of glasses is the presence of Ca and Al cations in the glass network of CAS and LSCAS which contribute to increase of the site symmetry inside the glassy phase. The Yb^{3+} 0-phonon line in CAS and LSCAS glasses is located at 977 nm, very close to the 975.5 nm in the silica glass of the glass ceramics, and consequently this comparison might be another proof of the assignment of the 975.5 nm peak to the silica glass phase in the glass ceramic sample.

Since the glass ceramic sample is composed of both the silicate phase and the nano-spinel crystalline phase, we also need some comparison with spectra of rare earth-doped spinel nano-crystalline phase. We have recently published the concentration dependences of the absorption spectra of Yb^{3+}-doped $MgAl_2O_4$ spinel nano-crystals in the range of 10–30 nm size, prepared by the sol-gel method and characterized by both structural and spectroscopic properties [31]. The first striking feature is the overlapping between the Yb^{3+} 0-phonon absorption line in the silica glass at 975.5 nm and the 0-phonon line of Yb^{3+} -doped $MgAl_2O_4$ spinel nano-crystals at 976 nm (Fig. 16.12).

Both of these two lines are very broad. In glasses this is due to the un-equivalent Yb^{3+} sites as usually expected in disordered amorphous materials and in $MgAl_2O_4$ spinel nano-crystals this is due to the following anti-site phenomenon:

$$MgAl_2O_4 \rightarrow (Mg_{1-2x}Al_{2x})\,[Mg_{2x}Al_{2-2x}]\,O_4.$$

Indeed, some Al^{3+} and Mg^{2+} cations are exchanged between the cation sublattices forming pairs of anti-site defects and thus a degree of inversion is created. This degree of inversion x ranges from almost 0 for the normal spinel up to 0.5 for the inverse spinel structure. Apparently, in the spinel nano-crystal phase in [19] the unresolved Yb^{3+} components indicate at large inverse spinel as a

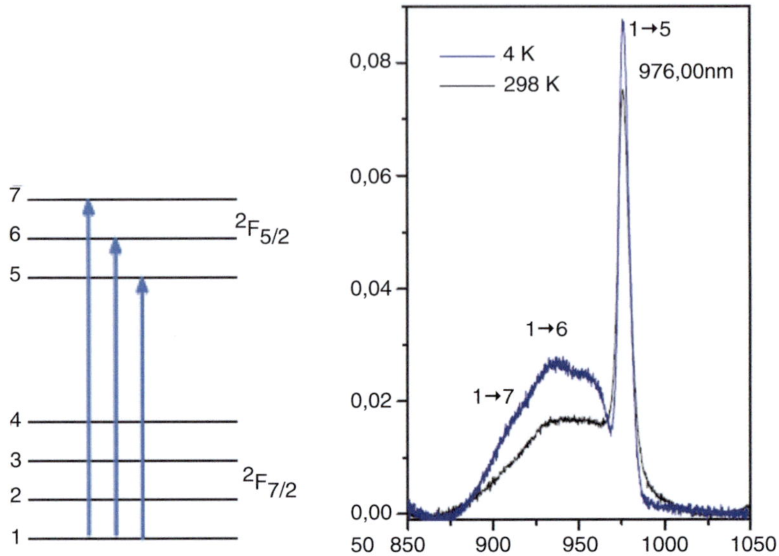

Fig. 16.12 Temperature dependence of absorption spectra of Yb^{3+}-doped $MgAl_2O_4$ spinel nanocrystals prepared by sol-gel method [31]

result of the multiplicity of Yb^{3+} sites observed either in octahedral or tetrahedral positions. Another effect increases the disorder by the substitution of Mg^{2+} cations. The compensation effect imposes to substitute three Mg^{2+} cations by two Yb^{3+} luminescent cations, creating one vacancy, pairs and aggregates even if the ionic radius of Yb^{3+} (85.8 pm) is much larger than those of Mg^{2+} (65 pm) and Al^{3+} (50 pm) [19]. This substitution was also proposed for the Yb^{3+}-Er^{3+}-co-doped $MgAl_2O_4$ phosphor powders prepared by the combustion method [32]. In addition, other perturbed Yb^{3+} ions exist due to the morphology of the surface of the grains in contrary with the regular structure of the internal grain. It means there are several reasons to increase disorder in spinel lattice and to observe large broadening of both absorption and emission spectra.

As a result, in the glass ceramic samples the line located at 975 nm can be assigned to the $^2F_{7/2}$ (1) $\rightarrow$ $^2F_{5/2}$ (5) transition either with Yb^{3+} ions in the glassy phase or with Yb^{3+} ions in the $MgAl_2O_4$ inverse spinel nano-crystalline phase. There is a surprising situation, adding difficulties for the final interpretation, never reported, with two Yb^{3+}-doped phases, one with SiO_2 glass, another one with inverse $MgAl_2O_4$ nano-crystalline phase, which show Yb^{3+} lines in coincidence.

Consequently we can assign the line at 975 nm of the glass ceramics to both the glass and the nano-crystalline phases but the higher population detected with TEM-EDX plays in favor of the detection first of the nano-crystalline phase and like a noisy signal for the glass phase.

At last we need to find new data to interpret the absorption line at 970 nm.

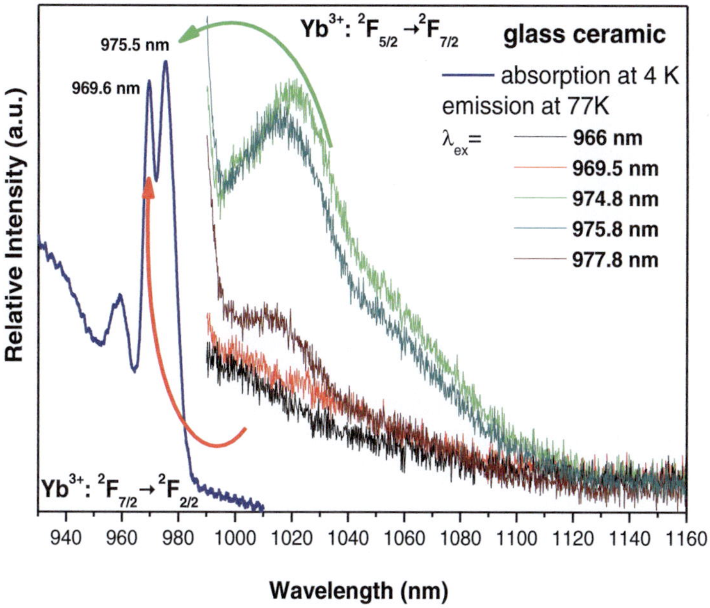

Fig. 16.13 Site selective laser excitation within the two broad $^2F_{7/2} \rightarrow ^2F_{5/2}$ 0-phonon lines of Yb^{3+} ions which were detected in absorption at RT and 4 K

16.3.3.4 Site Selective Spectroscopy as a Contribution to the Assignment of the Yb^{3+} Centres

Site selective laser excitation within the range 880–950 nm of Yb^{3+} ion into the two highest excited levels of the $^2F_{7/2} \rightarrow ^2F_{5/2}$ transitions of GC can be observed in Fig. 16.13 at 77 K.

The different excitations of the Ti-Sapphire laser pump are mentioned. The absorption spectrum of the sample at 4 K, almost the same as at RT, is indicated in blue as a reference. The first obvious feature is the absence of luminescence emission from Yb^{3+} ions characterized by the 0-phonon line at 970 nm. Only the Yb^{3+} 0-phonon lines around 977 nm associated to both SiO_2 glass and $MgAl_2O_4$ nano-crystal phases are the emitting centres shifted as expected from multi-sites in the glass ceramics. In addition we have not detected any energy transfer between the two centers responsible for these two broad absorption lines at 970 and 975 nm, respectively.

In Fig. 16.13, the site selective laser excitation at 77 K has been applied especially into the two broad $^2F_{7/2}$ (1) $\rightarrow ^2F_{5/2}$ (5) 0-phonon lines of Yb^{3+} ions which were detected in absorption with maxima at around 970 nm and 975 nm. Then, the different excitations are now in the range 966–980 nm. In red and black lines are reported the $^2F_{5/2}$ (5) $\rightarrow ^2F_{7/2}$ (1, 2, 3, 4) emission spectra under excitation at 966 nm and 969 nm, respectively into the new absorption peak present in the GC. In green, dark blue and brown colors are reported the ${}^2F_{5/2}$ (5) $\rightarrow {}^2F_{7/2}$ (1, 2, 3, 4) emission spectra under excitation at 974.8 nm, 975.8 nm and 977.8 nm, respectively, into the absorption peak already observed with glassy phase. We confirm only the Yb^{3+} lowest energy peak at 976 nm is giving radiative transitions and we see clearly the resolved Yb^{3+} emission spectra into excitation at 976 nm. A very weak emission is recorded under excitation at around 970 nm, due indeed to the excitation of the foot of the 976 nm absorption line since the overlapping of the two absorption peaks can't be avoided.

Consequently, the Yb^{3+} center associated with the line at 970 nm is only absorbing and not emitting. It means that the Yb^{3+} resolved emission spectra observed in Fig. 16.13 can be associated only to pumping of the broad line at around 976 nm and reflect mainly the intensity of Yb^{3+}-doped nano-spinel crystalline due to higher population of Yb^{3+} ions than in the glass.

At this point of the discussion, we interpret Yb^{3+} ions (and then associated neighbor Er^{3+} ions of similar ionic radii) in the two phases. Only Yb^{3+}-doped nano-crystalline phase is observed due to the highest population leading to higher fluorescence intensity. In the previous discussion on the emission properties we pointed out a radiation-less Yb^{3+} center associated to the 0-phonon absorption line at 970 nm only within the glass ceramic sample. Then, we think this is the result of strong perturbation acting on these centers due to the location of Yb^{3+} ions in vicinity of the surface of the nano-crystals. Such localization of this type of Yb^{3+} quencher ions on the surface close to the edges leads to stronger distortion by the crystal field than those of the regular ions inside the nano-crystal.

Another hypothesis is the presence of Yb^{3+} aggregates at the surface of nano-crystals so that migration of the energy by resonance between the 0-phonon lines at 970 nm might explain the quenching. However, TEM-EDX results did not detect any segregation of dopants at the interface of the nano-crystals and the glass and this is why we think the probability of the formation of aggregates is very weak.

16.4 Conclusion

The main goal of this work was to show how it is possible to characterize the location of Yb^{3+} and Er^{3+}rare earth ions and Co^{2+} transition metal ions by conjugating, what is rather rare, TEM-EDX and optical spectroscopy tools in a glass ceramics composed of $MgAl_2O_4$ spinel nano-crystals of 10–20 nm size embedded in SiO_2 glass. The application targeted is a compact self-Q-switched microchip laser for which spectroscopic properties are fulfilled. Here we have chosen this example at this School for pedagogical reasons to show the special way of assignment by applying the two independent tools of TEM-EDX and spectroscopy techniques.

The previous analysis had shown that Yb^{3+} and Er^{3+} rare earth ions seem to remain in SiO_2 glass matrix in the glass ceramics, while Co^{2+} occupy the tetrahedral sites in $MgAl_2O_4$ nano-crystals. The use of TEM-EDX technique led to the result that Yb^{3+}, Er^{3+} and Co^{2+} dopants are located into the two phases but preferentially located in the spinel nano-crystals. This important result has been clearly confirmed by the elemental mapping of Yb^{3+} and Er^{3+} rare earth ions and also by the spectroscopic results of Co^{2+} ions. TEM-EDX measurements was not be accurately applied to the very low Co^{2+} concentration whereas the assignment of Yb^{3+} ion spectroscopic properties was more difficult due to the overlapping of broad $^2F_{7/2}$ (1) $\rightarrow$ $^2F_{5/2}$ (5) 0-phonon lines belonging to both glassy and spinel nano-crystalline phases. Yb^{3+} ions localized either in the SiO_2 glass or in the $MgAl_2O_4$ spinel nano-crystalline are surprisingly absorbing at almost the same wavelength, around 975 nm with the characteristic of the greatest intensity line coming from the greatest population of the spinel nano-crystalline phase. The glassy phase was not really resolved at this wavelength in the glass ceramics due to the lower intensity of the 0-phonon line. A new radiationless Yb^{3+} centre most probably localized on the strongly perturbed area of the edge of the nano-crystal surface has been pointed out by the presence of an absorption line at 970 nm of comparable intensity with that of the regular Yb^{3+} ions inside the spinel nano-crystal structure.

Triply-doped glass ceramics composed of the spinel nano-crystals imbedded in a silicate glassy phase is a good example of material for which two complementary techniques like TEM-EDX and optical spectroscopy are necessary to assign the distribution of dopants. This is the reason why it has been exposed for the School. Undoubtedly, optimization of Yb^{3+}, Er^{3+} and Co^{2+} concentrations is expected in this glass ceramics before any development.

Acknowledgements We are grateful to the CLYM (Centre Lyonnais de Microscopie, www.clym.fr) for its guidance in the Ly-EtTEM (Lyon Environmental tomographic TEM) project, which was financially supported by the CNRS, The Région Rhône-Alpes, The 'Greater Lyon' and the French Ministry of Research and Higher Education. We wish to warmly thank the Faculty of Chemistry of the University of Wroclaw for the access of low temperature absorption measurements.

References

1. Maiman, T. (1960). Stimulated optical radiation in ruby. *Nature, 187*, 493.
2. Geusic, J. E., et al. (1964). Yttrium gallium and gadolinium garnets. *Applied Physics Letters, 4*(10), 182.
3. Boulon, G. (2012). 50 years of advances in solid-state laser materials. *Optical Materials, 34*(3), 499–512.
4. Pappalardo, R. G. (1987). Recent trends in laser material research. In: B. Di Bartolo, & E. Majorana (Eds.), *Spectroscopy of solid-state laser-type materials, international science series, physical sciences* (Vol. 30). New York: Plenum Press.
5. Ikesue, A., Furusato, I., & Kamata, K. (1995). Fabrication of polycrystalline, transparent YAG ceramics by a solid-state reaction method. *Journal of the American Ceramic Society, 78*, 225–228.
6. Ikesue, A., Kinoshita, T., Kamata, K., & Yoshida, K. (1995). Fabrication and optical properties of high-performance polycrystalline Nd:YAG ceramics for solid-state lasers. *Journal of the American Ceramic Society, 78*, 1033–1040.

7. Ikesue, A., Aung, Y. L., Yoda, T., Nakayama, S., & Kamimura, T. (2007). Fabrication and laser performance of polycrystal and single crystal Nd:YAG by advanced ceramic processing. *Optical Materials, 29*(10), 1289–1294.
8. Lu, J., Yagi, H., Takaichi, K., Uematsu, T., Bisson, J.-F., Feng, Y., Shirakawa, A., Ueda, K.-I., Yanagitani, T., & Kaminskii, A. A. (2004). 110 W ceramic Nd^{3+}: $Y_3Al_5O_{12}$ laser. *Applied Physics B, 79*, 25.
9. Yagi, H., Yanagitani, T., Takaichi, K., Ueda, K.-i., & Kaminskii, A. A. (2007). Characterizations and laser performances of highly transparent Nd^{3+}:$Y_3Al_5O_{12}$ laser ceramics. *Optical Materials, 29*(10), 1258–1262.
10. Kaminskii, A. A., Strek, W., Hreniak, D. (Eds.). (2007). *Proceedings of the 1st international laser ceramic symposium – E-MRS fall meeting 2005, 1st international laser ceramic symposium*, Warsaw, Poland, 05–06 September 2005. *Optical Materials, 29*(10), 1217–1294.
11. Sanghera, J., Kim, W., Villalobos, G., Shaw, B., Baker, C., Frantz, J., Sadowski, B., & Aggarwal, I. (2013). Ceramic laser materials: Past and present. *Optical Materials, 35*(4), 693–6991.
12. Ramirez, M. O., Wisdom, J., Li, H., Aung, Y. L., Stitt, J., Messing, G. L., Dierolf, V., Liu, Z., Ikesue, A., Byer, R. L., & Gopalan, V. (2008). Three-dimensional grain boundary spectroscopy in transparent high power ceramic laser materials. *Optics Express, 16*(9), 5965.
13. Zhao, W., Mancini, C., Amans, D., Boulon, G., Epicier, T., Min, Y., Yagi, H., Yanagitani, T., Yanagida, T., & Yoshikawa, A. (2010). Evidence of the inhomogeneous Ce^{3+} distribution across grain boundaries in transparent polycrystalline Ce^{3+}-doped $(Gd,Y)_3Al_5O_{12}$ garnet optical ceramics. *Japanese Journal of Applied Physics, 49*, 022602. 525, doi:10.1143/JJAP.49.022602.
14. Zhao, W., Anghel, S., Mancini, C., Amans, D., Boulon, G., Epicier, T., Shi, Y., Feng, X. Q., Pan, Y. B., Chani, V., & Yoshikawa, A. (2011). Ce^{3+} dopant segregation in $Y_3Al_5O_{12}$ optical ceramics. *Optical Materials, 33*, 684–687.
15. Esposito, L., Hostasa, J., Piancastelli, A., Toci G., Pirri, A., Vannini, M., Epicier, T., Malchére, A., Alombert-Goget, G., & Boulon, G. (2014). Fabrication and laser performance of multilayered YAG-Yb^{3+}:YAG ceramics . *Journal of Materials Chemistry C, 2*, 10138–10148.
16. Epicier, T., Boulon, G., Zhao, W., Guzik, M., Jiang, B., Ikesue, A., & Esposito, L. (2012). Spatial distribution of the Yb^{3+} rare earth ions in $Y_3Al_5O_{12}$ and Y_2O_3 optical ceramics as analyzed by TEM. *Journal of Materials Chemistry*, 22, 18221–18229.
17. Chani, V. I., Boulon, G., Zhao, W., Yanagida, T., & Yoshikawa, A. (2010). Correlation between segregation of rare earth dopants in melt crystal growth and ceramic processing for optical applications. *Japanese Journal of Applied Physics, 49*, 0756011–0756016.
18. Duan, X. L., Yuan, D. R., Cheng, X. F., Wang, Z. M., Sun, Z. H., Luan, C. N., Xu, D., & Lv, M. K. (2004). Absorption and photoluminescence characteristics of Co^{2+}:$MgAl_2O_4$ nanocrystals embedded in sol–gel derived SiO_2-based glass. *Optical Materials, 25*(1), 65.
19. Chen, L., Yu, C., Hu, L., & Chen, W. (2012). Preparation and spectroscopic properties of nanostructured glass-ceramics containing Yb^{3+}, Er^{3+} ions and Co^{2+}-doped spinel nanocrystals. *Solid State Sciences, 14*, 287.
20. Kalisky, Y., Kalisky, O., & Kokta, M. R. (2008). Passively Q-switched diode-pumped Nd: YAG and Nd:YVO_4 using (Cr^{4+}, Ca^{2+}):YAG and (Cr^{4+}, Mg^{2+}):YAG saturable absorbers. *Optical Materials, 30*, 1775.
21. Kalisky, Y., Ben Amar-Baranga, A., Shimony, Y., & Kokta, M. R. (1997). Cr^{4+} doped garnets: Novel laser materials and non-linear saturable absorbers. *Optical Materials, 8*, 129.
22. Kalisky, Y., Ben-Amar Baranga, A., Shimony, Y., Burshtein, Z., Pollack, S. A., & Kokta, M. (1996). Cr^{4+} doped garnets: Their properties as non-linear absorbers. *Optical Materials, 6*, 275.
23. Kalisky, Y. (2004). Passively Q-switched diode-pumped Nd:YAG and Nd:YVO 4 using (Cr^{4+}, Ca^{2+}):YAG and (Cr^{4+}, Mg^{2+}):YAG saturable absorbers. In R. Scheps, & T. Landsberg (Eds.), *Progress in quantum electronics* (Vol. 28, p. 552).
24. Chen, W., & Boulon, G. (2003). Growth mechanism of Cr: Forsterite laser crystal with high Cr concentration. *Optical Materials, 24*(1/2), 163–168.

25. Stultz, R. D., Camargo, M. B., Birnbaum, M. (1995). Divalent uranium and cobalt saturable absorber Q-Switches at 1.5 μm, OSA, *Processing of Advanced Solid-State Lasers*, *24*, 460.
26. Yumashev, K. V., Denisov, L. A., Posonov, N. N., Kuleshov, N. V., & Moncorge, R. (2002). Excited state absorption and passive Q-switch performance of Co^{2+} doped oxide crystals. *Journal of Alloys and Compounds, 341*, 366.
27. Li, P., Li, Y., Sun, Y., Hou, X., Zhang, H., & Wang, J. (2006). Passively Q-switched 1.34 μm Nd: $Y_xGd_{1-x}VO_4$ laser with Co^{2+}:$LaMgAl_{11}O_{19}$ saturable absorber. *Optics Express, 14*(17), 7730.
28. Boulon, G., Guyot, Y., Alombert-Goget, G., Guzik, M., Epicier, T., Blanchard, N., Chen, L., Hu, L., & Chen, W. (2014). Conjugation of TEM-EDX and optical spectroscopy tools for the localization of Yb^{3+}, Er^{3+} and Co^{2+} dopants in a laser glass ceramics composed of $MgAl_2O_4$ spinel nano-crystals embedded in SiO_2 glass. *Journal of Materials Chemistry C, 2*, 9385–9397.
29. Boulon, G., Guyot, Y., Guzik, M., Epicier, T., Gluchowski, P., Hreniak, D., & Strek, W. (2014). Yb^{3+} ions distribution in YAG nano-ceramics analyzed by both optical and TEM-EDX techniques. *Journal of Physical Chemistry C, 118*(28), 15474–15486.
30. Guyot, Y., Steimacher, A., Belançon, M. P., Medina, A. N., Baesso, L., Lima, S., Andrade, L. H. C., Brenier, A., Jurdyc, A. M., & Boulon, G. (2011). Spectroscopic properties, concentration quenching and laser investigations of Yb^{3+}-doped calcium aluminosilicate glasses. *JOSA B, 28*(10), 2510–2517.
31. Wiglusz, R. J., Boulon, G., Guyot, Y., Guzik, M., Hreniak, D., & Strek, W. (2014). Structural and spectroscopic properties of Yb^{3+}-doped $MgAl_2O_4$ nano-crystalline spinel. *Dalton Transactions, 43*, 7752–7759.
32. Singh, V., Kumar Rai, V., Watanabe, S., Gundu Rao, T. K., Badie, L., Ledoux-Rak, I., & Jho, Y.-D. (2012). Synthesis, Characterization, Optical absorption, luminescence and defect centres in Er and Yb co-doped $MgAl_2O_4$ phosphor. *Applied Physics B, 108*, 437.

Chapter 17
Nd^{3+}, Eu^{3+} and Yb^{3+} Ions as Structural Probes in the Scheelite-Type Cadmium Molybdates with Vacancies

Malgorzata Guzik, Janina Legendziewicz, Elżbieta Tomaszewicz, Yannick Guyot, and Georges Boulon

Abstract The main aim of this research is to demonstrate how important are the laser techniques like site selective spectroscopy and time-resolved spectroscopy at low temperature in determining the structure of compounds in complementarity of the usual XRD diffraction technique. Basing on the analyse of specific lines like the $^5D_0 \leftrightarrow {}^7F_0$ of Eu^{3+} ion, the $^4I_{9/2} \rightarrow {}^2P_{1/2}$, the $^4I_{9/2} \rightarrow {}^4F_{3/2}$ absorption transitions of Nd^{3+}, ion as well as the zero-phonon line of the $^2F_{5/2} \leftrightarrow {}^2F_{7/2}$ of Yb^{3+} ion we get precise information about the active metal site in the $CdMoO_4$ host lattice. As a good example we have chosen the series of rare earth -doped vacancied scheelite-type cadmium molybdate micro-crystals as $Cd_{1\text{-}3x}RE_{2x}\square_xMoO_4$ (where $RE = Nd^{3+}$, Eu^{3+} and Yb^{3+}) solid solutions prepared by a high-temperature solid state method. The monophasic samples crystallize in the tetragonal scheelite-type structure (the space group $I4_1/a$, with point symmetry S_4) when the concentration of optical activator is between 0 and 66.67 mol% of Nd^{3+} or Eu^{3+} ions and 33.36 mol% of Yb^{3+} ions. The substitution of divalent Cd^{2+} by trivalent RE^{3+} cations leads to the formation of cationic vacancies in the framework (which are denoted in the chemical formula as $\square$), due to the charge compensation: $3Cd^{2+} \rightarrow 2YRE^{3+} + \square$ vacancy. We check the relationship between structural and spectroscopic properties of Nd^{3+}, Eu^{3+} and Yb^{3+} rare earth ions as structural probes in the scheelite-type $CdMoO_4$ with vacancies mainly by confirming the presence of multi-sites with each dopant. We also analyze how the vacancy concentration dependence may enhanced

M. Guzik (✉) • J. Legendziewicz
Faculty of Chemistry, University of Wrocław, Joliot-Curie 14, 50-383 Wrocław, Poland
e-mail: goguzik@poczta.fm

E. Tomaszewicz
Department of Inorganic and Analytical Chemistry, West Pomeranian University of Technology, Al. Piastów 42, 71-065 Szczecin, Poland

Y. Guyot • G. Boulon
Univ Lyon, Université Claude Bernard Lyon 1, CNRS, Institut Lumière Matière, F-69622 Villeurbanne, France

© Springer Science+Business Media Dordrecht 2017
B. Di Bartolo et al. (eds.), *Nano-Optics: Principles Enabling Basic Research and Applications*, NATO Science for Peace and Security Series B: Physics and Biophysics, DOI 10.1007/978-94-024-0850-8_17

the optical parameters and brings some originality to this research program in optical materials. We show that RE^{3+} -doped $CdMoO_4$ investigated materials are interesting for potential applications like promising phosphors activated by Eu^{3+} ions under excitation by UV light with the very strong red luminescence, potential laser materials with Nd^{3+} and Yb^{3+} ions by the values of the absorption cross-section, the strong emission intensity and laser parameters.

These results selected for the School will be helpful in understanding how important is the role played by lanthanide ions not only as promising optical materials, both in the visible (Eu^{3+}) and near infrared spectral region (Nd^{3+}, Yb^{3+}) but also as structural probes for basics research.

17.1 Introduction

Solid-state lasers are subject of the intensive research since over half of the last century. For many years Nd^{3+}-doped YAG single crystals have been extensively studied and widely used for high-power lasers. However, we should not forget that the first continuously operating crystal laser doped with Nd^{3+} ions, reported in 1961, based on calcium tungstate ($CaWO_4$). Over the years it was discovered that especially alkali metal double tungstates and molybdates are very good matrices for laser materials and also for light generation or amplification in the near- infrared region when they are doped by Nd^{3+}, Dy^{3+}, Ho^{3+}, Er^{3+}, Tm^{3+} or Yb^{3+} rare earth ions [1–4]. Also, simply molybdates and tungstates belong to important families of inorganic materials used in the optical field due to their potential applications. Among tungstates, the most investigated groups are $CaWO_4$ and $MgWO_4$ (as phosphors), as well as $ZnWO_4$, $CdWO_4$ and $PbWO_4$ (as scintillators). As for molybdates $MMoO_4$ (M = Ca, Sr, Ba, and Pb, Cd) of tetragonal scheelite-type structure with space group $I4_1/a$ (No. 88) ($C^6{}_{4h}$), a form wide class of materials in different application fields, such as scintillator detectors, luminescence photoconductivity and photocatalysis [5–8]. Due to large X-ray absorption coefficient and efficient scintillation yield, they have been applied in the medical detecting X-rays and γ-rays [9]. The multi-purpose applications of the scheelite oxide systems are due to the possibility of their appearance in several phases with different structural types. They can crystallize in the scheelite structure $I4_1/a$, Z = 4 (No. 88), pseudoscheelite structure Pnma, Z = 4 (No.62), zircon structure $I4_1/amd$, Z = 4 (No. 141), M-fergusonite structure $I2/_a$, Z = 2 (No. 15) and M-fergusonite structure $P2_1/c$, Z = 2 (No. 14) [10]. Simple scheelite phase (*e.g.* $M^{II}M^{VI}O_4$, M^{VI} = Mo, W) has eight symmetry elements, body-centred tetragonal unit cell and four formula units (Z = 4). Each M^{VI} ion is surrounded by four equivalent oxygen ions in T_d symmetry and each $M^{II}O_8$ polyhedron shares corners with eight adjacent $M^{VI}O_4$ tetrahedra. It was shown previously that molybdate ($MoO_4{}^{2-}$) group is a good candidate as a host material in the scheelite related red phosphors because the molybdate are chemically stable due to their unique crystal structure, which is better than sulphide and oxysulfide red-emitting phosphors [11, 12].

RE^{3+}-doped tungstates with scheelite- and wolframite-type structures constitute a family of interesting optical materials. We should not forget that when RE^{3+} ions are incorporated in the $PbMO_4$ or $CdMO_4$ (M = Mo, W) lattice, on a regular lattice, it must be accompanied by a charge-compensating defect. The substitution of divalent M^{2+} ions by trivalent M^{3+} cations leads to the formation of cationic vacancies in a framework and some disordering what in consequence leads to improving of the luminescence intensity. This is why it is important to carry out in detail the relationship between structural and spectroscopic properties of Eu^{3+}, Nd^{3+}, Yb^{3+}- doped $CdMoO_4$.

Un-doped $CdMoO_4$ crystal is a candidate for cryogenic scintillation materials to investigate galactic Dark Matter, i.e. weakly interacting massive particles [13, 14] but its luminescence properties have been studied very little. Concerning RE^{3+} doped cadmium molybdates in the literature, one can find recent articles devoted to Ln^{3+}-doped $CdMoO_4$ (Ln = Pr^{3+}, Sm^{3+}, Eu^{3+}, Dy^{3+}, Ho^{3+} and Er^{3+}) but without presenting high spectral resolution, which allows to show any multisite character and any analysis of the site symmetry of the investigated compounds [15, 16].

Our previous studies have shown the existence of a new scheelite-type solid solutions described as $Cd_{1-3x}RE_{2x}\square_xMoO_4$ for RE^{3+} = Pr, Nd, Sm–Dy, and $0 < x \leq 0.25$, where $\square$ stands for cationic vacancy [17]. Structure and vibrational properties $Cd_{1-3x}Gd_{2x}\square_xMoO_4$ and $Cd_{1-3x}RE_{2x}\square_xMoO_4$ RE^{3+} = Sm-Dy have been investigated in details previously [18]. Recently dielectric and magnetic properties of $Cd_{0.958}Gd_{0.028}\square_{0.014}MoO_4$ single crystal were also investigated [19].

The main goal of this lecture presented at the School is to show how the spectroscopic analysis of RE^{3+} (RE^{3+} = Nd^{3+}, Eu^{3+}, Yb^{3+}) ions as dopants, in new $Cd_{1-3x}RE_{2x}\square_xMoO_4$ solid solutions with different concentrations of optically active ion (from 0.1 mol% to 66 mol%) and $\square$ vacancies, can bring complementary structural information to the usual X-Ray analysis of these optical materials [20–22].

17.2 Experimental Section

17.2.1 Sample Preparation

$CdMoO_4$ and $RE_2(MoO_4)_3$ were used as the starting materials. Cadmium molybdate was prepared by heating an equimolar mixture of CdO and MoO_3 (both reactants with purity 99.9 %, Aldrich) in the thermal conditions described previously [23]. RE (Eu^{3+}, Nd^{3+} and Yb^{3+}) molybdate $RE_2(MoO_4)_3$ were obtained by sintering of RE_2O_3 (Nd_2O_3 and Eu_2O_3-99.99 %, Stanford Materials and Alfa Asear, Yb_2O_3-99.998 %, Stanford Materials) and MoO_3 mixture with the 1:3 molar ratio, and by thermal treatment described previously [23, 28, 29]. $CdMoO_4$ and $Nd_2(MoO_4)_3$ mixed in appropriate ratios were sintered in static air, in corundum crucible. The following heating sequence was used: 1023 (12 h), 1073 K (12 h), 1123 K (12 h), 1173 K (12 h), and 1223 K (12 h). $CdMoO_4$ and $Yb_2(MoO_4)_3$ mixed in appropriate molar ratios were sintered in static air, in corundum crucibles, and the following heating treatment was used: 1123 K (12 h), 1223 K (12 h), 1273 K (12 h), 1323 K (12 h), and 1353 K (12 h). $CdMoO_4$ and $Eu_2(MoO_4)_3$ mixed in appropriate molar ratios were sintered in static air in the following heating treatment was used: 1123 K (12 h), 1223 K (12 h), 1273 K (12 h), and 1298 K (2×12h). Additionally, the mixtures with small concentration of Eu^{3+} ions, i.e. with the initial $Eu_2(MoO_4)_3$ content 10 mol% and less were heated in 1323 and 1348 K for 2×12 h. The preparation of samples under study can be described by the following equation:

$$\mathbf{(1{-}3x)\ CdMoO_{4(s)} + xRE_2(MoO_4)_{3(s)} = Cd_{1-3x}RE_{2x}\square_x MoO_{4(s)}}$$

After each heating period, all samples were cooled slowly down to ambient temperature, and for better reactivity ground in an agate mortar, and next examined for their content by powder XRD method. As an example in the Table 17.1 we show the chemical formulae of studied $Cd_{1\text{-}3x}Nd_{2x}\square_x MoO_4$ solid solutions and the corresponding concentration of Nd^{3+} ions. Dopant concentrations were quantified by ICP method. There is good agreement (within experimental error) between the starting and finally detected composition of the resulting compounds.

Table 17.1 The content of $CdMoO_4/Nd_2(MoO_4)_3$ initial mixtures, formula of a solid solution, calculated lattice parameters, calculated and experimental value of density of some $Cd_{1\text{-}3x}Nd_{2x}\square_x MoO_4$ solid solutions

Content of $Nd_2(MoO_4)_3$ in an initial $CdMoO_4/Nd_2(MoO_4)_3$ mixture [mol%]	Model of a solid solution, a value of x parameter	Concentr. of Nd^{3+} ions [mol%]	a [nm]	c [nm]	d_{cal} [g/cm^3]	d_{exp} [g/cm^3]
50.00	$Cd_{0.2500}Nd_{0.5000}\square_{0.2500}MoO_4$ x = 0.2500	66.67	0.52537	1.15413	5.42	5.12
25.00	$Cd_{0.4999}Nd_{0.3334}\square_{0.1667}MoO_4$ x = 0.1667	40.01	0.52027	1.13765	5.75	5.57
10.00	$Cd_{0.7483}Nd_{0.1678}\square_{0.0839}MoO_4$ x = 0.0839	18 .32	0.51754	1.12768	5.95	5.87
5.00	$Cd_{0.8635}Nd_{0.0910}\square_{0.0455}MoO_4$ x = 0.0455	9.53	0.51656	1.12437	6.01	5.97
2.57	$Cd_{0.9268}Nd_{0.0488}\square_{0.0244}MoO_4$ x = 0.0244	5.00	0.51609	1.12278	6.05	6.00
1.53	$Cd_{0.9556}Nd_{0.0296}\square_{0.0148}MoO_4$ x = 0.0148	3.00	0.51584	1.12210	6.06	6.02
0.05	$Cd_{0.9985}Nd_{0.0010}\square_{0.0005}MoO_4$ x = 0.0005	0.10	0.51559	1.12112	6.08	6.05
0.00	$CdMoO_4$ x = 0	0.00	0.51558	1.12109	6.08	6.05

17.2.2 Apparatus

XRD Phase Analysis Room temperature (RT) powder XRD method was used to identify the phase purity and the crystal structure of solid solutions. X-ray diffraction patterns were collected over the angular range $2\Theta = 10$–$90°$ in continuous scanning mode with the scan rate of 0.008° per step on a D8 ADVANCE powder diffractometer with nickel-filtered Cu $K\alpha_{1,2}$ radiation ($\lambda = 1.5418$ Å) and a Vantec detector, as well as on EMPYREAN II diffractometer (PANalytical) using Cu $K\alpha_{1,2}$ radiation ($\lambda = 1.5418$ Å) within $2\Theta = 10$–$90°$ range with a scan rate 0.008° per step and a counting time of 4 s per step. XRD patterns were analysed by *HighScore Plus 4.0* software and lattice parameters were calculated using the least squares refinement procedure by *POWDER* software [24].

Density Measurements The density of samples under study was measured on a Ultrapycnometer 1000 Quantachrome Instruments (model Ultrapyc 1200e, USA) using argon (99.999 %) as a pycnometric gas.

ICP Measurements The concentration of europium ions was determined by ICP-MS method (JY ULTRACE 238 spectrometer) after dissolving of $Cd_{1-3x}Eu_{2x}\square_xMoO_4$ samples with different Eu^{3+} concentrations *i.e.* 0.10 and 1.98 mol% in a hot and dilute aqueous hydrochloric acid solution. The Eu concentration was determined to be 0.056 and 1.073 wt%, respectively. These values correspond to the following formulas of solid solutions: $Cd_{0.9985}Eu_{0.0010}\square_{0.0005}MoO_4$ ($x = 0.0005$; 0.100 mol%) and $Cd_{0.9712}Eu_{0.0192}\square_{0.0096}MoO_4$ ($x = 0.0096$; 1.939 mol%), respectively. The experimental results are in very good agreement with the calculated data for: $Cd_{0.9985}Eu_{0.0010}\square_{0.0005}MoO_4$ (0.056 wt%; 0.100 mol%; $x = 0.0005$) and $Cd_{0.9706}Eu_{0.0196}\square_{0.0098}MoO_4$ (1.095 wt%; 1.969 mol%; $x = 0.0098$) and confirm our assumptions about the Eu^{3+} concentration in samples under study.

Scanning Electron Microscopy (SEM) Scanning electron microscopy studies were carried out on Hitachi S-3400 N equipped with an energy dispersive X-ray spectroscopy (EDX) EDAX analyser. The powders were coated with thin gold alloy layer to facilitate conductivity.

Emission and Excitation Measurements The steady state emission spectra were measured using a SpectraPro 750 monochromator, equipped with a Hamamatsu R928 photomultiplier and a 1200 l/mm grating blazed at 500 nm. A 450 W xenon arc lamp was used as the excitation source. It was coupled with a 275 mm excitation monochromator which used a 1800 l/mm grating blazed at 250 nm. The excitation spectra were corrected for the excitation light intensity while the emission spectra were corrected for the instrument response. The measurements were taken at room temperature (RT) and liquid nitrogen (77K) temperature.

Site Selective Excitation and Time Resolved Spectroscopy (or Fluorescence Line Narrowing) was performed at 77 K in a nitrogen cryostat (or quartz dewar). A tuneable laser (EKSPLA OPO NT342B, 7 ns, 10Hz) was used for scanning the excitation wavelength around 465 and 580 nm. The fluorescence was collected with an optical fiber and analysed with the help spectrometer equipped with a 1200 l/mm grating blazed at 500 nm, coupled to intensified charge coupled device (ICCD) camera (Shamrock 303 and iStar from Andor Technology).

Luminescence Decay Measurements The luminescence decay curves were recorded under pulsed laser excitation (OPO laser, EKSPLA NT342, 10Hz, 7 ns), the fluorescence intensity around 1.06 μm being detected with a R1767 Hamamatsu photomultiplier through a HRS1 Jobin-Yvon monochromator equipped with a 1 μm blazed grating and coupled to a LECROY LT 342 digital oscilloscope. The luminescence decay curves were recorded at RT and 77 K.

17.3 Results and Discussion

17.3.1 Morphology and Particle Size by SEM

To check crystallinity and morphology of obtained samples we used the scanning electron microscope (SEM) and transmission electron microscope (TEM) techniques. Here as an example in Fig. 17.1 we present only SEM micrograph of the

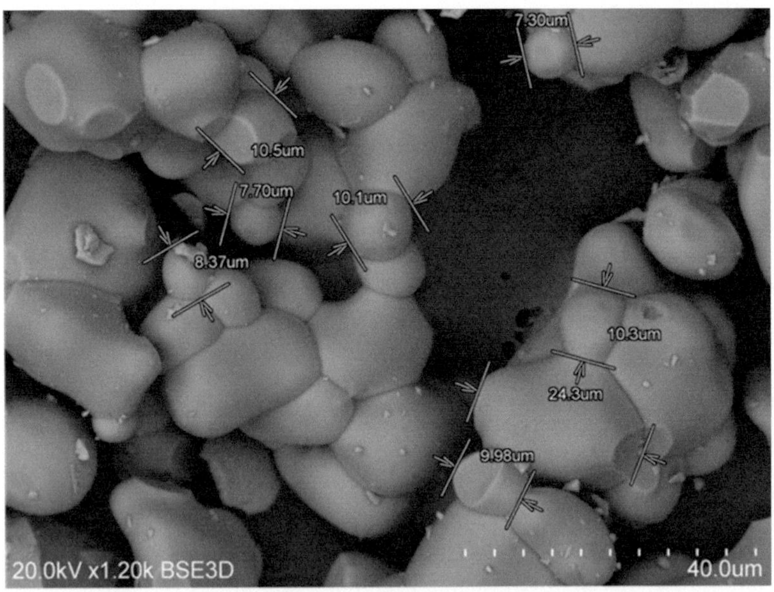

$Cd_{0.9985}Nd_{0.0010}\square_{0.0005}MoO_4$ (0.1% molNd^{3+})

Fig. 17.1 Scanning electron micrographs presenting the particle size of $Cd_{0.9985}Nd_{0.0010}\square_{0.0005}MoO_4$ (0.1 mol%) solid solution tetragonal cadmium molybdate doped with 0.1 mol% of Nd^{3+} ion. The images show that at a low Nd^{3+} ion concentration the particles of the cadmium and neodymium molybdate are homogenous, though high temperatures of the synthesis. The product crystallizes in a compact form with an average size of about 1–10 μm. The particles are well separated and a grain boundary between the microcrystals can be seen. The spherical-like shaped particles created loose agglomerates. With an increase of the active ions up to 66.67 mol%, the particles become more glued together to form bigger aggregates (not presented here).

The influence of applied annealing temperature on the morphology is well manifested for the solid solutions with Eu^{3+}, Yb^{3+} ions were the heating temperature was higher than that for the analogs with Nd^{3+} ones. This is immediately reflected in the morphology of the samples because the grain size is much larger than for Nd^{3+} -doped solid solutions (1–10 μm) [20]. The micrographs show homogenous particles in the compact form of an average size of about 2–25 μm for Eu^{3+}-doped samples [21] and 60–135 μm for Yb^{3+} [22]. The particles are well separated and the grain boundaries between the micro- crystals can be clearly seen. For Yb^{3+} we observe the grains of irregular with sharp ends and do not resemble to spherical-like particles which are observed for the Nd^{3+} and Eu^{3+}-doped samples.

The energy dispersive X-ray microanalysis (EDS) (not presented here) of the $Cd_{1\text{-}3x}RE_{2x}\square_xMoO_4$ phase exhibits the peaks related to Cd, Mo, O, and Nd, Eu and Yb elements and the intensity ratio of recorded peaks confirms the quantitative composition of the samples investigated.

17.3.2 Structural Analysis. Example of $Cd_{1\text{-}3x}Nd_{2x}\square_xMoO_4$ Solid Solutions

Room temperature XRD analysis was used for an identification of samples obtained after annealing of $CdMoO_4/Nd_2(MoO_4)_3$ mixtures. Figure 17.2 shows the representative X-ray powder diffraction patterns of pure $CdMoO_4$ compared with that from ICSD (#084455), as well as the materials obtained after the last annealing period of $CdMoO_4/Nd_2(MoO_4)_3$ mixtures comprising initially 0.0, 5, 10 and 50 mol% of $Nd_2(MoO_4)_3$. The obtained patterns were found to be similar to each other, as well as to that earlier reported by us for $Cd_{0.25}Nd_{0.50}\square_{0.25}WO_4$ [20, 25] and $Cd_{1\text{-}3x}Gd_{2x}\square_xMoO_4$ [18]. All diffraction lines were indexed with the tetragonal scheelie-type structure with the space group $I4_1/a$ (No.88) ($C_{4h}{}^6$), Z = 4. The insert in the Fig. 17.2 shows the XRD patterns in 2Θ range of 28–30° with the peaks, which correspond to [112] lattice plane. With increasing amounts of $Nd_2(MoO_4)_3$ in the initial mixture, a shifting of all diffraction peaks towards lower 2Θ angles is observed. The results indicate that the samples are of a single phase and contain $Cd_{1\text{-}3x}Nd_{2x}\square_xMoO_4$ solid solution. However, for the phases with the highest concentration of the Nd^{3+} ion (66.67 mol%), the additional small lines are observed at the lower 2Θ angles range (at 12.7, 17.1, 17.6, 21.9 and 22.3°). They may originate from the different phase which starts to form when the quantity of

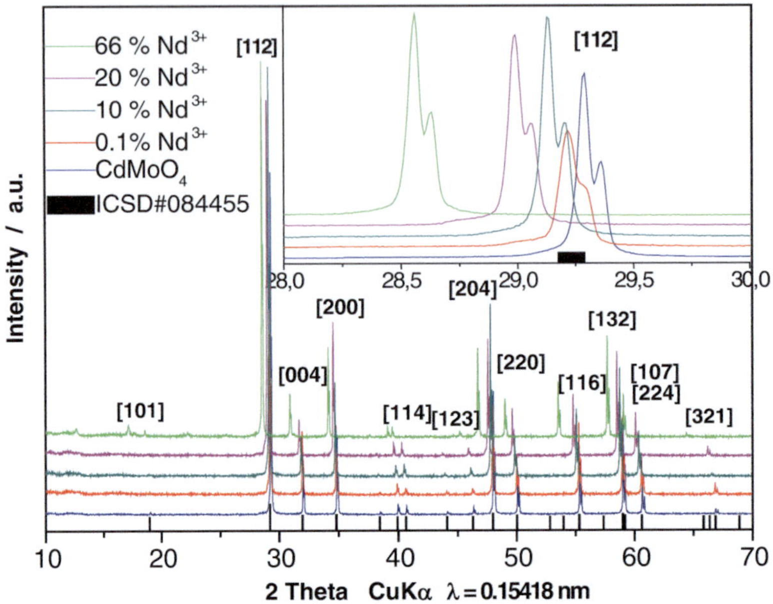

Fig. 17.2 X-ray powder diffraction patterns of $CdMoO_4$ and $Cd_{1\text{-}3x}Nd_{2x}\square_xMoO_4$ solid solutions with different content of the optically active ion. Simulated XRD pattern of tetragonal $CdMoO_4$ according to ICSD#084455 is presented by tick marks at the bottom of the figure

$Nd_2(MoO_4)_3$ is higher than 50 mol%. The maximum of $Nd_2(MoO_4)_3$ content in the initial $CdMoO_4/Nd_2(MoO_4)_3$ mixture, for which it is possible to obtain the single phase of $Cd_{1\text{-}3x}Nd_{2x}\square_xMoO_4$ is equal to 50 mol%.

The structure of $Cd_{1\text{-}3x}Nd_{2x}\square_xMoO_4$ is based on the scheelite-type structure of $CdMoO_4$. The calculated lattice parameters, observed and calculated values of density for each sample are given in Table 17.1. According to the Vegard law, the unit cell parameters of the $Cd_{1\text{-}3x}Nd_{2x}\square_xMoO_4$ solid solutions linearly decreased with increasing of $CdMoO_4$ content. On the other hand, the density of each solid solution was slightly increasing with the decreasing of a concentration of optically active ion. The difference between a value of the experimental and calculated density for each $Cd_{1\text{-}3x}Nd_{2x}\square_xMoO_4$ solid solution was also observed. This fact confirms partial occupancy of cationic positions by Nd^{3+} in the scheelite structure and presence of vacancies in the framework. It is worth to notice that using the Czochralski method we have been successfully obtained a single crystals for analogous solid solutions contained small concentrations of Eu^{3+}, Nd^{3+} and Yb^{3+} ions. Investigations are in progress.

Figure 17.3 presents the structure of $CdMoO_4$ with CdO_8 dodecahedra and MoO_4 tetrahedra. The scheelite–type structure is made up of CdO_8 polyhedra and MoO_4 tetrahedra connected by common vertices. CdO_8 polyhedra connect *via* the edges and form 3D framework. Each Mo^{6+} ion is surrounded by four equivalent O^{2-} ions at distances of 1.75 Å in approximately tetrahedral symmetry. Each Cd^{2+} atom is surrounded by eight O^{2-} ions at distances of 2.40 Å and

Fig. 17.3 The structure of $CdMoO_4$

2.44 Å in approximately octahedral symmetry. The Cd^{2+} cations in $CdMoO_4$ can be substituted by trivalent metal. The phases show a random distribution of Nd^{3+}. The structure of $Cd_{1-3x}Nd_{2x}\square_xMoO_4$ solid solution is shown in Fig. 17.4. The distance between the Cd^{2+}/ Nd^{3+} ions is equal to 3.908 Å. From Rietveld's analysis, knowing the Nd^{3+}–Nd^{3+} distance, we are able to predict the Eu^{3+}–Eu^{3+} bond distance, which is equal to 3.822 Å [22] and even shorter for Yb^{3+}–Yb^{3+} (3.66 Å) [21] which can affect the spectral behaviour of the system and lead to the creation of dimer states.

17.3.3 Spectroscopic Analysis of $Cd_{1-3x}Nd_{2x}\square_xMoO_4$

17.3.3.1 Evidence of Nd^{3+} Multi-sites

We are interested in the research of RE^{3+}- doped tungstates and molybdates for many years. We are studing already the creation of several symmetry sites with rare earth trivalent ions like Eu^{3+} and Nd^{3+} ions in the tungstates with scheelite – and wolframite-type structures but also in monoclinic and cubic molybdate host lattices [26–28]. Multisite character appeared in the wolframite-type tungstates such as $ZnY_4W_3O_{16}$, crystallizing in orthorhombic system, where more than one site is available for the doping of Eu^{3+} and Nd^{3+} ions and also in the molybdates of LAMOX type [29, 30]. In contrast, in the tetragonal $Cd_{0.25}Gd_{0.5}\square_{0.25}WO_4$ and monoclinic $CdY_2W_2O_{10}$, representing the scheelite-type structure, the Eu^{3+} or Nd^{3+} ions occupy only one crystallographic position [27, 31]. These doped

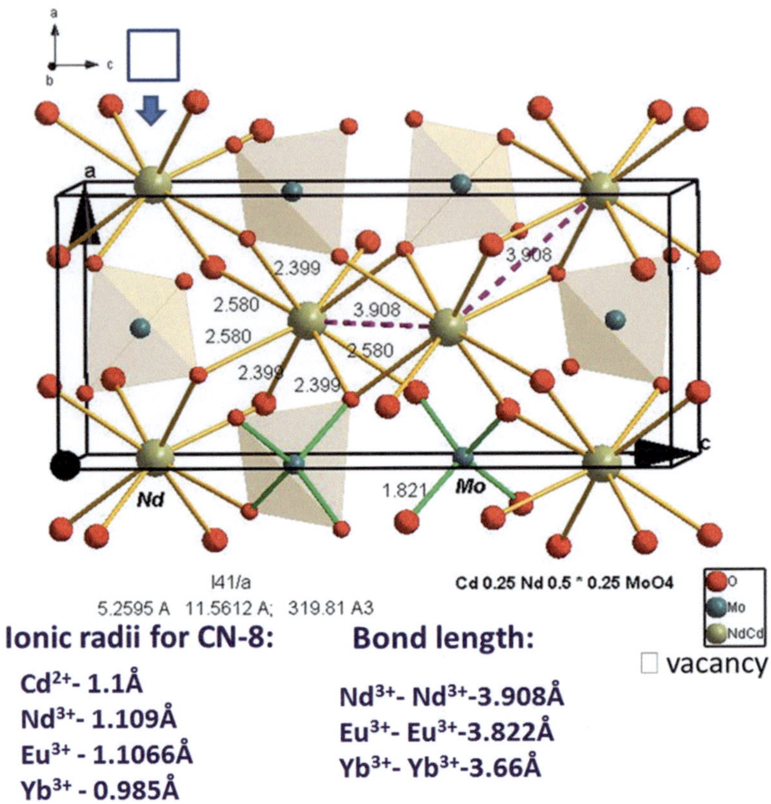

Fig. 17.4 The structure of $Cd_{0.25}Nd_{0.5}\square_{0.25}MoO_4$ solid solutions. The lengths of bonds are in Å

materials exhibit also structural disorder resulting in a significant inhomogeneous broadening of the spectral bands visible in the absorption and emission spectra.

In the new scheelite-type system under investigation, i.e. $Cd_{1\text{-}3x}Nd_{2x}\square_xMoO_4$ the substitution of trivalent RE^{3+} ions in the place of divalent Cd^{2+} ions is limited (until 66.67 mol%) and leads to the formation of cationic vacancies (□) in a framework, which can be connected also with creation of two different environments around Nd^{3+} ions. Detailed analysis of the structure of absorption and luminescence spectra recorded at cryogenic and room temperature allows us to confirm our supposition of more than one type of the activator center.

Confirmation from the Absorption Spectra

Figure 17.5 presents the absorption spectra at 4.2 K of $Cd_{0.9268}Nd_{0.0488}\square_{0.0244}MoO_4$ (5 mol% Nd^{3+}) in the spectral range from 400 to 900 nm (UV-vis and NIR). These spectra show 4f–4f transitions with crystal field (C.F.) fine structure in the range of all transitions.

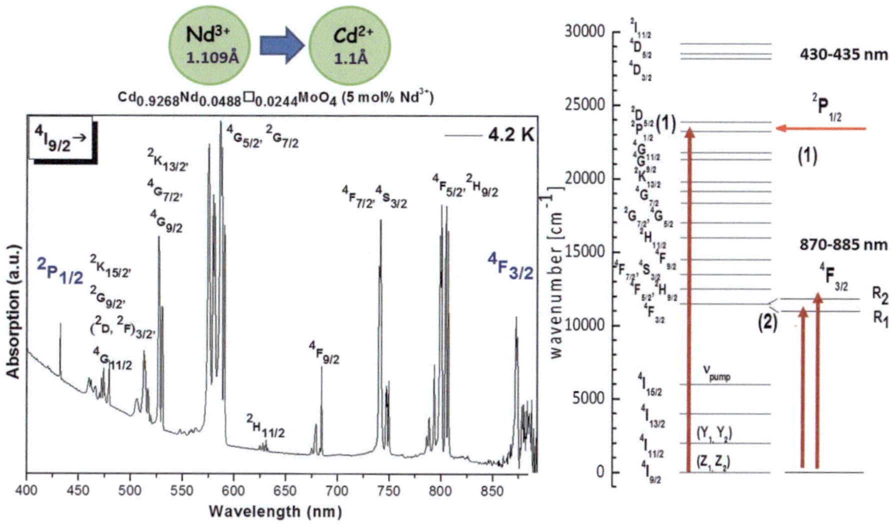

Fig. 17.5 Absorption spectra at 4.2 K of $Cd_{0.9268}Nd_{0.0488}\square_{0.0244}MoO_4$ (5 mol% Nd^{3+}) in the spectral range from 400 to 900 nm (UV–VIS and NIR with particular emphasis on the $^4I_{9/2} \rightarrow {}^2P_{1/2}$ and $^4I_{9/2} \rightarrow {}^4F_{3/2}$ transitions which simple Stark structure can play a role as a structural probe

At 4.2 K the absorption bands should be narrower than at room temperature and this behavior is well observed for $Cd_{1-3x}Nd_{2x}\square_xMoO_4$ solid solutions [20]. All the absorption lines are assigned to the transitions from the $^4I_{9/2}$ ground state to $^2P_{1/2}$ (433 nm), $^2G_{11/2} + {}^2F_{3/2} + {}^2D_{3/2} + {}^2G_{9/2}$ (480 nm), $^4G_{9/2}$, $^4G_{7/2}$, $^2K_{13/2}$ (530 nm), $^4G_{5/2}$, $^2G_{7/2}$ (580 nm), $^2H_{11/2}$ (630 nm), $^4F_{9/2}$ (683 nm), $^4F_{7/2} + {}^4S_{3/2}$ (750 nm), $^2H_{9/2} + {}^4F_{5/2}$ (807 nm), $^4F_{3/2}$ (875 nm).

From the structural point of view the careful analysis of the splitting of electronic transitions in the range of $^2P_{1/2}$ Kramers' doublet is important. Since at 4.2 K only the lowest Stark component of the ground state is populated, the number of components of $^4I_{9/2}$ $(Z_1, Z_2) \rightarrow {}^2P_{1/2}$ transition is consisted with the number of metal sites in the crystal structure. Figure 17.6 shows the absorption spectra at 4.2 K of $Cd_{0.9268}Nd_{0.0488}\square_{0.0244}MoO_4$ (5 mol% Nd^{3+}) in the spectral range of the $^4I_{9/2} \rightarrow {}^2P_{1/2}$ transition.

At room temperature three components at 432.5 nm (23 121 cm^{-1}), 435 nm (22 988 cm^{-1}), 437 nm (22 883 cm^{-1}) corresponding to the $^4I_{9/2} \rightarrow {}^2P_{1/2}$ transition are clearly visible. At 4.2 K the lines at 435 and 437 nm disappear because they represent the transitions from the higher energy Stark component (Z_2) of the $^4I_{9/2}$ ground state of Nd^{3+} ion. The line localized at 432.5 nm splits into two well resolved components at 432.4 nm (23 127 cm^{-1}) and 432.7 nm (23 110 $cm^{-1)}$, with a splitting of 17 cm^{-1} and a full width at half maximum (FWHM) of this absorption line equals to 27 cm^{-1}. This indicates clearly two main Nd^{3+} distributions of sites in the lattice with D_{2d} point symmetry with slightly lowering symmetry.

At this point we would like to draw attention and remind in a short recapitulation, how the line-widths of the $^4I_{9/2} \rightarrow {}^2P_{1/2}$ transition have been presented in another

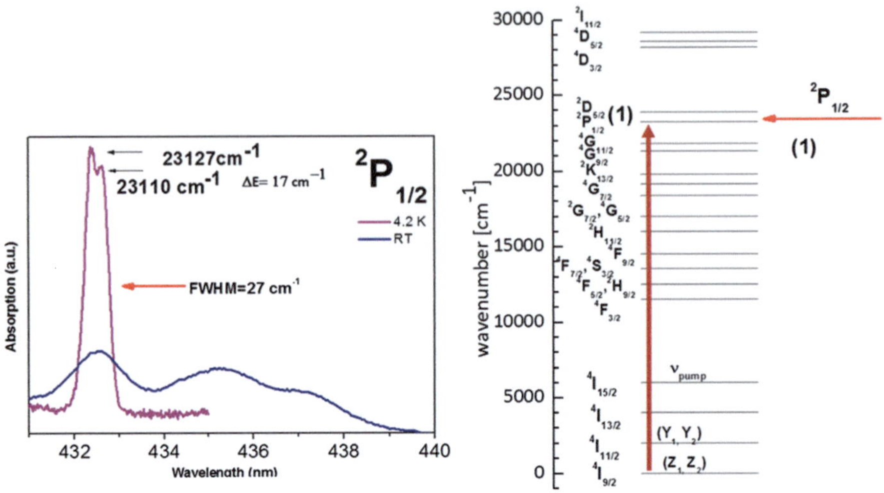

Fig. 17.6 Absorption spectra at 4.2 K of $Cd_{0.9268}Nd_{0.0488}\square_{0.0244}MoO_4$ (5 mol% Nd^{3+}) in the spectral range of the $^4I_{9/2} \rightarrow {}^2P_{1/2}$ transition

Nd^{3+}-doped tungstates investigated recently by us [26, 27]. For example in 5 % Nd^{3+}-doped $ZnY_4W_3O_{16}$ (orthorhombic system) the line-width of the 432.9 nm line was broad (FWHM ~ 130 cm^{-1}) as a consequence of Nd^{3+} multi-sites of slightly different crystal fields [26]. The differences in the widths of the bands at low and room temperatures were more noticeable for the tetragonal 5 % Nd^{3+} - doped $Cd_{0.25}Gd_{0.50}\square_{0.25}WO_4$ tungstates (FWHM = 76 cm^{-1}), where the structure was more ordered than in the monoclinic 5 % Nd^{3+}- doped $CdY_2W_2O_{10}$ compound (FWHM = 99 cm^{-1}) [27].

It is also important to notice that in all mentioned above compounds Nd^{3+} ions substituting Y^{3+} or Gd^{3+} ions, which were inserted (in 95 %) in the framework of the matrix, results in widespread disorder. As a consequence, the wide absorption lines are present at 4.2 K. In case of studied here $Cd_{1-3x}Nd_{2x}\square_xMoO_4$ phase the Nd^{3+} ions substitute only the Cd^{2+} ions, so that three Cd^{2+} are replaced by two Nd^{3+} ions with the creation of one vacancy, the FWHM = 27 cm^{-1} of the $^4I_{9/2} \rightarrow {}^2P_{1/2}$ absorption line is much narrower than in $Cd_{0.25}Gd_{0.50}\square_{0.25}WO_4$ and $CdY_2W_2O_{10}$ tungstates, what means that less disorder is observed in the lattice.

Another important observation is the easy substitution of Cd^{2+} ion (ionic radius = 1.1 Å for CN = 8) by Nd^{3+} ions (ionic radius 1.109 Å for CN = 8). In Fig. 17.7 other selected absorption lines like $^4I_{9/2} \rightarrow {}^4F_{3/2}$ of $Cd_{0.9268}Nd_{0.0488}\square_{0.0244}MoO_4$ (5 mol% Nd^{3+}) at room and 4.2 K are shown. The $^4I_{9/2} \rightarrow {}^4F_{3/2}$ transition also attracted our interest, because of two components expected at 4.2 K for one site. We observed two broad components, one at 873 nm (11 455 cm^{-1}) and the second one at 878.5 nm (11 382 cm^{-1}), that gives the value of $^4F_{3/2}$ state C.F. splitting equals to 73 cm^{-1}. We did not observe a subtle splitting of the $^4I_{9/2} \rightarrow {}^4F_{3/2}$ transition for the two types of expected sites but only the broadening of each component.

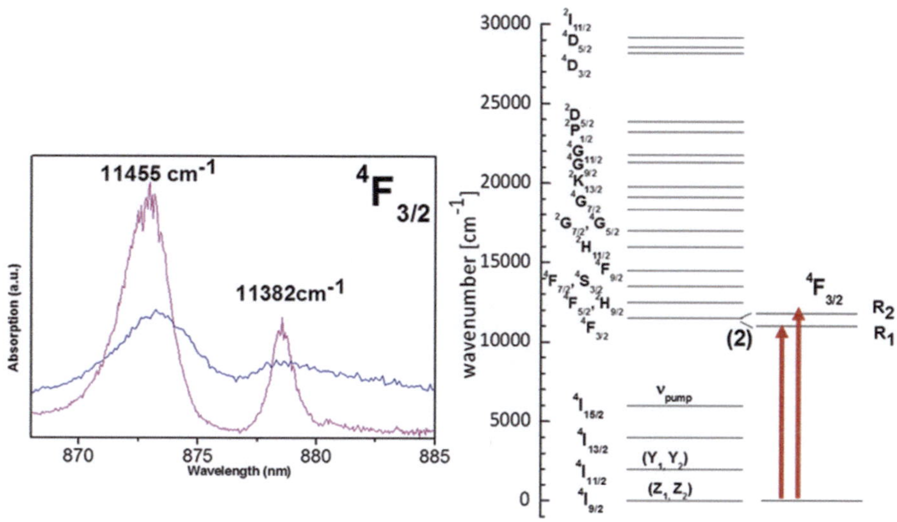

Fig. 17.7 Absorption spectra at 4.2 K of $Cd_{0.9268}Nd_{0.0488}\square_{0.0244}MoO_4$ (5 mol% Nd^{3+}) in the spectral range of the $^4I_{9/2} \rightarrow {}^4F_{3/2}$ transition

According to the Judd–Jorgensen statement, the f–f transitions, which obey $\Delta J = 2$, $\Delta L = 2$ selection rules are the most sensitive to the lanthanide ion environment changes, and are commonly referred as hypersensitive ones. The Nd^{3+} $^4I_{9/2} \rightarrow {}^4G_{5/2}$ transition satisfies the above criteria, and therefore its intensity is often used as a probe of structural changes, similar to the $^4I_{9/2} \rightarrow {}^4F_{5/2}$ transition in NIR region, which also partially (ΔJ) satisfies the selection rule for hypersensitive transition. In practice the energy of the $^2G_{7/2}$ term is very close to that of the $^4G_{5/2}$ one, thus at room temperature the $^4I_{9/2} \rightarrow {}^4G_{5/2}$, $^2G_{7/2}$ transitions must be analyzed together.

For the high symmetry systems, such as O_h and D_{4h}, the intensities of those two transitions in the IR region are about half of those of the $^4I_{9/2} \rightarrow {}^4G_{5/2}$, $^2G_{7/2}$, while in low symmetry environments the ratio is significantly higher. So, for the phases under investigations the highest intensity is observed for the hypersensitive $^4I_{9/2} \rightarrow {}^4G_{5/2}$, $^2G_{7/2}$ transitions because of lowering of occupied Nd^{3+} site symmetry. Thus on the base of the intensity ratio between two bands in the IR spectral range (725–778 nm) and (778–838 nm) and the $^4I_{9/2} \rightarrow {}^4G_{5/2}$, $^2G_{7/2}$ one (550–617 nm) presented in Fig. 17.8 one can conclude that symmetry of Nd^{3+} ion in our compound is rather high (S_4 or D_{2d}).

Figure 17.8 displays the $^4I_{9/2} \rightarrow {}^4F_{5/2}$, $^2H_{9/2}$ transition in the absorption spectra of $Cd_{0.9268}Nd_{0.0488}\square_{0.0244}MoO_4$ at room temperature. The broadening of the absorption lines in 780–830 nm spectral range reveals suitable absorption channels for the laser diode pumping. Important parameters for laser application are the absorption and emission cross-sections. It is well-known that large cross-section provides effective absorption of the excitation radiation. This parameter was determined by many authors to classify laser behavior of different Nd^{3+}-doped materials among

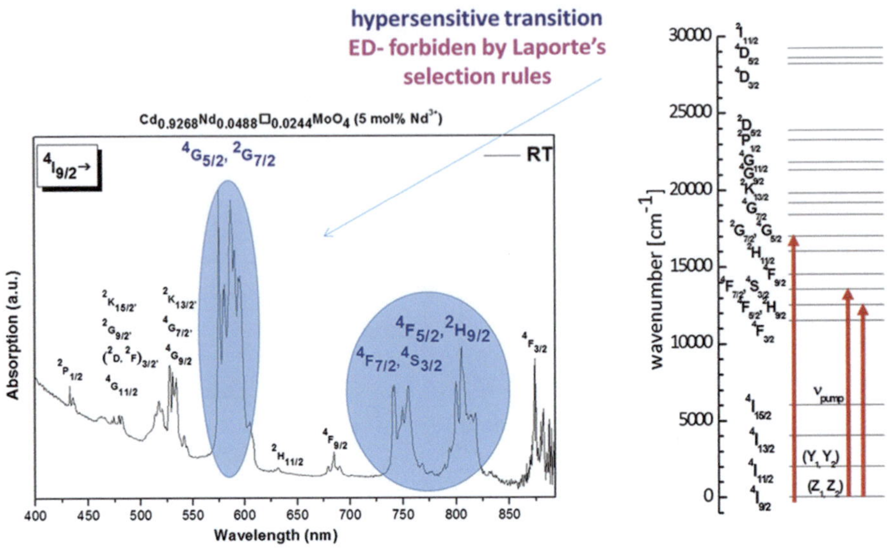

Fig. 17.8 Absorption spectra at 4.2 K of $Cd_{0.9268}Nd_{0.0488}\square_{0.0244}MoO_4$ (5 mol% Nd^{3+}) in the spectral range of 400–900 nm, with consideration of intensity ratio between two bands in the IR spectral range (725–778 nm) and (778–838 nm) and the $^4I_{9/2} \rightarrow ^4G_{5/2}$, $^2G_{7/2}$ one (550–617 nm) of the $^4I_{9/2} \rightarrow ^4F_{3/2}$ transition

which for well-known commercial laser materials. As some kind on standard one can use the absorption cross-section equal to 0.77×10^{-19} cm^2 for Nd^{3+} doped YAG commercial laser material [26, 27].

As in our previous papers on Nd^{3+}-doped tungstates we calculated the absorption coefficient and the absorption cross-section for diluted sample of $Cd_{0.9268}Nd_{0.0488}\square_{0.0244}MoO_4$ (5 mol% Nd^{3+}). The sample in the form of microcrystalline powder was pressed into the translucent pellet and then the absorption spectra were recorded. Basing on the absorption spectra the absorption coefficient and the absorption cross-section were calculated using formula:

$$\sigma = \mathbf{k}/\mathbf{N_0}$$

$\boldsymbol{\sigma}$ is the absorption cross-section, $\mathbf{N_0}$ the number of Nd^{3+} ions per volume ion/cm^3 and **k** is the absorption coefficient.

The assigned absorption cross-section at $\lambda = 804.8$ nm equal to 1.44×10^{-19} cm^2 for $Cd_{0.9268}Nd_{0.0488}\square_{0.0244}MoO_4$ (5 mol% Nd^{3+}), with number of Nd^{3+} ion per volume $N_0 = 6.5 \times 10^{20}$ ions/cm^3. Recently, the value of absorption cross-sections equal to 3×10^{-20} cm^2 for 10 mol% Nd^{3+}-doped $ZnY_4W_3O_{16}$ with $N_0 = 4.6 \times 10^{20}$ ion/cm^3 of Nd^{3+} ions per volume was determined [26], as well as for two types of cadmium tungstates, i.e. 5 mol% of Nd^{3+}–doped $Cd_{0.25}Gd_{0.50}\square_{0.25}WO_4$:$Nd^{3+}$ and 5 mol% Nd^{3+}-doped $CdY_2W_2O_{10}$. For cadmium

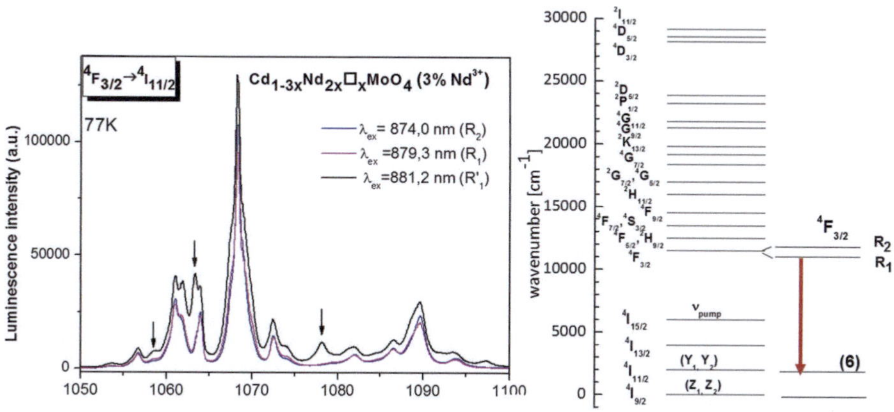

Fig. 17.9 Emission spectra of $Cd_{0.9556}Nd_{0.0296}\square_{0.0148}MoO_4$ (3 mol% of Nd^{3+}) recorded at 77 K under different excitation line of Ti-Sapphire laser

compounds the calculated values of these parameters were higher [27]. For the monoclinic 5 mol% Nd^{3+}-doped $CdY_2W_2O_{10}$, sample with number of Nd^{3+} ion per volume $N_0 = 5.13 \times 10^{20}$ ions/cm^3 the value of the absorption cross-section with correction for the single crystal is equal to 1.3×10^{-19} cm^2, while for the tetragonal 5 mol% Nd^{3+}-doped $Cd_{0.25}Gd_{0.50}\square_{0.25}WO_4$ with $N_0 = 1.35 \times 10^{20}$ ions/cm^3 is 2.5×10^{-19} cm^2.

Compared to YAG, the rare earth doped alkali metal double tungstates crystals like: $KY(WO_4)_2$ (KYW) and $KGd(WO_4)_2$ (KGdW) have an advantage of larger absorption cross section. For double tungstates these values are as follows: 1.28×10^{-20} cm^2 for Nd^{3+}-doped KGdW, while for Yb^{3+} – doped KGdW is 1.2×10^{-19} cm^2 and 1.33×10^{-19} cm^2 for Yb^{3+} – doped KYW [32, 33].

Confirmation from the Emission Spectra

Figure 17.9 presents emission spectra of $Cd_{0.9556}Nd_{0.0296}\square_{0.0148}MoO_4$ (3 mol% of Nd^{3+}) recorded at 77 K under different excitation lines of Ti-Sapphire laser. The number of Stark components vary when the excitation wavelength change, confirming the distribution of Nd^{3+} ion in more than one non-equivalent site in the lattice (additional components are indicated in the Fig. 17.9 by arrows).

From the group theory consideration the $^4F_{3/2}$ multiplet splits into two Kramers' doublets and the number of components correlates with $J + 1/2$ levels for one site. In the spectra recorded at room and liquid nitrogen temperature, in the band corresponding to the $^4F_{3/2} \rightarrow ^4I_{9/2}$ transition ($\lambda \approx 900$ nm) one can distinguish nine well split lines (for the complete removed degeneracy of the $^4I_{9/2}$ level should be more than five).

In the spectra recorded at 4.2 K in all spectral range the number of Stark components is also higher than that expected for the compound with one symmetry site due to the double distribution of Nd^{3+} ions.

Potential Application

To check the possibility to apply our compounds as laser materials in the future, it is useful to compare them to the well-known YAG: Nd^{3+} laser materials. A commercial LED emission spectrum with a maximum at 588 nm to pump the $^4I_{9/2} \rightarrow ^4G_{5/2}$, $^2G_{7/2}$ hypersensitive absorption transitions is well adapted for the excitation of the microcrystalline $Cd_{0.9556}Nd_{0.0296}\square_{0.0148}MoO_4$ (3 mol% of Nd^{3+}) and the powdered single crystal of YAG:Nd^{3+}. The comparison of the integral intensities of transitions for both types of compounds shows values eight and half times (for $^4F_{3/2} \rightarrow ^4I_{9/2}$ transition) and nine times (for $^4F_{3/2} \rightarrow ^4I_{11/2}$ transition) higher for the $Cd_{1\text{-}3x}Nd_{2x}\square_xMoO_4$ than for YAG:Nd^{3+}. The potential application of $Cd_{1\text{-}3x}Nd_{2x}\square_xMoO_4$ might come from this high intensity and also from the exploitation of wide bands especially promising for an ultra-short pulse laser generation [20].

17.3.4 Spectroscopic Analysis of $Cd_{1\text{-}3x}Yb_{2x}\square_xMoO_4$

17.3.4.1 Evidence of Yb^{3+} Multi-sites

The Yb^{3+} ions have only two Stark-split energy manifolds: the $^2F_{7/2}$ ground state and the $^2F_{5/2}$ excited state. In Fig. 17.10, Stark levels are distributed in these manifolds and labelled from 1 to 4 in the ground state and from 5 to 7 in the excited state from the lowest energy to the highest energy, respectively.

The spectroscopic studies performed for Nd^{3+} ions in this matrix have highlighted two main sites in the description of the local structure for $CdMoO_4$ doped by different quantities of neodymium ions. Each of such Nd^{3+} ions has not only Cd^{2+} atoms but also □ cationic vacancies in their neighborhood. This should give one type of local symmetry for all Nd^{3+} foreign ions if the □ vacancy was located always at the same place. But in case of these solid solutions it seems that the vacancy should be located mainly in two positions to create two main distributions of Nd^{3+} sites. Yb^{3+} ions, characterized by ionic radius of 0.985 Å, can substitute Cd^{2+} (1.10 Å) sites and a similar situation is observed now for $Cd_{1\text{-}3x}Yb_{2x}\square xMoO_4$ solid solutions where the number of components observed in the emission spectra suggests two types of local symmetry of Yb^{3+} ions. For both Nd^{3+} and Yb^{3+} ions the probability to create pairs and aggregates is high.

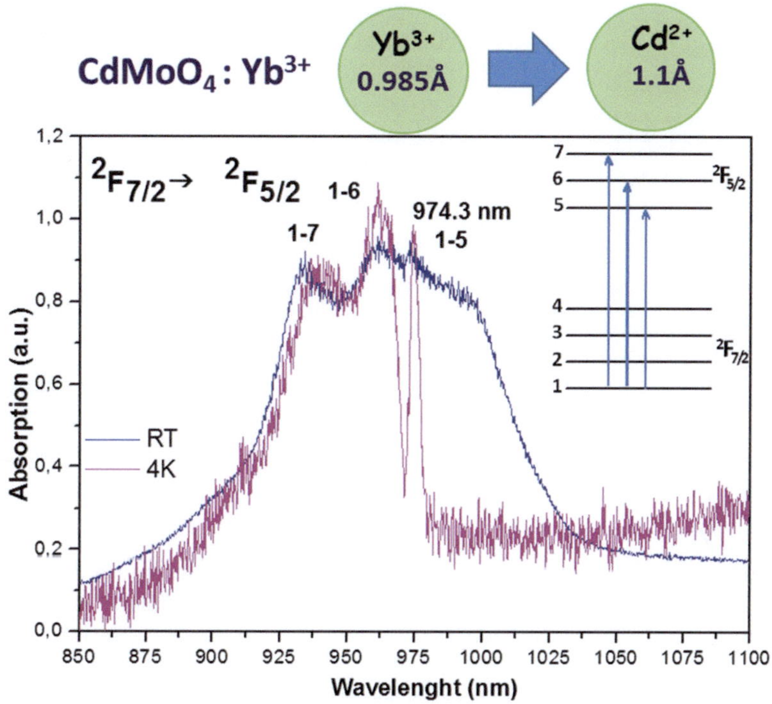

Fig. 17.10 Absorption spectra of $Cd_{0.8635}Yb_{0.1678}\square_{0.0839}MoO_4$ (9.53 mol%) solid solution recorded at room temperature and 4.2 K

17.3.4.2 Confirmation from the Absorption Spectra

Figure 17.10 presents the absorption spectra for 9.53 mol% Yb^{3+} in $Cd_{1\text{-}3x}Yb_{2x}\square_x MoO_4$ recorded at room and liquid helium temperature. At RT the spectra consist of broad weakly resolved bands located between 850 nm and 1050 nm corresponding to the transition from the $^2F_{7/2}$ ground state of Yb^{3+} to the three Stark components of the $^2F_{5/2}$ excited state. The absorption line corresponds to the so-called zero phonon line located at around 974.3 nm (10 264 cm^{-1}) and is not the strongest absorption line. This is the transition from the lowest Stark component of the $^2F_{7/2}(1)$ ground state to the lowest Stark level of the $^2F_{5/2}(5)$. Two others lines of comparable intensities are located at around 963 nm (10 384 cm^{-1}) corresponding to $^2F_{7/2}(1)\rightarrow{}^2F_{5/2}(6)$ and 933 nm (10 718 cm^{-1}) to $^2F_{7/2}(1)\rightarrow{}^2F_{5/2}(7)$ transitions. An additional components observed at RT at 993 nm (10 070.5 cm^{-1}) corresponds to the absorption from the higher energy Stark components of the ground state. The measurements performed at 4.2 K lead to narrowing and better resolution of the main three electronic Stark components. According to the crystal-field theory

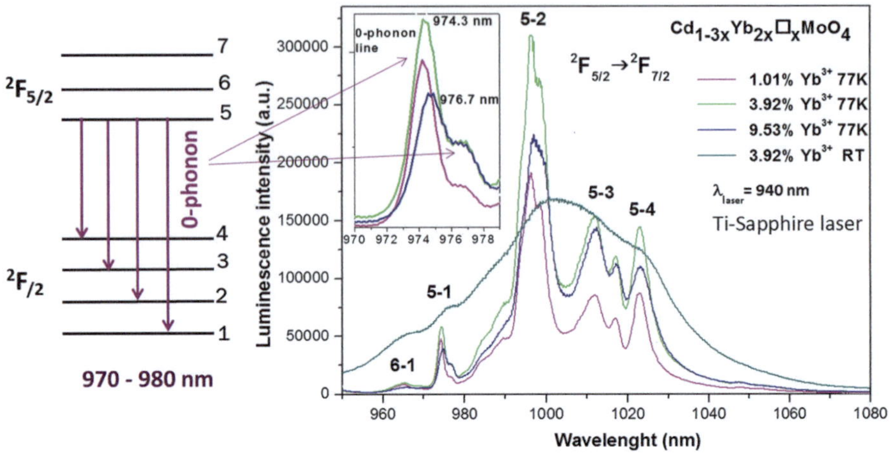

Fig. 17.11 Emission spectra of $Cd_{1\text{-}3x}Yb_{2x}\square_xMoO_4$ solid solutions at room temperature and 77 K

of Kramers' ions, the maximum of allowed components splitting for $J = 5/2$ state is three; therefore, for one symmetry site the absorption spectrum at liquid helium temperature should be resolved into three bands. The zero-phonon line is a sharp one and contains only one component at 4.2 K. The total splitting of the excited $^2F_{5/2}$ state 454 cm^{-1} is a smaller than that observed for oxide matrices (about 793 cm^{-1}) [34], $K_5Bi(MoO_4)_4$ molybdates [35], or double phosphates (750–800 cm^{-1}) [36, 37] and is similar to fluoride hosts (about 420 cm^{-1}) in which vacancies also occur [38]. The assigned absorption cross-section of the 1–5 transition at 974.3 nm is equal to 1.45×10^{-19} cm^2 with number of Yb^{3+} ions per volume $N_0 = 6.32 \times 10^{20}$ ions/cm^3 (for 9.32 mol%). This parameter was calculated in the same way as for analogue samples with Nd^{3+} [20] or Nd^{3+}-doped tungstates [27].

For double tungstates these values are as follows: 1.28×10^{-20} cm^2 for Nd^{3+}-doped KGdW, while for Yb^{3+}-doped KGdW is 1.2×10^{-19} cm^2 and 1.33×10^{-19} cm^2 for Yb^{3+}-doped KYW [32, 33].

17.3.4.3 Confirmation from the Emission Spectra

The selected emission spectra of the samples of $Cd_{1\text{-}3x}Yb_{2x}\square_xMoO_4$ registered at different temperatures and doped with different Yb^{3+} contents are plotted in Fig. 17.11. The studied materials exhibit the luminescence in the near-infrared region from the $^2F_{5/2} \rightarrow ^2F_{7/2}$ transitions of the Yb^{3+} ions. At RT the bands are broad and the distinction between vibronic components and Stark components is difficult to identify.

In the case of the Yb^{3+} ion, one should observe transitions into four Stark components of the ground $^2F_{7/2}$ split by the crystal field. It corresponds to the transitions from the lowest Stark level of the $^2F_{5/2}$ excited state (5) to each of the four Stark levels of the $^2F_{7/2}$ state (1, 2, 3, 4) respectively. The (5 → 1) component is the

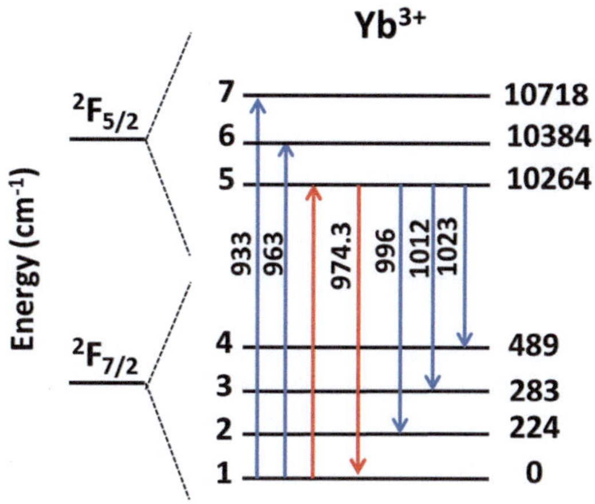

Fig. 17.12 Low temperature Stark splitting levels of Yb^{3+} ions in $Cd_{0.8635}Yb_{0.1678}\square_{0.0839}MoO_4$ (9.53 mol% of Yb^{3+}) calculated from the experimental data of absorption and emission

so-called zero-phonon line at around 975–977 nm. In case of $Cd_{1\text{-}3x}Yb_{2x}\square_xMoO_4$ solid solutions at 4.2 K the intensity of this line for all investigated concentrations of Yb^{3+} ions is not very high due to the well- known re-absorption effect by energy transfer between resonant zero-phonon lines. The charge compensation by association of two Yb^{3+} cations in place of three Cd^{2+} cations perturbs the local structure of optical centres since a vacancy has been created. The slight shift of the zero-phonon emission line seen only in luminescence spectra of the obtained samples when Yb^{3+} concentration increases is probably connected to the multisite distribution of $Cd_{1\text{-}3x}Yb_{2x}\square_xMoO4$. This line localized at 974.3 nm splits into two well resolved components at 974.3 nm (10 264 cm^{-1}) and 976.7 nm (10 238.5 cm^{-1}), with a splitting of 25.5 cm^{-1} and the full width at half maximum (FWHM) equals to 21 cm^{-1}. This indicates clearly two main Yb^{3+} distributions of sites in the lattice of D_{2d} point symmetry with slightly lower symmetry (see insert Fig. 17.12). The line did not shift with the Yb^{3+} content. The position of zero-phonon line from the absorption spectra stays in good agreement with that from the emission. Contrary to emission spectra at RT high resolution emission spectra are reached at low temperature (4.2 K) and we observe doubling of almost each line, so there are more than four components expected for one symmetry site. In fact seven intense components are clearly resolved. The main bands corresponding to the four Stark emission components are located at 974.3 nm (10 264 cm^{-1}) ($5 \rightarrow 1$), 996 nm (10 040 cm^{-1}) ($5 \rightarrow 2$), 1012 nm (9 881 cm^{-1}) ($5 \rightarrow 3$) and 1 023 nm (9 775 cm^{-1}) ($5 \rightarrow 4$), respectively. In the Yb^{3+}-doped molybdates under investigation, not only the lowest level 5 of the $^2F_{5/2}$ excited state is emitting, but one can observe the $6 \rightarrow 1$ transition at 962.4 and 964 nm in Figs. 17.10 and 17.11.

Additional weak components may be a result of electron-phonon coupling with M-O modes. Among all compounds under investigation, the highest emission intensity is shown by the material containing 4 mol% of Yb^{3+} ions. The comparison of the luminescence intensity of all samples was performed by recording the spectra under the same conditions.

Figure 17.11 presents emission spectra of $Cd_{0.9424}Yb_{0.0384}\square_{0.0192}MoO_4$ (3.92 mol% of Yb^{3+}) recorded at 77 K under different excitation lines of the Ti-Sapphire laser. The number of zero-phonon line components varies when the excitation wavelength changes, confirming the distribution of Yb^{3+} ions in more than one non-equivalent site in the lattice [21].

17.3.4.4 Energy Level Diagram of Yb^{3+} Ions in $Cd_{1\text{-}3x}Yb_{2x}\square_xMoO_4$

Figure 17.12 represents a summary of the spectroscopic results by combining experimental absorption and emission spectra of the highest population of $Cd_{0.8635}Yb_{0.1678}\square_{0.0839}MoO_4$ solid solution (9.53 mol% of Yb^{3+} ions). An energy level diagram can be derived from low temperature measurements because at RT the spectra were not structured enough to assign the transitions.

Yb^{3+} concentration dependences of the $^2F_{5/2}$ experimental decay time were analyzed in order to attempt the understanding of the concentration quenching mechanism and estimate the main parameters useful for a theoretical approach of laser potential.

17.3.4.5 Potential Application to Laser Source

We have applied on $Cd_{1\text{-}3x}Yb_{2x}\square_xMoO_4$ solid solutions the model proposed earlier on concentration dependence of the life time [39] for Yb^{3+}-doped cubic oxides such as: Y_2O_3, Sc_2O_3, and Lu_2O_3 sesquioxides, YAG, GGG and LuAG garnets, and Yb^{3+}-doped cubic fluorides (CaF_2 and KY_3F_{10}). The variation is interpreted in term of self-trapping and self-quenching processes [40]. The optimized concentrations for gain have been found 6 mol% at 77 K and 3 mol% at RT [21].

17.3.5 *Spectroscopic Analysis of* $Cd_{1\text{-}3x}Eu_{2x}\square_xMoO_4$

17.3.5.1 Evidence of Eu^{3+} Multisites Under Site Selective Excitation

Figure 17.13 present the emission spectra of $Cd_{1\text{-}3x}Eu_{2x}\square_xMoO_4$ with 0.05 mol% of Eu^{3+} ions recorded at 77 K. We have applied the laser site-selective excitation into$^7F_0 \rightarrow {}^5D_2$ transition of Eu^{3+} ion, by changing excitation lines every 0.1 nm in the spectral range from 465.0 nm to 466.4 nm. In the figures we present the selected emission spectra, which clearly show the differences in the number of components and position of the lines under different excitation. The most significant differences can be observed for $^5D_0 \rightarrow {}^7F_0$ transition and $^5D_0 \rightarrow {}^7F_1$ one. Emission spectra recorded at 77 K presents for $^5D_0 \rightarrow {}^7F_0$ transition three different positions of maxima *i.e.* at 580.65 nm (under 465.4 nm excitation), 580.8 nm (under 465.5 nm

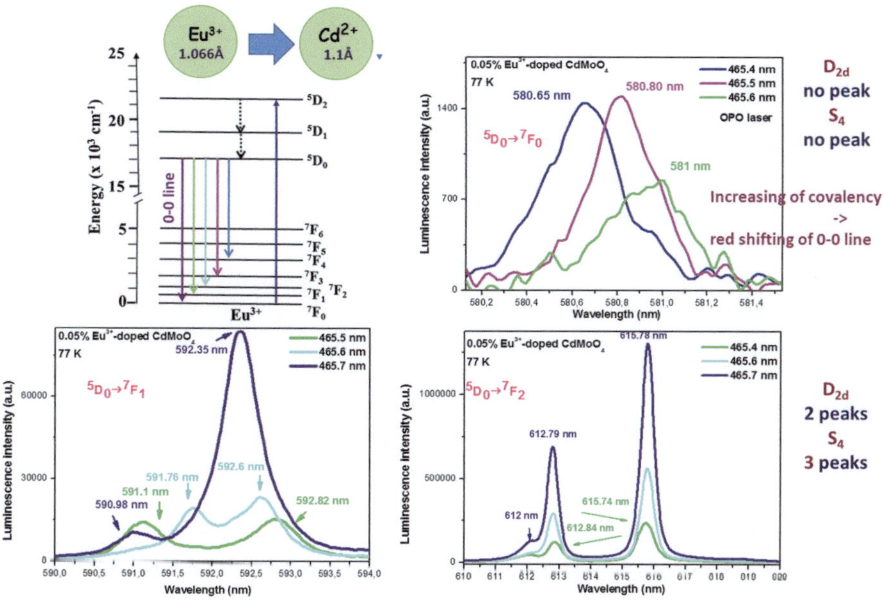

Fig. 17.13 Site selective emission spectra of $Cd_{1-3x}Eu_{2x}\square_xMoO_4$ with 0.05 mol% of Eu^{3+} ions recorded at 77 K presenting luminescence from the $^5D_0$ state under different excitations of OPO laser into $^7F_0 \rightarrow {}^5D_2$ transition

excitation) and 581 nm (under 465.6 nm excitation). It is worth noting that the excitation lines differ by only 0.1 nm, what lead us to the conclusion that there are three non-equivalent symmetry sites, very close to each other due to the distinct difference with the subtle change of excitation line.

The multisite character was also well illustrated by the $^5D_0 \rightarrow {}^7F_1$ transition, where the peak positions vary depending on the excitation line. It could be again related to the existence of three main distributions of Eu^{3+} ions. Indeed, the comparative analysis reveals that under $\lambda_{ex} = 465.5$ nm we observe two components at 591.1 nm and 592.82 nm. The change the excitation step of only 0.1 nm leads to changing the peaks positions to 591.76 nm and 592.6 nm. The variation of the excitation line by another 0.1 nm results the maxima at 590.98 nm and 592.35 nm. The last example is the excitation at $\lambda_{ex} = 466$ nm with the emission peaks at 591.85 nm and 592.59 nm.

For $^5D_0 \rightarrow {}^7F_2$ transition the spectra are dominated by the peak at 615.78 nm, which only slightly varies to 615.74 nm with changing of excitation line. The second line located at 612.79 nm shifts to 612.84 nm. The third, additional component appears at 612 nm and is the most intense under 465.7 nm excitation line. For the $^5D_0 \rightarrow {}^7F_2$ transition the differences under site selective excitation are not as spectacular as in case of $^5D_0 \rightarrow {}^7F_0$ and $^5D_0 \rightarrow {}^7F_1$ transitions due to $^5D_2 \rightarrow {}^5D_1$ non-radiative relaxation processes. Presence of third components between 611 nm and 617 nm allow us to confirm rather S_4 symmetry than D_{2d} symmetry, for which three peaks corresponding to $^5D_0 \rightarrow ^5F_2$ transition should be observed. According to the selection rules, for the D_{2d} symmetry, which was supposed to be the case for the phases under investigation, two peaks should be observed for the $^5D_0 \rightarrow ^7F_1$ transition, no peak should be observed for the $^5D_0 \rightarrow ^7F_0$ transition and one or two peaks should be observed for the $^5D_0 \rightarrow ^7F_2$ transition. If the symmetry is slightly different, for example S_4, the expected number of components for the $^5D_0 \rightarrow ^7F_0$ and $^5D_0 \rightarrow ^7F_1$ transitions is similar to the symmetry D_{2d}, but for the $^5D_0 \rightarrow ^7F_2$ transition, it increases to three.

Our spectroscopic studies provide at least three main sites in the description of local structures for $CdMoO_4$ doped by different quantities of europium ions. Similarly as in case of Nd^{3+} and Yb^{3+} in this matrix, each of such Eu^{3+} ions have not only Cd^{2+} atoms but also cationic vacancies in their neighborhood. This should give one type of local symmetry for all Eu^{3+} foreign ions if the vacancy was always located at the same place. It means that the vacancies □ should be located mainly in three positions to create three main distributions of Eu^{3+} sites.

17.3.5.2 Time Resolved Emission Spectra

Figure 17.14 shows time resolved emission spectra of $Cd_{1-3x}Eu_{2x}\square_xMoO_4$ with 0.05 mol% of Eu^{3+} ion, recorded at 77 K under excitation into $^7F_0 \rightarrow ^5D_2$ level of Eu^{3+} ion at 465.7 nm by an OPO laser. The measurements allowed well distinguish

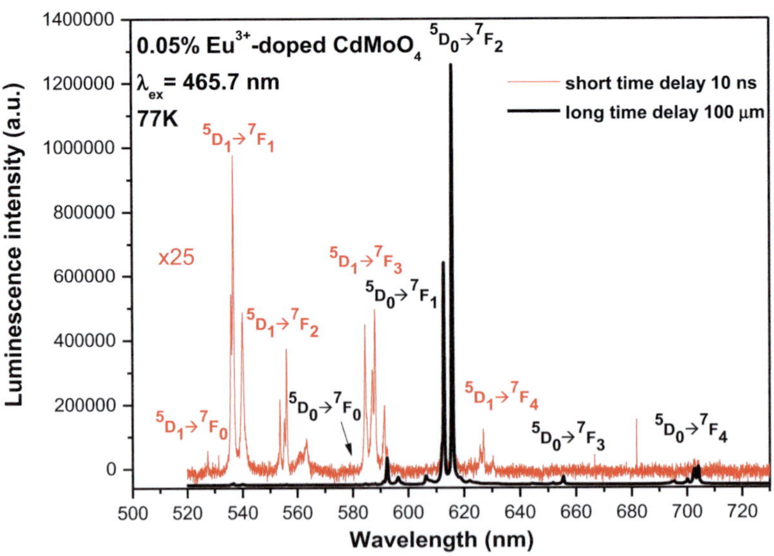

Fig. 17.14 Time resolved emission spectra of $Cd_{1-3x}Eu_{2x}\square_xMoO_4$ with 0.05 mol% of Eu^{3+}, recorded at 77 K presenting luminescence from the $^5D_1$ and $^5D_0$ states under excitation into $^7F_0 \rightarrow ^5D_2$ level of Eu^{3+} ion at 465.7 nm of OPO laser

the lines corresponding to the transitions from the $^5D_1$ and $^5D_0$ states to ^7F_J (J = 0, 1, 2, 3, 4) and to define contributions of the individual emission lines in the steady state spectrum. The spectra obtained with short time delay (10 ns) consist of several lines belongs to fast (~11 μs) fluorescence transitions from $^5D_1 \rightarrow ^7F_0$ located at 527 nm, $^5D_1 \rightarrow ^7F_1$ at 535.9 nm, 536.7 nm and 540 nm, $^5D_1 \rightarrow ^7F_2$ at 556.0 nm, 560.9 nm, 563.24 nm, 553.58 nm and 555.24 nm, $^5D_1 \rightarrow ^7F_3$ at 584.5 nm, 585 nm, 587 nm, 588 nm, 591.5 nm and 592. 25 nm, as well as from $^5D_1 \rightarrow ^7F_4$ at 625.78 nm and 626.88 nm, respectively. The emission spectra obtained with pulse delay equal to 100 μs present the long luminescence (~315 μs) originating from the $^5D_0$ level of excited state, *i.e.* $^5D_0 \rightarrow ^7F_0$ at 580.8 nm, $^5D_0 \rightarrow ^7F_1$ at 591, 592.35, 596.3 nm, $^5D_0 \rightarrow ^7F_2$ at 606.3 nm, 608 nm, 612.2, 612.8 nm, 615.8 nm, $^5D_0 \rightarrow ^7F_3$ at 644.3 nm, 650.8 nm, 651.83 nm, 655.5 nm, $^5D_0 \rightarrow ^7F_4$ at 695.1 nm, 699.36 nm, 700.15 nm, 702.71 nm and 704 nm [22].

17.3.5.3 Color Points and Colorimetric Characterization

Similar as in our previous papers on Eu^{3+}-doped tungstates, we have analysed also the ability of $Cd_{1\text{-}3x}Eu_{2x}\square_xMoO_4$ solid solution as candidates for phosphor application. To characterize the luminescence (λ_{ex} =310 nm, 395 and 466.5 nm) of the systems under investigation with respect to their emission colour at 293 K, the CIE (x, y) coordinates were calculated [41]. A coordinate system (x, y) to characterize colors has been defined by the *Commission Internationale de l'Eclairage (CIE)* [42]. For the Eu^{3+} doped samples generally used in applications as red phosphors, the color points x and y have to satisfy the following conditions: $x>0.65$ and $y<0.35$. The CIE chromatic diagram with the points indicating the emission colour for $Cd_{1\text{-}3x}Eu_{2x}\square_xMoO_4$ solid solution with 0.05 mol% of Eu^{3+} recorded at room temperature under excitation at 310 nm (x = 0.474, y = 0.343), 395 nm (x = 0.652, y = 0.348) and 466 nm (x = 0.63, y = 0.369). We have shown by using the Commission Internationale de l'Éclairage 1931 x-y chromatic diagram, that this system presents an emission very close to the ideal red light [22].

17.4 Conclusion

In this study we investigated the structural and photoluminescence properties of Nd^{3+}, Eu^{3+} and Yb^{3+} ions in tetragonal $CdMoO_4$ molybdate host lattice in the presence of a precise amount of cationic vacancies in the framework, which are the consequence of the substitution of divalent Cd^{2+} by trivalent RE^{3+} cations. Disordering is manifested by broad emission lines and by the presence of multisites that are characterized by slight variation of the local symmetries due to the position of the next nearest vacancies. The studies performed on Nd^{3+}, Eu^{3+} and Yb^{3+} ions as structural probes by using site-selective laser spectroscopy at low temperature have clearly indicated two unequivalent sites for Nd^{3+} and Yb^{3+} ions and three unequivalent sites for Eu^{3+} ions, according to the main nearest positions of □ cationic vacancies around trivalent rare earth pairs. Eu^{3+}, and most probably Nd^{3+}, Yb^{3+} major ions, possess the S_4 point symmetry. The three nonequivalent symmetry sites are very close to each other due to the subtle differences observed in the emission spectra. Site-selective laser spectroscopy has allowed to define the energy level schemes for the main site of each of all three rare earth ions in this host lattice.

Finally, we have check the possibility to apply these compounds either as phosphors (samples emitting in the visible spectral region doped with Eu^{3+} ions very close to the ideal red light) or as laser materials (samples emitting in the near infrared spectral region doped with Nd^{3+} and Yb^{3+} ions). Especially for Nd^{3+} dopant, a potential application might come from both large value of the absorption cross-section, high integral intensity of luminescence and also the exploitation of wide bands for an ultra-short pulse laser generation. Investigation is in progress with rare earth-doped single crystals grown by the Czochralski method.

We hope these results selected for the School are helpful in understanding how important is the role played by lanthanide ions not only as promising optical materials, both in the visible (Eu^{3+}) and near infrared spectral region (Nd^{3+}, Yb^{3+}) but also as structural probes for basics research.

Acknowledgements We wish to thank the Minister of Science and Higher Education in Poland and in France for the Grant POLONIUM for scientific exchange between Institute Light Matter (ILM), UMR5306 CNRS-University Lyon1, University of Lyon, France and Faculty of Chemistry, University of Wrocław in Poland. We also wish to thank the National Science Center of Poland for the grant HARMONIA No. UMO013/08/M/ST5/007484700/PB/WCH/13 and the French Embassy in Warsaw for a French Government scholarship for research stage of M. Guzik in Lyon. These financial supports are gratefully acknowledged.

References

1. Pujol, M. C., Mateos, X., Solé, R., Massons, J., Gavaldà, J., Solans, X., Díaz, F., & Aguiló, M. (2002). Structure, crystal growth and physical anisotropy of $KYb(WO_4)_2$, a new laser matrix. *Journal of Applied Crystallography, 35*, 108.
2. Boulon, G., Metrat, G., Muhlstein, N., Brenier, A., Kokta, M. R., Kravchik, L., & Kalisky, Y. (2003). Efficient diode-pumped Nd: $KGd(WO_4)_2$ laser grown by top nucleated floating crystal method. *Optical Materials, 24*, 377.
3. Brenier, A., Bourgeois, F., Metrat, G., Muhlstein, N., & Boulon, G. (2001). Spectroscopic properties at 1.351 μm of Nd^{3+}-doped $KY(WO_4)_2$ and $KGd(WO_4)_2$ single crystals for Raman conversion. *Optical Materials, 16*, 207.
4. Lu, J., Ueda, K., Yagi, H., Yanagitani, T., Akiama, Y., & Kaminski, A. A. (2002). Neodymium doped yttrium aluminum garnet ($Y_3Al_5O_{12}$) nanocrystalline ceramics a new generation of solid state laser and optical materials. *Journal of Alloys and Compounds, 341*, 220–225.
5. Minowa, M., Itakura, K., Moriyama, S., & Otani, W. (1992). Measurement of the property of cooled lead molybdate as a scintillator. *Nuclear Instruments and Methods in Physics Research Section A, 320*, 500–503.

6. Sczancoski, J. C., Bomio, M. D. R., Cavalcante, L. S., Joya, M. R., Pizani, P. S., Varela, J. A., Longo, E., Siu Li, M., & Andrés, J. A. (2009). Morphology and blue photoluminescence emission of $PbMoO_4$ processed in conventional hydrothermal. *Journal of Physical Chemistry C, 113*, 5812.
7. Hizhnyi, Y. A., Nedilko, S. G., & Lumin, J. (2003). Investigation of the luminescent properties of pure and defect lead tungstate crystals by electronic structure calculations. *Journal of luminescence, 102–103,* 688–693.
8. Kudo, A., Steinberg, M., Bard, A. J., Campion, A., Fox, M. F., Mallouk, T. E., Webber, S. E., & White, J. M. (1990). Photoactivity of ternary lead-group VIB oxides for hydrogen and oxygen evolution. *Catalysis Letters, 5*, 61.
9. Millers, D., Grigorjeva, L., Chernov, S., Popov, A., Lecoq, P., & Auffray, E. (1997). The temperature dependence of scintillation parameters in $PbWO_4$. *Physica Status Solidi B, 203*, 585–589.
10. Godlewska, P., Tomaszewicz, E., Macalik, L., Hanuza, J., Ptak, M., Tomaszewski, P., Mączka, M., & Ropuszyńska-Robak, P. (2013). Correlation between the structural and spectroscopic parameters for$Cd_{1-3x}Gd_{2x}\square_xMoO_4$ solid solutions where □ denotes cationic vacancies. *Materials Chemistry and Physics, 139*, 890–896.
11. Wang, W. S., Zhen, L., Xua, C. Y., Shao, W. Z., & Chen, Z. L. (2012). Eu^{3+}-doped $CdMoO_4$ red phosphor synthesized through an aqueous solution route at room temperature. *Journal of Alloys and Compounds, 529*, 17.
12. Zhang, J., Zhang, N. Zou, L., Gan, S. (2014). Formation mechanism and optical properties of $CdMoO_4$ and $CdMoO_4$:Ln^{3+} (Ln = Pr, Sm, Eu, Dy, Ho and Er) microspheres synthesized via a facile sonochemical route. *RSC Advances, 4*, 38455.
13. Itoh, M. (2012). Luminescence study of self-trapped excitons in $CdMoO_4$. *Journal of Luminescence, 132*, 645–651.
14. Mikhailik, V. B., & Kraus, H. (2010). Performance of scintillation materials at cryogenic temperatures. *Physica Status Solidi B, 247*, 1583.
15. Zhang, J., Zhang, N., Zou, L., & Gan, S. (2014). Formation mechanism and optical properties of $CdMoO_4$ and $CdMoO_4$:Ln^{3+} (Ln = Pr, Sm, Eu, Dy, Ho and Er) microspheres synthesized via a facile sonochemical route. *RSC Advances, 4*, 38455.
16. Lin, J., Zeng, Z., & Wang, Q. (2013). $CdMoO_4$:Eu^{3+} micro-sized luminescent particles synthesis and photo-catalytic performance. *Inorganica Chimica Acta, 408*, 59.
17. Tomaszewicz, E., Filipek, E., Fuks, H., & Typek, J. (2014). Thermal and magnetic properties of new scheelite type $Cd_{1-3x}\square_xGd_{2x}MoO_4$ ceramic materials. *Journal of the European Ceramic Society, 34*, 1511.
18. Godlewska, P., Tomaszewicz, E., Macalik, L., Hanuza, J., Ptak, M., Tomaszewski, P.E., & Ropuszyńska-Robak, P. (2013). Structure and vibrational properties of scheelite type $Cd_{0.25}RE_{0.5}\square_{0.25}MoO_4$ solid solutions where □ is the cationic vacancy and RE = Sm-Dy. *Journal of Molecular Structure, 1037*, 332.
19. Groń, T., Tomaszewicz, E., Berkowski, M., Duda, H., Kukuła, Z., Pawlus, S., Mydlarz, T., Ostafin, T., & Kusz, J. (2014). Dielectric and magnetic properties of $CdMoO_4$:Gd^{3+} single crystal. *Journal of Alloys and Compounds, 593*, 230.
20. Guzik, M., Tomaszewicz, E., Guyot, Y., Legendziewicz, J., & Boulon, G. (2015). Structural and spectroscopic characterizations of new $Cd_{1-3x}Nd_{2x}\square_xMoO_4$ scheelite-type molybdates with vacancies as potential optical materials. *Journal of Materials Chemistry C, 3*, 4057.
21. Guzik, M., Tomaszewicz, E., Guyot, Y., Legendziewicz, J., & Boulon, G. (2016). Spectroscopic properties, concentration quenching and Yb^{3+} site occupations in vacancied scheelite-type molybdates. *Journal of Luminescence, 169*, 755–764.
22. Guzik, M., Tomaszewicz, E., Guyot, Y., Legendziewicz, J., & Boulon, G. (2015). Eu^{3+} luminescence from different sites in a scheelite-type cadmium molybdate red phosphor with vacancies. *Journal of Materials Chemistry C, 3*, 8582.
23. Tomaszewicz, E., Kaczmarek, S. M., & Fuks, H. (2010). New cadmium and rare-earth metal molybdates with scheelite-type structure. *Materials Chemistry and Physics, 122*, 595.

24. Taupin, D. (1973). A powder-diagram automatic-indexing routine. *Journal of Applied Crystallography, 6*, 380.
25. Tomaszewicz, E., Kaczmarek, S. M., & Fuks, H. (2009). New cadmium and rare earth metal tungstates with the scheelite type structure. *Journal of Rare Earths, 27*, 569.
26. Guzik, M., Cybinska, J., Tomaszewicz, E., Guyot, Y., Legendziewicz, J., Boulon, G., & Strek, W. (2011). Spectroscopic behavior of Nd^{3+} in a new microcrystalline $ZnY_4W_3O_{16}$ tungstate. *Optical Materials, 34*, 487–495.
27. Guzik, M., Tomaszewicz, E., Guyot, Y., Legendziewicza, J., & Boulon, G. (2012). Structural and spectroscopic characterizations of two promising Nd-doped monoclinic or tetragonal laser tungstates. *Journal of Materials Chemistry, 22*, 14896.
28. Tomaszewicz, E., Guzik, M., Cybińska, J., & Legendziewicz, J. (2009). Spectroscopic investigation of the Europium(3+) ions in a new $ZnY_4W_3O_{16}$ matrix. *Helvetica Chimica Acta, 92*, 2274–2290.
29. Guzik, M., Bieza, M., Tomaszewicz, E., Guyot, Y., Zych, E., & Boulon, G. (2015). Nd^{3+} dopant influence on the structural and spectroscopic properties of microcrystalline $La_2Mo_2O_9$ molybdate. *Optical Materials, 41*, 21–31.
30. Guzik, M., Bieza, M., Tomaszewicz, E., Guyot, Y., & Boulon, G. (2014). Development of Nd^{3+} -doped monoclinic dimolybdates $La_2Mo_2O_9$ as optical materials. *Zeitschrift fur Naturforschung B: Journal of Chemical Sciences, 69b*, 193–204.
31. Guzik, M., Tomaszewicz, E., Kaczmarek, S. M., Cybińska, J., & Fuks, H. (2010). Spectroscopic investigations of $Cd_{0.25}Gd_{0.50}\square_{0.25}WO_4{:}Eu^{3+}$ – a new promising red phosphor. *Journal of Non-Crystalline Solids, 35*, 1902–1907.
32. Specification of Nd:KGW, Dohrer Electrooptic, www.doehrer.com
33. Specification of Yb:KGW and Yb:KYW, Eksma Optics, www.eksmaoptics.com
34. Boulon, G., Laversenne, L., Goutaudier, C., Guyot, Y., & Cohen-Adad, M. T. (2003). Radiative and non-radiative energy transfers in Yb^{3+}-doped sesquioxide and garnet laser crystals from a combinatorial approach based on gradient concentration fibers. *Journal of Luminescence, 102–103*, 417.
35. Canibano, H., Boulon, G., Palatella, L., Guyota, Y., Brenier, A., Voda, M., Balda, R., & Fernandez, J. (2003). Spectroscopic properties of new Yb^{3+}-doped $K_5Bi(MoO_4)_4$ crystals. *Journal of Luminescence, 102–103*, 318.
36. Legendziewicz, J., Cybinska, J., Guzik, M., Boulon, G., & Meyer, G. (2008). Comparative study of crystal field analysis in Pr^{3+} and Yb^{3+}-doped K_2LaX_5 ($X = Cl^-$, Br^-) ternary halides and Yb^{3+}-doped $A_3Lu(PO_4)_2$ ($A = Na^+$, Rb^+) double phosphates. Charge transfer band observations of Yb^{3+}-doped systems. *Optical Materials, 30*, 1655.
37. Legendziewicz, J., Guzik, M., & Szuszkiewicz, W. (2008). Charge transfer and f–f emission of trivalent ytterbium observed in double phosphates $M^IM^{III}(PO_4)_2$ ($M^I = Na$, Rb; $M^{III} = Lu$, Y). *Journal of Alloys and Compounds, 451*, 165.
38. Bensalah, A., Guyot, Y., Brenier, A., Sato, H., Fukuda, T., & Boulon, G. (2004). Spectroscopic properties of Yb^{3+}: $LuLiF_4$ crystal grown by the Czochralski method for laser applications and evaluation of quenching processes: A comparison with Yb^{3+}: $YLiF_4$. *Journal of Alloys and Compounds, 380*, 15.
39. Auzel, F., Baldacchini, G., Laversenne, L., & Boulon, G. (2003). Radiation trapping and self-quenching analysis in Yb^{3+}, Er^{3+}, and Ho^{3+} doped Y_2O_3. *Optical Materials, 24*, 103.
40. Boulon, G., Guyot, Y., & Yoshikawa, A. (2009). Optimization of the gain in Yb^{3+}-doped cubic laser crystals of 99.99 % purity. *Journal of Rare Earths, 27*, 616.
41. Judd, D. B., & Wyszecki, G. (1962). *Color in business, science and industry* (p. 2). New York: Wiley.
42. Shionoya, R., & Yen, W. M. (1998). Phosphor handbook, phosphor research society, SRC Press, 459.

Chapter 18
Medical Applications of Nanomaterials

Anna Vedda and Irene Villa

Abstract This chapter is devoted to an introduction to the fascinating field of nanomedicine. In particular, the role of material scientists in this interdisciplinary area will be discussed, by considering how the structural, morphological, and functional properties of nanoparticles can be designed for specific medical applications. Due to the large extent of the field, focus will be given to the optical applications of inorganic nanoparticles in medicine. These include *in vitro* and *in vivo* imaging, self-lighting photodynamic therapy, hyperthermia, and thermometry. Moreover some specific issues like self-absorption and autofluorescence, which turn out to be critical for the functionalization of luminescent nanoparticles, will be also described and discussed.

18.1 Introduction

Nanomedicine includes all applications of nanotechnology for treatment, diagnosis, monitoring, and control of biological systems [1]. Diagnostics and therapy applications are also sometimes merged in the *theranostics* field where nanoparticles are developed in order to play simultaneously both a diagnostics and a therapeutic role. Specific interesting characteristics of nanoparticles are their large surface area useful to accommodate functional groups, as well as the possibility to be up-taken by cells. This last aspect depends on nanoparticles dimensions, their surface charge, and surface functionalization with proper ligands. This makes nanomedicine a very interdisciplinary research area that requires the interaction between medicine and chemistry, physics, biology, engineering.

During the last years, the scientific community of materials scientists has carried out great efforts in the field of the nanotechnology applied to medicine focusing on the design, synthesis, characterization, and application of nanometer-size particles.

A. Vedda (✉) • I. Villa
Department of Materials Science, University of Milano-Bicocca, Via Cozzi 55, 20125 Milan, Italy
e-mail: anna.vedda@mater.unimib.it

© Springer Science+Business Media Dordrecht 2017
B. Di Bartolo et al. (eds.), *Nano-Optics: Principles Enabling Basic Research and Applications*, NATO Science for Peace and Security Series B: Physics and Biophysics, DOI 10.1007/978-94-024-0850-8_18

Starting with early communications in the 90ies (see for example Ref. [2]) the number of publications in nanomedicine has constantly grown, reaching nowadays more than one thousand per year.

Metal, semiconductor, and insulating inorganic systems, as well as polymeric compounds have been considered [3–9]. For a successful outcome in biomedical application, biocompatibility, biodegradability, and water dispersibility are indispensable requirements. Nanoparticles in the size range of 2–100 nm possessing tunable optical properties have been proposed in specific cancer treatments. A further employment of properly designed nanostructures concerns their use as heat generators for hyperthermia treatments, aimed to damage tumor cells by the locally induced temperature increase, which can also be monitored by changes of the optical features of selected nanoparticles. Moreover, luminescent nanoparticles can be employed also for direct imaging of biological systems. Fluorescence bioimaging can be an excellent non-invasive technique, relatively cheap and displaying a high spatial resolution (hundreds of nm), thus potentially allowing tumor detection at the very early stages.

The text is organized as follows. Section 18.2 is devoted to optical imaging. Both imaging by photoluminescence and by persistent luminescence are described. Section 18.3 concerns nanoparticles in therapy, where two kind of therapeutic modalities with nanoparticles are described, namely the self-activated photodynamic therapy and the photothermal therapy.

18.2 Optical Imaging

18.2.1 Optical Imaging by Photoluminescence

The kind of interaction of the electromagnetic spectrum with the human body depends on its energy and this offers a variety of possibilities to investigate it and control diseases. The high energy ionizing X and gamma rays are exploited for example in Computed Tomography and Positron Emission Tomography respectively, while low energy MHz waves are used in Magnetic Resonance Imaging. The visible, and especially the near infrared region (NIR) of the electromagnetic spectrum, can also be exploited for imaging [10]. Luminescent nanoparticles, if properly localized close to the regions of interest (like tumors), can be excited optically and provide an optical image of the body. The most important specific requirements of nanoparticles for imaging, especially *in vivo*, are (i) controlled dimensions, (ii) absence of toxicity, (iii) excitation and emission in the near infrared, (iv) high luminescence efficiency, and (v) emission far from autofluorescence signals. The need of infrared emitters particularly for *in vivo* imaging can be easily understood by the inspection of Fig. 18.1 that reports the extinction coefficient of light by the human tissues. Biological media (water, blood, haemoglobin, melanin and lipids) significantly absorb and simultaneously scatter light [11]. However it can be seen that human tissues are partially transparent in two regions, called the "first biological window" from approximately 700 to 950 nm, and the "second biological

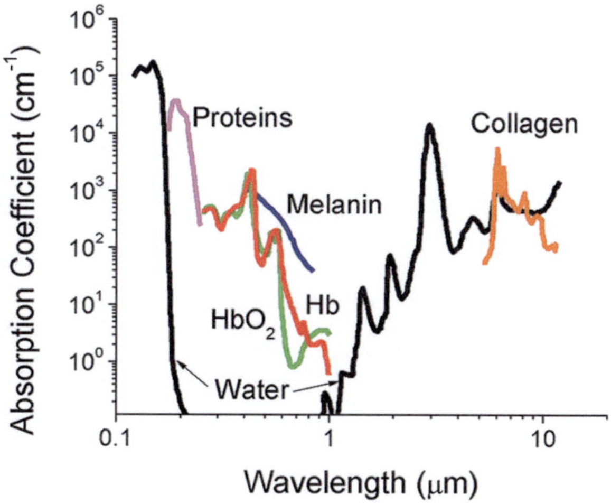

Fig. 18.1 Optical absorption of various tissue and blood components from 200 nm to 10 μm (Reprinted with permission from Ref. [12]. Copyright (2012) American Chemical Society)

window" from 1000 to 1500 nm. These regions are limited by water absorptions at about 980 and 1500 nm. They are therefore the spectral regions of interest for imaging purposes.

Among the various types of luminescent nanoparticles are organic dyes, carbon nanotubes and nanodots, as well as gold nanoparticles (nanorods, nanocages, nanospheres), quantum dots (CdS, CdSe, ZnS, ZnSe, PbSe, PbS). Also rare earth activated insulating crystals like $NaGdF_4$, $NaYF_4$, CaF_2, LaF_3, SrF_2 are the subject of intense studies and their favorable properties will be specifically discussed here. These systems are mostly optically inactive unless they are doped with a luminescent activator. Indeed it is found that they are good hosts for rare earth ions (RE), a class of elements characterized by a great variety of optical transitions that make them very attractive as luminescent activators in many optical applications including lighting devices, lasers, solar energy materials, ionizing radiation sensors and, finally, in nanomedicine. After doping, the RE energy levels can lie within the host band-gap and confer an optical activity to otherwise transparent systems. The low phonon energies of fluoride crystals like those listed above favor a high luminescence efficiency of RE-related transitions in these hosts.

When doping nanoparticles with such luminescent activators, a number of fundamental spectroscopic questions are worth to be addressed. Several ones are related to the fact that, even if nanoparticles have a crystalline structure, their degree of disorder is always relatively high due to their large surface areas. This characteristic can have various and opposite effects on luminescence. On one hand, disorder can lower the luminescence efficiency by introducing non-radiative channels. On the other hand, crystal field fluctuations can limit energy migration and non-radiative recombination. Moreover symmetry reduction in the local field can improve fluorescence of forbidden transitions of RE by increasing the radiative recombination probability of the involved electronic excited states.

Several cases of infrared imaging studies can be mentioned from the recent literature [12]. For example (see Fig. 18.2) core-shell structures consisting of a

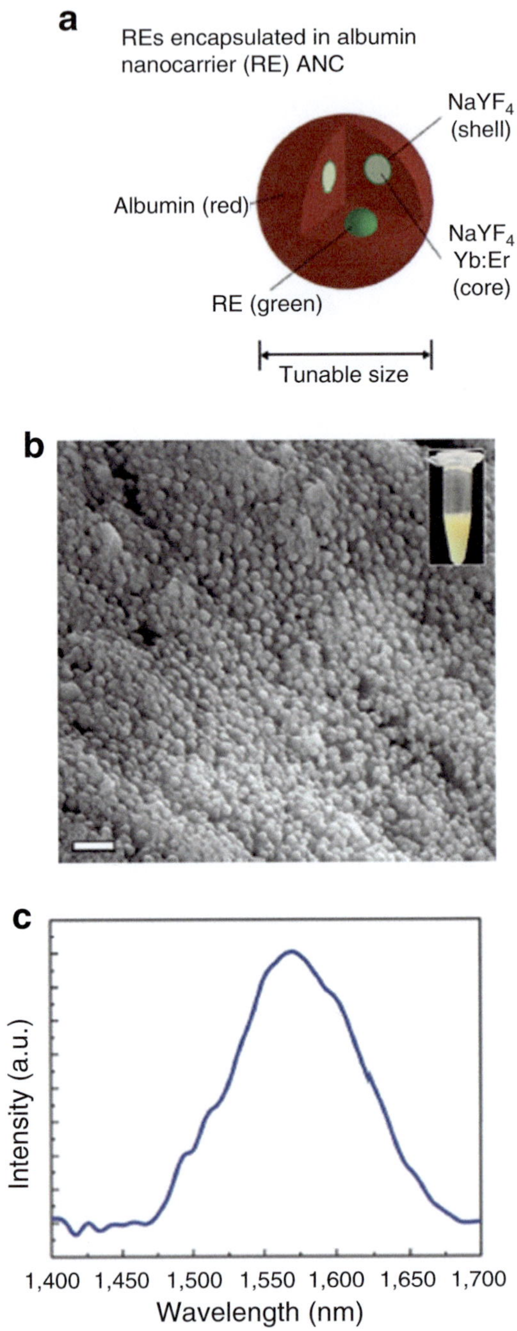

Fig. 18.2 Rare-earth encapsulated albumin nanocarriers (**a**) consist of REs encapsulated by a coating of human serum albumin (HSA) to form an inorganic–organic nanocomposite, which can be tuned in size. SEM images (**b**) of (RE)ANCs show uniform sub-100 nm spherical particles (scale bar = 200 nm). (RE)ANCs retain the infrared emission of the encapsulated REs (**c**), exhibiting peak emission between 1550–1600 nm following 980 nm excitation (Reprinted by permission from Macmillan Publishers Ltd: Nature Communications (Ref. [13]), copyright (2013))

$NaYF_4$:Er core (featuring an emission peaking at 1525 nm, even beyond the second biological window) surrounded by an undoped $NaYF_4$ shell were found to possess a higher luminescence efficiency and detection sensitivity with respect to other IR emitters. Moreover, after controlled albumin encapsulation the targeting of such NPs to malignancies was evidenced thus demonstrating the good potential of such structures for *in vivo* imaging [13].

Among other problems, *in vivo* imaging can suffer from the occurrence of tissue autofluorescence due to optical excitation, that can lie in the same spectral region as the luminescence of the nanoparticles and give rise to a detrimental background increase. Autofluorescence can originate also from the animal diet. The problem of autofluorescence can be solved by carefully selecting the spectral position of the nanoparticle emission to be separated from it. Nd-doped SrF_2 and LaF_3 nanoparticles were found to be particularly useful to this purpose [14]. In fact the Nd^{3+} emission line peaking at 1330 nm is well separated from autofluorescence of the animal diet that occurs at shorter wavelengths (see Fig. 18.3). In this way a reliable biodistribution study of the nanoparticle in the animal organs could be performed.

18.2.2 Optical Imaging by Persistent Luminescence

The imaging by photoluminescence above described requires continuous excitation of nanoparticles by an optical source. In order to excite nanoparticles in the body, not only emission, but also excitation must occur in the transparency window of tissues. This requirement, together with the possible simultaneous autofluorescence excitation, limits to some extent the applicability of photoluminescence to medical imaging. Alternatively, excitation of the nanoparticles before their introduction in the body can be foreseen. This is possible for those systems that display persistent luminescence, i.e. a luminescence signal that lasts for a very long time (minutes, or even hours) after excitation. Such a phenomenon allows to excite the nanoparticles before their introduction in living tissues, therefore removing the restrictions regarding the excitation source, and obtaining a completely autofluorescence-free signal. Persistent phosphors are crystals or amorphous materials that are usually been developed for other applications fields like for example night signals. For the emission to last for a so long time, the recombination mechanism does not involve only an optical transition in a single luminescent ion. An energy level of a distinct center is always also involved, named "trap" because of its ability to keep a carrier (an electron or a hole) for a long time. The recombination is then called "persistent luminescence" or also sometimes "phosphorescence" (but this term is also often used to indicate forbidden recombinations). The process is sketched in Fig. 18.4. Excitation with an energy greater than the band gap of the material is necessary.

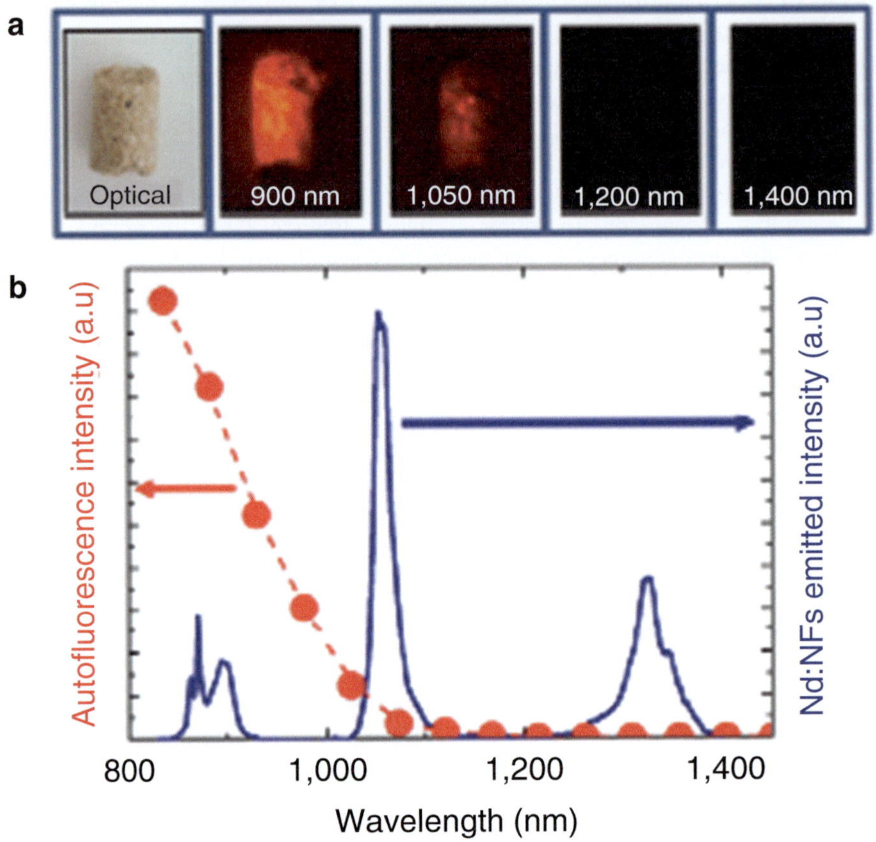

Fig. 18.3 (**a**) Optical image of a mouse food pellet and fluorescence images of the same pellet as obtained using different longpass filters when excited with a 808 nm laser. (**b**) Wavelength dependence of the infrared autofluorescence intensity generated by a mouse food pellet. Dots are experimental data and dashed line is a guide for the eyes. The figure also includes the room temperature emission spectrum (uncorrected by the detection system spectral response) of a colloidal solution of Nd:SrF_2 nanoparticles (Reprinted from Ref. [14] with permission of Springer)

In this way, besides recombining promptly at luminescence centers, carriers can also be trapped by localized levels in the material band gap due to point defects or impurities. In the figure such levels are traps for electrons, but a similar mechanism can be sketched for holes.

After trapping, carriers can be released again towards the conduction band by thermal activation, and recombine with holes at the recombination centers giving rise to a luminescence signal. The decay turns out to be exponential, with a decay time

$$\tau = 1/s \exp(-E/kT) \tag{18.1}$$

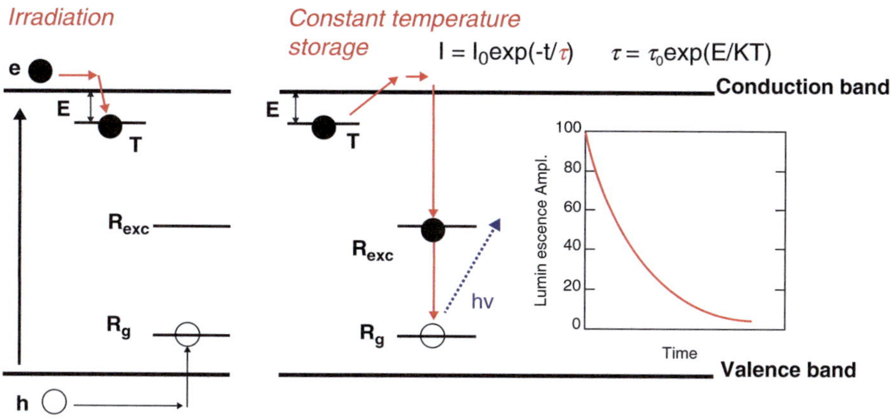

Fig. 18.4 Sketch of the persistent luminescence occurring in a material after irradiation with ionizing radiations. T is a trap, R_{exc} and R_g are the excited and ground state levels of the recombination center that emits light after electron capture by the excited level

where s is a constant, E is the thermal depth of the trap, k is the Boltzmann's constant and T is the absolute temperature [15]. Depending on trap parameters s and E, the decay time can have very different values. By a proper engineering of the material and of its defects, a suitable decay time of the order of several minutes or hours can be obtained. A possible issue in this mechanism is that the decay strongly depends on temperature and small temperature variations can give rise to signal instabilities. A similar, but more suitable mechanism involves athermal tunneling of the trapped electron towards the excited state of the luminescent center. For such a mechanism to occur, it is necessary that the trap lies at a similar level as the excited state of the center, and that the trap-center distance is very small (typically of a few angstrom) [16]. In such a case the decay has still an exponential course, but the decay time does no more depend on temperature. The decay time can be written as:

$$\tau = 1/A_R W \tag{18.2}$$

where A_R is the radiative transition probability of the luminescent center and W is the trap-center tunneling probability, that depends exponentially from the trap-to-center distance.

An even more favourable, and also frequent, situation occurs if traps and centers are randomly distributed in the material giving rise to a continuous distribution of trap-center distances, because in that case the decay follows a t^{-1} power law. Such a decay is slow and this characteristic is particularly favourable for signal detection for long times [16].

Such considerations demonstrate that persistent phosphors are also worth to be developed for medicine. Excitation is perfomed with ionizing radiation or above band-gap light before NP insertion in the body. Red and IR emitters lying in the

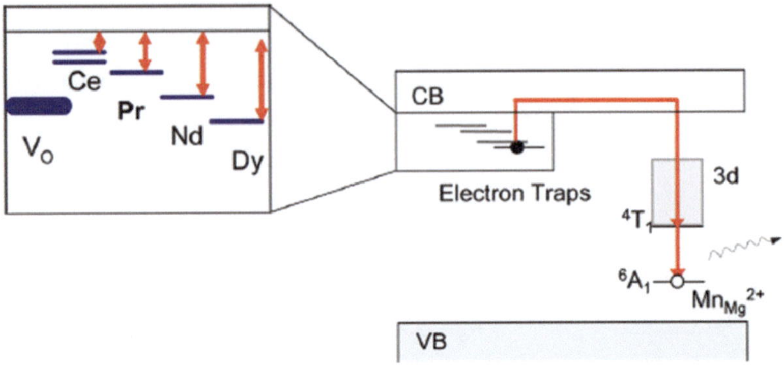

Fig. 18.5 Schematic energy level diagram of Mn^{2+} and RE^{3+} in $CaMgSi_2O_6$:Mn^{2+}, RE^{3+}. The main hole traps are Mn^{2+} ions, while electrons are trapped by oxygen vacancies (V_O) and RE^{3+} ions. TSL and persistent luminescence occur by thermally activated electron release and capture by Mn^{3+} ions, giving the Mn^{2+} emission. The insert shows the relative positions of electron trap levels with respect to the conduction band edge (Reprinted with permission from Ref. [20]. Copyright (2011) American Chemical Society)

transparency window of the body are employed [17–27]. An interesting example is represented by the work of Maldiney et al. [20] in which several RE ions were introduced in Mn^{2+}-doped diopside-based persistent luminescence nanoparticles. Such RE (Dy, Pr, Ce, Nd) play the role of electron traps with variable depth, while Mn^{2+} plays the role of recombination center. Figure 18.5 shows a scheme of the levels involved in the process. Pr-doping turns out to give rise to an optimal trap, allowing to produce an intense long persistent luminescence following X-ray or UV light irradiation. After intravenous injection of previously excited nanoparticles, the emitted light detection through living tissues was also realized successfully. A second very interesting example is represented by near-IR (from 650 to 1000 nm) emitting Cr^{3+}- doped zinc gallogermanates nanoparticles [21]. The long persistent luminescence occurs *via* a tunneling process as it is demonstrated by its t^{-1} decay law, allowing signal detection for as many as 360 h. Figure 18.6 displays the emission spectrum and the luminescence decay, while Fig. 18.7 depicts the localized trap-center recombination process.

18.3 Luminescent Nanoparticles in Medical Therapy

Cancer is a group of diseases characterized by the uncontrolled growth and spread of abnormal cells. Although regular screening and surveillance programs as well as early intervention are the best ways to improve the outcome and survival, efforts need to be devoted at finding better cancer therapeutic options that are

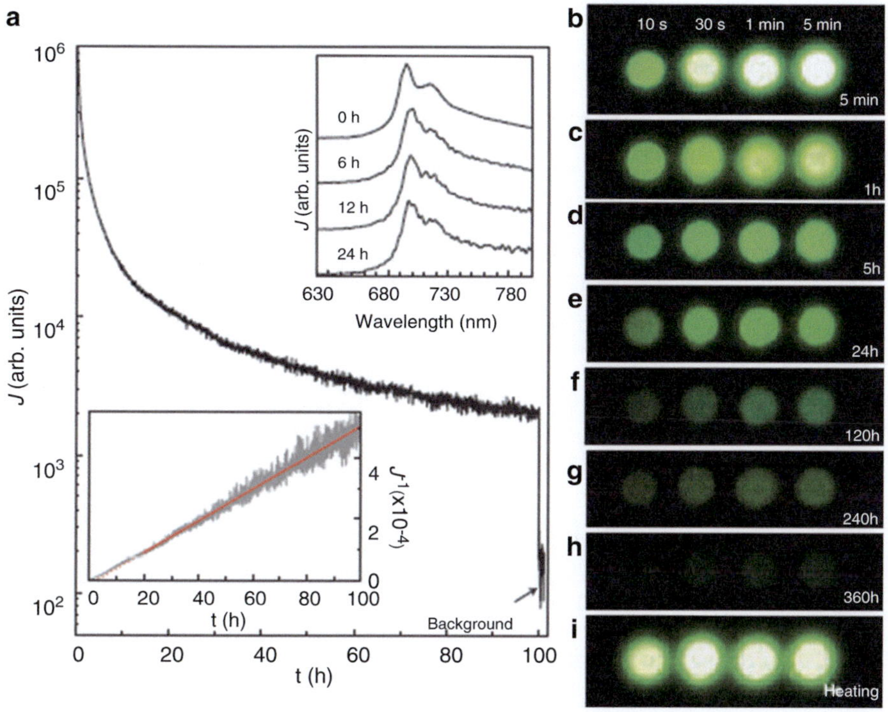

Fig. 18.6 NIR afterglow decay of $Zn_3Ga_2Ge_2O_{10}$:0.5%Cr^{3+} discs irradiated by a 4 W 365 nm ultraviolet lamp. (**a**), Afterglow intensity (I) monitored at 713 nm as a function of time (t). The sample was irradiated for 5 min. The bottom inset shows the same data plotted as I^{-1} versus t. The upper inset shows four afterglow spectra recorded at 0 h, 6 h, 12 h and 24 h after the end of the irradiation. (**b–h**), NIR images of four phosphor discs taken at different afterglow times (5 min to 360 h) after irradiation by a 365 nm lamp for 10 s to 5 min (Reprinted by permission from Macmillan Publishers Ltd: Nature Materials (Ref. [21]), copyright (2012))

effective, efficient, affordable, and acceptable to patients. The conventional cancer treatment options are chemotherapy, radiotherapy, and surgery, and more recently small molecule-based therapies and immunotherapy, along with a combination of these strategies, are being experimented.

Chemotherapy is often associated with systemic side effects. High recurrence rate is associated with surgical resection of tumors, while radiation therapy is limited by the cumulative radiation dose, since the use of high intensity doses increases the possibility of damaging healthy tissues. For this reason, the desire for safer therapies has increased interest in novel methods, leading to the refinement of the conventional cancer treatment modalities. Nowadays, the research on developing alternate treatment modalities that are safe, effective, and of low cost is very active and challenging.

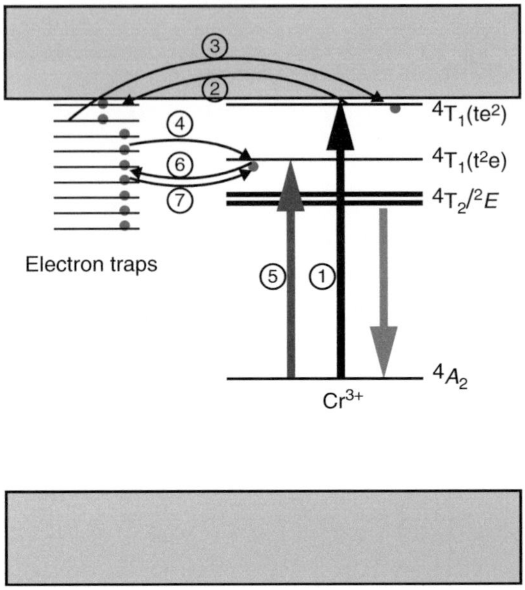

Fig. 18.7 A schematic representation of the persistent NIR luminescence mechanism in $Zn_3Ga_2Ge_2O_{10}$:0.5%Cr^{3+}. The straight-line arrows and curved-line arrows represent optical transitions and electron transfer processes, respectively (Reprinted by permission from Macmillan Publishers Ltd: Nature Materials (Ref. [21]), copyright (2012))

In the following, the role of luminescent nanoparticles in therapy treatments will be described particularly for the cases of self-activated photodynamic therapy and thermal therapy, including temperature monitoring.

18.3.1 Self-Activated Photodynamic Therapy

Photodynamic therapy (PDT) is an alternative tumor-ablative and function-sparing oncologic intervention. Since its inception and its first modern demonstration [28] PDT has undergone extensive investigations and has emerged as a site-specific disease treatment modality. For instance, PDT is used against bladder cancers; esophageal and gastric cancers; brain cancers; both primary site and metastatic breast cancers; skin cancers; gynecological malignancies; colorectal cancers; thoracic malignancies; oral, head, and neck tumors [29]. PDT involves three essential components: light, oxygen, and a phototosensitizer (PS). Photosensitizers, often pharmacologically inactive without illumination, can be activated by light of a specific wavelength [30, 31] as depicted in Fig. 18.8. The PS absorbs a light photon and becomes activated from the ground state to a short-lived excited singlet state (S_1^*). The excited PS can undergo intersystem crossing whereby the spin of its excited electron inverts to form a relatively long-lived triplet state (T_1^*).

The triplet excited PS can then directly interact with a substrate, for instance the cell membrane, and transfer a proton or an electron to form a radical anion or

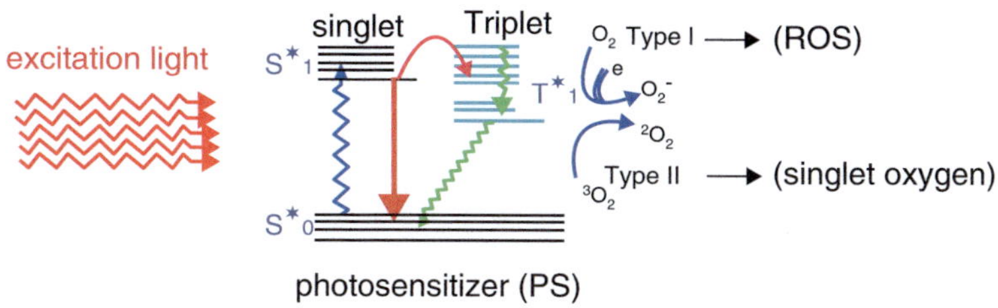

Fig. 18.8 Scheme of the production of singlet oxygen and radicals by light excited porphyrin (Reproduced from Ref. [31] with permission of the Royal Society of Chemistry)

cation, respectively, which then reacts with oxygen (Type I reaction) to produce oxygenated products such as superoxide anion radicals, hydroxyl radicals, and hydrogen peroxides (reactive oxygen species or ROS). These ROS have been shown to destroy tumors by direct cytotoxicity towards tumor cells and producing necrosis. Alternatively, the PS in the triplet state can react subsequently with molecular oxygen ($^3O_2$) *via* energy transfer to produce singlet oxygen $^1O_2$ (Type II reaction), a highly reactive and transient molecular species. This last reaction is considered to be more effective and it plays a dominant role in PDT. The products formed as a result of the reactions are responsible for the cell killing and therapeutic effect in PDT [32].

PDT has specific merits compared to the conventional treatment methods due to its minimal invasiveness, repeatability without cumulative toxicity, excellent functional results, and improved quality of life of the patients. However, the effectiveness of photodynamic therapy is largely determined by the efficiency of singlet oxygen production. Many factors influence how efficiently singlet oxygen is generated in a PDT process, including the photosensitizer used, light intensity and wavelength, and oxygen concentration. Porphyrins, chlorines, and bacteriochlorins are among the most useful photosensitizers, and also quantum dots have been recently considered as promising new generation photosensitizers [33].

Despite the widespread and rapidly growing applications, PDT has yet to gain clinical acceptance as a first-line oncological intervention due to certain limitations. Although it has the potential to be a stand-alone modality for the treatment and management of cancers at different stages, currently it is only being used in the treatment of superficial and flat lesions, that are accessible through endoscopes or as a surgical adjuvant. It is considered challenging or impossible for conventional PDT to treat deep or large volume tumors. This is primarily caused by the tissue penetration depth limit of the visible light required by most of the conventional PS for its activation as shown in Fig. 18.1. Thus, effectiveness of PDT greatly decreases with tissue thickness due to strong attenuation of visible light with increasing tissue depth, leading to incomplete treatment and tumor relapse.

As explained above, for a deeper penetration of light into tissues, wavelength of the irradiated beam should be in the NIR region between 700 and 1300 nm.

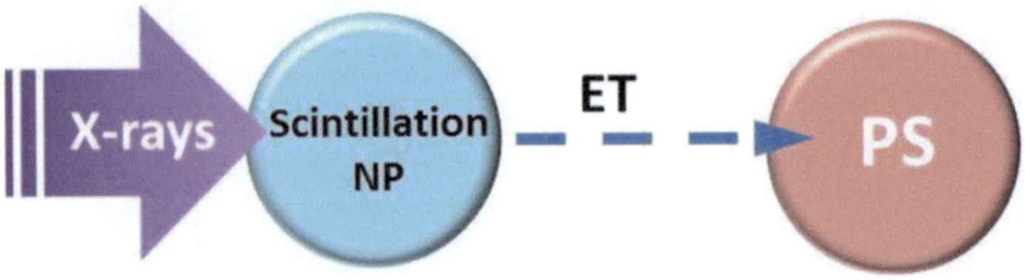

Fig. 18.9 Outline of the X-ray photoactivated PDT therapy mechanism. The scintillating nanoparticle transfers the energy absorbed from the incident ionizing radiation to the PDT photosensitizer, without limits on the excitation penetration depth in biological tissues

Recently, there have been exciting developments of novel PDT derivatives, such as two-photon PDT or up-conversion nanoparticle mediated PDT, which aim at minimizing tissue interference and improve penetration depth. Although the light penetration depth can be improved to some extent in this way, it is still limited to not more than 15 mm [34].

A potential solution for this major issue is the use of scintillating nanoparticles that can be exploited to activate the singlet oxygen sensitizers (i.e. porphyrins). The large penetration depth of X-ray removes the limit for the application of the photodynamic therapy in localized deep tissues. The X-ray activated PDT mechanism is outlined in Fig. 18.9.

Upon excitation by ionizing radiation such as X-rays, light is emitted by the nanoparticles and it activates the photosensitizers through energy transfer to produce singlet oxygen. This newly proposed PDT is called self-lighting photodynamic therapy (SLPDT) [30]. Therefore, the radiation and photodynamic therapies are combined and occur simultaneously, leading to a more efficient tumor destruction.

Scintillating nanoparticles are useful in this process due to their intense luminescence under X-ray irradiation.

The nanoparticles selected for SLPDT must meet the following requirements:

(1) matching of the emission spectrum with the photosensitizer absorption spectrum to guarantee an efficient production of singlet oxygen. We remark that all porphyrin-derived PDT compounds approved for clinical uses have a strong absorption band near 400 nm called the Soret band. Therefore, it is clear that the most efficient excitation light for photodynamic therapy lies in the UV-blue spectral region. Other weak satellite absorption peaks (Q-bands) between 500 nm and 800 nm can also be considered;
(2) high scintillation efficiency;
(3) possibility to be linked to photosensitizers;
(4) water solubility, stability in biological environments, absence of toxicity.

X-ray excited optical luminescence of rare earths is presently been considered for biological purposes. RE-doped fluoride or oxide nanoparticles such as LaF_3:Tb/Ce [34, 35] and Tb_2O_3 or Gd_2O_3 [36, 37] have been proposed as scintillators in PDT.

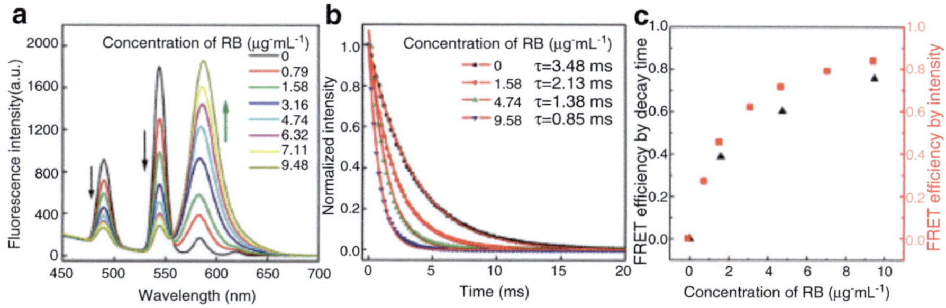

Fig. 18.10 Fluorescence analysis of LaF_3:Tb-photosensitizers nanocomposites. (**a**) Successive change of emission spectra by increasing Rose Bengal (RB) loading amount; (**b**) fluorescence decay curves of nanocomposites monitored at 544 nm by increasing RB loading amount; (**c**) Forster resonant energy transfer (FRET) efficiency as a function of RB concentration (Reprinted with permission from Ref. [34]. Copyright (2015) American Chemical Society)

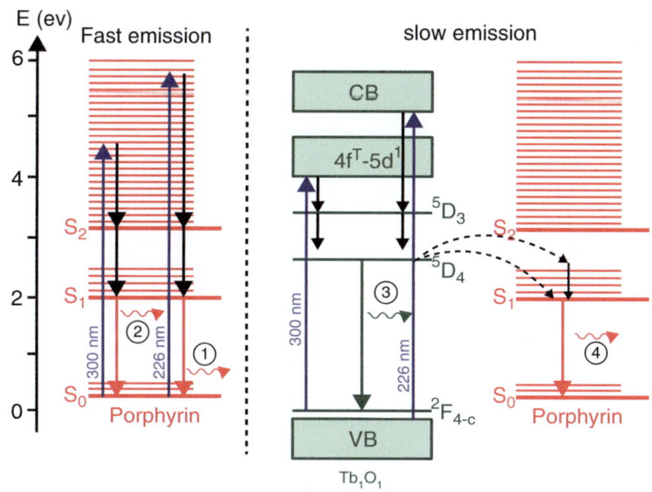

Fig. 18.11 The *left side* of the scheme represents direct excitation of porphyrin by 226 or 300 nm pulsed laser radiation. For both excitations, porphyrin is excited in highly energetic vibrational levels and relaxes in a nonradiative way to the S_2 and then the S_1 level. From S_1 to S_0, a radiative de-excitation leads to visible emissions (1) and (2). The *right* part of the Figure represents the transfer between Tb_2O_3@SiO_2 and porphyrin. Under a 300 nm radiation, Tb is excited from $4f_8$ to $4f_7-5d_1$ band. It relaxes non radiatively to the $^5D_4$ level. From this level, Tb can either relax radiatively, emitting visible light (3), or transfer energy to porphyrin, leading to a red emission (4) (Reprinted with permission from Ref. [36]. Copyright (2013) American Chemical Society)

Besides, compounds like ZnS:Cu,Co have been also studied [38]. Figure 18.10 reports spectroscopic data demonstrating energy transfer from Tb^{3+} in LaF_3:Tb and Rose Bengal used as photosensitizer [34]; Fig. 18.11 depicts the energy transfer mechanism between Tb_2O_3 and porphyrin [36].

18.3.2 Photothermal Therapy and Temperature Monitoring

The interest towards the use of heating to treat tumors arose rather long time ago as a consequence of different clinical observations. More recently an increasing number of investigations are conducted in this field due to the need of exploring novel therapeutic techniques alternatives to surgery and chemotherapy.

Indeed, it was found in several cases that a relatively light and controlled heating of the tumor mass can lead to killing of tumor cells. The reasons for such a positive effect are not completely understood and they are the subject of intense investigations. The range of temperatures for an effective therapeutic effect is rather limited, from approximately 38 to 60 °C. According to the specific temperature interval employed and the effect obtained such treatments are called "diathermia" (38–41 °C) or "hyperthermia" (41–48 °C). Above 48 °C, thermal treatments cause also irreversible injury to surrounding tissues with several collateral effects. Thermal treatments are often coupled to other therapies.

In all cases, a controlled and very localized heating of the tumor is needed. In the past, external heating procedures were employed using for example radio-frequencies, microwaves or ultrasounds. Nowadays thanks to the advent of nanotechnology, it seems possible to direct heating to the specific target by using appropriately functionalized nanoparticles. Besides magnetic heating nanoparticles, also luminescent ones are considered good candidates for this purpose. This is true especially for those systems in which the excitation energy lies in the transparency biological regions, in order to allow deep and intense tissue excitation by infrared laser beams.

For thermal therapy with rare-earth doped nanoparticles, the mechanism exploited is just the one complementary to radiative de-excitation, i.e. non radiative losses that generate heat. Nanoparticles with a low quantum yield, that are discarded for imaging purposes, can in principle find application in this field.

To the purpose of effective heating, the nanoparticles and their doping levels are designed in order to display strongly phonon assisted transitions, and to enhance all those effects that induce thermal losses like energy migration at high concentrations, or cross-relaxation. A very rich amount of papers recently appeared on this topic [39]. A significant example (Fig. 18.12) is represented by the case of $NaYF_4$:Nd allowing both hyperthermia and *in vivo* imaging functions [40].

Moreover, a crucial problem in thermal therapy is the direct and local temperature measurement that allows a good control of the treatment. It was recently demonstrated that this further requirement can be fulfilled by the analysis of luminescence spectral shapes of Nd in fluoride hosts [41–43] since the nanoparticles emission spectrum is strongly sensitive to temperature. The case of LaF_3:Nd is depicted in Fig. 18.13. By this way it is possible to simultaneously perform the hyperthermia treatment and read the local temperature.

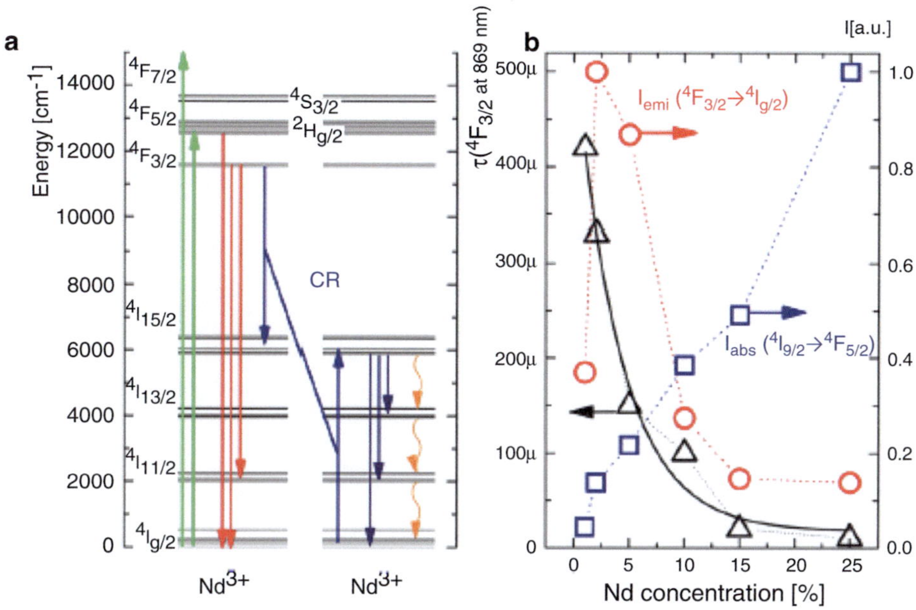

Fig. 18.12 (**a**) A diagram of Nd^{3+} energy levels. The *arrows* indicate excitation (*green*), luminescence (*red*), depopulating cross relaxation (*blue*), IR emission (*navy blue*) and nonradiative de-excitation (*orange wave arrows*). (**b**) A comparison of Nd^{3+} concentration dependent $^4F_{3/2}$ luminescence lifetimes (*black*), normalized relative to $^4F_{3/2} \rightarrow {}^4I_{9/2}$ emission intensity (*red*) and integrated absorption of $^4I_{9/2} \rightarrow {}^4F_{5/2}$ (*blue*) (Reproduced from Ref. [40] with permission from Springer)

18.4 Conclusions

The role of inorganic luminescent nanoparticles in medicine, for some selected imaging modalities and therapy treatments, has been described. This lecture gives only a partial view of the numerous and very wide applications that nanoparticles of many kinds, organic and inorganic, can have in medicine. These include among all for example drug delivery, not treated here. In any case, the examples reported testify the very vivid character of the present research. Nanomedicine proves to be a challenging field for spectroscopists who desire to explore a really interdisciplinary topic that requires a tight collaboration between scientists of different areas.

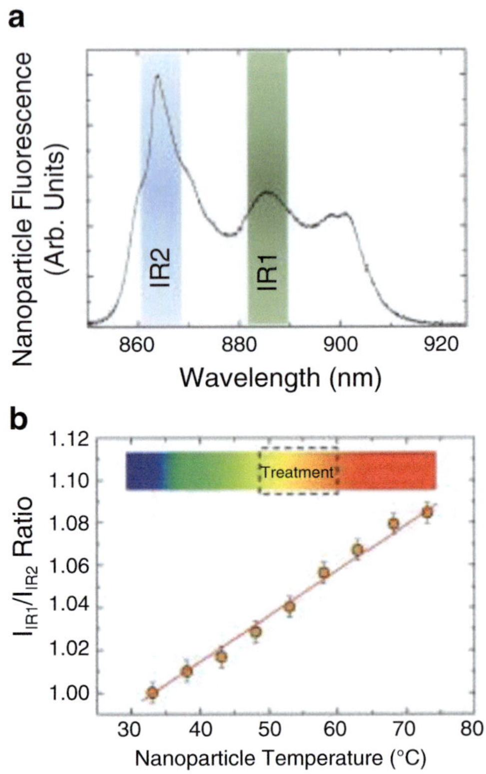

Fig. 18.13 (**a**) Near infrared (850–930 nm) room temperature emission spectrum of heavily (5.6 at%) doped Nd:LaF_3 NPs. The two spectral ranges used for ratiometric fluorescence thermal sensing (IR1 and IR2) are schematically indicated. (**b**) Temperature dependence of the IR1/IR2 intensity ratio. *Dots* are experimental data and solid line is the best linear fit. The temperature range required for efficient photo-thermal treatment of tumors is schematically indicated (Reprinted with permission from Ref. [43]. Copyright (2013) John Wiley and Sons)

Acknowledgements The authors are very grateful to Professors José Garcia Solé, Daniel Jaque, M. Bettinelli, and A. Speghini for the fruitful collaboration and discussions.

References

1. Li, C. (2014). A targeted approach to cancer imaging and therapy. *Nature Materials, 13*, 110.
2. Fahi, G. (1992). *Nanotechnology in medicine*. U.S. Pharmacopeial Convention, https://www.foresight.org/Updates/Update16/Update16.1.html#anchor576239
3. Cheng, C. J., Tietjen, G. T., Saucier-Sawyer, J. K., & Saltzman, W. M. (2015). A holistic approach to targeting disease with polymeric nanoparticles. *Nature Reviews, 14*, 239.
4. Cheng, Z., Al Zaki, A., Hui, J. Z., Muzykantov, V. R., & Tsourkas, A. (2012). Multifunctional nanoparticles: Cost versus benefit of adding targeting and imaging capabilities. *Science, 38*, 903.
5. Daniel, M.-C., & Astruc, D. (2004). Gold nanoparticles: Assembly, supramolecular chemistry, quantum-size-related properties, and applications toward biology, catalysis, and nanotechnology. *Chemical Reviews, 104*, 293.
6. Eustis, S., & El-Sayed, M. A. (2006). Why gold nanoparticles are more precious than pretty gold: Noble metal surface Plasmon resonance and its enhancement of the radiative and nonradiative properties of nanocrystals of different shapes. *Chemical Society Reviews, 35*, 209.

7. Park, J.-H., Gu, L., von Maltzahn, G., Ruoslahti, E., Bhatia, S. N., & Sailor, M. J. (2009). Biodegradable luminescent porous silicon nanoparticles for in vivo applications. *Nature Materials, 8*(331).
8. Welsher, K., Sherlock, S. P., & Dai, H. (2011). Deep-tissue anatomical imaging of mice using carbon nanotube fluorophores in the second near-infrared window. *PNAS, 108*, 8943.
9. Weissleder, R., Nahrendorfand, M., & Pittet, M. J. (2014). Imaging macrophages with nanoparticles. *Nature Materials, 13*, 125.
10. Hilderbrandt, S. A., & Weissleder, S. (2010). Near-infrared fluorescence: application to in vivo molecular imaging. *Current Opinion in Chemical Biology, 14*, 71.
11. Jacques, S. L. (2013). Optical properties of biological tissues: a review. *Physics in Medicine and Biology, 58*, R37.
12. Pansare, V. J., Hejazi, S., Faenza, W. J., & Prud'homme, R. K. (2012). Review of long-wavelength optical and NIR imaging materials: Contrast agents, fluorophores, and multifunctional nano carriers. *Chemistry of Materials, 24*, 812.
13. Naczynski, D. J., Tan, M. C., Zevon, M., Wall, B., Kohl, J., Kulesa, A., Chen, S., Roth, C. M., Riman, R. E., & Moghe, P. V. (2013). Rare-earth-doped biological composites as in vivo shortwave infrared reporters. *Nature Communications, 4*, 2199. doi:10.1038/ncomms3199.
14. Villa, I., Vedda, A., Xochilt Cantarelli, I., Pedroni, M., Piccinelli, F., Bettinelli, M., Speghini, A., Quintanilla, M., Vetrone, F., Rocha, U., Jacinto, C., Carrasco, E., Sanz Rodríguez, F., Juarranz de la Cruz, Á., Haro Gonzalez, P., García Solé, J., & Jaque García, D. (2015). 1.3 μm emitting $SrF_2:Nd^{3+}$ nanoparticles for high contrast in vivo imaging in the second biological window. *Nano Research, 8*, 649.
15. Mc Keever, S. W. S. (1985). *Thermoluminescence of solids*. Cambridge: Cambridge University Press.
16. Avouris, P., & Morgan, T. N. (1981). A tunneling model for the decay of luminescence in inorganic phosphors: The case of Zn_2SiO_4:Mn. *Journal of Chemical Physics, 74*, 4347.
17. Caratto, V., Locardi, F., Costa, G. A., Masini, R., Fasoli, M., Panzeri, L., Martini, M., Bottinelli, E., Gianotti, E., & Miletto, I. (2014). NIR persistent luminescence of lanthanide Ion-doped rare-earth oxycarbonates: The effect of dopants. *ACS Applied Materials and Interfaces, 6*, 17346.
18. Lecointre, A., Viana, B., LeMasne, Q., Bessière, A., Chaneac, C., & Gourier, D. (2009). Red long-lasting luminescence in clinoenstatite. *Journal of Luminescence, 129*, 1527.
19. Lecointre, A., Bessiere, A., Viana, B., & Gourier, D. (2010). Red persistent luminescent silicate nanoparticles. *Radiation Measurements, 45*, 497.
20. Maldiney, T., Lecointre, A., Viana, B., Bessièere, A., Bessodes, M., Gourier, D., Richard, C., & Scherman, D. (2011). Controlling electron trap depth to enhance optical properties of persistent luminescence nanoparticles for in vivo imaging. *Journal of the American Chemical Society, 133*, 11810.
21. Pan, Z., Lu, Y.-Y., & Liu, F. (2012). Sunlight-activated long-persistent luminescence in the near-infrared from Cr^{3+}-doped zinc gallogermanates. *Nature Materials, 11*, 58.
22. le Masne de Chermont, Q., Chanéac, C., Seguin, J., Pellé, F., Maîtrejean, S., Jolivet, J.-P., Gourier, D., Bessodes, M., & Scherman, D. (2007). Nanoprobes with near-infrared persistent luminescence for in vivo imaging. *PNAS, 104*, 9266.
23. Singh, S. K. (2014). Red and near infrared persistent luminescence nano-probes for bioimaging and targeting applications. *RSC Advances, 4*, 58674.
24. Sun, M., Li, Z.-J., Liu, C.-L., Fu, H.-X., Shen, J.-S., & Zhang, H.-W. (2014). Persistent luminescent nanoparticles for super-longtime in vivo and in situ imaging with repeatable excitation. *Journal of Luminescence, 145*, 838.
25. Ueda, J., Shinoda, T., & Tanabe, S. (2013). Photochromism and near-infrared persistent luminescence in Eu^{2+}-Nd^{3+}-co-doped $CaAl_2O_4$ Ceramics. *Optical Materials Experimental, 3*, 787.
26. Wang, X.-J., Jia, D., & Yen, W. M. (2003). Mn^{2+} activated green, yellow, and red long persistent phosphors. *Journal of Luminescence, 102–103*, 34.

27. Maldiney, T., Bessière, A., Seguin, J., Teston, E., Sharma, S. K., Viana, B., Bos, A. J. J., Dorenbos, P., Bessodes, M., Gourier, D., Scherman, D., & Richard, C. (2014). Thein vivo activation of persistent nanophosphors for optical imaging of vascularization, tumors and grafted cells. *Nature Materials, 13*, 418.
28. Dougherty, T. J., Grindey, G. B., Fiel, R., Weishaupt, K. R., & Boyle, D. G. (1975). Photoradiation therapy. II. Cure of animal tumors with hematoporphyrin and light. *Journal of the Nature Cancer Institute, 55*, 115.
29. Chen, W., & Zhang, J. (2006). Using nanoparticles to enable simultaneous radiation and photodynamic therapies for cancer treatment. *Journal of Nanoscience and Nanotechnology, 6*, 1159.
30. Chen, H., Wang, G. D., Chuang, Y.-J., Zhen, Z., Chen, X., Biddinger, P., Hao, Z., Liu, F., Shen, B., Pan, Z., & Xie, J. (2015). Nanoscintillator-mediated X-ray inducible photodynamic therapy for in vivo cancer treatment. *Nano Letters, 15*, 2249.
31. St Denis, T. G., Aziz, K., Waheed, A. A., Huang, Y.-Y., Sharma, S. K., Mroz, P., & Hamblin, M. R. (2011). Combination approaches to potentiate immune response after photodynamic therapy for cancer. *Photochemical and Photobiological Sciences, 10*(5), 792.
32. Robertson, C. A., Hawkins Evans, D., & Abrahamse, H. (2009). Photodynamic therapy (PDT): A short review on cellular mechanisms and cancer research applications for PDT. *Journal of Photochemistry and Photobiology. B, 96*, 1.
33. Juzenas, P., Chen, W., Sun, Y.-P., Coelho, M. A. N., Generalov, R., Generalova, N., & Christensen, I. L. (2008). Quantum dots and nanoparticles for photodynamic and radiation therapies of cancer. *Advanced Drug Delivery Reviews, 60*, 1600.
34. Tang, Y., Hu, J., Elmenoufy, A. H., & Yang, X. (2015). Highly efficient FRET system capable of deep photodynamic therapy established on X-ray excited mesoporous LaF_3:Tb scintillating nanoparticles. *ACS Applied Materials and Interfaces, 7*, 12261.
35. Liu, Y., Chen, W., Wang, S., Joly, A. G., Westcott, S., & Woo, B. K. (2008). Investigation of water-soluble x-ray luminescence nanoparticles for photodynamic activation. *Applied Physics Letters, 92*, 043901.
36. Bulin, A.-L., et al. (2013). X-ray-induced singlet oxygen activation with nanoscintillator-coupled porphyrins. *Journal of Physical Chemistry C, 117*, 21583.
37. Bulin, A.-L., Vasil'ev, A., Belsky, A., Amans, D., Ledoux, G., & Dujardin, C. (2015). Modelling energy deposition in nanoscintillators to predict the efficiency of the X-ray-induced photodynamic effect. *Nanoscale, 7*, 5744.
38. Ma, L., Zou, X., Bui, B., Chen, W., Song, K. H., & Solberg, T. (2014). X-ray excited ZnS:Cu, Co afterglow nanoparticles for photodynamic activation. *Applied Physics Letters, 105*, 013702.
39. Jaque, D., Martınez Maestro, L., del Rosal, B., Haro-Gonzalez, P., Benayas, A., Plaza, J. L., Rodrıguez, E. M., & Solé, J. G. (2014). Nanoparticles for photothermal therapies. *Nanoscale, 6*, 9494.
40. Bednarkiewicz, A., Wawrzynczyk, D., Nyk, M., & Strek, W. (2011). Optically stimulated heating using Nd^{3+} doped $NaYF_4$ colloidal near infrared nanophosphors. *Applied Physics B, 103*, 847.
41. Wawrzynczyk, D., Bednarkiewicz, A., Nyk, M., Strek, W., & Samoc, M. (2012). Neodymium(III) doped fluoride nanoparticles as non-contact optical temperature sensors. *Nanoscale, 4*, 6959.
42. Rocha, U., da Silva, C. J., Silva, W. F., Guedes, I., Benayas, A., Maestro, L. M., Elias, M. A., Bovero, E., van Veggel, F. C. J. M., Solé, J. A. G., & Jaque, D. (2013). Subtissue thermal sensing based on neodymium-doped LaF_3 nanoparticles. *ACS Nano, 7*, 1188.
43. Carrasco, E., del Rosal, B., Sanz-Rodríguez, F., Juarranz de la Fuente, Á., Gonzalez, P. H., Rocha, U., Kumar, K. U., Jacinto, C., Solé, J. G., & Jaque, D. (2015). Intratumoral thermal reading during photo-thermal therapy by multifunctional fluorescent nanoparticles. *Advanced Functional Materials, 25*, 615.

Chapter 19
Emission Cross Section, Füchtbauer-Ladenburg Equation, and Purcell Factor

Markus Pollnau* **and Marc Eichhorn***

Abstract We review the physics underlying the process of spontaneous emission into a resonator mode. We verify the fundamental modal dimensions, present the spectral mode profile, the Füchtbauer-Ladenburg equation, and the Purcell factor. Furthermore, we obtain the relation between peak emission cross section, radiative lifetime, and emission linewidth.

19.1 Introduction

In our work [1], we review the physics underlying the process of spontaneous emission into a resonator mode. In a straight-forward manner we demonstrate that spontaneous emission can be interpreted as stimulated emission driven by the one vacuum photon per optical mode and polarization. We present the relationship between spectral mode profile, coherence time, and Q-factor and derive the fundamental modal dimensions. We then obtain the relation between the radiative lifetime and the emission cross section, called the Füchtbauer-Ladenburg equation, introduce the Purcell factor to this equation, and discuss its influence on different types of lasers. We derive the relation between peak emission cross section, radiative lifetime, and emission linewidth and show that semiconductor and solid-state laser

*Author contributed equally with all other contributors.

M. Pollnau (✉)
Department of Materials and Nano Physics, School of Information and Communication Technology, KTH—Royal Institute of Technology, Electrum 229, Isafjordsgatan 22-24, 16440 Kista, Sweden
e-mail: M.Pollnau@utwente.nl

M. Eichhorn
French-German Research Institute of Saint-Louis ISL, 5 rue du Général Cassagnou, BP 70034, 68301 Saint Louis CEDEX, France

© Springer Science+Business Media Dordrecht 2017
B. Di Bartolo et al. (eds.), *Nano-Optics: Principles Enabling Basic Research and Applications*, NATO Science for Peace and Security Series B: Physics and Biophysics, DOI 10.1007/978-94-024-0850-8_19

materials exhibit a similar product of radiative lifetime and peak emission cross section, because the atomic de-coherence of the excited state is on the same order of magnitude in both gain materials.

19.2 Vacuum Photon and Spontaneous Emission

We interpret spontaneous emission of photons from upper to lower laser level as stimulated emission driven by vacuum fluctuations. Since vacuum fluctuations create energy only within the related time uncertainty interval, they trigger only spontaneous emission [2]. Spontaneous absorption would lead to a long-lasting creation of energy, thereby violating the uncertainty principle [3, 4].

We assume that light at frequency ν propagates with speed $c = c_0/n_r(\nu)$, where c_0 is the speed of light in vacuum, in the active medium of refractive index $n_r(\nu)$, which homogeneously fills the mode volume V_{mode} of the resonator. With the wavenumber k and wavelength λ in the active medium,

$$\frac{k}{2\pi} = \frac{1}{\lambda} = \frac{\nu}{c}, \tag{19.1}$$

when taking into account that in a three-dimensional hohlraum resonator each mode extends over the round-trip length, i.e., forward (k) and backward ($-k$) propagating waves belong to the same mode, and the same mode volume V_{mode} is shared by two modes with orthogonal polarizations, in free space the mode density $M(\nu)$ per unit volume up to the frequency ν and the spectral mode density $\tilde{M}(\nu)$ are derived by the volume integral over the whole k-sphere:

$$M(\nu) = 2\int_0^k\int_0^\pi\int_0^{2\pi} \frac{d^3k'}{(2\pi)^3} = 2\int_0^\nu\int_0^\pi\int_0^{2\pi} \frac{d^3\nu'}{c^3} = \frac{4\pi}{c^3}\int_0^\nu\int_0^\pi \nu'^2 \sin(\vartheta)\, d\vartheta\, d\nu' = \frac{8\pi\nu^3}{3c^3}$$
$$\Rightarrow \tilde{M}(\nu) = \frac{dM(\nu)}{d\nu} = \frac{8\pi\nu^2}{c^3}. \tag{19.2}$$

In this work, all parameters that are defined per frequency interval, i.e., in units of the parameter divided by the frequency unit, are denoted by a tilde.

Exploiting Einstein's rate-equation description [2] of blackbody radiation, the radiative lifetime $\tau_{21,rad}$ and rate constant or Einstein coefficient A_{21} of spontaneous emission are connected to the Einstein coefficient B_{21} of stimulated emission by

$$\frac{1}{\tau_{21,rad}} = A_{21} = \frac{8\pi h\nu^3}{c^3} B_{21} = \tilde{M}(\nu)\, h\nu B_{21}, \tag{19.3}$$

where h is Planck's constant. Interpreted as stimulated emission triggered by the vacuum spectral energy density $\tilde{u}_0(\nu)$ generated by the number φ_0 of vacuum photons of energy $h\nu$ per mode,

$$\tilde{u}_0(\nu) = \tilde{M}(\nu)\, h\nu\varphi_0, \tag{19.4}$$

the spontaneous-emission rate constant becomes

$$\frac{1}{\tau_{21,rad}} = A_{21} = \tilde{u}_0(\nu)\, B_{21} = \tilde{M}(\nu)\, h\nu\varphi_0 B_{21}. \tag{19.5}$$

By comparison of Eqs. (19.3) and (19.5) one obtains a fundamental result of quantum electrodynamics that

$$\varphi_0 = 1 \tag{19.6}$$

vacuum photon is present per mode as a result of vacuum fluctuations. In the present derivation, this result obtains from the information contained in Einstein's Eq. (19.3), see e.g. Ref. [5]. The spectrum of blackbody radiation obeys Planck's equation [6], because Eq. (19.6) holds true.

19.3 Resonator Modes

In this section, we will introduce the spectral mode profile and derive the fundamental modal dimensions. For an optical resonator of geometrical length ℓ and the Gaussian TEM_{00} beam with beam waist w_0 as the resonator mode, the mode volume over which this resonator mode is distributed is

$$V_{mode} = \ell\pi w_0^2. \tag{19.7}$$

All energy densities, the population densities N_2 and N_1 of upper and lower laser level, respectively, as well as all rates considered in this work are averaged over the mode volume.

19.3.1 Spectral Mode Profile, Coherence Time, and Q-Factor

A number $\varphi(t_0)$ of photons at frequency ν_0 that are present inside a resonator at the initial time t_0 decay out of the resonator with time t as a result of resonator losses which are quantified by the photon decay time τ_c (where the index c stands for "cavity", which means the resonator):

$$\varphi(t) = \varphi(t_0)\, e^{-t/\tau_c} \quad \text{for} \quad t \geq t_0 = 0. \tag{19.8}$$

Fourier transformation of the amplitude of this exponential decay to frequency space and calculation of its absolute value squared,

$$\tilde{\gamma}_c(\nu) = \frac{1}{\tau_c}\left|\frac{1}{\sqrt{\varphi(t_0)}}\int_{t_0}^{\infty}\sqrt{\varphi(t)}e^{-i2\pi(\nu-\nu_c)t}dt\right|^2$$

$$= \frac{1}{\tau_c}\left|\frac{2}{\pi}\frac{1}{i4(\nu-\nu_c)+(\pi\tau_c)^{-1}}\right|^2 = \frac{1}{\pi^2\tau_c}\frac{1}{4(\nu-\nu_c)^2+(2\pi\tau_c)^{-2}}, \quad (19.9)$$

provides the spectral intensity distribution of this emission as a function of frequency ν. It has a normalized Lorentzian spectral profile

$$\tilde{\gamma}_c(\nu) = \frac{2}{\pi}\frac{\Delta\nu_c}{4(\nu-\nu_c)^2+\Delta\nu_c^2}, \quad \text{with} \quad \int \tilde{\gamma}_c(\nu)\,d\nu = 1, \quad (19.10)$$

which is centered at frequency ν_c and of full-width-at-half-maximum (FWHM) linewidth

$$\Delta\nu_c = \frac{1}{2\pi\tau_c} = \frac{1}{\pi\tau_c^{coh}}. \quad (19.11)$$

The photon decay time τ_c is half the coherence time τ_c^{coh} of light emitted from the resonator, which, unlike an exponential decay, is defined over the interval from $t = -\infty$ to $t = \infty$ [7].

The quality of an optical resonator is expressed by its intrinsic Q-factor, originally introduced as the "coil dissipation constant" of a resonant electric circuit [8] and later generalized [9] as the energy stored in the resonator, E_{stored}, divided by the energy lost per oscillation cycle, E_{lost},

$$Q_c := 2\pi\frac{E_{stored}(t)}{E_{lost}(t)} = 2\pi\frac{\varphi(t)}{-\frac{1}{\nu_c}\frac{d}{dt}\varphi(t)} = 2\pi\nu_c\tau_c, \quad (19.12)$$

where $E = h\nu_c\varphi$. Equations (19.12) and (19.11) relate the Q-factor and photon decay time to the linewidth, which can be measured by injecting continuous-wave (cw) light into the passive resonator, thus establishing a steady state in which the injected energy $E_{injected} = E_{lost} = E_{stored}/(\nu_c\tau_c)$, and observing its spectral transmission:

$$Q_c = 2\pi\left.\frac{E_{stored}}{E_{injected}}\right|_{cw} = 2\pi\left.\frac{\varphi}{-\frac{1}{\nu_c}\frac{\delta\varphi}{\delta t}}\right|_{cw} = 2\pi\nu_c\tau_c = \frac{\nu_c}{\Delta\nu_c}. \quad (19.13)$$

$\delta\varphi$ is the photon number lost during the time interval δt and replaced by injecting photons. The injection must take place via an existing loss channel of the resonator (e.g. one of the resonator mirrors) in order not to alter its Q-factor, and the injected signal must be optically isolated from its source to avoid feedback between the light-source resonator and the resonator under investigation.

Fourier transformation according to Eq. (19.9) and its Fourier back-transformation unambiguously transform the amplitudes of an exponential decay in the time domain and a Lorentzian line function in the frequency domain into each other, thereby relating the exponential decay time to the Lorentzian linewidth according to Eq. (19.11), i.e., the underlying physical process is independent of its description in frequency or time domain. Hence, the right-hand side of Eq. (19.13) serves as an equivalent definition of the Q-factor. In other words, the Q-factor of a passive resonator does not change with the way we feed light into it or the way we observe the outcoupled light. The Q-factor solely depends on the resonator losses.

Accordingly, in a passive Fabry-Pérot resonator [10] with a free spectral range of

$$\Delta\nu_{FSR} = \frac{c}{2\ell} = \frac{1}{t_{RT}}, \tag{19.14}$$

where t_{RT} is the photon round-trip time in the resonator, each transverse-fundamental resonator mode with its longitudinal-mode index q, centered at frequency ν_q, where $\nu_0 = c/\lambda_0 = c/(2\ell)$ represents the lowest frequency, has an individual photon decay time of $\tau_q(\nu_q)$ given by

$$\frac{1}{\tau_q(\nu_q)} = -\frac{c}{2\ell} \ln\left\{\left[1 - L_{RT}(\nu_q)\right]\left[1 - T_{out}(\nu_q)\right]\right\}, \tag{19.15}$$

where $L_{RT}(\nu_q)$ and $T_{out}(\nu_q)$ quantify the intrinsic round-trip losses and the out-coupling losses at the frequency ν_q, respectively. τ_q results in a Q-factor of Q_q according to Eqs. (19.12) or (19.13), a linewidth of $\Delta\nu_q$ according to Eq. (19.11), and a finesse of

$$\mathscr{F}(\nu_q) = \frac{\Delta\nu_{FSR}}{\Delta\nu_q} = \frac{\Delta\nu_{FSR}}{\nu_q} Q_q = \frac{\lambda_q}{2\ell} Q_q = \frac{Q_q}{q+1}. \tag{19.16}$$

Each mode exhibits a normalized Lorentzian spectral profile equivalent to Eq. (19.10),

$$\tilde{\gamma}_q(\nu) = \frac{2}{\pi} \frac{\Delta\nu_q}{4(\nu - \nu_q)^2 + \Delta\nu_q^2}. \tag{19.17}$$

The sum over all longitudinal modes results in a spectral Airy function [11, 12]. The non-averaged spectral mode density of a single resonator mode with longitudinal-mode index q is given by

$$\tilde{M}_q(\nu) = \frac{\tilde{\gamma}_q(\nu)}{V_{mode}}. \tag{19.18}$$

In case the Lorentzian line $\tilde{\gamma}_q(\nu)$ is narrow compared to the free spectral range of the resonator, the spectral mode density averaged over the free spectral range obtains as

$$\tilde{M}_{\text{mode}}(\nu) \approx \frac{1}{\Delta\nu_{FSR}} \int_{\nu_q-\Delta\nu_{FSR}/2}^{\nu_q+\Delta\nu_{FSR}/2} \tilde{M}_q(\nu)\, d\nu$$
$$= \frac{1}{\Delta\nu_{FSR}} \int_{\nu_q-\Delta\nu_{FSR}/2}^{\nu_q+\Delta\nu_{FSR}/2} \frac{\tilde{\gamma}_q}{V_{mode}} d\nu \approx \frac{1}{\Delta\nu_{FSR}} \frac{1}{V_{mode}}. \tag{19.19}$$

In average, there exists one mode, hence also one vacuum photon φ_0, per polarization state and free-spectral range $\Delta\nu_{FSR}$. Whereas $\tilde{M}_{\text{mode}}(\nu)$ is only approximately equal to the finite integral in Eq. (19.19), which cuts off the wings of the Lorentzian profile $\tilde{\gamma}_q(\nu)$ of the single resonator mode of Eq. (19.18), the missing amount is exactly compensated by the wings of other resonator modes that reach into this free spectral range, hence the very left-hand and right-hand terms of Eq. (19.19) are equal to each other. Consideration of the solid angle of the resonator modes,

$$\Delta\Omega_{\text{mode}} = 2 \times 2\pi\,[1 - \cos(\theta)] \approx 4\pi\left[1 - \left(1 - \frac{1}{2}\theta^2\right)\right]$$
$$= 2\pi\theta^2 = 2\frac{\lambda^2}{\pi w_0^2} = 2\frac{c^2}{\nu^2}\frac{1}{\pi w_0^2}, \tag{19.20}$$

where θ is the divergence angle of the Gaussian beam and the factor of 2 takes both propagation directions into account, and executing the volume integral of Eq. (19.2) only over this solid angle,

$$M_{\text{mode}} = \int_0^{\nu} \int_{\Delta\Omega_{\text{mode}}} \frac{d^3\nu'}{c^3} = \int_0^{\nu} \int_{\Delta\Omega_{\text{mode}}} \frac{\nu'^2 d\Omega d\nu'}{c^3}$$
$$= \int_0^{\nu} \frac{\nu'^2 \Delta\Omega_{\text{mode}} d\nu'}{c^3} = 2\frac{\nu}{c}\frac{1}{\pi w_0^2}, \tag{19.21}$$

results in the same spectral mode density,

$$\tilde{M}_{\text{mode}}(\nu) = \frac{dM_{\text{mode}}}{d\nu} = \frac{2}{c}\frac{1}{\pi w_0^2} = \frac{2\ell}{c}\frac{1}{\pi w_0^2 \ell} = \frac{1}{\Delta\nu_{FSR}}\frac{1}{V_{\text{mode}}}, \tag{19.22}$$

thereby proving that in Eq. (19.19) the very left-hand and right-hand terms are indeed equal.

19.3.2 Fundamental Modal Dimensions

When taking into account that in a one-dimensional linear resonator of length ℓ each mode extends over the round-trip length 2ℓ, i.e., forward (k) and backward ($-k$) propagating waves belong to the same mode, and the same mode volume is shared by two modes with orthogonal polarizations, the one-dimensional mode density $M^{1D}(\nu)$ per unit length up to a frequency ν and the one-dimensional spectral mode density $\tilde{M}^{1D}(\nu)$ along the resonator direction are given by

$$M^{1D}(\nu) = 2\int_{-k}^{k} \frac{dk'}{2\pi} = 2\int_{0}^{\nu} \frac{2d\nu'}{c} = \frac{4\nu}{c}$$

$$\Rightarrow \quad \tilde{M}^{1D}(\nu) = \frac{dM^{1D}(\nu)}{d\nu} = \frac{4}{c} = \frac{2}{\Delta\nu_{FSR}}\frac{1}{\ell}. \tag{19.23}$$

Although the volume integral in Eq. (19.2) ranges only from 0 to k, the integrated sphere includes negative and positive k values in all three dimensions. To cover the equivalent k-space, the one-dimensional integral in Eq. (19.23) must range from $-k$ to k. By inserting Eq. (19.14) to the right-hand side of Eq. (19.23), we find that two modes with orthogonal polarizations exist per free spectral range of the resonator of length ℓ, in agreement with Eq. (19.19). When considering the frequency range up to a maximum frequency ν, for transverse-fundamental-mode operation the three-dimensional spectral mode density inside the resonator amounts to

$$\frac{\tilde{M}^{1D}(\nu)}{\pi w_M^2} = \frac{4}{c}\frac{1}{\pi w_M^2} = \frac{2}{\Delta\nu_{FSR}}\frac{1}{\ell}\frac{1}{\pi w_M^2} = \tilde{M}(\nu) = \frac{8\pi\nu^2}{c^3}, \tag{19.24}$$

allowing us to determine the fundamental modal area πw_M^2 and radius w_M that are, in average, occupied by two resonator modes with orthogonal polarizations:

$$\pi w_M^2 = \frac{\tilde{M}^{1D}(\nu)}{\tilde{M}(\nu)} = \frac{c^2}{2\pi\nu^2} = \frac{\lambda^2}{2\pi} \quad \Rightarrow \quad w_M = \frac{\lambda}{\sqrt{2}\pi}. \tag{19.25}$$

Inserting this result into Eq. (19.20) yields

$$\frac{\Delta\Omega_{\text{mode}}}{4\pi} = \frac{\pi w_M^2}{\pi w_0^2}, \tag{19.26}$$

i.e., a mode confined to a beam waist w_0 that equals the fundamental modal radius w_M leads to divergence of the light propagation, in both directions, into the whole sphere, whereas a larger beam waist results in an accordingly smaller divergence.

The fundamental modal length that allows these two modes with orthogonal polarizations to exist inside the linear resonator equals one half sine cycle of light,

$$\ell_M = \frac{\lambda}{2}. \tag{19.27}$$

Physically interpreted, it encompasses one complete sine cycle over the round-trip length 2ℓ, leading to constructive interference of the wave with itself. As expected, with Eq. (19.14) the fundamental modal length results in a fundamental free spectral range of

$$\Delta\nu_{FSR,M} = \frac{c}{2\ell_M} = \nu. \tag{19.28}$$

The fundamental modal volume V_M that two linear-resonator modes with orthogonal polarizations would occupy if this volume did not spatially overlap with the volumes of any other modes then obtains with Eqs. (19.25) and (19.27) as

$$V_M = \ell_M \pi w_M^2 = \frac{\lambda^3}{4\pi}. \tag{19.29}$$

This result for the fundamental modal volume is independent of the existence of a resonator, hence is the same as directly derived from the mode density in free space of Eq. (19.2) and the fundamental free spectral range of Eq. (19.28) by use of Eq. (19.1):

$$V_M = \frac{1}{\Delta\nu_{FSR,M}} \frac{2}{\tilde{M}(\nu)} = \frac{\lambda^3}{4\pi}. \tag{19.30}$$

If the volume V_{mode} is smaller than the fundamental modal volume V_M of Eq. (19.29), no resonator mode exists up to the maximum frequency ν. If V_{mode} equals $V_M(\nu)$, two orthogonal modes exist. If V_{mode} is larger than this limit, the number of modes and, hence, also the number of vacuum photons that, in average, share the volume V_{mode} of the lasing resonator mode increases stepwise to

$$\varphi_0(V_{mode}) = 2\frac{V_{mode}}{V_M(\nu)}. \tag{19.31}$$

These vacuum photons have largely different frequencies.

19.4 Transition Cross Section and Emission Rates

The strengths of emission from the upper and absorption from the lower level of a transition can be described in terms of the effective emission and absorption cross sections $\sigma_e(\nu)$ and $\sigma_a(\nu)$, respectively, which are defined via the interaction of light at frequency ν with spectral intensity $\tilde{I}(\nu, z)$ propagating along the z direction and interacting with a gain medium:

$$\frac{d\tilde{I}(\nu,z)}{dz} = \sigma_e(\nu) N_2(z)\tilde{I}(\nu,z) \quad \text{and}$$
$$\frac{d\tilde{I}(\nu,z)}{dz} = -\sigma_a(\nu) N_1(z)\tilde{I}(\nu,z). \tag{19.32}$$

In an active medium with crystal-field splitting of upper and lower electronic multiplets, $\sigma_e(\nu)/b_2 = \sigma_a(\nu)/b_1 = \sigma(\nu)$, where b_2 and b_1 are the Boltzmann factors of upper and lower crystal-field level of the specific transition at frequency ν, respectively, and $\sigma(\nu)$ is the atomic cross section of this transition (for simplicity, the degeneracies of the crystal-field levels are assumed to be equal).

19.4.1 Füchtbauer-Ladenburg Equation and Purcell Factor

Spontaneous emission that is interpreted as stimulated emission triggered by the vacuum spectral energy density $\tilde{u}_0(\nu)$, by use of Eq. (19.5), gives rise to a spectral fluorescence power $\tilde{P}_f(\nu)$ generated per unit volume V of

$$\frac{d\tilde{P}_f(\nu)}{dV} = -\frac{dN_2}{dt} h\nu\tilde{\rho}_f(\nu) = B_{21}\tilde{u}_0(\nu) N_2 h\nu\tilde{\rho}_f(\nu). \tag{19.33}$$

The spectral fluorescence distribution $\tilde{\rho}_f(\nu)$ is normalized to

$$\int \tilde{\rho}_f(\nu)\, d\nu = 1. \tag{19.34}$$

From the spectroscopic point of view, the same spectral fluorescence power $\tilde{P}_f(\nu)$ or spectral fluorescence intensity $\tilde{I}_f(\nu)$ can be regarded as a result of stimulated emission triggered by an equivalent plane-wave spectral vacuum photon intensity $\tilde{I}_0(\nu)$ generated per unit length z, penetrating the area A_{xy}, which, by use of Eq. (19.32), can be expressed as

$$\frac{d\tilde{P}_f(\nu)}{dV} = \frac{\partial\tilde{P}_f(\nu)}{\partial A_{xy}\partial z} = \frac{d\tilde{I}_f(\nu)}{dz}$$
$$= \sigma_e(\nu) N_2\tilde{I}_0(\nu) = c\sigma_e(\nu) N_2\tilde{u}_0(\nu). \tag{19.35}$$

Equality of Eqs. (19.33) and (19.35) yields

$$\sigma_e(\nu) = \frac{h\nu}{c} B_{21}\tilde{\rho}_f(\nu). \tag{19.36}$$

With Eq. (19.3), the emission cross section obtains as

$$\sigma_e(\nu) = \frac{1}{c}\frac{1}{\tau_{21,rad}}\frac{1}{\tilde{M}(\nu)}\tilde{\rho}_f(\nu). \tag{19.37}$$

Exploiting Eqs. (19.34) and (19.2) results in the Füchtbauer-Ladenburg equation in free space, which relates the effective emission cross section $\sigma_e(\nu)$ to the radiative lifetime $\tau_{21,rad}$:

$$\begin{aligned}
&\frac{1}{\tau_{21,rad}}\tilde{\rho}_f(\nu) = c\sigma_e(\nu)\tilde{M}(\nu) \\
&\quad\Rightarrow \int \frac{1}{\tau_{21,rad}}\tilde{\rho}_f(\nu)\,d\nu = \int c\sigma_e(\nu)\tilde{M}(\nu)\,d\nu \\
&\quad\Rightarrow \frac{1}{\tau_{21,rad}} = A_{21} = \int c\sigma_e(\nu)\tilde{M}(\nu)\,d\nu = \frac{8\pi}{c^2}\int \sigma_e(\nu)\nu^2 d\nu.
\end{aligned} \tag{19.38}$$

Accordingly, with Eqs. (19.18), (19.2), and (19.17) the spontaneous-emission rate into a single resonator mode becomes

$$\begin{aligned}
\frac{1}{\tau_{21,rad}^{\text{mode}}} &= A_{21,rad}^{\text{mode}} = \int c\sigma_e(\nu)\tilde{M}_q(\nu)\,d\nu \\
&= \int c\sigma_e(\nu)\frac{\tilde{\gamma}_q(\nu)}{V_{mode}}d\nu = \int c\sigma_e(\nu)F_q(\nu)\tilde{M}(\nu)\,d\nu.
\end{aligned} \tag{19.39}$$

Therein,

$$\begin{aligned}
F_q(\nu) &= \frac{\tilde{M}_q(\nu)}{\tilde{M}(\nu)} = \frac{c^3}{8\pi\nu^2}\frac{\tilde{\gamma}_q(\nu)}{V_{mode}} \\
&= \frac{c^3}{4\pi^2\nu^2}\frac{1}{V_{mode}}\frac{\Delta\nu_q}{4\left(\nu-\nu_q\right)^2+\Delta\nu_q^2}
\end{aligned} \tag{19.40}$$

quantifies the difference between the spontaneous-emission rate into a single resonator mode and that into free space.

In the limit of an emitter with a very narrow linewidth, $\Delta\nu_e \to 0$, if the resonator is tuned such that the center frequency ν_q of one of its modes matches the frequency ν_e of the peak cross section, the integral in Eq. (19.39) delivers a value different from zero only at frequencies $\nu \approx \nu_q = \nu_e$, equivalent to wavelengths $\lambda \approx \lambda_q = \lambda_e$ in the active medium, hence by use of Eq. (19.13) we can approximate Eq. (19.40) as

$$F_q(\nu) \overset{\substack{\Delta\nu_e \to 0 \\ \nu = \nu_q = \nu_e}}{\approx} \frac{c^3}{4\pi^2\nu_q^2} \frac{1}{V_{mode}} \frac{1}{\Delta\nu_q} = \frac{c^3}{4\pi^2\nu_q^3} \frac{1}{V_{mode}} Q_q = F_q(\nu_q)$$

$$\Rightarrow \quad F_q(\lambda_q) = \frac{\lambda_q^3}{4\pi^2} \frac{1}{V_{mode}} Q_q. \tag{19.41}$$

$F_q(\nu_q)$ and $F_q(\lambda_q)$ can be considered as a one-dimensional Purcell factor. Adding up this factor in all three dimensions results in the Purcell factor [13] that is defined for an emitter placed inside a three-dimensional hohlraum resonator:

$$3F_q(\lambda_q) = F_P(\lambda_q) = \frac{3\lambda_q^3}{4\pi^2} \frac{1}{V_{hohlraum}} Q_q. \tag{19.42}$$

The radiative decay into the single resonator mode of Eq. (19.39) is then altered to

$$\frac{1}{\tau_{21,rad}^{mode}} = \int c\sigma_e(\nu) F_q(\nu) \tilde{M}(\nu)\, d\nu \overset{\substack{\Delta\nu_e \to 0 \\ \nu = \nu_q = \nu_e}}{\approx} F_q(\nu_q) \int c\sigma_e(\nu) \tilde{M}(\nu)\, d\nu = F_q(\nu_q) \frac{1}{\tau_{21,rad}}. \tag{19.43}$$

Inserting Eqs. (19.41), (19.40), (19.30), and (19.31) into Eq. (19.43) displays the underlying physics:

$$\frac{1}{\tau_{21,rad}^{mode}} \overset{\substack{\Delta\nu_e \to 0 \\ \nu = \nu_q = \nu_e}}{\approx} \frac{\tilde{M}_q(\nu_q)}{\tilde{M}(\nu_q)} \frac{1}{\tau_{21,rad}} = \frac{V_M}{2V_{mode}} \frac{2}{\pi} Q_q \frac{1}{\tau_{21,rad}} = \frac{1}{\varphi_0(V_{mode})} \frac{2}{\pi} Q_q \frac{1}{\tau_{21,rad}}. \tag{19.44}$$

Compared to the spontaneous-emission rate $1/\tau_{21,rad}$ into free space, the spontaneous-emission rate $1/\tau_{21,rad}^{mode}$ into the single resonator mode is, on the one hand, reduced by the factor $V_M/(2V_{mode})$, because out of the number $\varphi_0(V_{mode})$ of vacuum photons that in average occupy the mode volume V_{mode} only one triggers spontaneous emission into this mode. On the other hand, $1/\tau_{21,rad}^{mode}$ is enhanced by the factor Q_q, the resonant enhancement due to a high resonator quality, because the vacuum photon of this mode is better confined to the mode volume V_{mode} by this factor. Only when the latter factor is larger than the former, the spontaneous-emission rate into a single resonator mode exceeds that into free space.

In the other limit, if the effective emission cross-section $\sigma_e(\nu)$ varies insignificantly over the Lorentzian profile of the resonator mode, $\Delta\nu_e >> \Delta\nu_q$, Eqs. (19.40) and (19.39) can be approximated as

$$F_q\left(\nu_q\right) \overset{\Delta\nu_e >> \Delta\nu_q}{\approx} \frac{1}{\tilde{M}\left(\nu_q\right)} \frac{1}{\Delta\nu_{FSR}} \frac{1}{V_{mode}} = \frac{c^3}{8\pi\nu_q^2} \frac{1}{\Delta\nu_{FSR}} \frac{1}{V_{mode}}, \tag{19.45}$$

$$\begin{aligned} \frac{1}{\tau_{21,rad}^{mode}} &= \int c\sigma_e(\nu) \frac{\tilde{\gamma}_q(\nu)}{V_{mode}} d\nu \\ &\overset{\Delta\nu_e >> \Delta\nu_q}{\approx} \frac{c\sigma_e\left(\nu_q\right)}{V_{mode}} \int \tilde{\gamma}_q(\nu)\, d\nu = \frac{c\sigma_e\left(\nu_q\right)}{V_{mode}}. \end{aligned} \tag{19.46}$$

For certain lasers with narrow spectral luminescence linewidth, $\Delta\nu_e << \Delta\nu_q$, e.g. far-infrared and sub-millimeter-wave lasers [14], as well as for solid-state and semiconductor lasers with resonator lengths that are not significantly longer than the laser wavelength, leading to a large free spectral range in Eq. (19.14), a short photon decay time τ_q in Eq. (19.15), and an accordingly broad linewidth $\Delta\nu_q$ that can be on the order of $\Delta\nu_e$, see Fig. 19.1a–c, the integral in Eq. (19.39) must be considered, and experimentally the resonator length must be adjusted, such that ν_q equals the peak emission frequency ν_e, to ensure a large spectral overlap between the two profiles $\sigma_e(\nu)$ and $\tilde{\gamma}_L(\nu)$ of the lasing resonator mode [15].

In bulk lasers with resonator lengths that are long compared to the emission wavelength, $\ell >> \lambda_e$, one finds an accordingly small free spectral range in Eq. (19.14), a long photon decay time τ_q in Eq. (19.15), and a narrow linewidth $\Delta\nu_q$. In, e.g., rare-earth-ion-doped near-infrared or visible solid-state bulk lasers, phonon perturbation of the excited-state wave function, resulting in a virtual decay, as well as phononic re-distribution within the crystal-field multiplet, resulting in a real decay and re-population among the individual crystal-field levels, lead to atomic decoherence with an atomic coherence time τ_{atom}^{coh}. This decay occurs with a lifetime $\tau_{atom} = ½\, \tau_{atom}^{coh}$, on the time scale of a picosecond, thus usually being significantly faster than the effective decay time $\tau_{2,eff}$ and the radiative lifetime $\tau_{21,rad}$ of the excited state, as well as the photon decay time τ_q in bulk lasers,

$$\begin{aligned} &\frac{1}{2}\tau_{atom}^{coh} = \tau_{atom} << \tau_{2,eff}, \tau_{21,rad}, \tau_q \\ &\quad \Rightarrow \quad \Delta\nu_e = \frac{1}{2\pi}\left(\frac{1}{\tau_{atom}} + \frac{1}{\tau_{2,eff}}\right) \approx \frac{1}{2\pi\tau_{atom}} >> \Delta\nu_q. \end{aligned} \tag{19.47}$$

thereby forming a key linewidth broadening mechanism. Consequently, with Eq. (19.11), one obtains $\Delta\nu_e >> \Delta\nu_q$ in Eq. (19.47), hence the transition cross sections $\sigma_e(\nu)$ and $\sigma_a(\nu)$ are broader than, and vary insignificantly over, the Lorentzian profile $\tilde{\gamma}_q(\nu)$ of the resonator mode, see Fig. 19.1d, such that Eq. (19.39) can be approximated by Eq. (19.46).

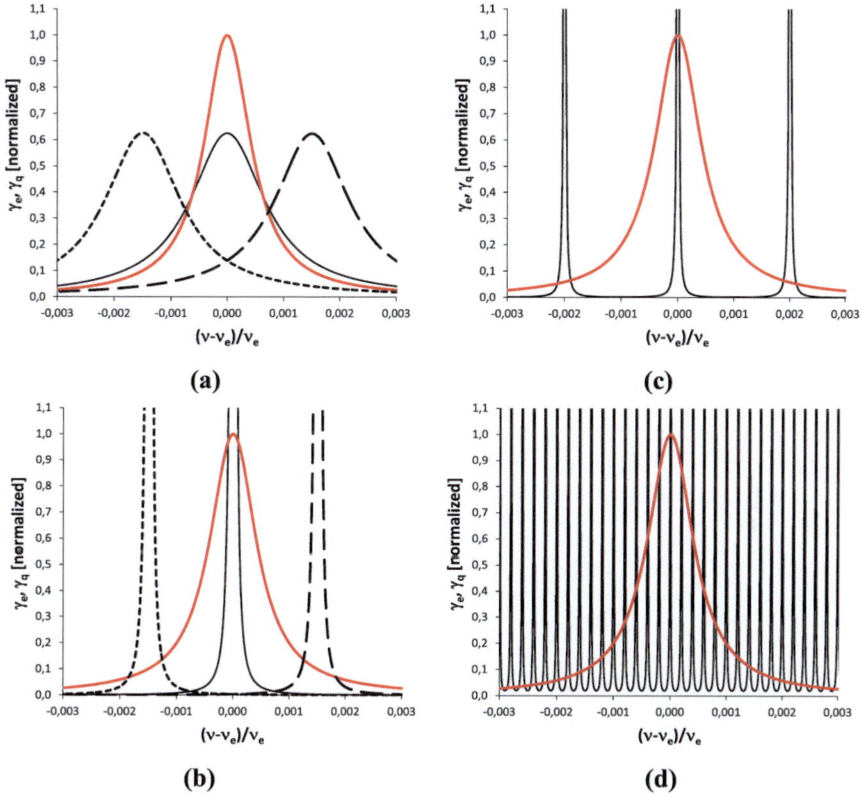

Fig. 19.1 Overlap of Lorentzian-shaped emission-line spectral profile $\tilde{\gamma}_e(\nu)$ (*red*) with Lorentzian-shaped resonator-mode spectral profiles $\tilde{\gamma}_q(\nu)$ (*black*), normalized to the peak of the emission-line spectral profile, for a resonator with $n_r = 1$, $\lambda_e = 1\ \mu\text{m}$, $\Delta\lambda_e = 1$ nm, $T_{out} = 0.01$, and $L_{RT} = 0$, for resonator lengths of (**a**) $\ell = 1 \times \lambda_e/2$, showing the resonator mode with index $q = 0$ for ℓ (*solid line*) and detuning of ℓ by $\pm 0.15\,\%$ (*dotted* and *dashed lines*), (**b**) $\ell = 50 \times \lambda_e/2$, showing the resonator mode with index $q = 49$ for ℓ (*solid line*) and detuning of ℓ by $\pm 0.15\,\%$ (*dotted* and *dashed lines*), (**c**) $\ell = 500 \times \lambda_e/2$, showing the 3 resonator modes with mode indices around $q = 499$, and (**d**) $\ell = 5000 \times \lambda_e/2$, showing the 31 resonator modes with mode indices around $q = 4999$ (Figure extended from Ref. [1])

19.4.2 Relation Between Peak Emission Cross Section, Radiative Lifetime, and Emission Linewidth

For a single atomic emission line, centered at frequency ν_e, with a Lorentzian-shaped profile $\tilde{\gamma}_e(\nu)$ of linewidth $\Delta\nu_e$ given by Eq. (19.10), the emission cross section becomes

$$\sigma_e(\nu) = \sigma_e(\nu_e)\,\frac{\pi}{2}\Delta\nu_e\tilde{\gamma}_e(\nu) = \sigma_e(\nu_e)\,\frac{\Delta\nu_e^2}{4(\nu-\nu_e)^2 + \Delta\nu_e^2}. \tag{19.48}$$

When assuming a narrow emission linewidth, $\Delta\nu_{e,\min} << \nu_e$, resulting in a large peak cross section $\sigma_{e,\max}(\nu_e)$, Eq. (19.38) yields

$$\frac{1}{\tau_{21,rad}} \overset{\Delta\nu_e << \nu_e}{=} \frac{8\pi}{c^2}\int \sigma_{e,\max}(\nu_e)\,\frac{\pi}{2}\Delta\nu_{e,\min}\tilde{\gamma}_e(\nu)\,\nu_e^2 d\nu$$
$$= \frac{4\pi^2\nu_e^2}{c^2}\sigma_{e,\max}(\nu_e)\,\Delta\nu_{e,\min}. \tag{19.49}$$

Exploiting Eqs. (19.11) and (19.1),

$$\Delta\nu_{e,\min} = \frac{1}{2\pi\tau_{21,rad}} = \frac{c}{\lambda_e^2}\Delta\lambda_{e,\min}, \tag{19.50}$$

where $\Delta\lambda_{e,\min}$ is the minimum emission linewidth in wavelength units inside the active medium, one finds from Eq. (19.49) that the largest possible peak cross section, called the natural emission cross section, equals the fundamental modal area of Eq. (19.25):

$$\sigma_{e,\max}(\nu_e) = \frac{c^2}{4\pi^2\nu_e^2\tau_{21,rad}}\frac{1}{\Delta\nu_{e,\min}} = \frac{c^2}{2\pi\nu_e^2} = \frac{\lambda_e^2}{2\pi} = \pi w_M^2. \tag{19.51}$$

With Eqs. (19.48) and (19.51), the integral emission cross section Σ_e, i.e., the area underneath the spectral emission-cross-section curve, which is proportional to the product of peak emission cross section and emission linewidth, is solely determined by the radiative lifetime of spontaneous emission:

$$\Sigma_e = \int \sigma_e(\nu_e)\,\frac{\pi}{2}\Delta\nu_e\tilde{\gamma}_e(\nu)\,d\nu = \sigma_e(\nu_e)\,\frac{\pi}{2}\Delta\nu_e$$
$$= \sigma_{e,\max}(\nu_e)\,\frac{\pi}{2}\Delta\nu_{e,\min} = \frac{c^2}{8\pi\nu_e^2\tau_{21,rad}}. \tag{19.52}$$

Consequently, the product of peak emission cross section and radiative lifetime cannot exceed the value of

$$\sigma_e(\nu_e)\,\tau_{21,rad} = \frac{c^2}{4\pi^2\nu_e^2}\frac{1}{\Delta\nu_e} = \frac{\lambda_e^4}{4\pi^2 c}\frac{1}{\Delta\lambda_e} \le \sigma_{e,\max}(\nu_e)\,\tau_{21,rad}$$
$$= \frac{c^2}{4\pi^2\nu_e^2}\frac{1}{\Delta\nu_{e,\min}} = \frac{\lambda_e^4}{4\pi^2 c}\frac{1}{\Delta\lambda_{e,\min}}. \tag{19.53}$$

Transitions within the 4*f* sub-shell of rare-earth ions are parity forbidden, leading to long radiative lifetimes on the order of $\tau_{21,rad} \approx 10^{-3}$ s. Assuming a host material with a refractive index of $n_r = 2$, resulting in $c = c_0/n_r$, and an emission wavelength of $\lambda_e^{vac} = 1$ μm emitted from the material, resulting in λ_e

$= \lambda_e^{vac}/n_r = 500$ nm, the emission linewidth expected from Eq. (19.50) would be as narrow as $\Delta\lambda_{e,\min} = \lambda_{e,\min}^{vac}/n_r = 2.6 \times 10^{-19}$ m. From Eq. (19.51) one obtains a maximum peak emission cross section of $\sigma_{e,\max}(\lambda_e) = 4.0 \times 10^{-14}$ m^2. These values would satisfy Eq. (19.53). However, in rare-earth-ion-doped solid-state laser materials phonon perturbation of the excited-state wave function as well as phononic re-distribution within the crystal-field multiplet lead to atomic de-coherence with an atomic coherence time τ_{atom}^{coh} according to Eq. (19.47) on the time scale of $\tau_{atom}^{coh} \approx 10^{-12}$ s, orders-of-magnitude faster than $\tau_{21,rad}$, resulting in the peak emission cross section being decreased and the emission linewidth $\Delta\lambda_e$ inside the active medium being increased by a factor of

$$f_\sigma = \frac{\Delta\nu_{e,\min}}{\Delta\nu_e} = \frac{\Delta\lambda_{e,\min}}{\Delta\lambda_e} = \frac{\sigma_e(\nu_e)}{\sigma_{e,\max}(\nu_e)} \overset{\tau_{atom} << \tau_{21,rad}}{\approx} \frac{\tau_{atom}}{\tau_{21,rad}}, \qquad (19.54)$$

on the order of $f_\sigma \approx 10^{-9}$, thereby leaving the integrated emission cross section of Eq. (19.52) and, thus, the radiative lifetime $\tau_{21,rad}$ unchanged. The fast atomic de-coherence significantly broadens the emission linewidth to $\Delta\lambda_e \approx 0.5$ nm, equal to $\Delta\lambda_e^{vac} \approx 1$ nm, and accordingly decreases the natural emission cross section of Eq. (19.52) to $\sigma_e(\nu_e) \approx 2 \times 10^{-23}$ m^2, thereby again satisfying Eq. (19.53). Thus, the significantly smaller experimental peak emission cross-sections observed in rare-earth-ion-doped host materials compared to the natural emission cross section obtained from Eq. (19.51) are not, as sometimes assumed, a direct consequence of the long lifetimes due to the parity rule, but of the ratio between these long lifetimes and the fast de-coherence of excited atomic states, as quantified by Eq. (19.54).

The allowed inter-band transitions in III-V semiconductor materials as well as the allowed transitions in laser dyes lead to significantly shorter radiative lifetimes on the order of $\tau_{21,rad} \approx 10^{-7}$ s. Assuming that de-coherence of the excited states in these materials occurs at a similar time scale of $\tau_{atom}^{coh} \approx 10^{-12}$ s as in rare-earth-ion-doped materials, the linewidth-broadening and peak-decreasing factor f_σ is by a ratio of $10^{-7}/10^{-3} \approx 10^{-4}$ smaller in the semiconductor materials, leading to accordingly larger peak emission cross sections on the order of $\sigma_e \approx 10^{-20} - 10^{-19}$ m^2 in III-V semiconductor materials, compared to $\sigma_e \approx 10^{-24} - 10^{-23}$ m^2 in rare-earth-ion-doped materials [16]. In fact, the product of peak emission cross section $\sigma_e(\nu_e)$ and radiative lifetime $\tau_{21,rad}$ in Eq. (19.53) is similar in these different classes of materials [16], because the de-coherence time is on the same order of magnitude. The situation is illustrated in a simplified manner in Fig. 19.2.

Taking the example of a host material, monoclinic potassium double tungstate, that is known for providing comparatively large emission cross sections to rare-earth ions such as Yb^{3+} [17, 18], its refractive index in the near-infrared spectral region is $n_r \approx 2$, resulting in $c \approx ½\, c_0$, and at the central emission line of $\lambda_e^{vac} = 981$ nm, equal to $\lambda_e = 490$ nm in the active medium, the measured emission cross section is $\sigma_e(\lambda_e) \approx 1.5 \times 10^{-23}$ m^2, and the emission linewidth is $\Delta\lambda_e^{vac} \approx 4.0$ nm, equal to $\Delta\lambda_e \approx 2.0$ nm in the active medium. The luminescence lifetime is $\tau_2 \approx 261$ μs [19]. However, the upper and lower state exhibit a crystal-field

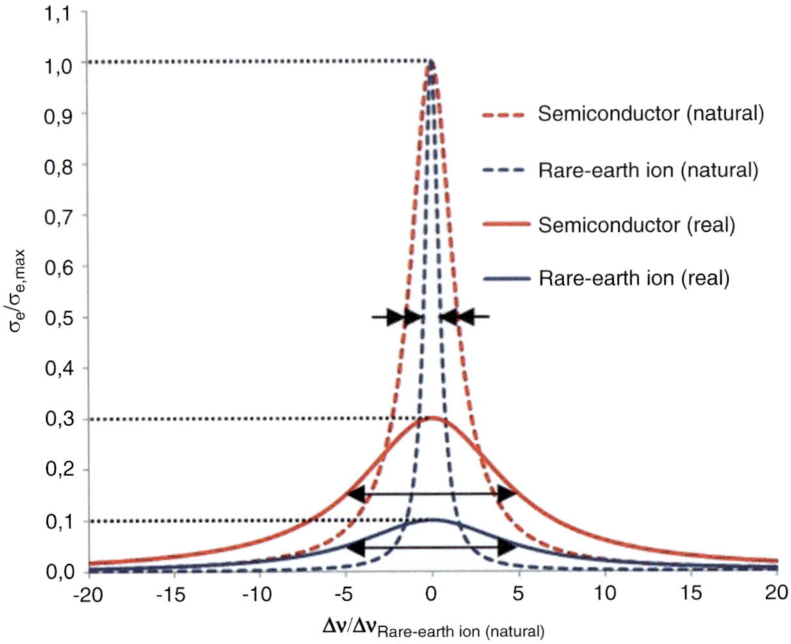

Fig. 19.2 Emission cross sections in a semiconductor and a rare-earth-ion-doped material, calibrated to the natural emission cross section, as a function of frequency detuning from the emission peak, calibrated to the FWHM linewidth of the natural emission line of the rare-earth ion. For the purpose of clarity of illustration in a single figure, it is assumed that the excited-state lifetime of the semiconductor is only 3 times (instead of orders-of-magnitude) shorter than that of the rare-earth ion, whereas the atomic de-coherence time is the same for both excited states and only 10 times (instead of orders-of-magnitude) shorter than the excited-state lifetime of the rare-earth ion. It results in $f_\sigma = 10$ for the rare-earth ion, $f_\sigma = 3.33$ for the semiconductor, and an experimentally observed (real) peak emission cross section which is 3 times larger for the semiconductor compared to the rare-earth ion. The *arrows* indicate the FWHM linewidths. The areas underneath the natural and experimentally observed (real) semiconductor emission lines are equal to each other. The same holds for the areas underneath the two rare-earth-ion emission lines (Figure taken from Ref. [1])

splitting into 3 and 4 crystal-field levels, respectively, resulting in 12 individual crystal-field transitions. Within each crystal-field multiplet, the population density is thermalized by fast phononic intra-multiplet transitions, resulting in a Boltzmann distribution. The radiative lifetime of spontaneous emission from upper to lower crystal-field multiplet reflects the decay on all 12 crystal-field transitions, whereas the area underneath the central line at 981 nm occupies only ~25 % of the complete emission spectrum [19]. Hence, assuming that the decay from the upper crystal-field multiplet is entirely radiative, the spontaneous-emission rate on the central line corresponds to a radiative lifetime that is approximately four times the intrinsic luminescence lifetime, $\tau_{21,rad} \approx 4\tau_2 \approx 1.044$ ms.

Equivalent to Eq. (19.50), $\Delta\lambda_e \approx 2.0$ nm results in $\tau_{atom} \approx 1.3 \times 10^{-13}$ s, thereby assuming that the measured emission line-width is entirely due to atomic decoherence and additional inhomogeneous linewidth-broadening mechanisms are negligible. Therefore, $f_\sigma = \tau_{atom}/\tau_{21,rad} \approx 1.2 \times 10^{-10}$. The decrease in peak emission cross section leads to $f_\sigma = \sigma_e(\lambda_e)/\sigma_{e,\max}(\lambda_e) = 1.5 \times 10^{-23}$ $m^2/3.8 \times 10^{-14}$ $m^2 = 3.9 \times 10^{-10}$, and the broadening of emission line-width results in $f_\sigma = \Delta\lambda_{e,\min}/\Delta\lambda_e = 2.4 \times 10^{-19}$ m$/2.0 \times 10^{-9}$ m $= 8.2 \times 10^{-9}$. Consequently, the experimental values of $\sigma_e(\lambda_e) \approx 1.5 \times 10^{-23}$ m^2, $\tau_{21,rad} \approx 1.044$ ms, and $\Delta\lambda_e \approx 2.0$ nm satisfy Eq. (19.53) only with an error of a factor of five, which may be owing partly to experimental inaccuracies when measuring these values, a spectral overlap of the central emission line with other nearby crystal-field transitions, additional inhomogeneous linewidth broadening, a certain amount of non-radiative decay from the upper crystal-field multiplet, which decreases the intrinsic luminescence lifetime from its radiative value to the measured value of $\tau_{2,eff}$, and considering only one of the three directions of the optical indicatrix.

19.5 Summary

We have fundamentally derived the Füchtbauer-Ladenburg equation, the Purcell factor, and the relation between peak emission cross section, radiative lifetime, and emission linewidth.

References

1. Eichhorn, M., & Pollnau, M. (2015). *IEEE Journal of Selected Topics in Quantum Electronics, 21*, 9000216.
2. Einstein, A. (1917). *Physikalishce Zeitschrift, 18*, 121.
3. Heisenberg, W. (1927). *Zeitschrift für Physik, 43*, 172.
4. Robertson, H. P. (1929). *Physics Review, 34*, 163.
5. Lengyel, B. A. (1971). *Lasers* (2nd ed., pp. 81 ff). New York: Wiley.
6. Planck, M. (1901). *Annals of Physics, 309*, 553.
7. Saleh, B. E. A., & Teich, M. C. (2007). *Fundamentals of Photonics* (2nd ed., pp. 408–409). Hoboken: Wiley-Interscience.
8. Johnson, K. S. (1931). *Transmission circuits for telephonic communication*, (5th ed., Ch. XV, pp. 174–179). New York: D. Van Nostrand, Inc., 1st ed. published by Western Electric (1924).
9. Hansen, H. H. (1938). *Journal of Applied Physics, 9*, 654.
10. Fabry, C., & Pérot, A. (1899). *Annales de chimie et de physique, 16*, 115.
11. Airy, G. B. (1838). *Transactions of the Cambridge Philosophical Society, 6*, 379.
12. Svelto, O. (2010). *Principles of lasers* (5th ed., pp. 142–146). New York: Springer.
13. Purcell, E. M. (1946). *Physics Review, 69*, 681.
14. Affolter, E., & Kneubühl, F. K. (1981). *IEEE Journal of Quantum Electronics, 17*, 1115.
15. Schwaller, P., Steffen, H., Moser, J. F., & Kneubühl, F. K. (1967). *Applied Optics, 6*, 827.
16. Geskus, D., Aravazhi, S., García-Blanco, S. M., & Pollnau, M. (2012). Giant optical gain in a rare-earth-ion-doped microstructure. *Advanced Materials, 24*, OP19.

17. Kuleshov, N. V., Lagatsky, A. A., Shcherbitsky, V. G., Mikhailov, V. P., Heumann, E., Jensen, T., Diening, A., & Huber, G. (1997). *Applied Physics B: Lasers and Optics, 64*, 409.
18. Lagatsky, A. A., Kuleshov, N. V., & Mikhailov, V. P. (1999). *Optics Communications, 165*, 71.
19. Aravazhi, S., Geskus, D., van Dalfsen, K., Vázquez-Córdova, S. A., Grivas, C., Griebner, U., García-Blanco, S. M., & Pollnau, M. (2013). *Applied Physics B: Lasers and Optics, 111*, 433.

Chapter 20
Surface Plasmon Enhanced Fluorescence of Glycine-Dimer-Functionalized Silver Nanoparticles

Vira Kravets and Anatoliy Pinchuk

Abstract We present experimental results on the fluorescence of glycine-dimers-functionalized silver nanoparticles. Silver nanoparticles were synthesized by using a solid state thermal reduction of silver nitrate in a glycine matrix yielding polydisperse silver nanoparticles. For particles of average core diameter of 23 nm the fluorescence quantum yield was determined to be 5.2+/−0.1 % and depended non-monotonically on the size of the nanoparticles. The fluorescence emission of the glycine dimers was enhanced by the strong local electric field due to the surface plasmon resonance in silver nanoparticles. The application of small fluorescent silver nanoparticles of average diameters 5 +/− 4 nm as imaging markers for yeast cells was demonstrated. The nanoparticles were able to penetrate yeast cell membranes and accumulate inside the vesicles within the cell. The luminescence was observed after 12 h of incubation inside the nuclei of about 10% of the cells suggesting that silver nanoparticles penetrated into the nuclei of the cells.

20.1 Introduction

Noble metal nanoparticles have recently attracted much interest because of their potential use in photovoltaic, drug delivery, and bioimaging, to name just a few [1–3]. The unique optical and electronic properties of noble metal nanoparticles, such as tunable excitations of collective coherent oscillations of conducting electrons, known as surface plasmon resonance (SPR), make them promising for a number of applications in surface enhanced Raman spectroscopy (SERS), biosensing and bioimaging, [2, 3] photothermal therapy, [4] and as plasmonic rulers [5]. The environmental impact of nanoparticles on living organisms has been studied extensively in the recent years. [3, 4] In addition to biological cell viability

V. Kravets • A. Pinchuk (✉)
Physics Department and Center for Biofrontiers Institute, University of Colorado Springs, 1420, Austin Bluffs Parkway, Colorado Springs, CO 80918, USA
e-mail: apinchuk@uccs.edu

© Springer Science+Business Media Dordrecht 2017

B. Di Bartolo et al. (eds.), *Nano-Optics: Principles Enabling Basic Research and Applications*, NATO Science for Peace and Security Series B: Physics and Biophysics, DOI 10.1007/978-94-024-0850-8_20

studies, investigation of nanoparticles' effect on cell functionality and morphology is important. In such studies, the detection of metal nanoparticles within a live cell is possible by using confocal fluorescent microscopy. It is important that the fluorescent probes exhibit bright and stable fluorescence along with the possibility to penetrate the membrane of a cell.

The unique electronic and optical properties of noble metal nanoparticles result from their size dependent electron band structure and the surface-to-volume ratio. The local SPR induced electric field near the surface of a nanoparticle depends on the size of the particle. This local electric field may enhance the fluorescence of the nearby molecule if properly placed near the surface of the nanoparticle [6]. The size-dependent enhancement of the photoluminescence has not been well studied and it's mechanism remains questionable [7].

In this work, we report the synthesis, size separation, optical characterization, and application as biomarkers of glycine-dimer-functionalized silver nanoparticles. The strong local SPR-enhanced electric field near the surface of silver nanoparticles is hypothesized to enhance the photoluminescence of glycine-dimer ligands. The fluorescent silver nanoparticles might potentially be used as robust, non-toxic, stable, and bright fluorescent biomarkers.

20.2 Results

One-to-ten ratio of silver nitrate and glycine was mixed in distilled deionized water. The water was then rotor-evaporated and the solid mixture was heated in a silicon bath at 445°C for 5 min. The product of reaction was re-dissolved in water and sonicated for 24 h. The polydisperse solution with nanoparticles was size- separated by using centrifugation. The size separated silver nanoparticles were then characterized by measuring their optical extinction and fluorescence, as well as the quantum yield. Here on figure one we present selected results of two samples, one containing small nanoparticles (~5 nm diameter), and another containing larger nanoparticles (~60 nm diameter).

Figure 20.1 exhibits photographs of the solutions of silver nanoparticles and their extinction spectra. Large nanoparticles predominantly scatter light (Fig. 20.1b, left), while small nanoparticles mostly absorb light. Smaller nanoparticles enhance the emission of glycine dimers, which results in yellow light emission when excited at 450 nm wavelength (Fig. 20.1b, right).

The formation of glycine dimers was proved by the Fourier Transformation Infrared (FTIR) spectra of the glycine-silver nitrate mixtures treated at different temperatures reported in ref.[13]. The FTIR spectra revealed characteristic glycine-dimer (peptide-bond) peak near 1730 inverse centimeters, [9] but it was only present at the reaction temperatures 170–190 °C, and only in the presence of silver nitrate. No such peak was observed when pure glycine was treated under identical conditions.

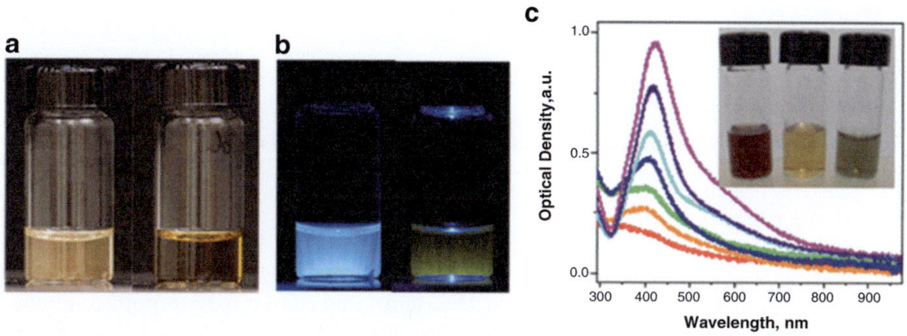

Fig. 20.1 Comparison of the aqueous solutions of silver nanoparticles of *large* (~60 nm) diameters (*right*) and *small* (~5 nm) diameters (*left*) under (**a**) daylight illumination, (**b**) blue illumination of the wavelength 450 nm. (**c**) Optical density of size-separated silver nanoparticles. Inset: Water solutions of glycine-dimer-coated nanoparticles of different sizes

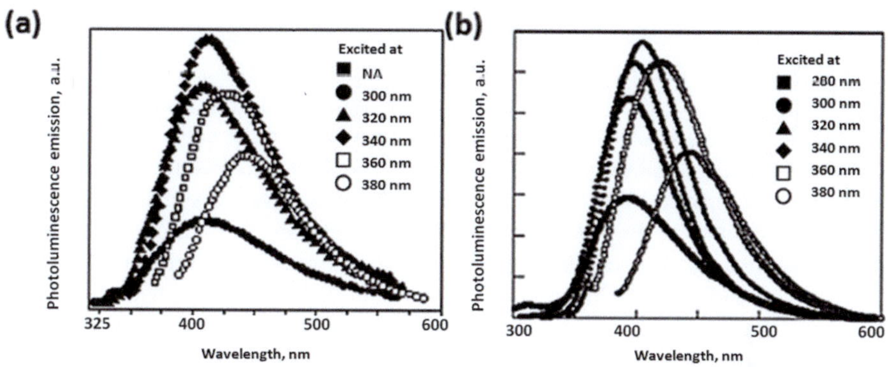

Fig. 20.2 Fluorescence emission spectra of water solutions of the glycine-dimer-coated silver nanoparticles for different excitation wavelengths. (**a**) our silver-nitrate-mediated synthesis, (**b**) synthesized without silver nitrate, reported in the literature [8]

Previously, the formation of fluorescent molecules from glycine dimers was proposed by Futamura et.al. [8] The emission spectra reported in their work are identical to the emission spectra of our glycine-silver nanoparticles shown in Fig. 20.2. It is interesting to note that in previous studies the glycine dimers were synthesized without silver nitrate and under different temperature, pressure and treatment conditions [8]. The spectra shown in Fig. 20.2 have the same scale and almost the same legends for comparison. One can see that when excited at 340 nm (diamond-shaped scatter points) both spectra have the same maximum emission wavelength and intensity. The figure on the right has one extra curve (excitation 280 nm). The emission spectrum for 320 nm excitation on the right has higher intensity than the one the left. Despite these two differences the spectra are very much alike.

The synthesis of glycine dimers has been previously reported by using wetting-drying of the silica surface [9] as well as in super-critical water [8]. The reported

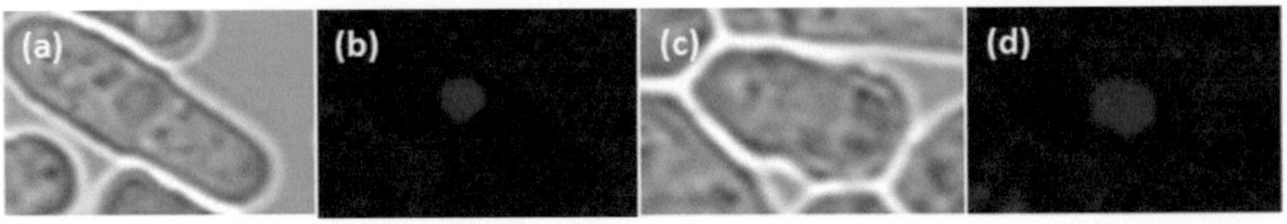

Fig. 20.3 Bio-imaging of the yeast cells (**a**, **c**) optical transmission microscopy; (**b**, **d**) fluorescence confocal microscopy

yield of the reaction was very low. We note that previously reported photoluminescence of atomic-like silver nanoclusters [10] is most likely is due to photoluminescence of glycine dimer molecules enhanced by the SPR from silver nanoparticles.

To demonstrate the potential application of glycine-dimer-coated silver nanoparticles as bio-imaging markers, we incubated nanoparticles of average diameters 5+/−4 nm with yeast cells in the cell media and monitored penetration of nanoparticles through the cell membranes and into the cell nuclei over the course of 12 h (Fig. 20.3). The transmission and fluorescence confocal microscopy images of individual yeast cells are shown in Fig. 20.3. Within the first hour of incubation fluorescent signal was observed mostly from the surface of the cells. After an hour, the fluorescence was observed from the vesicles inside the cells, and finally after 12 h exposure, about 10% of the cells expressed fluorescence signal inside their nuclei. This can be attributed to the smaller nanoparticles penetrating into the cell nuclei membranes. The fact that only in some cells fluorescence was observed within the nuclei could be rationalized by the fact that our sample had +/− 4 nm deviation in the particle diameter, and the spatial distribution of small nanoparticles (which are more likely to penetrate into the nuclei) was non-uniform.

Based on the confocal fluorescence images we suggest that nanoparticles can penetrate through the yeast cell membranes, as well as inside the nuclei. Possible membrane penetration passive mechanisms may be via membrane alternation (via interaction of the nanoparticles surface groups with the membrane), or less likely, via membrane deformation (this pathway has been reported for larger nanoparticles). Previous reports indicate that nanoparticle's ligands can soften the lipid membrane allowing nanoparticles penetration without the membrane damage [11]. In the cell media and inside the cell, nanoparticles potentially can react with the proteins, lipids and sugars, which leads to formation of the "protein corona". This, in turn, affects nanoparticles' ability to penetrate the cell membrane and when inside the cell it affects nanoparticles interactions with cell organelles and it's digestion.

Alternatively, the active transport of nanoparticles through the membrane may be via clathrin-mediated endocytosis [12]. Cells use endocytosis for transport of large macro-molecules (of size 2–10 nm), such as proteins, carbohydrates, lipids, nucleic acids. Small nanoparticles are on the same scale as biomolecules and the cells may uptake nanoparticles. Note that sometimes paraformaldehide causes partial permeabilization of the cell membranes, which could have been one of the mechanism for the nanoparticles penetration into the cells. After nanoparticles penetrate the cell membrane, they may be found in the nuclei as can be seen in Fig. 20.3. It has been reported that small ~2 nm diameter gold nanoparticles can penetrate into the nucleus [11, 12]. Another possibility of the penetration into the nuclei is release of the silver ions, or parts of the nanoparticles, digested within the cell.

The ability of small nanoparticles to penetrate the cell membrane and to accumulate in the nuclei suggests potential toxic effects of such nanoparticles. Previously toxicity of these nanoparticles has been investigated towards rat basophilic leukemia cells, and it was found that they are toxic in the 10–100 micro-molar range of silver atoms concentration [13].

In this study we have not performed high resolution transmission electron microscopy, which would allow us to tell if the nanoparticles themselves or their ions, or possibly metabolized remnants were penetrating the yeast nuclei. This may be an interesting question to address in the future studies, relevant to the nanoparticle's toxicity issue, since it is known, that the main mechanism of the nanoparticles toxicity is release of silver ions. It is proposed that silver ions when penetrating into the cell bind to the thiol-containing bio-molecules, which causes multiple toxic effects.

20.3 Conclusions

Highly fluorescent, glycine-dimer-functionalized silver nanoparticles were synthesized, size-separated by using centrifugation, and characterized. Based on the fluorescence emission spectra and FTIR spectroscopy the origin of fluorescence was attributed to the cyclic glycine dimers. Quantum yield of 5.2+/−0.1 % was determined for glycine-dimer-functionalized silver nanoparticles of the average size 23+/−4 nm. We utilized small (average of 5+/− nm in diameter) nanoparticles in bioimaging of live yeast cells and observed accumulation of the nanoparticles in the nuclei of the cells after 12 h of exposure to nanoparticles solution. The fluorescent silver nanoparticles might be used as bioimaging markers for the biological cell imaging. Further research of the potential of glycine-dimer-coated silver nanoparticles in bioimaging may benefit from study of size-dependence of the nanoparticles internalization, mechanisms of the interaction and toxicity, and their further distribution within the cell.

Acknowledgements The authors acknowledge partial support from the Center for Biofrontiers Institute, the Butcher Foundation seeding grant and NATO (SFPP-984617) grant.

References

1. Haes, A., Zou, S., Schatz, G., & Van Duyne, R. (2004). Nanoscale optical biosensor: Short range distance dependence of the localized surface plasmon resonance of silver and gold nanoparticles. *Journal of Physical Chemistry B, 108*, 6961–6968.

2. Haynes, C., Yonzon, C., Zhang, X., & Van Duyne, R. (2005). Surface-enhanced Raman sensors: Early history and the development of sensors for quantitative biowarfare agent and glucose detection. *Journal of Raman Spectroscopy, 36*, 471–484.
3. Braun, G., Friman, T., Pang, H., Pallaoro, A., De Mendoza, T., Willmore, A., Kotamraju, V., Mann, A., She, Z., Sugahara, K., Reich, N., Teesalu, T., & Ruoslahti, E. (2014). Etchable plasmonic nanoparticle probes to image and quantify cellular internalization. *Nature Materials, 13*, 904–911.
4. O'Neal, D., Hirsch, L., Halas, N., Payne, J., & West, J. (2004). Photothermal tumor ablation in mice using near infrared absorbing nanoshells. *Cancer Letters, 209*, 171–176.
5. Kravets, V., Ocola, L., Khalavka, Y., & Pinchuk, A. (2015). Polarization and distance dependent coupling in linear chains of gold nanoparticles. *Applied Physics Letters, 106*, 053104.
6. Lakowicz, J., Geddes, C., Gryczynski, I., Malicka, J., Gryczynski, Z., Aslan, K., Lukomska, J., Matveeva, E., Zhang, J., Badugu, R., & Huang, J. (2004). Advances in surface-enhanced fluorescence. *Journal of Fluorescence, 14*(4), 425–441.
7. Yeschenko, O., Dmitruk, I., Alexeenko, A., Losytskyy, M., Kotko, A., & Pinchuk, A. (2009). Size-dependent surface-plasmon-enhanced photoluminescence from silver nanoparticles embedded in silica. *Physical Review B, 79*, 235438.
8. Futamura, Y., Yahara, K., & Yamamoto, K. (2007). Evidence for the production of fluorescent pyrazine derivatives using supercritical water. *Journal of Supercritical Fluids, 41*(2), 279–284.
9. Meng, M., Stievano, L., & Lambert, J. F. (2004). Adsorption and thermal condensation mechanisms of amino acids on oxide supports. 1. Glycine on silica. *Langmuir, 20*(3), 914–923.
10. Kravets, V., Culhane, K., Dmitruk, I., & Pincuk, A. (2002). Glycine-coated photoluminescent silver nanoclusters. *Proceedings SPIE, 8232*(1).
11. Shang, L., Nienhaus, K., & Nienhaus, G. (2014). Engineered nanoparticles interacting with cells: size matters. *Journal of Nanobiotechnology, 12*(5).
12. Wu, Y., Putcha, N., Ng, K., Leong, D., Lim, C., Loo, S., & Chen, X. (2013). Biophysical responses upon the interaction of nanomaterials with cellular interfaces. *Accounts of Chemical Research, 46*, 782–791.
13. Kravets, V., Almemar, Z., Jiang, K., Culhane, K., Machado, R., Hagen, G., Kotko, A., Dmitruk, I., Spendier, K., & Pinchuk, A. (2016). Imaging of biological cells using luminescent silver nanoparticles. *Nanoscale Research Letters, 11*, 30. doi:10.1186/s11671-016-1243-x.

Part II
Interdisciplinary Lecture

Chapter 21
Andrea Pozzo: The Art of Perspective

Elpidio Silvestri

Abstract The advent of the Jesuits on the Roman environment caused all kinds of fears, debates, quarrels and, above all, accuses of lack of concern toward Art. Nothing could be more misleading. In this presentation I will illustrate how they, instead, used Art to create the right and proper environment to attract the people to enter the Church, feel welcomed, at ease and willing to pay attention to the word of God. The Jesuits went insofar as to make use of the artists within their own Religious Order. The most famous one goes by the name of Andrea Pozzo whose highly ingenious talent achieved the highest illusionistic effects ever created in Art. We will talk about two of the most famous example: The Rooms of St. Ignatius in the novices' residence and the cupola of the Church of St. Ignatius in Rome.

21.1 Introduction

Andrea Pozzo was born, in 1642, in Trento, the major city of the region of South Tyrol, in Northern Italy, and died in Vienna in 1709. His parents sent him to the Jesuits school in Trento, but he was not very profitable since he "preferred to draw and scribble". Therefore, when he was seventeen, his father sent him to learn painting in various painters' studios in the area. In 1665 he moved to Milan where he joined the Jesuit to became a brother within the Religious Order (Fig. 21.1). This basically means that he is not a Priest, although he has taken the same wows, but he shares the religious life, exercising the daily duties that are usually carried out by layman. At the same time he completed the artistic studies.

His first official painting was titled *Escape to Egypt*, for a church in Trento. Today that work of art is part of a collection in Turin. In 1667 he moved his residence within the residence of the Jesuits priests, next to the Church of St. Fedele. In this environment he also started producing scenographic works for the religious celebrations of the *Quarantore*, which translates to *the Forty Hours*. According to Christian belief, Christ resurrected 40 h after his death on the cross. During the

E. Silvestri (✉)
Pontifical Gregorian University, Piazza della Pilotta, 4 00187 Roma, Italy
e-mail: elpidio.silvestri@alice.it

© Springer Science+Business Media Dordrecht 2017
B. Di Bartolo et al. (eds.), *Nano-Optics: Principles Enabling Basic Research and Applications*, NATO Science for Peace and Security Series B: Physics and Biophysics, DOI 10.1007/978-94-024-0850-8_21

Fig. 21.1 Andrea Pozzo, *Self Portrait*, Church of Jesus [1]

following few years, in Milan, Pozzo focused his studies so as to acquire a very substantial know-how of *perspective painting* and architecture. Thus, he was soon able to apply it to the works that he was called to do in the Jesuits churches in Northern Italy. The best example of his works, in those years, is in the small town of Mondoví, in the Piedmont region. In 3 years he restructured the Church of St. Francis Xavier, expressing all his capacities of a complete artist: Architect, painter, scenographer and maker of machines to enliven the scenography.

His first action was to modify the internal architectural structure. He achieved his goal by making use of his architectural knowledge and various chromatic elements, which multiplied the decorative effects of the nave, such as the painting of the column as if they were made of marble.

In the four pendentives, at the base of what appears as a cupola, he painted four allegories of the four continents. In reality he painted the deceitful cupola as an octagonal tambour which opens toward the sky (Figs. 21.2 and 21.3). Thus he deceived the observer who does not realize that, in reality, the ceiling is flat!

This is the perspective that he wants. But he has not finished with the surprises. He then moved to the main altar and, behind it, he constructed a scenic machinery. Through its use it is possible to direct the apparatus which creates a theatrical levitation of the statue of St. Francis Xavier (Fig. 21.5).

An in depth analysis of the apparatus reveals that the theatrical effect of the church architecture, both real and illusory, has been thought as a function of this

Fig. 21.2 Andrea Pozzo, Nave of church St. Francis Xavier, Mondoví (particular) [1]

scenic machine-apparatus which produces the theatrical effect of the levitation. According to the Jesuit Father H. Pfeiffer, Andrea Pozzo has thought of such triumphal altar because the Papal Bull regarding the canonization of Francis Xavier mentions his levitation as the main miracle. This altar apparatus is a true theatrical apparatus made of wood and cardboard. Notwithstanding the very poor and easily perishable materials, the apparatus is still perfectly functioning today. In addition, it is coherent with the scenography of the church, its proportions and the architecture order.

In Fig. 21.4 are shown two section designs of this church. These two designs are an attempt to reproduce the imaginary vision of Pozzo, who inserted this unique experience of *open vision* into the real architecture and, thus, transforming

Fig. 21.3 Andrea Pozzo, *St. Francis Xavier ascending to heaven*, Church of St. Francis Xavier, Mondoví [1]

it into spectacular architectural scenery. To achieve this goal he configured multiple *observation zone*, based on the mobility of the observer within the church. The object of this project was therefore to localize these focal points, having established that the vision would never be zenithal, but it is always a function of oblique visual approach which has to be directed to the depicted scenes. The first point of view, at the entrance, is naturally directed toward the main altar and the painting on the apse. But, as the observer moves the vision toward the higher part of the apse, it is also drawn backward toward the perspective vision of the *dome vault* with its colonnade architecture. It will take some time to realize that there is no dome vault: it is, again, an illusionistic painting of architecture, on a flat surface. In this illusionistic *open sky*, above the colonnades, that he created, Pozzo painted the figure of St. Francis, escorted by angels, as he rise toward heaven (Fig. 21.3). The illusion will be confirmed in the following views (Fig. 21.4) of the open ceilings between the arcades. Only at point 3 (top part of slide) the observer can correctly vision the scene of the offer of a crown by the angels to St. Francis. This is the point of view of the young Andrea Pozzo, at the time, only 35 years old. But he continued his studies of *quadratura*, as this type of painting are indicated in Italian. The literary meaning of the term *quadratura* is that the painting is enclosed into a square, in Italian *quadrato*. Therefore, the painted architecture becomes the unifying mean between the real space and the imaginary one. In the latter it is placed the represented scene.

Pozzo felt that to achieve the best knowledge he had to go to Rome, because there were the best artists and the most advanced techniques were taught. Since he

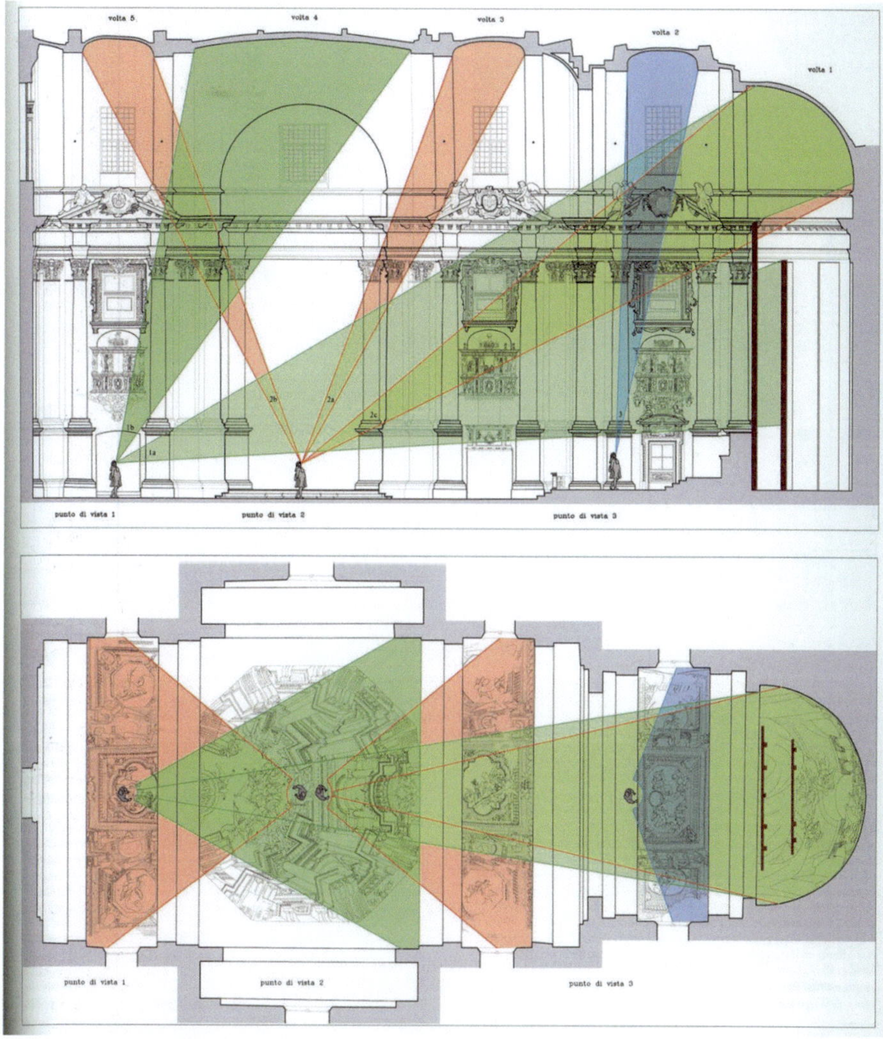

Fig. 21.4 (*Top*) Project & Relief by G. Dardanello – University of Turin, (*Bottom*) Relief & Graphical Reinstatement by M. Boetti – Geomar – Mondoví [1]

was a “brother” in the Jesuits Order he decides to write to Father Oliva, at the time the Provost General and, to better explain his goal, he sent two small paintings with the letter. Father Oliva (Fig. 21.6), had the paintings examined by his friend Carlo Maratta, the most famous painter in Rome at the time.

Maratta expressed a very positive judgment and told father Oliva to let the young painter come at the earliest. But, although Father Oliva wanted him in Rome, many obstacles came about. Eventually Brother Pozzo made it to Rome, but Father Oliva

Fig. 21.5 Statue of St. Francis Xavier, main altar Church of St. Francis Xavier, Mondoví [1]

died right after his arrival. Since he was unknown to the others in charge, he was temporarily relegated to the kitchens. Eventually he started to work for the Church of Jesus in Rome, but the Duke of Savoy requested his services to the new Provost General of the Jesuits. Pozzo was upset and so decided to ask the highest person in charge: he wrote to the Pope asking him to emit a specific order prohibiting Brother

Fig. 21.6 Anonymous, *Father Oliva*, sec XVII [1]

Pozzo to service lay persons. The Pope Innocent XI gladly emitted a decree as per the wishes of Brother Pozzo. At this point he was free to dedicate himself to work for the churches of the Jesuits. His first important work was for the Casa Professa of the Church of Jesus, were St. Ignatius had inhabited the last years of his life. After him his successors had inhabited the same rooms. Father Oliva had thought of asking Pozzo to decorate these rooms. His successor did commission the work to Pozzo. He received the commission around June 1682 and completed the work at the beginning of 1685. Right after, he also received the commission for the false cupola for the Church of St. Ignazio. He completed the painting within the same year.

21.2 The Painting of Casa Professa

Pozzo started to study the space which he had to decorate. It was a long corridor which terminated with a slanted wall, as we can see in the next figure (Fig. 21.7). The space of interest can be restricted to the *Corridor*. Originally the entrance to the chambers was placed on the room identified with the letter "E" and the entrance door is identified with the letter "D". This room was the personal bed room of St. Ignatius, while the next room, identified with the roman number "I", was used as his studio/office, were he received the visitors. This room was connected directly to the antechamber, identified with the roman number "II", were the visitors waited to be received by Ignatius.

The area where Pozzo worked is the above mentioned Corridor, identified with the letter "X". The then external wall was characterized by five windows, strongly tapered toward the outside. The entrance was then placed on the slanted wall, identified with the small letter "y". Father Oliva, before getting in touch with Pozzo had already tried to decorate this Corridor, but he was uncertain as to who could best do the work. He decided to try with a painter who was well known to him: Jacques Courtois, who was also a Jesuits Brother. Courtois was very famous at the time, because of his paintings of battle scenes. So Father Oliva commissioned to Courtois the painting of all the splay area of the five windows and the area below the window sill. He almost completed the work, before he died. So when Pozzo was commissioned he started the work on the last uncompleted decoration. But before starting the actual decoration he did some rearranging work in the Corridor.

The latter was used, at the time, as a "short-cut" from people coming and going from/to other floors in the building. Thus a very annoying confusion was created. Pozzo, therefore, decided to close this entrance (identified with the letter "y" in the Fig. 21.7), and modified the slanted wall transforming it into a small apse. The opposite side was actually wide open and Pozzo, very wisely, thought, instead, of building a wall with a door, after the fourth window, thus transforming the interested area into a some kind of seclusion (Fig. 21.7). In addition he opened a new communication door with the Chambers, placing it above a small double ramp staircase of three stairs on each side ("N" in Figs. 21.7 and 21.8). The final conclusion was that the chambers became a memento to Saint Ignatius.

The main problem that Pozzo tries to solve, with all these changes in the arrangement of the room, is very well known and very simple: light in the room. The only source of real light in the seventeenth century was the "sun". This is the reason why windows, in palaces and churches, are usually numerous, tall and with large splay area. This is why the communication door with the rooms of St. Ignatius is placed on the south wall, right across the fourth window from the left ("L" in

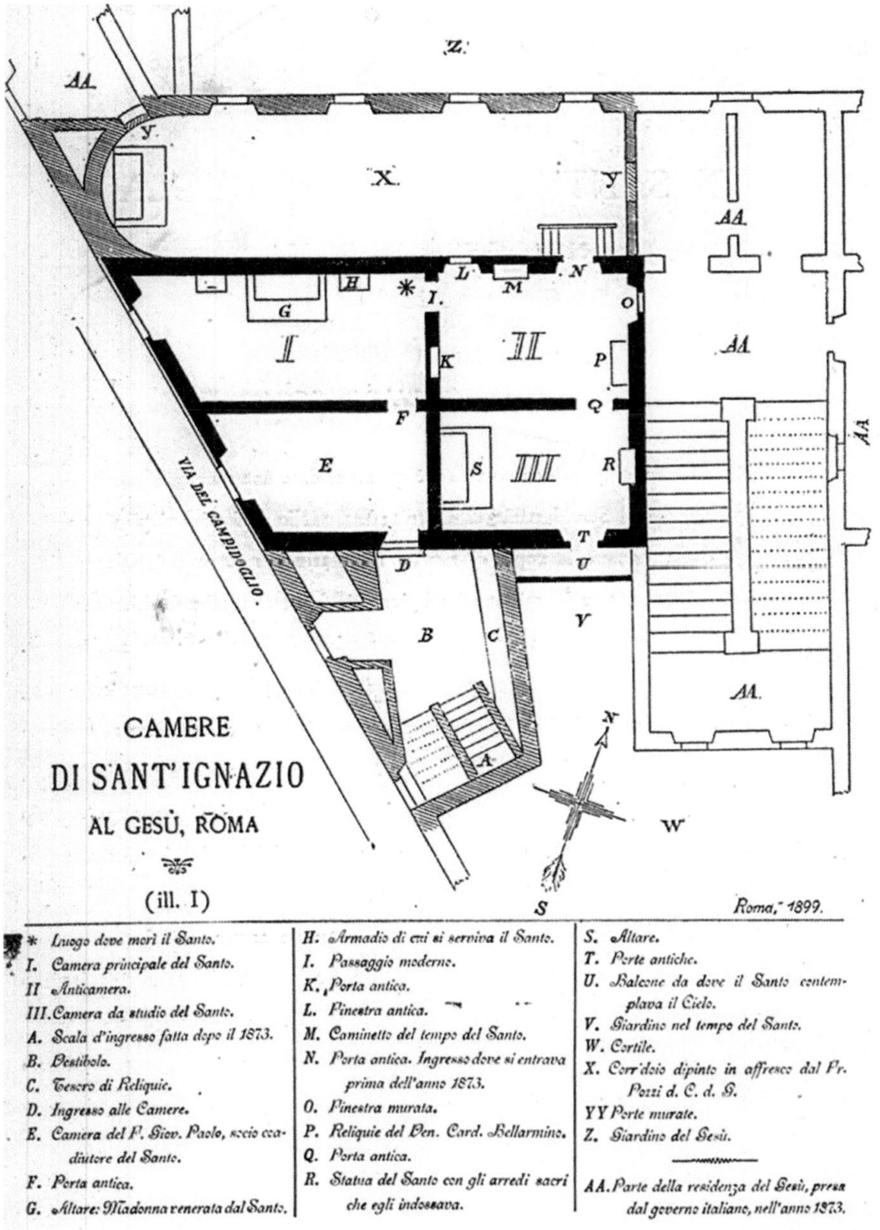

Fig. 21.7 Plan of the first floor of the Casa Professa with the Chambers of St. Ignatius and the Corridor [2]

Fig. 21.8 Pozzo's new communication door. (*left*) View from the Corridor. (*right*) Grated window references as "L" in Fig. 21.7 [2]

Fig. 21.7). This way, the light will penetrate into the room through the door and the adjacent small window (Fig. 21.8 (right)).

Having solved the above problems, the next steps were the decoration of the walls and ceiling of the Corridor. Let us now examine the physical characteristic of the Corridor. It is a rectangular shaped room having the following measures: the wall with windows is 13.50 m long; the opposite wall is 11.50 m long; the slanted end wall is 3.75 m long; the entrance wall is 3.50 m large. On the end wall (Fig. 21.9) it is depicted a mighty arch supported by four pairs of columns with gilded capitals. In the two external niches of the arch, are depicted two persons possibly identifiable with Evangelists, since they are writing.

Behind, (figuratively) the columns it is visible a space where there is an altar with, again, double pairs of columns of light green marble, surmounted by golden capitals, which support a broken tympanum with two allegoric statue placed on the edges. The altar table is evidenced with straight lines and, at the center of the front marble slab, there is plate with the inscription: S. Ignazio di Loyola. Next, with great surprise, the painting above the altar is a canvas. It is encapsulated in real golden ribs and represents the glory of St. Ignatius. It is painted with the technique of *tempera on canvas* and it is about a meter high.

The theme of Pozzo's paintings on the walls of the corridor concern the life of St. Ignatius, the founder of the Jesuits order. Pozzo was a Jesuit Brother and it is quite obvious that he chose this corridor, linking the various rooms of the apartment that St. Ignatius inhabited during the last years of his life, as the best place were to recount the miracles of St. Ignatius. As it can be seen in Fig. 21.10 Pozzo decided to mix the normal painting technique with that of *anamorphosis*. The reason for this choice is the fact that Pozzo wanted to *narrate* Ignatius' life using his painting abilities. Thus, he decided to put the pages of his *painted book* on the walls of

Fig. 21.9 South East end wall of the Corridor [2]

Fig. 21.10 (*left*) North wall of the Corridor. (*right*) South wall of the Corridor [2]

the corridor. Each individual *page – panel* would be configured by the two different techniques mentioned above. Thus, the *reading – observation* would, necessarily, be done by standing on a precise spot, placed in the center of the corridor and evidenced by a circle incorporated in a rhomboid design. Standing there the observer is able see a perfect image, while outside of it the anamorphic deformations are evidenced.

Pozzo does not want the observer to deambulate in front of each painting, but stay still and concentrate on the theme of the painting. At this stage, Pozzo decides to supply what is usually required to increase the gratification of mind: soft music. Hence, at the end of the corridor, he painted two angels playing musical instruments. They sit in front of the altar over which towers the figure of St. Ignatius (Fig. 21.9). They are seating at the entrance, under the arch, and facing each other. The one on the left, dressed in white with a red mantle, plays a violin (or *viola da braccio*), while the other, dressed in light violet with a green mantle, plays a *viola da gamba* with eight strings. An unusual musical instrument that brings to mind the Parnaso fresco, by Raffaello, in the Stanza della Segnatura, in the Vatican. Both Raffaello and Pozzo added strings to the instruments which usually carried seven. Raffaello added two, supposedly because he wanted to pair the Nine Muses, Nine Poets, Nine Writers, depicted in the fresco, while Pozzo adds just one. Most probably Pozzo was aware of the claim by the English musician William Young of having invented the eight string *viola da gamba*. This is confirmed by the correspondence that Young had with Athanasius Kircher who had erroneously attributed it to somebody else. Pozzo and Kircher did not knew each other, since Kircher died in Rome on the 28 November 1680, while Pozzo arrived in Rome on June 1681. But both were Jesuits and both had their laboratories at the Collegio Romano.

The next area that it is to be analyzed concerns the two sides of the Corridor (Fig. 21.10). For this type of decoration, as stated, Pozzo decided to use the technique of *anamorphosis*, but only for certain figures. In fact the big scenes of the life of Ignatius that he paints in the spaces between windows are realized with the traditional linear perspective. Very practically, as we see, Pozzo places the *proper point of view* in the center of the Corridor, implementing the unique point of view from where the observer will have a correct view of the figures. Needless to say that to implement such painting for individual figures, anywhere in the room, was a very painstaking enterprise due to the exact calculations that he had to do. (No electronic computing at that time!).

Pozzo invents an ambient where the architecture is completely fictitious, since it is totally the result of painting and the *anamorphosis* increases this illusory sensation. Thus the real space is merged with the imaginary one. *Anamorphosis*, is a perspective virtuosity. Pozzo, who bases all his works on perspective, exalted the *anamorphosis* and personalized it. The motivation that drives Pozzo is due to his firm belief in the correct application of the perspective painting. The fixed point, from which the visitor is forced to look at the frescos, allows also the visitor to ascertain and verify, in a practical way, the profound divergence between the inanity of the ephemeral reality and the rock stability of the true vision.

All of the above is stated clearly, by Pozzo himself in his exhaustively illustrated treatise "Perspectiva Pictorum et Architectorum" where he wrote his famous advice: "tirar sempre tutte le linee [. . .] al vero punto dell'occhio che é la gloria divina", which freely translates into: "direct all the lines from the true point in the eye, where the divine glory resides". Pozzo believes that the technique of *anamorphosis* lets you see all the painted structures and persons (e.g. architectures, angels, etc,) in a distorted form, but when you "correct" your personal observation by putting yourself in the proper point of view, you see all the same things in its proper form, as created by God. Therefore the *source* of view, the eye, coincides with the residence of the Creator, or God.

21.3 The Side Paintings

We will proceed in looking at the side walls paintings proceeding from the entrance of the Corridor toward the end. The first fresco, on the south wall, (Fig. 21.10 (right)) regards the miracles of the oil in the lamp of St. Ignatius in the Manresa grotto. Ignatius, in 1522, decided to terminate his military life and decided to donate the arms, that he used as knight, to the Madonna celebrated in the Chapel of Montserrat. After that he retired, for a time, in a grotto in nearby Manresa. After his death, to remember this fundamental period of his life, a painting of him was placed in the grotto and an oil lamp in front of it. Very soon, sick people who visited the grotto claimed recovery from their illness due to the fact that they were anointed with the oil from that lamp. So what the fresco is celebrating is a series of miracles that occurred in that grotto if the sick person was anointed with the oil from the lamp. The writing on the top of the frame says: "Contactu olei Sancto Ignatio sacri, sanitas restituitur ac vita" ("The good health and life are returned if anointed with the sacred oil of St. Ignatius").

The second fresco, on the same side of the Corridor, concerns a miracle occurred in Rome, on the street in front of the Casa Professa, on April 19, 1541. A young man who was serving in Casa Professa, became possessed by devil and was acting accordingly. Notwithstanding the generous help of all the people around, he could not be refrained. But as soon as Ignatius came back from his errand in the city, he pointed the finger to the young man and the devil flew away from him. This is actually the report of the people who participated or just witnessed the happening. The fresco emphasizes the architecture of the place, but for scenic purpose only. But one thing is hidden in the painting. Right behind Ignatius there is a *novizio* who is leaning out of a window talking to two kids. His face is very familiar: it is the face of Andrea Pozzo!

The third fresco, on the same side of the Corridor, does not concern a miracle, but the important theme of the Sacred Family. Pozzo paints in fresco the painting and the its golden frame. Pozzo again plays with the observer: this time, the observer thinks that is in front of real painted canvas, like at the end of the Corridor, but Pozzo has inverted the play. This time is a real fresco! Pozzo's version of the Sacred Family is filled with a description of what is reciprocal love in a family. The look that Mother and Son exchange is something that one expect in real life, but Pozzo is capable of reproducing it. But also the look of Joseph reveals his feeling of participation in this exchange of love between Mother and Son. Artistically speaking it is to be noticed the choice of the colors: while the face of the Madonna is outlined and the flesh color is brightened up together with hair. This is in contrast with classical iconography of Joseph, as an old man with a white beard and covered with a dark brown mantle. This enter in a chromatic contrast with the bright and clear colors of Mary and Jesus. The uniqueness of this fresco is also evidenced by the fact that on top cross bar does not carries a painted plaque with a theme description. In addition, in the bottom part there are not two Putti, but two young Angels which attend to the vase of flowers.

The fourth fresco, on the same side of the Corridor, illustrates the theme of a painting of St. Ignatius realized by an angelic hand. *Sancti Ignatii effigies, coelesti manu depicta*. On the top left corner it is visible a green curtain kept tied to the top, thus leaving wide open a room where, at the center, there is a young person wearing a dress whose top is white and the rest light blue. He has blond hair and the face has very delicate traits and he is seating at the center of the room engaged in painting, "caelesti manu", a picture of a man. It is easy to recognize that the man in the painting is Ignatius and the painter is an angel who, as good artist keeps the painter's palette in the left hand, while hold the bristle brush in the right hand. Around him there are four persons who look upon this miraculous event enraptured. The scene does not represent a specific event with connection to the life of Ignatius, but it refers to the fact that Ignatius refused, during all his life, to be portrayed. With this painting, Pozzo wishes to make a "spiritual portray" of the Saint who, after all, was regarded as Saint, even during his lifetime. As such it was impossible to portray him, in his lifetime, when his sainthood was so evident, only with his human reality.

The message that Pozzo sends to the observer of his works, at any age in time, is very simple: "*Delectare Docendo*" ("Delight while Teaching").

References

1. Insolera, L. S., & Boesel, R. (2010). *Andrea Pozzo*. Roma: Artemide.
2. Insolera, L. S. (2014). *Andrea Pozzo e il Corridoio di S. Ignazio*. Roma: Artemide.

Part III
Short Seminars

Chapter 22
Effective Oscillator Strengths of Tb^{3+} Ions in a Garnet Crystal Determined from Low Temperature Magneto-Optic Rotations

Muhammad Sabieh Anwar

Abstract In the presentation, I quickly reviewed, from a quantum mechanical perspective, how electric dipole transitions can contribute to the rotation of the plane of polarization of light in a gyrotropic medium. This is called the magneto-optic Faraday effect. This was followed by an investigation of ultra-large Faraday rotations in the garnet single crystal at cryogenic temperatures. A Fourier analysis coupled with Mueller calculus allows one to unambiguously determine the Faraday rotation and ellipticity. Finally, from rotary dispersion data, it is possible to determine the effective oscillators strengths of effective electric dipole transitions.

M.S. Anwar (✉)
Department of Physics, Lahore University of Management Sciences, Lahore, Pakistan
e-mail: sabieh@gmail.com

© Springer Science+Business Media Dordrecht 2017

B. Di Bartolo et al. (eds.), *Nano-Optics: Principles Enabling Basic Research and Applications*, NATO Science for Peace and Security Series B: Physics and Biophysics, DOI 10.1007/978-94-024-0850-8_22

Chapter 23
Analysis of Surface Layer Properties of Evaporating Microdroplet of Aqueous SiO_2 Nanospheres Suspension with Sodium Dodecyl Sulfate

Justice Archer, Maciej Kolwas, Genadij Derkachov, Mariusz Woźniak, Daniel Jakubczyk, and Krystyna Kolwas

Abstract We studied the evolution of interference patterns and intensities of polarized and depolarized scattered laser light on a single evaporating microdroplet of suspension composed of silica (SiO_2) spherical nanospheres (225 nm radius) and sodium dodecyl sulfate (SDS) dispersed in diethylene glycol (DEG). Our experimental set-up description and procedures can be found elsewhere in (Jakubczyk D, Derkachov G, Kolwas M, Kolwas K, J Quant Spectrosc Radiat Trans 126:99–104, 2012; Kolwas M, Kolwas K, Derkachov G, Jakubczyk D, Phys Chem Chem Phys 17:6881–6888, 2015; Jakubczyk D, Kolwas M, Derkachov G, Kolwas K, Zientara M, Acta Phys Pol-Ser A Gen Phys 122:709, 2012).

In our experiments, the refractive indices of the liquid (diethylene glycol (DEG): $n_{D20} = 1.45$), the silica (SiO_2: 225 nm radius nanospheres: $n_{D20} = 1{:}46$), and the sodium dodecyl sulfate (SDS: $n_{D20} = 1{:}46$) were nearly equal. Therefore, the suspension microdroplet can be treated as optically homogeneous. As the liquid evaporates, we observe successive stages of drying processes manifested in the light scattering properties as a result of the presence of the SDS and the SiO_2 nanosphere inclusions. We see: (1) Mie scattering patterns typical for homogeneous sphere with well-pronounced whispering gallery modes (WGMs), then (2) a sudden appearance of short lived but intense "spiky modes" followed by (3) a mixture of fluctuating scattered modes (speckles) from the dried microobject. The observed changes in the interference patterns and the scattered laser light intensities (Ivv, Ihh, Ivh, and Ihv) originate from the evolving surface morphologies of the forming microobject during the slow drying process (Langmuir 31:7860–7868, 150627110639005, 2015). Figure 23.1 presents the evolution of the scattered light intensities and snapshots of interference patterns captured by CCD camera from video recording

J. Archer (✉) • M. Kolwas • G. Derkachov • M. Woźniak • D. Jakubczyk • K. Kolwas
Institute of Physics of the Polish Academy of Sciences, Al Lotników 32/46, 02-668 Warsaw, Poland
e-mail: archer@ifpan.edu.pl

© Springer Science+Business Media Dordrecht 2017

B. Di Bartolo et al. (eds.), *Nano-Optics: Principles Enabling Basic Research and Applications*, NATO Science for Peace and Security Series B: Physics and Biophysics, DOI 10.1007/978-94-024-0850-8_23

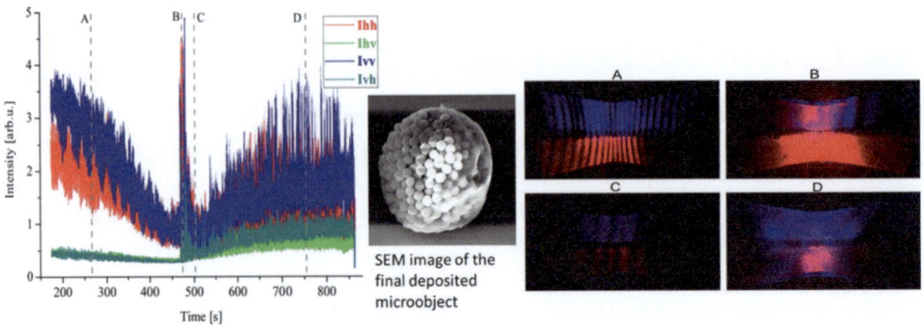

Fig. 23.1 Evolution of the scattered light intensities and snapshots of interference patterns captured by CCD camera from video recording of the evaporation process and a SEM image of the final deposited quasi- spherical dried microobject

of the evaporation process and a SEM image of the final deposited quasi- spherical dried microobject.

We found also the temporal evolution of the effective droplet radius on the basis of Mie scattering theory with the analysis of whispering gallery mode (WGMs) and the Mie Scattering Look-up Table Method (MSLTM) (Jakubczyk D, Derkachov G, Kolwas M, Kolwas K, J Quant Spectrosc Radiat Trans 126:99–104, 2012; Appl Spectros 52:284–291, 1998). The evolution of the effective droplet radius made possible the determination of the SDS surface volume concentrations near the droplet surface. This study serves as the basis for wide range of novel surface diagnostic methods for configurational changes in complex systems of nano – microparticles evolving at the sub-wavelength scale.

Acknowledgment This work was supported by the National Science Center, Poland under grant number 2014/13/D/ST3/01882

References

1. Jakubczyk, D., Derkachov, G., Kolwas, M., & Kolwas, K. (2012). Combining weighting and scatterometry: Application to a levitated droplet of suspension. *Journal of Quantitative Spectroscopy and Radiative Transfer, 126*, 99–104.
2. Kolwas, M., Kolwas, K., Derkachov, G., & Jakubczyk, D. (2015). Surface diagnostics of evaporating droplets of nanosphere suspension: Fano interference and surface pressure. *Physical Chemistry Chemical Physics, 17*, 6881–6888.
3. Jakubczyk, D., Kolwas, M., Derkachov, G., Kolwas, K., & Zientara, M. (2012). Evaporation of micro-droplets: The radius-square-law revisited. *Acta Physica Polonica-Series A General Physics, 122*, 709.
4. Woźniak, M., Derkachov, G., Kolwas, K., Archer, J., Wojciechowski, T., Jakubczyk, D., et al. (2015). Formation of highly-ordered spherical aggregates from drying microdroplets of colloidal suspension. *Langmuir, 31*, 7860–7868. 150627110639005.
5. Popp, J., Lankers, M., Trunk, M., Hartmann, I., Urlaub, E., & Kiefer, W. (1998). High-precision determination of size, refractive index, and dispersion of single microparticles from morphology-dependent resonances in optical processes. *Applied Spectroscopy, 52*, 284–291.

Chapter 24
Compressed Sensing Techniques Applied to the Reconstruction of Magnetic Resonance Images

Francesco Baldacchini

Abstract Magnetic resonance Imaging (MRI) is nowadays a central technique in medical diagnostics but, in order to acquire a high resolution image of the human body, long acquisition times are still needed. MRI scanners have traditionally been limited to imaging static structures over a short period, and the patient has been instructed to hold his or her breath. However, the image can now be treated as a sparse signal in space and time, and MRI scanners have begun to overcome the previous limitations and produce images, for example, of a beating heart. Compressive sensing, and more generally the possibility of efficiently capturing sparse and compressible signals, using a relatively small number of measurements, paves the way for a number of possible applications. New physical sampling devices may be designed that directly record discrete low-rate incoherent measurements of the analog signal, which is needed for the completeness of the signal itself. This should be especially useful in situations where large collections of samples may be costly, difficult or impossible to obtain. A digital camera newly developed by Richard Baraniuk and Kevin Kelly at Rice University (see dsp.rice.edu/cs/cscamera) provides a particularly interesting example of successful implementation of compressive sensing methodology. In the detector array of a conventional digital camera, each pixel performs an analog-to-digital conversion; for example, the detector on a 5-megapixel camera produces 5 million bits for each image. This large amount of data is then dramatically reduced through a compression algorithm (using wavelet or other techniques), so as not to overburden typical storage and transfer capacities. Rather than collect 5 million pixels for an image, the new camera samples only 200,000 single-pixels that provide an immediate 25-fold savings in data collected compared with 5 megapixels. However, CS-MRI is still in its infancy, and many crucial issues remain unsettled. These include:

- optimizing sampling trajectories,
- developing improved sparse transforms that are incoherent to the sampling operator,

F. Baldacchini (✉)
Via Guglielmo Quattrucci 246, 00046 Grottaferrata, Roma, Italy
e-mail: frankbald@gmail.com

© Springer Science+Business Media Dordrecht 2017
B. Di Bartolo et al. (eds.), *Nano-Optics: Principles Enabling Basic Research and Applications*, NATO Science for Peace and Security Series B: Physics and Biophysics, DOI 10.1007/978-94-024-0850-8_24

- studying reconstruction quality in terms of clinical significance,
- improving the speed of reconstruction algorithms.

Therefore, there are still fascinating theoretical and practical research problems, promising substantial payoffs in improved medical care. In conclusion, a lot of work is still waiting ahead, but signal acquisition and processing field are finally off the constrictions imposed by previous theoretical and practical limits.

Chapter 25
FDTD Method and HPC for Large-Scale Computational Nanophotonics

Antonino Calà Lesina, Alessandro Vaccari, Pierre Berini, and Lora Ramunno

Abstract From 2010 to 2014 the number of papers using the Finite-Difference Time-Domain (FDTD) method for nanophotonics has doubled (from more than 700 to 1430 papers – Google Scholar, searching "FDTD and nanophotonics"). The simulation of the light-matter interaction at the nano-scale requires very fine mesh-sizes to capture the physics, as for example in nanoplasmonics. Large computational resources are needed to achieve a good accuracy, and not taking care of that could result in non-convergent or inaccurate calculations. The availability of a high-performance computing (HPC) system (a massively parallel IBM Blue Gene/Q of the Southern Ontario Smart Computing Innovation Platform – SOSCIP) allows us to perform high-resolution simulations otherwise impossible using an in-house parallel 3D-FDTD code. Here we demonstrate the convergence of the FDTD method simulating the interaction of a linearly-polarized plane-wave with a gold nanosphere of radius 60 nm in the range 400–700 nm. The full study is reported in (Calà Lesina et al., On the convergence and accuracy of the FDTD method for nanoplasmonics, Opt Exp 23(8), 2015). The problem of the scattering of a plane-wave by a sphere is analytically known from the Mie theory. Varying the mesh size down to the limit allowed by the supercomputer, which has 65,536 cores, the numerical results are compared to the analytical ones. This study can be considered as a guideline for a proper FDTD simulation setup.

In Fig. 25.1 we use per-component mesh, double-precision, and Drude+2CP. A mesh size of 0.5 nm is required to get an average percentage error <*%err*> less than 1 %. In Fig. 25.2 we use uniform staircase mesh; we observe a slower convergence. In Fig. 25.3 we use single-precision; we observe divergences due to accumulating round-off errors. In Fig. 25.4 we use the standard Drude model which do not model the interband transitions.

A. Calà Lesina (✉) • P. Berini • L. Ramunno
Department of Physics and Centre for Research in Photonics, University of Ottawa, Ottawa, ON, Canada
e-mail: antonino.calalesina@uottawa.ca

A. Vaccari
Centre for Materials and Microsystems, Fondazione Bruno Kessler, Trento, TN, Italy
e-mail: vaccari@fbk.eu

© Springer Science+Business Media Dordrecht 2017
B. Di Bartolo et al. (eds.), *Nano-Optics: Principles Enabling Basic Research and Applications*, NATO Science for Peace and Security Series B: Physics and Biophysics, DOI 10.1007/978-94-024-0850-8_25

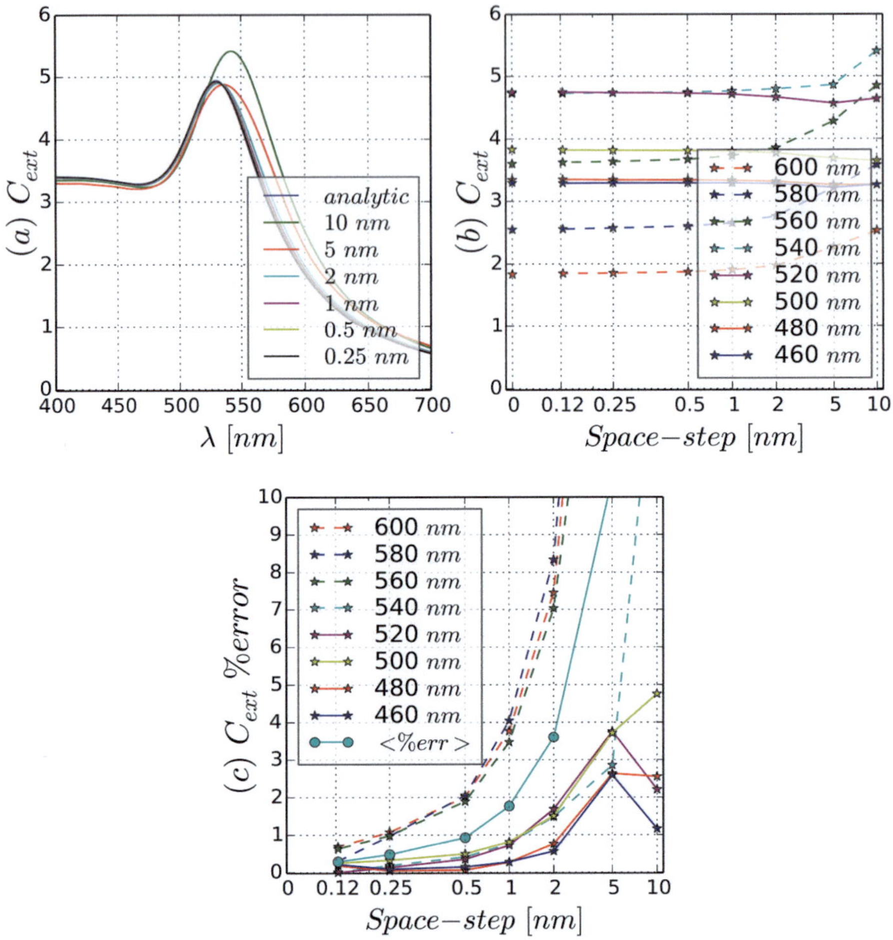

Fig. 25.1 Gold sphere C_{ext} convergence study: (**a**) and (**b**) spectra, (**c**) percentage error. Reference simulation setup: per-component mesh, double precision, and Drude+2CP model

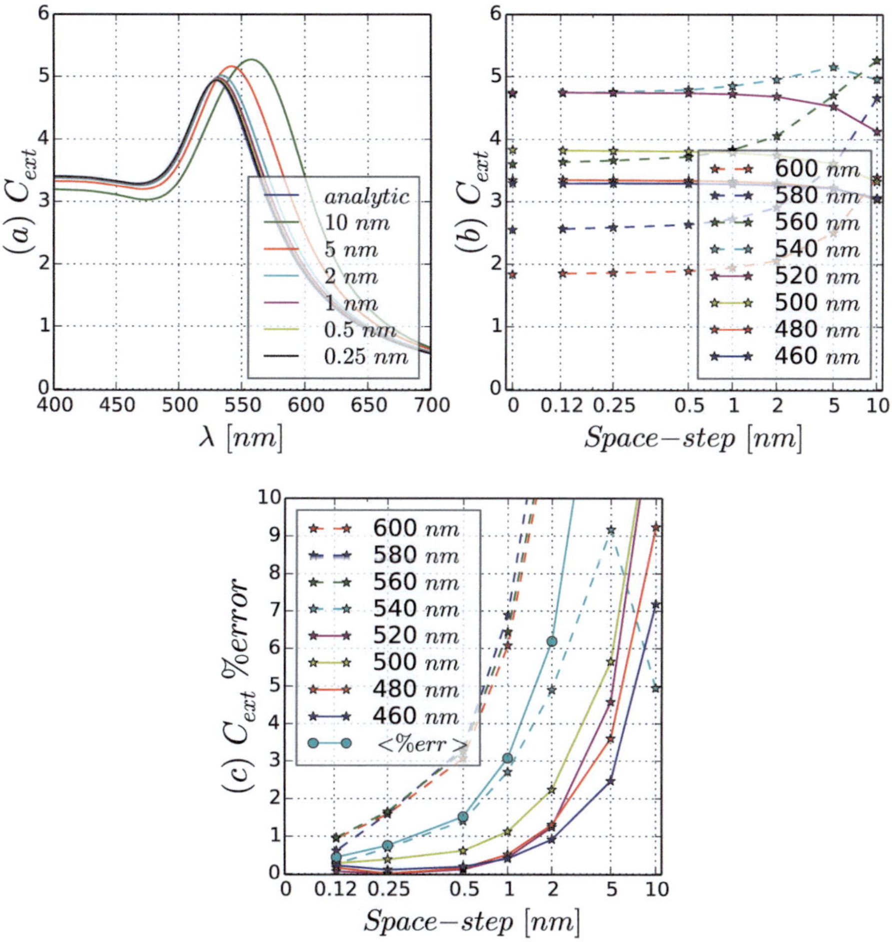

Fig. 25.2 Gold sphere C_{ext} convergence: (**a**) and (**b**) spectra, (**c**) percentage error. Simulation setup as in Fig. 25.1, except uniform staircase

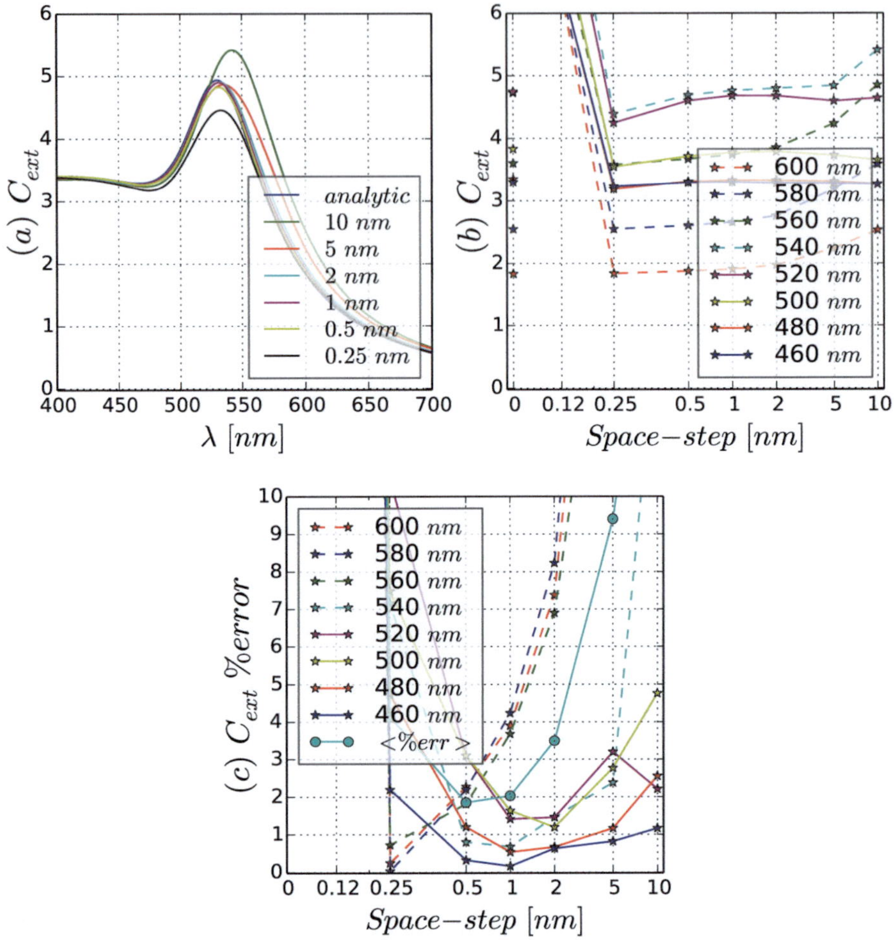

Fig. 25.3 Gold sphere C_{ext} convergence: (**a**) and (**b**) spectra, (**c**) percentage error. Simulation setup as in Fig. 25.1, except single-precision

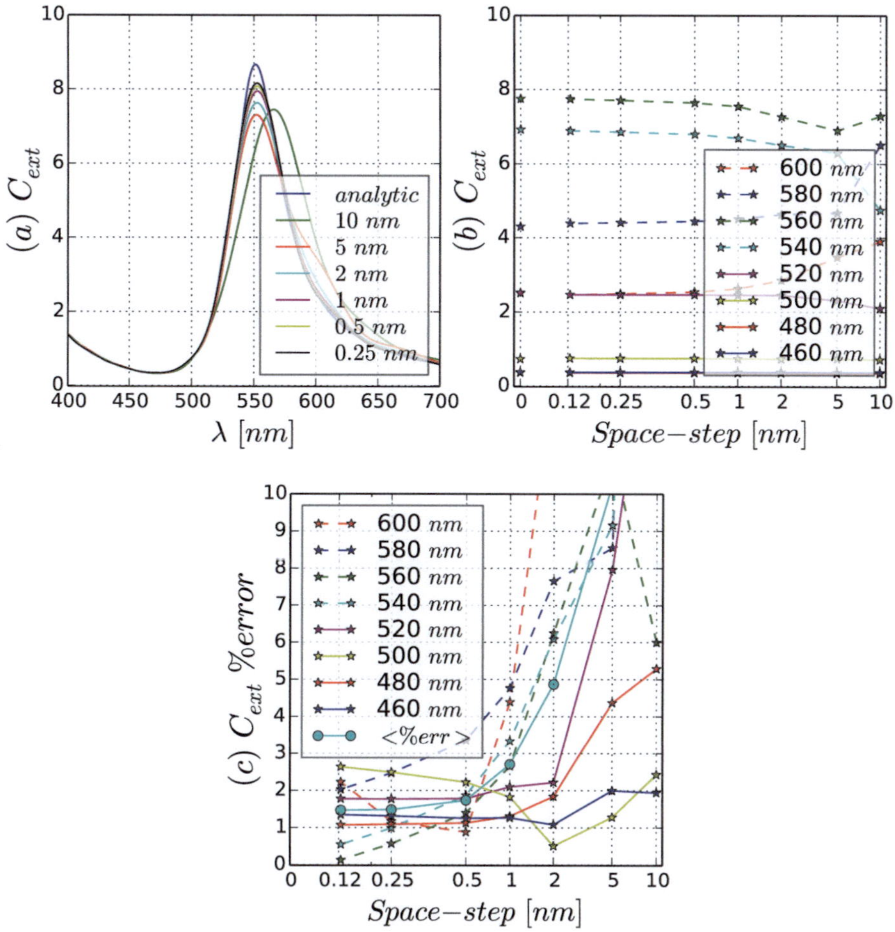

Fig. 25.4 Gold sphere C_{ext} convergence: (**a**) and (**b**) spectra, (**c**) percentage error. Simulation setup as in Fig. 25.1, except standard Drude model

Chapter 26
Investigation of the Luminescence Spectral Profiles and the Efficiencies of Yb^{3+}, Nd^{3+}, Tm^{3+} Doped Y_2O_3-SiO_2 Nano-phosphors

Hatun Cinkaya, Gokhan Bilir, Murat Erdem, and Gonul Eryurek

Abstract Among various developed UC luminescence materials, rare earth doped yttrium silicates (YSO) exhibit high chemical durability and thermal stability which are suitable for many applications such as lighting, lasers, optical sensors and biological applications. The reduction to nano size of such materials aims to investigate the issue of how the optical properties affected, and thus to further extend the use of these materials in photonic industry. The studies about the effect of nanoscale to the lattice- electron interaction is not yet enough to understand the mechanisms. For this reason, the influence of the various synthesis techniques to the size and the effect of the size of those nano-materials on the optical and thermal properties are basic research areas for various photonic device design and their developments.

In most cases, the rare earth ions replace other ions of similar size and same valence state in the host medium. The dopant ions (Tm^{3+}, Nd^{3+} and Yb^{3+}) replace the Y^{3+} ions in the host, randomly. Because of the dopant ions have a different ionic radius than the yttrium ions, they distort the lattice around them. As a consequence, the local environment of the different dopants gives different absorption and radiation energies for each single ion in the host. The selection of host material change the luminescent effects on the emission properties such as obtaining different color in emitted light.

Apart from choosing the host material and rare earth dopants, the synthesis technique also affects the optical and structural properties. Sol-gel method has great significance because the technique allows to control the during all process (Boilot JP, Gacoin T, Perruchas S, C R Chim 13:186, 2010). There are lots of advantages such as devices and materials that are used in sol-gel technique are very simple. The needed temperature for processes without densification is low such as room

H. Cinkaya (✉) • G. Bilir • G. Eryurek
Department of Physics Engineering, Faculty of Science and Letters, Istanbul Technical University, Maslak, 34469 Istanbul, Turkey
e-mail: cinkayahatun@itu.edu.tr

M. Erdem
Department of Physics, Marmara University, 34722 Istanbul, Turkey

© Springer Science+Business Media Dordrecht 2017
B. Di Bartolo et al. (eds.), *Nano-Optics: Principles Enabling Basic Research and Applications*, NATO Science for Peace and Security Series B: Physics and Biophysics, DOI 10.1007/978-94-024-0850-8_26

temperature. Energy saving, the synthesis of materials with high homogeneity, the controllability of the microstructure of the material and not entering the pollution to the lattice are other advantages of the sol-gel method (Mohammad MR, Fray DJ, Particuology 9:235–242, 2011).

In this study, luminescent yttrium silicate Y_2O_3-SiO_2 host material doped with Yb^{3+}, Nd^{3+} and Tm^{3+} rare-earth ions in various proportions were obtained by the sol–gel method and the samples were annealed at 1150 °C during 12 h. After the annealing process, the powder sample form was obtained. Structural and luminescence properties of the prepared nanophosphors were studied experimentally by using different analytical techniques. For the structural characterization Fourier Infrared spectroscopy (FTIR), X-Ray Diffraction (XRD), Scanning Electron microscopy (SEM) were performed. FTIR spectra of un-doped yttrium silicate and triple doped with Yb^{3+}, Nd^{3+} and Tm^{3+} rare-earth ions were recorded for un-annealed gel and annealed powder samples. In the range of 3000–4000 cm^{-1}, OH stretching vibrations of water and OH groups appear and 400–1100 cm^{-1} where the Si–O and Y–O vibrations occurred (Rezvani Nikabadi H, Shahtahmasebi N, Rezaee Rokn-Abadi M, Bagheri Mohagheghi M, Goharshadi EK, Phys Scr 87:025802, 2013).

XRD results showed that the formation and stability of high-temperature crystalline phases were affected by the rare earth ions and mostly α-$Y_2Si_2O_7$ structure is obtained for un-doped and triple doped nanophosphor powders (according to JCPDS 38-0223 card) (Becerro AI, Escudero A, Phase Transit Multinatl J 77(12):1093–1102, 2004).

SEM observations of the sample were investigated at low (50,000×) and high magnification (200,000×) for annealed powder samples and confirmed the average particle sizes estimated as 80–100 nm with XRD results.

The optical absorption and emission measurements were performed to determine spectral properties of yttrium silicate Y_2O_3-SiO_2 host material doped with Yb^{3+}, Nd^{3+} and Tm^{3+} rare-earth ions and the energy transfer processes between dopant lanthanide ions were studied at room temperature. The band gap energies for both direct and indirect possible transitions energies were measured from the absorption spectrum of each sample. The effect of the dopant concentrations on the spectral features were investigated at room temperature in detail Table. 26.1.

Table 26.1 Direct and indirect transition energies of yttrium silicates doped with rare earth ions

Sample name and composition (Mole %)	Thickness (cm)	Direct transition E_{opt} (eV)	Indirect transition E_{opt} (eV)
SY1 (undoped YSO)	0.34	4.1815	3.0534
HT1 (% 0.5 Tm^{3+} doped YSO)	0.31	4.2379	3.3865
HY1 (% 10Yb^{3+} doped YSO)	0.115	4.2554	3.4519
HN1 (% 1Nd^{3+} doped YSO)	0.295	4.2398	3.1077
HNYT1 (triple doped YSO)	0.12	4.1918	3.1852

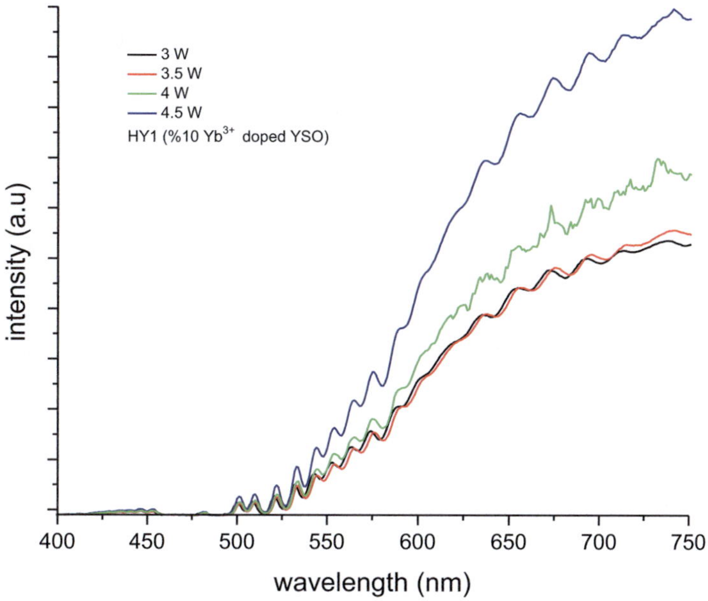

Fig. 26.1 White light generation of Y_2O_3-SiO_2 host material doped with %10 (per mole) Yb^{3+} ion with varying power intensity

Both the emission and upconversion processes were analyzed for samples by using laser spectroscopy technique excitation under 975 nm. The emission spectra in the range of 850–1700 nm and up conversion spectra in the range of 350–800 nm of the samples (yttrium silicates doped with Tm^{3+}, Nd^{3+} and Yb^{3+} ions) were performed at room temperature in detail. The production of white light was achieved by illuminating the yttrium silicates doped with Yb^{3+} nano-powder sample with the beam of the laser diode operating at 975 nm with an output power of at least 3 W. The white light spectrum (440–870 nm) obtained in our measurements is given in Fig. 26.1.

As seen from figure we observed a notable change of the white light intensity with the varying pumping power (Bilir G, Ozen G, Collins J, Cesario M, Di Bartolo B, IEEE Photonics J 6:8200518, 2014).

References

1. Boilot, J.-P., Gacoin, T., & Perruchas, S. (2010). Synthesis and sol-gel assembly of nanophosphers. *Comptes Rendus Chimie, 13*, 186–198.
2. Mohammad, M. R., & Fray, D. J. (2011). Sol-gel drived nanocrystalline and mesoprous barium strontium titanate prepared at room temperature. *Particuology, 9*, 235–242.

3. Rezvani Nikabadi, H., Shahtahmasebi, N., Rezaee Rokn-Abadi, M., Bagheri Mohagheghi, M., & Goharshadi, E. K. (2013). Gradual growth of gold nanoseeds on silica for SiO2 gold homogeneous nano core/shell applications by the chemical reduction method. *Physica Scripta, 87*, 025802.
4. Becerro, A. I., & Escudero, A. (2004). Revision of the crystallographic data of polymorphic $Y_2Si_2O_7$ and Y_2SiO_5 compounds. *Phase Transitions: A Multinational Journal, 77*(12), 1093–1102.
5. Bilir, G., Ozen, G., Collins, J., Cesario, M., & Di Bartolo, B. (2014). Luminescent properties of nanophosphors. *IEEE Photonics Journal, 6*, 8200518.

Chapter 27
Surface Electromagnetic Waves Guided by Non-metallic Interfaces

Muhammad Faryad

Abstract Surface electromagnetic waves guided by interfaces with metal are called surface plasmon-polariton (SPP) waves, and are much more famous than their counterparts that are guided by interfaces formed by dielectric materials. The surface wave electromagnetic waves can be guided if at least one of the partnering dielectric materials is periodically nonhomogeneous normal to the direction of propagation, and/or one of the partner is anisotropic. The surface can still exist if both partnering dielectric materials are isotropic and homogeneous, provided that at least one of them is lossy. These surface waves find applications in optical sensing of changes in the refractive index of fluids, and could potentially be used for on chip communication.

M. Faryad (✉)
Department of Physics, Lahore University of Management Sciences, Lahore, Pakistan
e-mail: muhammad.faryad@lums.edu.pk

© Springer Science+Business Media Dordrecht 2017

B. Di Bartolo et al. (eds.), *Nano-Optics: Principles Enabling Basic Research and Applications*, NATO Science for Peace and Security Series B: Physics and Biophysics, DOI 10.1007/978-94-024-0850-8_27

Chapter 28
Metal-Enhanced Fluorescence in Plasmonic Waveguides

Alexandre Grégoire and Denis Boudreau

Abstract The field known as plasmonics has seen a tremendous growth in interest from the research community. One of the more popular topics is the surface plasmon waveguide. As an alternative to solid linear nanostructures, a chain of plasmonic nanoparticles can also be used to propagate a traveling wave of surface charges. Experimentally, plasmonic nanoparticles lend themselves well to the fabrication of plasmonic waveguides via the self-assembly of nanoparticles into linear structures with very small inter-particle separation. In the present work, we used wrinkled PDMS stamps to fabricate large (i.e., several microns in length) one-dimensional nanoparticle assemblies. Combining this technique with the use of tunable, metal dielectric, core-shell nanoparticles [1] allows the fabrication of linear assemblies with adjustable inter-particle separation. The observation of isolated nanoparticle assemblies by single particle microscopy/spectroscopy (epifluorescence, dark-field scattering, fluorescence lifetime microscopy) and scanning electron microscopy allows us to gather optical and structural information in order to derive interesting structure-properties correlations – in particular the interaction between propagating dark plasmons with the emission behavior from nearly fluorophores.

Reference

1. Asselin, J., Legros, P., Grégoire, A., & Boudreau, D. (2016). Correlating metal-enhanced fluorescence and structural properties in Ag@SiO2 core-shell nanoparticles. *Plasmonics, 5*, 1–8.

A. Grégoire (✉) • D. Boudreau
Département de chimie et Centre d'optique, photonique et laser (COPL), Université Laval, Québec (QC), G1V 0A6, Canada
e-mail: alexandre.gregoire.1@ulaval.ca

© Springer Science+Business Media Dordrecht 2017

B. Di Bartolo et al. (eds.), *Nano-Optics: Principles Enabling Basic Research and Applications*, NATO Science for Peace and Security Series B: Physics and Biophysics, DOI 10.1007/978-94-024-0850-8_28

Chapter 29
Nonlinear Optics in TiO_2 Nanoscale Waveguides

Sarah Griesse-Nascimento, Orad Reshef, Michael Moebius, Christopher Evans, and Eric Mazur

Abstract We propose polycrystalline anatase TiO_2 nanoscale waveguides as a platform for efficient, on-chip generation of triplet photons via third-order spontaneous parametric down-conversion (TOSPDC). TiO_2 has a high nonlinear index of refraction (0.16×10^{18} m^2/W at 1565 nm), a wide transparency window (absorption edge at 400 nm) and negligible two-photon absorption above 800 nm, making it an advantageous material for integrated photonics. Efficient generation of three-photon entanglement on an integrated platform is necessary for many experiments in quantum optics, with applications in quantum metrology and communication. We couple simulations to experiments to determine and confirm optimal waveguide geometries for meeting TOSPDC phase-matching constraints and maximizing the efficiency of triplet-photon generation in TiO_2 nanoscale devices.

S. Griesse-Nascimento (✉) • O. Reshef • M. Moebius • E. Mazur
School of Engineering and Applied Sciences, Harvard University, Cambridge, MA, USA
e-mail: sgriesse@seas.harvard.edu

C. Evans
Kavli Institute, Cornell University, Ithaca, NY, USA

© Springer Science+Business Media Dordrecht 2017

B. Di Bartolo et al. (eds.), *Nano-Optics: Principles Enabling Basic Research and Applications*, NATO Science for Peace and Security Series B: Physics and Biophysics, DOI 10.1007/978-94-024-0850-8_29

Chapter 30
Mapping the Local Density of States of Periodic Plasmonic Nanostructures with Stochastic Super-resolution

Ke Guo, Marc Verschuuren, and Femius Koenderink

Abstract Periodic plasmonic nanostructures have been found promising in controlling the directivity and efficiency of solid-state lighting devices, in particular LED-pumped phosphors. Due to the inhomogeneous spatial distribution of the photonic resonances of periodic plasmonic nanostructures, their influence on the emission is strongly dependent on the position of emitters relative to the nanostructures. Therefore, mapping the local dependence of directivity and efficiency can be a key to understand and optimize the devices. Here we introduce a method of mapping local directivity and efficiency of emitters coupled to periodic nanostructures based on stochastic super-resolution. As an example, we show measurements of local density of states (LDOS) induced by a hexagonal lattice of aluminum nanoantennas with a spatial resolution of $\sim$40 nm. With FDTD simulation, we demonstrate that the variation of the decay rate of the emitters in the investigated sample is hardly influenced by the lattice modes and mainly governed by single-particle LDOS variations.

K. Guo (✉) • F. Koenderink
Center for Nanophotonics, FOM Institute AMOLF, Science Park 104, 1098 XG Amsterdam, The Netherlands
e-mail: guo@amolf.nl, f.koenderink@amolf.nl

M. Verschuuren
Philips Research Laboratories, High Tech Campus 4, 5656 AE Eindhoven, The Netherlands
e-mail: marc.verschuuren@philips.com

© Springer Science+Business Media Dordrecht 2017
B. Di Bartolo et al. (eds.), *Nano-Optics: Principles Enabling Basic Research and Applications*, NATO Science for Peace and Security Series B: Physics and Biophysics, DOI 10.1007/978-94-024-0850-8_30

Chapter 31
Design of Optical Nanobiosensors for Detection of Toxic Compounds and Pharmaceutical Products

Ahmed Hassaan, S. Yaneva, and L. Yotova

Abstract Demands for developing optical nanobiosensors for controlling the quality of food [1–4], pharmaceuticals [5], environmental activities [6] and industrial processes [7] have encountered an enormous increase over the last decade. A lot of effort was devoted to design highly sensitive and selective methods for that purpose [8]. Sol-gel approach has rapidly become a fascinating new field of research in materials science. The use of nanosystems and organic molecules in the gel formation process to influence the dimensions of the formed pores represents another strategy to construct optical nanobiosensors and increase the immobilized enzyme activity for more efficient sensing. The aim of our work is a construction of optical nanobiosensors based on horseradish peroxidase (HRP) for quantitative and qualitative detection of toxic compounds, contaminants and pharmaceutical products. Different categories of hybrid matrices were fabricated contained silica nanopartices, cellulose derivatives and Poly (amido amine) dendrimers (PAMAM) as perspective carriers for covalent immobilization. Horseradish peroxidase (HRP) was used as a model enzyme. Conditions were optimized, kinetic parameters, pH and temperature optimums were determined. Constructed nanobiosensors were implemented to optically detect compounds like phenol, resorcinol, epinephrine and acetaminophen. Results showed that the highest parameters were achieved by TMOS/CAB/PAMAM based matrices, where records of relative activity reached 92 % while having a shift for the optimum pH that recorded 6.5 and the optimum temperature that recorded 40 °C. The applied system demonstrated enhanced operational potential towards designing biosensor for medical, pharmaceutical, food industry as well as environmental monitoring purposes.

A. Hassaan (✉)
Department of Biotechnology, University of Chemical Technology and Metallurgy, Ohrdiski 8, 17563 Sofia, Bulgaria
e-mail: ahmed.hassaan@ymail.com

S. Yaneva • L. Yotova
Department of Fundamentals of Chemical Technologies, University of Chemical Technology and Metallurgy, Blvd. Kl. Ohrdiski 8, 17563 Sofia, Bulgaria

© Springer Science+Business Media Dordrecht 2017

B. Di Bartolo et al. (eds.), *Nano-Optics: Principles Enabling Basic Research and Applications*, NATO Science for Peace and Security Series B: Physics and Biophysics, DOI 10.1007/978-94-024-0850-8_31

References

1. Willets, K. A., Hall, W. P., Sherry, L. J., Zhang, X., Zhao, J., & Van Duyne, R. P. (2007). *Nanoscale localized surface Plasmon resonance biosensors* (pp. 159–173). Weinheim: Wiley-VCH Verlag GmbH & Co. KGaA.
2. Abdullah, J., Ahmad, M., Heng, L. Y., Karuppiah, N., & Sidek, H. (2006). Chitosan-based tyrosinase optical phenol biosensor employing hybrid nafion/sol-gel silicate for MBTH immobilization. *Talanta, 70*, 527–32.
3. Huang, H., Tang, C., Zeng, Y., Yu, X., Liao, B., Xia, X., Yi, P., & Chu, P. K. (2009). Label-free optical biosensor based on localized surface plasmon resonance of immobilized gold nanorods. *Colloids and Surfaces B: Biointerfaces, 71*, 96–101.
4. Ozsoz, M., & Wang, J. (1991). Tomato seed-based amperometric sensor for the determination of alcohols. *Electroanalysis, 3*, 655–658.
5. Deshpande, M. V., & Hall, E. A. (1990). An electrochemically grown polymer as an immobilisation matrix for whole cells: Application in an amperometric dopamine sensor. *Biosensors and Bioelectronics, 5*, 431–448.
6. Erdem, A., Altıniğne, N., Kılınç, E., Gokgunnec, L., Dalbastı, T., & Ozsoz, M. (1998). Amperometric biosensor based on mushroom tissue tyrosinase for the determination of phenolic compounds. *Journal of Pharmaceutical Society of Ankara, 23*, 1–6.
7. Kilinc, E., Erdem, A., Gokgunnec, L., Dalbasti, T., Karaoglan, M., & Ozsoz, M. (1998). Buttermilk based Cobalt Phthalocyanine dispersed Ferricyanide mediated amperometric biosensor for the determination of Xanthine. *Electroanalysis, 10*, 273–275.
8. Kaushik, A., Arya, S. K., Vasudev, A., & Bhansali, S. (2013). Recent advances in detection of Ochratoxin-A. *Open Journal of Applied Biosensor, 2*, 1–11.

Chapter 32
Random Nanocomposites: Fundamental Properties and Application for Harmful Agents Detection

Ivan Karbovnyk

Abstract Versatile composite nanostructured materials with random structure have significant potential in chemical sensing, radiation detection and shielding. Below, two types of such materials are discussed: metal-dielectric films with random morphology and nanocomposites fabricated by random dispersion of nanoparticles or nanotubes in the insulating host matrix. The influence of chemical or radiological factors on these materials changes structural, electrical and optical properties allowing such structures to serve the purpose of detecting and, in some cases, neutralization of harmful agents.

A possible way of obtaining metallic (in particular, silver and gold) random films is a vacuum deposition of metal on dielectric (*e.g.* glass or silicon) substrates. Using such technique one can grow films with an average thickness ranging from several nanometers to hundreds of nanometers. Film morphology can be influenced by adjusting the parameters of the deposition process.

Electromagnetic response of nanostructured films can be studied by using traditional spectroscopic techniques, for example, impedance spectroscopy or optical absorption spectroscopy. Important information can also be extracted by simulation of the electromagnetic wave – film surface interaction. This is effectively performed by reconstructing the surface morphology using statistical simulations followed by finite difference time domain modeling of the electromagnetic properties.

Since the surrounding of the film surface largely determine the optical properties, optical absorption spectrum of the nanostructured film can be used as an indicator of the presence of specific contaminants in the environment. For example, silver film with 5 nm mass thickness deposited on glass shows the absorption peak around 500 nm (Karbovnyk I, Collins J, Bolesta I, Stelmashchuk A, Kolkevych

---

I. Karbovnyk (✉)
Department of Electronics, Ivan Franko National University of Lviv, 107 Tarnavskogo str., 79017 Lviv, Ukraine
e-mail: ivan_karbovnyck@yahoo.com

© Springer Science+Business Media Dordrecht 2017

B. Di Bartolo et al. (eds.), *Nano-Optics: Principles Enabling Basic Research and Applications*, NATO Science for Peace and Security Series B: Physics and Biophysics, DOI 10.1007/978-94-024-0850-8_32

A, Velupillai S, Klym H, Fedyshyn O, Tymoshuk S, Kolych I, Nanoscale Res Lett. doi:10.1186/s11671-015-0855-x). After the short time exposure in CO_2 atmosphere, the blue shift of the absorption peak is observed and the absorption intensity is decreased. Controlling the film morphology may allow to maximize the absorption shift in the presence of designated gas with respect to the peak position at ambient condition.

Another type of probabilistic nanostructures is formed when certain type of nanofiller is dispersed in the host insulating matrix. There are numerous examples of such structures and it has been demonstrated that composites reinforced with Boron, Nitrogen or Carbon elements in the form of nanostructures dispersed in a matrix can provide radiation shielding for different ranges of energies and without the generation of harmful secondary particles.

A series of tests performed on polymer/carbon nanotubes and epoxy/boron nitride (BN) microparticles composites showed that electrical and luminescence properties of these structures are sensitive to the radiation.

The variation of resistance of hybrid polymer/carbon nanotubes structures after exposing them to gamma rays coming from ^{226}Ra source depends on the loading level of nanotubes. Typically, as the absorbed dose varies in the range 10–100 nGy, polymer/carbon nanotubes composite with 5 % loading shows 15–20 % increase in resistance.

Adding BN microparticles to the epoxy resin makes it more resistant to incoming harsh rays. At the same time, monitoring how the light emission from such composites change after the exposure to radiation makes this system useful for detecting purposes. A series of spectroscopic measurements have shown that the intensity of 470 nm luminescence band in epoxy/boron microparticles composite decreases after the irradiation with an average photon energy of 0.19 MeV is absorbed (see Fig. 32.1).

Increasing the concentrations of reinforcing particles makes the influence of gamma rays on the emission properties much less pronounced.

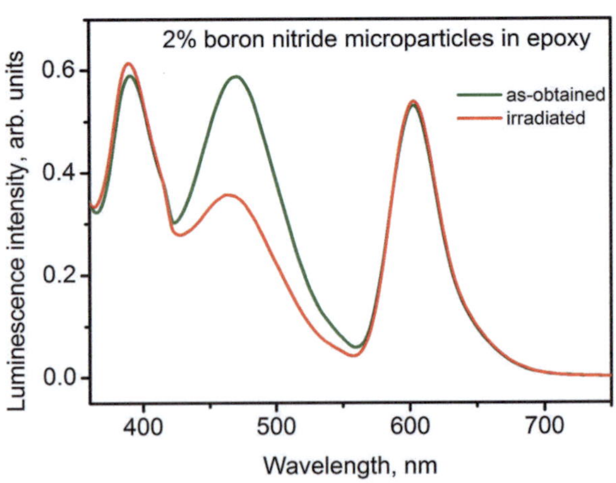

Fig. 32.1 Luminescence from BN microparticles/epoxy resin composite before and after irradiation with 0.19 MeV gamma rays

Reference

1. Karbovnyk, I., Collins, J., Bolesta, I., Stelmashchuk, A., Kolkevych, A., Velupillai, S., Klym, H., Fedyshyn, O., Tymoshuk, S., & Kolych, I. *Nanoscale Research Letters*. doi:10.1186/s11671-015-0855-x.

Chapter 33
Hall Effect Sign-inversion and Parallel Hall Effect in Single-constituent 3D Metamaterials

Christian Kern, Muamer Kadic, Robert Schittny, Tiemo Bückmann, and Martin Wegener

Abstract A few years ago, Briane and Milton have shown the existence of three-dimensional, isotropic, periodic metamaterials with a Hall coefficient that is negative with respect to that of its constituents using homogenization theory [1]. They also gave an example of such a metamaterial structure. Recently, we simplified their blueprint decisively and numerically demonstrated this sign-inversion which turned out to be controllable by a structure parameter [2].

We extend this work to single-constituent, anisotropic structures exhibiting a Hall voltage which is *parallel* to an external magnetic field. Furthermore, our numerical analysis reveals that we can combine orthogonal and parallel Hall voltages and achieve full control over the direction of the Hall voltage in the plane perpendicular to the imposed current by adjusting two structure parameters. On the experimental side we fabricate metamaterial Hall bar samples using three-dimensional direct laser writing [3]. Since the obtained polymer structure is electrically non-conducting we coat it with ZnO using atomic layer deposition [4], a deposition technique featuring outstanding conformality and thickness control.

References

1. Briane, M., & Milton, G. W. (2009). Homogenization of the three-dimensional Hall effect and change of sign of the Hall coefficient. *Archive for Rational Mechanics and Analysis, 193*, 715.
2. Kadic, M., Schittny, R., Bückmann, T., Kern, C., & Wegener, M. (2015). Hall-effect sign inversion in a realizable 3D metamaterial. *Physical Review X, 5*, 021030.

C. Kern (✉) • M. Kadic • R. Schittny • T. Bückmann
Institute of Applied Physics, Karlsruhe Institute of Technology (KIT), 76128 Karlsruhe, Germany
e-mail: christian.kern@kit.edu

M. Wegener
Institute of Applied Physics, Karlsruhe Institute of Technology (KIT), 76128 Karlsruhe, Germany

Institute of Nanotechnology, Karlsruhe Institute of Technology (KIT), 76344 Eggenstein-Leopoldshafen, Germany

© Springer Science+Business Media Dordrecht 2017

B. Di Bartolo et al. (eds.), *Nano-Optics: Principles Enabling Basic Research and Applications*, NATO Science for Peace and Security Series B: Physics and Biophysics, DOI 10.1007/978-94-024-0850-8_33

3. von Freyman, G., Ledermann, A., Thiel, M., Staude, I., Essig, S., Busch, K., & Wegener, M. (2010). Three-Dimensional Nanostructures for Photonics. *Advanced Functional Materials, 20*, 1038.
4. Tynell, T., & Karppinen, M. (2014). Atomic layer deposition of ZnO: A review. *Semiconductor Science and Technology, 29*, 043001.

Chapter 34
Organization of Metallic Nanoparticles in Block Copolymer Ultra-Thin Films for Optical Devices

Jean-François Lemineur and Anna Ritcey

Abstract Metallic nanoparticles (NPs) with sub-wavelength dimensions can interact with light because of the collective oscillation of conduction band electrons. This phenomenon, the so-called localized surface plasmon resonance (LSPR), offers significant promise for optical applications such as biosensors. Importantly, the LSPR is a function of several parameters, such as the size and shape of the particles. Moreover, the plasmonic properties of NPs assemblies differ from those of isolated particles because of interparticle coupling. Block copolymer Langmuir-Blodgett films have been investigated as templates for the organization of NPs. Lemineur JF, Saci N, Ritcey AM, Colloids Surf A. 498:88 97, 2016. Different kinds of self-assembled structures have been observed including isolated particles, clusters, rings and strands (Fig. 34.1). With this approach, the size, the shape and the periodicity of the domains used for assemblies can be systematically tuned. In addition, the particle diameter was also modifiable via an in situ seeded growth procedure. The optical properties of the resulting assemblies are currently under investigation and

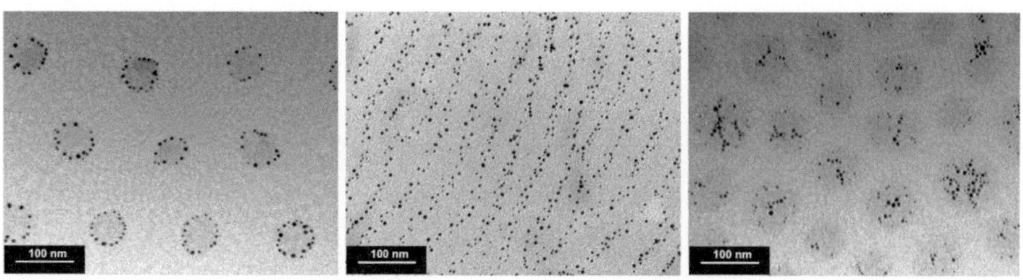

Fig. 34.1 Transmission electron microscopy images of self-assembled structures obtained by the Langmuir-Blodgett deposition of block copolymers and gold nanoparticles

J.-F. Lemineur (✉) • A. Ritcey
Department of Chemistry and CERMA, Université Laval, Pavillon Alexandre-Vachon, 1045 avenue de la Médecine, G1V 0A6 Québec, Canada
e-mail: jean-francois.lemineur.1@ulaval.ca

© Springer Science+Business Media Dordrecht 2017

B. Di Bartolo et al. (eds.), *Nano-Optics: Principles Enabling Basic Research and Applications*, NATO Science for Peace and Security Series B: Physics and Biophysics, DOI 10.1007/978-94-024-0850-8_34

theoretical calculations (based on the discrete dipole approximation) were carried out to guide the interpretation of experimental results. Finally, the monolayers were evaluated as potential substrates for SERS and as bio-sensors.

Reference

1. Lemineur, J.-F., Saci, N., & Ritcey, A. M. (2016). Impact of concentration and capping ligand length on the organization of metal nanoparticles in Langmuir-Blodgett surface micelles and nanostrands. *Colloids and Surfaces A: Physicochemical and Engineering Aspects, 498*, 88–97.

Chapter 35
Self-Assembled Laser-Activated Plasmonic Substrates for High-Throughput, High-Efficiency Intracellular Delivery

Marinna Madrid, Nabiha Saklayen, Marinus Huber, Nicolas Vogel, and Eric Mazur

Abstract Delivering material into cells is important for a diverse range of biological applications, including gene therapy, cellular engineering and imaging. We presented a plasmonic substrate for delivering membrane-impermeable material into cells at high throughput and high efficiency while maintaining cell viability. The substrate fabrication is based on an affordable and fast colloidal self-assembly process. When illuminated with a femtosecond laser, the light interacts with the electrons at the surface of the metal substrate, creating localized surface plasmons that form bubbles via energy dissipation in the surrounding medium. These bubbles come into close contact with the cell membrane to form transient pores and enable entry of membrane-impermeable material via diffusion. We performed proof of principle experiments using both a femtosecond laser and a nanosecond laser system. We used fluorescence microscopy to verify delivery of membrane-impermeable dye into HeLa CCL-2 cells and to verify cell viability after laser treatment. Our findings indicate that self-assembled plasmonic substrates may be an affordable, flexible tool for high-throughput, high-efficiency delivery of material into mammalian cells.

M. Madrid (✉) • E. Mazur
School of Engineering and Applied Sciences, Harvard University, Cambridge, MA, USA
e-mail: marinnamadrid@gmail.com; mazur@physics.harvard.edu

N. Saklayen
Physics Department, Harvard University, Cambridge, MA, USA
e-mail: nsaklayen@gmail.com

M. Huber
Max Planck Institute, Munich, Germany

N. Vogel
University of Erlangen, Erlangen, Germany

© Springer Science+Business Media Dordrecht 2017

B. Di Bartolo et al. (eds.), *Nano-Optics: Principles Enabling Basic Research and Applications*, NATO Science for Peace and Security Series B: Physics and Biophysics, DOI 10.1007/978-94-024-0850-8_35

Chapter 36
Single-Crystal vs Polycrystalline Gold: A Non-linear-Optics Analysis

Régis Méjard, Olivier Demichel, Anthonin Verdy, Marlène Petit, Alexandre Bouhelier, and Benoît Cluzel

Abstract Standard gold in the field of plasmonics is obtained by evaporation or sputtering and therefore is polycrystalline. Yet, this gold presents numbers of drawbacks such as roughness, grains and ill-defined electronic band diagrams in addition to the lack of reproducibility from one instrument to another. It is, thus, beneficial to turn to a metal production that can enable well-defined and controlled gold parameters. To that end, we have explored the wet synthesis of gold nanoplates which represents a simple and robust means of obtaining single-crystal gold (Guo Z, Zhang Y, DuanMu Y, Xu L, Xie S, Gu N, Colloids Surf A 278:33 38, 2006). The synthesized nanoplates are from 50 to less than 100 nm in thickness and can span over micrometers in lateral dimensions corresponding to areas of several hundreds of μm^2. They can thus be considered as thin film material perfectly suitable for plasmonic applications.

We have carried out non-linear, confocal, hyperspectral, optical measurements of such synthesized nanoplates and compared them to those of gold-evaporated nanoplates of similar dimensions. The single-crystal gold nanoplates are brighter at the second harmonic generation (SHG) wavelength than at multi-photon absorption luminescence (MPL) wavelengths whereas the opposite is observed for the polycrystalline gold nanoplates. Surprisingly, the single-crystal gold, which is a centrosymmetric medium, emits as much as 20-fold more SHG signal than the polycrystalline nanoplates despite an expected reduction of the SHG (Fig. 36.1). Moreover, a reduction of 2.5-fold in the MPL emission has been observed for the single-crystal compared to the polycrystalline gold.

Comparing the two samples allowed us to preclude certain causes of the origin of SHG, namely bulk centrosymmetry breakup, roughness of the film and electronic defects. However, the results indicate that $\chi^{(2)}_{surface}$ can be a pivotal parameter

R. Méjard (✉) • O. Demichel • A. Verdy • M. Petit • A. Bouhelier • B. Cluzel
Laboratoire Interdisciplinaire Carnot de Bourgogne, UMR 6303 CNRS, Université de Bourgogne – Franche-Comté, DIJON Cedex, France
e-mail: Regis.Mejard@u-bourgogne.fr

© Springer Science+Business Media Dordrecht 2017
B. Di Bartolo et al. (eds.), *Nano-Optics: Principles Enabling Basic Research and Applications*, NATO Science for Peace and Security Series B: Physics and Biophysics, DOI 10.1007/978-94-024-0850-8_36

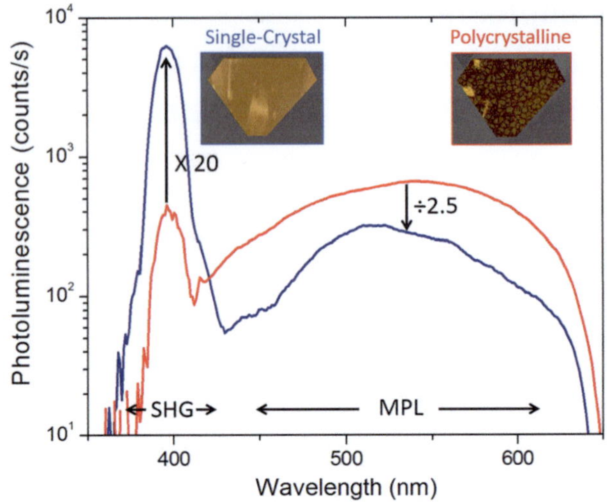

Fig. 36.1 Non-linear emission spectra of the center of the single- and multi-crystal gold nanoplates

(Wang X, Rodríguez F, Albers W, Ahorinta R, Sipe JE, Kauranen M, Phys Rev B 80:233402, 2009) and also that bulk quadrupoles and magnetic dipoles could play a part in the SHG emission mechanism (Suresh S, Arivuoli D, Rev Adv Mater Sci 30:243–253, 2012).

Regarding the MPL mechanism, since single-crystal gold features less electronic defects than its polycrystalline counterpart, the cutback in the emission of the former material shows that MLP is directly associated with electronic defects.

Last, we have etched single-crystal nanoantennas in the single-crystal nanoplates. We have performed an auto-correlation measurement on these nanoantennas and have extracted electron-phonon scattering times below those of polycrystalline nanoantennas. This finding is another support to the argument that MPL strength depends on the amount of electronic defects.

References

1. Guo, Z., Zhang, Y., DuanMu, Y., Xu, L., Xie, S., & Gu, N. (2006). *Colloids Surface A, 278*, 33–38.
2. Wang, X., Rodríguez, F., Albers, W., Ahorinta, R., Sipe, J. E., & Kauranen, M. (2009). *Physical Review B, 80*, 233402.
3. Suresh, S., & Arivuoli, D. (2012). *Reviews on Advanced Materials Science, 30*, 243–253.

Chapter 37
Transport of Light Through White-LED Phosphor Plates

Maryna Meretska, H. Thyrrestrup, A. Lagendijk, T. W. Tukker, A. P. Mosk, W. L. Ijzerman, and W. Vos

Abstract Energy efficient generation of white light has become an important societal issue in recent years. The technology of white-light emitting diodes (LEDs) is one of the main directions (Akasaki I, Amano H, Nakamura S, Blue LEDs – filling the world with new light, http://www.nobelprize.org/, 2014; Schubert EF, Light emitting diodes. Cambridge: Cambridge University, 2006). Key challenges in the white LED usage are understanding scattering, absorption and emission from ab-initio, and extracting the transport properties in the region where both emission and absorption overlap. Physical understanding of multiple light scattering in the LED provides tools to extract optical parameters of this system, and greatly simplify the LED design process. In this work we have been able to measure the total transmission, using a novel technique, in the region where emission and absorption overlap, and to extract transport parameters in the whole visible range.

We have studied transport of light in the wavelength range 400–700 nm, through polycarbonate diffuser plates with different phosphor concentrations of YAG:Ce3+, that are used for commercial LEDs. By using a broadband light source we have previously measured the total transmission in the whole wavelength range except for the blue-green region 490–520 nm, where emission and absorption overlap (Vos WL, Tukker TW, Mosk AP, Lagendijk A, IJzerman WL, Appl Opt 52:2602, 2013; Leung VYF, Lagendijk A, Tukker TW, Mosk AP, IJzerman WL, Vos WL, Opt Express 22:8320–8332, 2014). We used diffusion theory to extract from these data transport ltr and absorption lab mean free paths. To extract lab in the overlap region, however, we cannot use anymore broadband white light source, as the contributions to the signal from emission and absorption cannot be separated.

M. Meretska (✉) • H. Thyrrestrup • A. Lagendijk • A.P. Mosk • W. Vos
MESA+ Institute for Nanotechnology, Complex Photonic Systems (COPS), University of Twente, P.O. Box 217, 7500 AE Enschede, The Netherlands
e-mail: m.meretska@utwente.nl

T.W. Tukker
Philips Research, 34 High Tech Campus, 5656 AE Eindhoven, The Netherlands

W.L. Ijzerman
Philips Lighting, 44 High Tech Campus, 5656 AE Eindhoven, The Netherlands

© Springer Science+Business Media Dordrecht 2017

B. Di Bartolo et al. (eds.), *Nano-Optics: Principles Enabling Basic Research and Applications*, NATO Science for Peace and Security Series B: Physics and Biophysics, DOI 10.1007/978-94-024-0850-8_37

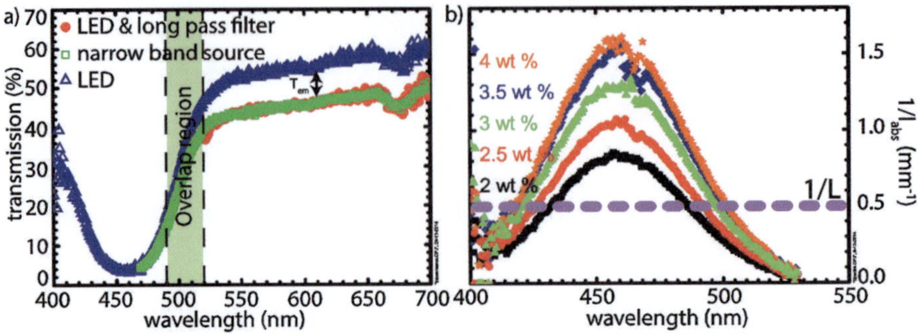

Fig. 37.1 (**a**) The total transmission measurement performed with the narrow band and the broad band light sources (phosphor concentration: 4 wt%). The *green* area indicates region where emission and absorption of the YAG:Ce^{3+} overlaps. (**b**) The absorption mean free path deduced from the total transmission measurements for different concentrations of YAG:Ce^{3+}

To overcome this ambiguity we have developed a new technique to measure light transport of LEDs in the range where emission and absorption overlap. A narrow band light source is employed to measure the total transmission (see Fig. 37.1a). In this case we observe a clear separation in wavelength between transmitted and re-emitted signals. We have therefore managed to extract the total transmission in the overlap region. These data allow us to deduce the lab in the previously inaccessible overlap region. By combining these results with the results obtained with the broadband light source, we have managed to extract the full range of transport parameters for phosphor plates in the visible wavelength range using nanophotonic theory (see Fig. 37.1b). In our presentation we elaborated on the consequences for real devices.

References

1. Akasaki, I., Amano, H., & Nakamura, S. (2014). Blue LEDs – Filling the world with new light. http://www.nobelprize.org/. Online Document.
2. Schubert, E. F. (2006). *Light emitting diodes*. Cambridge: Cambridge University.
3. Vos, W. L., Tukker, T. W., Mosk, A. P., Lagendijk, A., & IJzerman, W. L. (2013). Broadband mean free path of diffuse light in polydisperse ensembles of scatterers for white LED lighting. *Applied Optics, 52*, 2602.
4. Leung, V. Y. F., Lagendijk, A., Tukker, T. W., Mosk, A. P., IJzerman, W. L., & Vos, W. L. (2014). Interplay between multiple scattering, emission, and absorption of light in the phosphor of white light-emitting diode. *Optics Express, 22*, 8320–8332.

Chapter 38
Direct Laser Writing of 3D Nanostructures Using a 405 nm Laser Diode

Patrick Mueller, Michael Thiel, and Martin Wegener

Abstract Direct laser writing (DLW) is a well-known and established technology for fabricating 3D micro- and nanostructures. Usually, red femtosecond laser sources with wavelengths around 800 nm are used. Here, we use a laser diode with a wavelength of 405 nm as the exciting laser source and thus improve structures in terms of decreasing feature size and line distance by exploiting the linear wavelength dependence of the Sparrow resolution limit. A nonlinear multi-photon polymerization process is necessary for manufacturing true 3D structures. We investigate different photoresists and measure their nonlinearities by variation of the electronic pulse scheme of the laser. We observe an adequately high nonlinearity in a resist system based on the monomer pentaerythritol triacrylate. To benefit from the improved theoretical resolution of the smaller wavelength, it is necessary to achieve a close to diffraction-limited focal spot which we have confirmed by measuring the point spread function of the objective lens and comparing it to numerical simulations. In order to prove the performance of the system, we fabricate benchmark structures and characterize them with different experimental methods. Line gratings and point arrays are written to investigate 2D resolution and feature sizes. To characterize the capabilities in 3D, we fabricate woodpile photonic crystals that show a photonic stop band in the visible. The achievable lattice constants in both 2D and 3D are considerably smaller than in previous work with red femtosecond lasers, proving the success of the wavelength-reduction approach. Previous work using the conceptually diffraction-unlimited STED technology is also outperformed.

P. Mueller (✉)
Institute of Nanotechnology, Karlsruhe Institute of Technology (KIT), 76128 Karlsruhe, Germany
e-mail: pat.mueller@kit.edu

M. Thiel
Nanoscribe GmbH, 76344 Eggenstein-Leopoldshafen, Germany

M. Wegener
Institute of Applied Physics and DFG-Center for Functional Nanostructures (CFN), Karlsruhe Institute of Technology (KIT), 76128 Karlsruhe, Germany

© Springer Science+Business Media Dordrecht 2017
B. Di Bartolo et al. (eds.), *Nano-Optics: Principles Enabling Basic Research and Applications*, NATO Science for Peace and Security Series B: Physics and Biophysics, DOI 10.1007/978-94-024-0850-8_38

Chapter 39
Luminescent Labeling of Nanoparticles: SiO_2@$LaPO_4$

Jacobine van Hest and Andries Meijerink

Abstract Nanomaterials are embedded in a variety of products we use daily, e.g. silica nanocrystals are applied in rubber and food. However, more research has to be performed on nano-enabled products to learn more about their environmental and health risks. The demand for guidelines for nanomaterials asks for model-nanosystems that can be traced during their life cycle. In this project, luminescent $LaPO_4$ core particles with sizes ranging from 4 to 8 nm [1] were incorporated into silica nanocrystals. The luminescence properties of the nanocrystal were varied by changing or combining the nature of the lanthanide dopants in the $LaPO_4$ nanocrystal in order to obtain a variety of unique luminescent labels. Silica was grown around the luminescent $LaPO_4$ cores using the reverse micelle method (see Fig. 39.1a) [2]. Silica spheres with sizes between 35 and 85 nm could be obtained by changing the polarity of the dispersion medium of the $LaPO_4$ nanocrystals. The nanocrystals preserved their luminescence properties after silica coating (see Fig. 39.1b). The fluorescent nanoparticles make it possible to use combined fluorescence and electron microscopy for the analysis of the nanoparticles down to the single particle level – and is in progress.

J. van Hest (✉) • A. Meijerink
Universiteit Utrecht, Princetonplein 1, 3584 CC Utrecht, The Netherlands
e-mail: J.J.H.A.vanHest@uu.nl; A.Meijerink@uu.nl

© Springer Science+Business Media Dordrecht 2017
B. Di Bartolo et al. (eds.), *Nano-Optics: Principles Enabling Basic Research and Applications*, NATO Science for Peace and Security Series B: Physics and Biophysics, DOI 10.1007/978-94-024-0850-8_39

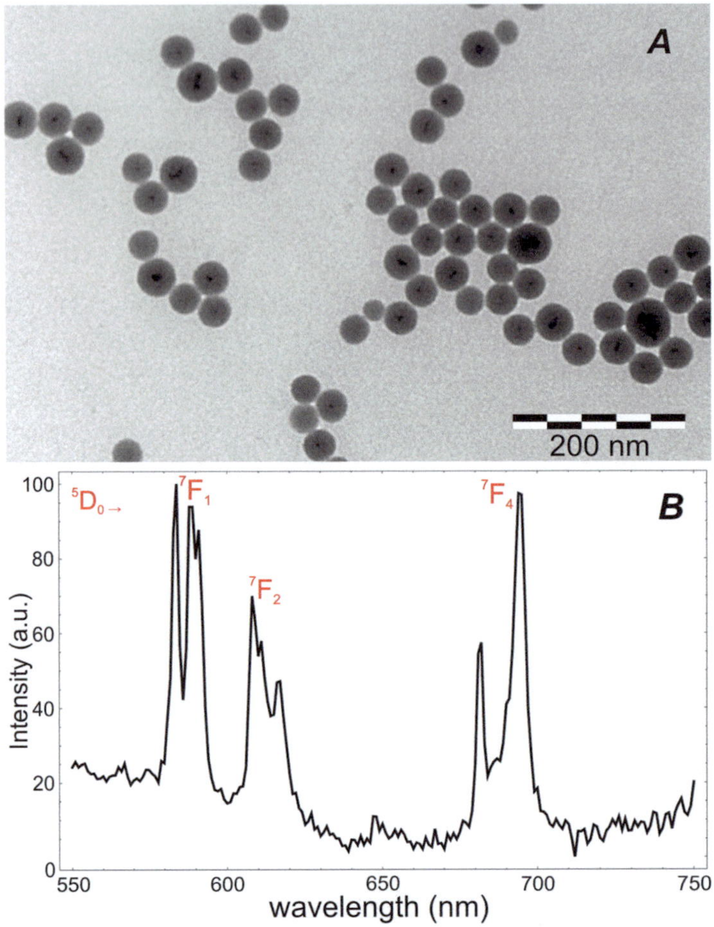

Fig. 39.1 **(a)** TEM image of $LaPO_4:Eu^{3+}$ nanocrystals incorporated into silica spheres of 36.9 ± 3.9 nm. **(b)** emission spectrum of these particles upon excitation at 280 nm

References

1. Hickmann, K., Kömpe, K., Hepp, A., et al. (2008). The role of amines in the growth of terbium(III)-doped cerium phosphate nanoparticles. *Small, 4*, 2136–2139.
2. Koole, K., van Schooneveld, M., Hilhorst, J., et al. (2008). On the incorporation mechanism of hydrophobic quantum dots in silica spheres by a Reverse Micelle Method. *Chemistry of Materials, 20*, 2503–2512.

Chapter 40
Integrated Super-Couplers Based on Zero-Index Metamaterials

Daryl I. Vulis, Orad Reshef, Philip Muñoz, Shota Kita, Yang Li, Marko Lončar, and Eric Mazur

Abstract There has been strong interest in the confinement of electromagnetic energy in sub-diffraction limit waveguide configurations. Such an achievement would offer applications in in telecommunications, subwavelength imaging, optical memory storage, and on-chip photonic processes. Materials with a refractive index of zero have been considered as strong contenders for such "super-coupling" applications (Silveirinha MG, Engheta N, Phys Rev B 76:245109, 2007; Engheta N, Science 340:286–287, 2013). Though ε-near-zero and μ-near-zero metamaterials, where zero index is obtained by tuning the effective electric permittivity or permeability to zero, have been proposed as a possible candidate, their infinite or zero impedance causes large reflections which pose a challenge for coupling applications.

Recently, metamaterials with simultaneously zero effective permittivity and permeability have been demonstrated in both out-of-plane and on-chip configurations (Moitra P, Yang Y, Anderson Z, Kravchenko II, Briggs DP, Valentine J, Nat Photonics 7:791, 2013; Li Y, Kita S, Muñoz P, Reshef O, Vulis D, Lončar M, Mazur E, On-chip zero index metamaterials. Manuscript in Preparation, 2015). Both of these configurations take advantage of an accidental Dirac cone at the center of the Brillouin zone and an all-dielectric design that offers both impedance matching and low losses (Huang X, Lai Y, Hang ZH, Zheng H, Chan CT, Nat Mat 10:582–586, 2011). An on-chip implementation of such an ε-and-μ-zero metamaterial offers a suitable platform for exploring supercoupling. This design consists of an array of silicon pillars fabricated using a standard CMOS-compatible process. We have previously fabricated a prism consisting of this material through which we are able to observe an unambiguous demonstration of zero effective index (Moitra P, Yang Y, Anderson Z, Kravchenko II, Briggs DP, Valentine J, Nat Photonics 7:791, 2013).

D.I. Vulis (✉) • O. Reshef • P. Muñoz • S. Kita • Y. Li • M. Lončar • E. Mazur
School of Engineering and Applied Sciences, Harvard University, 9 Oxford Street, Cambridge, MA 02138, USA
e-mail: dvulis@seas.harvard.edu; mazur@physics.harvard.edu

© Springer Science+Business Media Dordrecht 2017
B. Di Bartolo et al. (eds.), *Nano-Optics: Principles Enabling Basic Research and Applications*, NATO Science for Peace and Security Series B: Physics and Biophysics, DOI 10.1007/978-94-024-0850-8_40

In this work, we used an ε-and-μ-zero metamaterial to achieve demonstrable super-coupling. Using FDTD simulations, we explored arbitrary angle waveguide bends and powersplitter configurations of super-coupling structures between two separate waveguides for wavelengths within the telecom regime.

References

1. Silveirinha, M. G., & Engheta, N. (2007). Theory of supercoupling, squeezing wave energy, and field confinement in narrow channels and tight bends using ε near-zero metamaterials. *Physical Review B, 76*, 245109.
2. Engheta, N. (2013). Pursuing near-zero response. *Science, 340*, 286–287.
3. Moitra, P., Yang, Y., Anderson, Z., Kravchenko, I. I., Briggs, D. P., & Valentine, J. (2013). Realization of an all-dielectric zero-index optical metamaterial. *Nature Photonics, 7*, 791.
4. Li, Y., Kita, S., Muñoz, P., Reshef, O., Vulis, D., Lončar, M., & Mazur, E. (2015). *On-chip zero index metamaterials*. Manuscript in Preparation.
5. Huang, X., Lai, Y., Hang, Z. H., Zheng, H., & Chan, C. T. (2011). Dirac cones induced by accidental degeneracy in photonic crystals and zero-refractive-index materials. *Nature Materials, 10*, 582–586.

Chapter 41
3D Micro-printing of Optical Temperature Probes

Andreas Wickberg, Jonathan B. Mueller, Yatin J. Mange, Thomas Nann, and Martin Wegener

Abstract We present printable optical temperature probes to monitor the temperature with a spatial precision on the micrometer scale. Our approach is based on the temperature-dependent upconversion fluorescence from NaYF4:Yb3+, Er3+ co-doped nanocrystals. These nanoparticles are dispersed in a standard photoresist for direct laser writing, allowing for spatially resolved micro-printing of single or multiple probe spots. For demonstration, we decapsulate a fully operational integrated circuit and print temperature probes directly on the semiconductor chip to monitor its local heating. The printability of the probes facilitates an easy integration into diverse systems, especially when aiming at integrated optics or lab-on-a-chip systems.

A. Wickberg (✉) • J.B. Mueller • M. Wegener
Institute of Applied Physics, Karlsruhe Institute of Technology (KIT), 76128 Karlsruhe, Germany
e-mail: andreas.wickberg@kit.edu

Y.J. Mange • T. Nann
Ian Wark Research Institute, University of South Australia, Adelaide, SA 5095, Australia

© Springer Science+Business Media Dordrecht 2017
B. Di Bartolo et al. (eds.), *Nano-Optics: Principles Enabling Basic Research and Applications*, NATO Science for Peace and Security Series B: Physics and Biophysics, DOI 10.1007/978-94-024-0850-8_41

Chapter 42
Evaporation-Driven Aggregation of Nanoparticles in a Free Droplet: Spherical Symmetry in Nanostructured Material

M. Woźniak, G. Derkachov, K. Kolwas, J. Archer, D. Jakubczyk, T. Wojciechowski, and M. Kolwas

Abstract We have investigated the evaporation-driven aggregation of nanoparticles during the slow-drying process of single colloidal droplets levitating in the electrodynamic trap. The final, spherical quasi-crystal composed of nanoparticles is an interesting nanostructured material of wide range of possible applications (plasmonics, metamaterials, colloidal crystals).

The experimental setup [1] consist of the linear quadrupole electrodynamic trap with 4 rod electrodes in vertical alignment, the climatic chamber, the laser system and 2 CCD cameras. The electrodes provide the alternating (AC) electric field in a quadrupole configuration. The annular electrodes placed around the vertical ones provide static (DC) field balancing particles (droplets) weight. The droplet has been illuminated by the laser beam along the vertical axis of the trap which enables stabilization of the vertical position using CCD cameras and a PID-type loop driving DC voltage of annular electrodes. The geometry of the trap allows to progressively reduce DC field in order to deposit the aggregated structures on the silicon substrate at the bottom of the chamber to further investigate them with SEM.

We have investigated the successive aggregation stages using the static light scattering and the electrostatic weighting. Combining modeling with experimental observations, we have been able to examine and describe consecutive stages of the aggregate formation. An assembly of nanoparticles develops from water suspension in spherical symmetry [2], mainly due to the surface tension of evaporating droplet [3]. At some stages of evaporation a film of nanoparticles is formed

M. Woźniak (✉) • G. Derkachov • K. Kolwas • J. Archer • D. Jakubczyk
• T. Wojciechowski • M. Kolwas
Institute of Physics, Polish Academy of Sciences, Al.Lotników 32/46, 02-668 Warsaw, Poland
e-mail: mwozniak@ifpan.edu.pl

© Springer Science+Business Media Dordrecht 2017
B. Di Bartolo et al. (eds.), *Nano-Optics: Principles Enabling Basic Research and Applications*, NATO Science for Peace and Security Series B: Physics and Biophysics, DOI 10.1007/978-94-024-0850-8_42

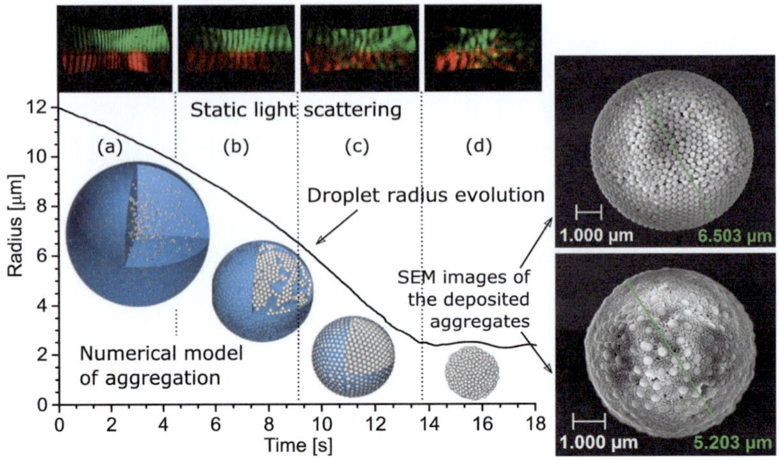

Fig. 42.1 Evolution of the radius of evaporating droplet together with the scattering patterns from the static light scattering measurements and the snapshots of the numerical modeling showing the stages of aggregation: **(a)** evaporation of homogeneous droplet, **(b)** formation of a film of nanoparticles at the interface, **(c)** formation of highly-ordered shell, **(d)** slow drying of the aggregate

at the liquid-air spherical interface. Subsequently, depending on the aggregation parameters, it either collapses or rearranges to form the shell-like structure. The morphology of the final structures significantly depends on the drying rate. We observed the process of evaporation, which, from the thermodynamic point of view, was stationary and close to equilibrium. Deposition of the aggregates at the silicon substrate allows us to analyze the final aggregates with SEM and compare the results with the optical scattering measurements [2, 4, 5]. Therefore, we were able to confirm the proposed scenario of aggregation. Figure 42.1 presents evolution of the radius of evaporating droplet together with the scattering patterns from the static light scattering measurements and snapshots of the numerical modeling showing the stages of aggregation.

Acknowledgements This work was supported by the National Science Center, Poland under grant number 2014/13/D/ST3/01882.

References

1. Woźniak, M., Derkachov, G., Kolwas, K., Archer, J., Wojciechowski, T., Jakubczyk, D., & Kolwas, M. (2015). Formation of highly-ordered spherical aggregates from drying microdroplets of colloidal suspension. *Langmuir*. doi:10.1021/acs.langmuir.5b01621.
2. Derkachov, G., Kolwas, K., Jakubczyk, D., Zientara, M., & Kolwas, M. (2008). Drying of a microdroplet of water suspension of nanoparticles: from surface aggregates to microcrystal. *Journal of Physical Chemistry C, 112*(43), 16919–16923.

3. Hołyst, R., Litniewski, M., Jakubczyk, D., Kolwas, K., Kolwas, M., Kowalski, K., Migacz, S., Palesa, S., & Zientara, M. (2013). Evaporation of freely suspended single droplets: experimental, theoretical and computational simulations. *Reports on Progress in Physics, 76*, 034601–034620.
4. Jakubczyk, D., Kolwas, M., Derkachov, G., & Kolwas, K. (2009). Surface states of microdroplet of suspension. *Journal of Physical Chemistry C, 113*(24), 10596–10602.
5. Kolwas, M., Kolwas, K., Jakubczyk, D., & Derkachov, G. (2015). Surface diagnostics of evaporating droplet of nanospheres suspension; Fano interference and surface pressure. *Physical Chemistry Chemical Physics, 17*, 6881–6888.

Part IV
Posters

Chapter 43
Ultrafast Optical Spectroscopy Techniques Applied to Colloidal Nanocrystals

Mauro Aresti, Michele Saba, Francesco Quochi, Andrea Mura, and Giovanni Bongiovanni

Abstract Colloidal semiconductor nanocrystals are promising materials for applications in solution-processable optoelectronic devices including light-emitting diodes, photodetectors, photocatalysts and photovoltaic cells. Chemical synthesis enables refined control over dimension and shape of nanocrystals, allowing for unique functionalities and novel physical properties.

We applied ultrafast optical spectroscopy techniques to investigate the photophysics of different nanostructured materials. In particular, these novel nano-materials will emit or absorb light efficiently so that they can be used in optoelectronics applications. In this experimental work we investigated several optical and electronic properties of the nanocrystals, in particular, we focused our studies on *CdSe*/*CdS* core/shell nanocrystals in order to investigate dark and weakly emitting states in a model known for high luminescence quantum yield and reduced blinking.

The experiment is designed to identify the fundamental mechanism of photoluminescence quenching, based on transient nonlinear photoluminescence spectroscopy applied to an ensemble of nanocrystal quantum dots.

We have introduced variable pulse rate photoluminescence as an experimental technique to investigate long-lived states in colloidal nanocrystals (Fig. 43.1). We demonstrate that dark and weakly emitting states, originating both from charges or traps, can be reversibly turned on and off by light [1].

In the second part of this experimental work we studied the properties of *CdSe*/*CdS* octapod nanocrystals decorated with platinum nanoparticles, we investigated the consequences of different decorations on photophysical properties, focusing on capture of photoexcited electrons by Pt domains.

The photophysics of Pt-decorated nanocrystals is studied with ultra-fast optical spectroscopy, in particular femtosecond transient absorption, a technique sensitive to the dynamics of the excited state (Fig. 43.2).

M. Aresti (✉) • M. Saba • F. Quochi • A. Mura • G. Bongiovanni
Dipartimento di Fisica, Universita' di Cagliari, Cittadella universitaria, I-09042 Monserrato, Italy
e-mail: mauro.aresti@dsf.unica.it

© Springer Science+Business Media Dordrecht 2017

B. Di Bartolo et al. (eds.), *Nano-Optics: Principles Enabling Basic Research and Applications*, NATO Science for Peace and Security Series B: Physics and Biophysics, DOI 10.1007/978-94-024-0850-8_43

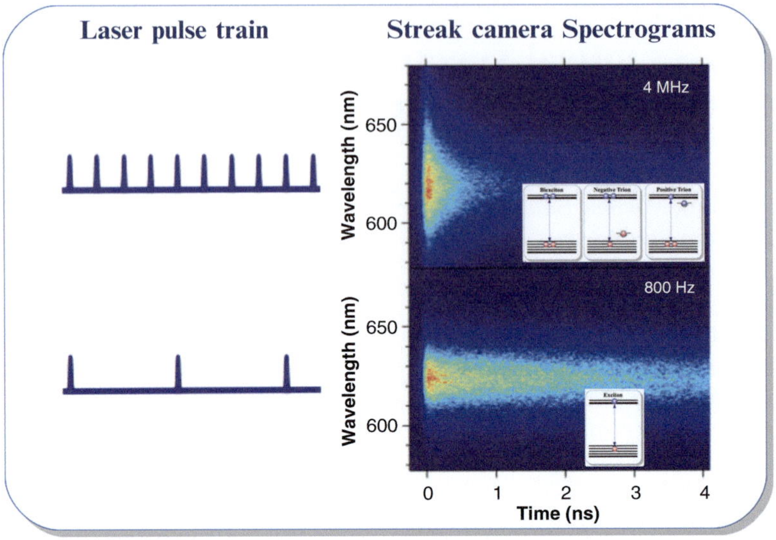

Fig. 43.1 Variable pulse rate photoluminescence spectroscopy measures, two streak camera photoluminescence spectrograms acquired under different conditions (with two different pulse repetition rates, but with the same energy per pulse) show the potential of this technique to investigate emission quenching and to distinguish light emission from neutral, positively, and negatively charged quantum dots

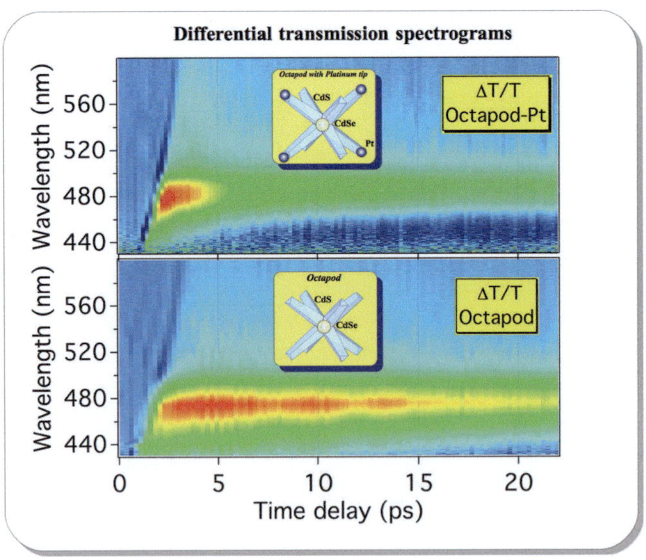

Fig. 43.2 Transient Absorption Spectroscopy measures, two differential transmission spectrograms for CdSe/CdS octapod nanocrystals and CdS/CdSe octapod nanocrystals decorated with tips of platinum are shown here. The shortening of the differential transmission decay in Pt-decorated nanocrystals can be safely attributed to electron capture by Pt, electrons can be trapped at the tips in a time scale of few picoseconds

Our measurements show that electron capture by Pt tips occurs in ∼200 ps, providing just a perturbation to the excited state dynamics, while in Pt-covered nanocrystals electron capture is much faster, ∼1 ps, and competes with energy relaxation and Auger recombination, causing the capture of hot electrons [2].

References

1. Saba, M., Aresti, M., Quochi, F., Marceddu, M., Loi, M. A., Huang, J., Talapin, D. V., Mura, A., & Bongiovanni, G. (2012). Light-induced charged and trap states in colloidal nanocrystals detected by variable pulse rate photoluminescence spectroscopy. *ACS Nano, 7*, 229–238.
2. Conca, E., Aresti, M., Saba, M., Casula, M. F., Quochi, F., Mula, G., Loche, D., Kim, M. R., Manna, L., Corrias, A., Mura, A., & Bongiovanni, G. (2014). Charge separation in Pt-decorated CdSe@ CdS octapod nanocrystals. *Nanoscale, 6*, 2238–2243.

Chapter 44
Directivity Based Nanoscopic Position Sensing

Ankan Bag, Martin Neugebauer, Pawel Woźniak, Gerd Leuchs, and Peter Banzer

Abstract Precise position sensing of a nanoparticle or a biomolecule is of paramount importance for the field of photonics, specifically in medicine and biophysics. This is a fundamental step towards several super-resolution imaging techniques, such as fluorescence based photoactivated localization microscopy (PALM) [1]. In the last decade, using different nonlinear or linear techniques, a localization precision down to few nanometers, even Ångström has been achieved [2]. Here, we present a new concept of position sensing enabling Ångström accuracy, based on strongly directional light emission off a single subwavelength dielectric scatterer. To realize the strong directional emission, we take advantage of a high refractive index dielectric silicon nanosphere, which supports both electric as well as magnetic resonances in the visible spectra [3]. As a probe beam, we use a radially polarized vector beam, which upon tight focusing provides an inhomogeneous field distribution, with a strong longitudinal electric field component present on-axis [4]. While, the transverse electric field components vanish on-axis, but increase linearly with radial distance (linearity holds in close vicinity to optical axis, around 50 nm). Using this tailored electromagnetic field, electric and magnetic dipoles resonances can be induced in the dielectric scatterer, when it is located off-axis; and interference of those dipole emissions may yield strong directional emission (see Fig. 44.1). By appropriately choosing the wavelength of the probe beam, this directivity has been maximized to get a strong position dependence. With proper calibration of this strong position dependent directivity, it was possible to show that a displacement of 5 nm can

A. Bag (✉) • M. Neugebauer • P. Woźniak
Max Planck Institute for the Science of Light, D-91058 Erlangen, Germany

Institute of Optics, Information and Photonics, Department of Physics,
Friedrich-Alexander-University Erlangen-Nuremberg, D-91058 Erlangen, Germany
e-mail: ankan.bag@mpl.mpg.de

G. Leuchs • P. Banzer
Max Planck Institute for the Science of Light, D-91058 Erlangen, Germany

Institute of Optics, Information and Photonics, Department of Physics,
Friedrich-Alexander-University Erlangen-Nuremberg, D-91058 Erlangen, Germany

Department of Physics, University of Ottawa, 25 Templeton, Ottawa, Ontario, K1N 6N5 Canada

© Springer Science+Business Media Dordrecht 2017

B. Di Bartolo et al. (eds.), *Nano-Optics: Principles Enabling Basic Research and Applications*, NATO Science for Peace and Security Series B: Physics and Biophysics, DOI 10.1007/978-94-024-0850-8_44

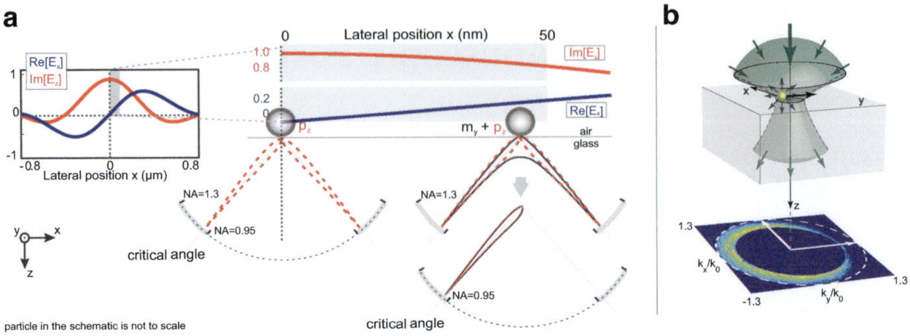

Fig. 44.1 (**a**) Electric field components along x-direction of a tightly focused radially polarized beam ($\lambda = 652\,\text{nm}$). Close to beam axis *(gray area)*, the transverse field amplitude $\text{Re}[E_x]$ is linearly dependent on the position, while the longitudinal field $\text{Im}[E_z]$ is approximately constant. Electric dipole p_z is induced when the dielectric silicon particle (on a glass substrate) is located on-axis, emitting symmetrically in the far-field. Both electric p_z and magnetic dipoles m_y are generated when the particle is off-axis, and interference of these dipole emissions give directionality. (**b**) A nanoparticle sitting on a glass substrate in the focal plane of a tightly focused light beam, is slightly shifted from the beam-axis; thus asymmetrically scattering. This is indicated by the backfocal plane image below (angular spectrum above NA = 1 is considered)

be easily distinguished whereas with further statistical analysis, it was possible to resolve smaller displacement with position uncertainty of 0.2 nm. This fast, easy to calibrate, linear technique can be very much useful for high resolution spatial and temporal particle localization and several other applications, also might constitute an alternative pathway towards linear high resolution imaging.

References

1. Betzig, E., Patterson, G. H., Sougrat, R., Lindwasser, O. W., Olenych, S., Bonifacino, J. S., Davidson, M. W., Lippincott-Schwartz, J., & Hess, H. F. (2006). Imaging intracellular fluorescent proteins at nanometer resolution. *Science*. doi:10.1126/science.1127344.
2. Nugent-Glandorf, L., & Perkins, T. T. (2004). Measuring 0.1-nm motion in 1 ms in an optical microscope with differential back-focal-plane detection. *Optics Letters*. doi:10.1364/OL.29.002611.
3. Woźniak, P., Banzer, P., & Leuchs, G. (2015). Selective switching of individual multipole resonances in single dielectric nanoparticles. *Laser & Photonics Reviews*. doi:10.1002/lpor.201400188.
4. Quabis, S., Dorn, R., Eberler, M., Glöckl, O., Leuchs, G. (2000). Focusing light to a tighter spot. *Optical Communication*. doi:10.1016/S0030-4018(99)00729-4.

Chapter 45
5d-4f Radioluminescence in Pr^{3+}-doped $K_3Y_xLu_{1-x}(PO_4)_2$

I. Carrasco, K. Bartosiewicz, F. Piccinelli, M. Nikl, and M. Bettinelli

Scintillating materials show *5d-4f* electric dipole allowed transitions located in the UV and visible regions, which make them widely used in various fields for the detection of X-ray and γ-ray photons or accelerated particles. Nowadays, most of the used materials are based on Ce^{3+} activated systems, but in the last years much attention has been focused on the study of the *5d-4f* emission of Pr^{3+} in wide band-gap hosts. The emitting lowest *5d* state of Pr^{3+} ions is located at higher energy than the one for Ce^{3+} in the same host, and its emission lifetime is about two or three times shorter (Ronda C, Gondek J, Goirand E, Jüstel T, Bettinelli M, Meijerink A, Mater. Res. Soc. Symp.Proc 1111:1111- D08-01, 2009; Korzhik M,. Fedorov A, Annenkov A, Borissevitch A, Dossovitski A, Missevitch O, Lecoq P, Nucl Instrum Methods A 571:122, 2007). As a result, the choice of Pr^{3+} as an activator ion for wide band-gap hosts may constitute a promising alternative in the development of new fast scintillators.

Among the many possible hosts, double phosphates appear as a promising option. It has been found that, in several double phosphates constituted of lanthanide ions, the absorption edge is located at rather high energy (Ivanovskikh K, Meijerink A, Ronda C, Piccinelli F, Speghini A, Bettinelli M, Opt Mater 34:419–423; Schwarz L, Finke B, Kloss M, Rohmann A, Sasum U, Haberland D, Journal of Luminescence,

I. Carrasco (✉) • F. Piccinelli • M. Bettinelli
Department of Biotechnology, University of Verona, Verona, Italy
e-mail: irene.carrascoruiz@univr.it

K. Bartosiewicz
Department of Optical Materials, Institute of Physics ASCR, Prague, Czech Republic

Faculty of Nuclear Sciences and Physical Engineering, Czech Technical University, Prague, Czech Republic

M. Nikl
Department of Optical Materials, Institute of Physics ASCR, Prague, Czech Republic

© Springer Science+Business Media Dordrecht 2017
B. Di Bartolo et al. (eds.), *Nano-Optics: Principles Enabling Basic Research and Applications*, NATO Science for Peace and Security Series B: Physics and Biophysics, DOI 10.1007/978-94-024-0850-8_45

72–74:257-259, 1997), making these compounds suitable for various optical applications including fast scintillators and vacuum ultraviolet phosphors. In particular, the $K_3Lu(PO_4)_2$ double phosphate is interesting due to the strong energy transfer from host-to-ion that it shows upon VUV synchrotron radiation excitation (Trevisani M, Ivanovskikh KV, Piccinelli F, Bettinelli M, Journal of Luminescence, 152:2–6, 2014). This opens the possibility of its application as a fast scintillator.

In this work, several materials belonging to the Pr^{3+} doped $K_3Y_xLu_{1-x}(PO_4)_2$ crystal family have been studied. The synthesis and structural characterization of these doped materials will be described in detail. Luminescence emission spectra measured using X-ray excitation will be presented and discussed.

Acknowledgments We would like to thank the European Commission for funding through the Marie Curie Initial Training network LUMINET, grant agreement No. 316906.

References

1. Ronda, C. Gondek, J. Goirand, E. Jüstel, T. Bettinelli M. & Meijerink, A. (2009). *Materials Research Society Symposium Proceedings, 1111,* 1111-D08-01.
2. Korzhik, M., Fedorov, A., Annenkov, A., Borissevitch, A., Dossovitski, A., Missevitch, O., & Lecoq, P. (2007). *Nuclear Instruments and Methods A, 571*, 122.
3. Ivanovskikh, K., Meijerink, A., Ronda, C., Piccinelli, F., Speghini, A., & Bettinelli, M. (2011). *Optical Materials, 34,* 419–423.
4. Schwarz, L., Finke, B., Kloss, M., Rohmann, A., Sasum, U., & Haberland, D. (1997). *Journal of Luminescence, 72–74*, 257–259.
5. Trevisani, M., Ivanovskikh, K. V., Piccinelli, F., & Bettinelli, M. (2014). *Journal of Luminescence, 152*, 2–6.

Chapter 46
Biosensing on a Chip: Study of Plasmonic Nanostructures Integrated in Microfluidic Devices

Juan Castro Arias, Andrea Cattoni, Dominique Decanini, Stéphane Collin, and Anne-Marie Haghiri-Gosnet

Abstract Arrays of plasmonic nanocavities with very low volumes, down to $\lambda^3/1000$, have been fabricated by soft UV nanoimprint lithography. It leads to high refractive index sensitivity (368 nm/RIU) and figure of merit (~46) and offers new perspectives for efficient biosensing experiments in ultralow volumes.

Surface plasmon resonance (SPR) based sensors are a well-established technology utilized for label-free biochemical sensing in different applications, from immunoassay and medical diagnostics, to environmental monitoring and food safety. The two main limitations of this technique is the wide probe volume ($>\lambda^3$) and the complex detection scheme (optical excitation through a prism in Kretschmann geometry). Localized SPR (LSPR) based sensors, constitute a promising alternative due to the possibilities for sensor miniaturization offered by nanoscale plasmon localization combined with efficient and simpler microspectroscopy. The performance of the biosensors are usually characterized by the bulk sensitivity $S_{\mathrm{bulk}} = \Delta\lambda/\Delta n$ (where n is the refractive index of the medium around the nanostructure), or through the figure of merit $\mathrm{FOM} = \mathrm{S}_{\mathrm{bulk}}/\mathrm{FWHM}$ (where FWHM is the full width at half-maximum of the plasmon resonance). SPR have higher sensitivity and FOM (typical values are $S_{\mathrm{bulk}} \sim 3000$ nm/RIU and $\mathrm{FOM} \sim 50$), nevertheless SPR fails in the detection of analytes of a few nanometers, because the effective variation of the index refraction is small. This can be understood considering the much long field decay length of propagating plasmons (~300 nm) that is proportional to the detection volume. In contrast, LSPR even though its FOM is much smaller ($S_{\mathrm{bulk}} \sim 200$ nm/RIU and $\mathrm{FOM} \sim 2$), its effective volume of detection is of the same order of magnitude of small analytes (~30 nm) which leads to a bigger effective variation of the refractive index (Svedendahl et al., Nano Lett 9 :4428, 2009). In this work we present a LSPR biosensor based on $\lambda^3/1000$ plasmonic nanocavities fabricated by soft UV nanoimprint lithography.

J. Castro Arias (✉) • A. Cattoni • D. Decanini • S. Collin • A.-M. Haghiri-Gosnet
Laboratoire de Photonique et de Nanostructure (LPN), 91460 Marcoussis, France
e-mail: juan_manuel.castro_arias@lpn.cnrs.fr

© Springer Science+Business Media Dordrecht 2017
B. Di Bartolo et al. (eds.), *Nano-Optics: Principles Enabling Basic Research and Applications*, NATO Science for Peace and Security Series B: Physics and Biophysics, DOI 10.1007/978-94-024-0850-8_46

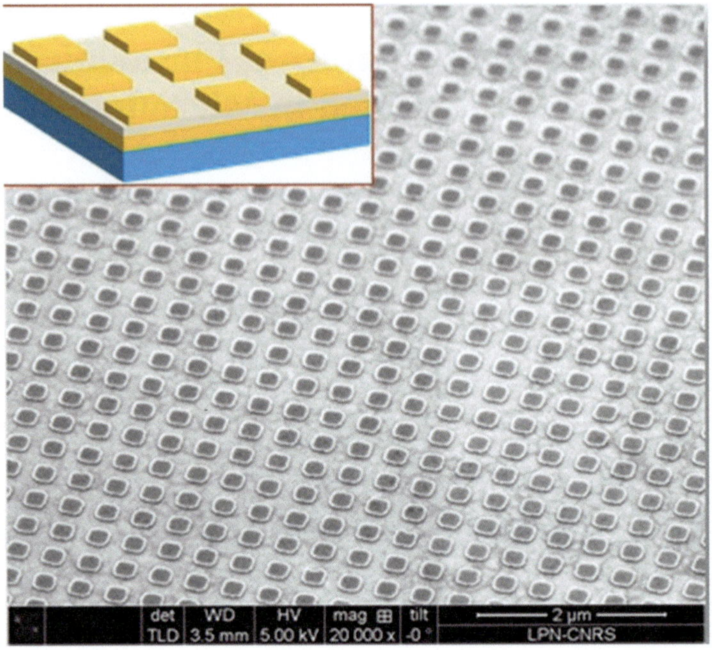

Fig. 46.1 SEM images and schematics of 1D and 2D arrays of $\lambda^3/1000$ plasmonic nanocavities

Following the former considerations, we have designed optical nanocavities with strong and efficient confinement of the incoming photons, and an improved FOM. Our method takes advantage of the strong light confinement of a Localized SPR in order to increase the limit of detection of small biological analytes. Nearly perfect omnidirectional absorption (3–70°) is demonstrated for the fundamental mode of the cavity ($\lambda = 1{,}15\ \mu$m). The second-order mode exhibits a sharper resonance with strong angular dependence and total optical absorption when the critical coupling condition is fulfilled. It leads to high refractive index sensitivity (368 nm/RIU) and FOM ($\sim$46) and offers new perspectives for efficient biosensing experiments in ultralow volumes (Cattoni et al., Nano Lett 11:3557, 2011) (Fig. 46.1).

References

1. Svedendahl et al. (2009). *Nano Letters, 9*, 4428.
2. Cattoni et al. (2011). *Nano Letters 11*, 3557.

Chapter 47
New Directions in Tip-Enhanced Near-Field Optical Microscopy

Nicolás Coca-López, Nina Mauser, Tobia Mancabelli, Alberto Comin, and Achim Hartschuh

Tip-enhanced near-field optical microscopy is used since more than 10 years in different configurations on an increasing variety of sample materials. The key advantage of the method is its combination of nanoscale spatial resolution and strong signal enhancement making it possible to study the weak optical response, such as Raman scattering or photoluminescence, of tiny sample volumes [1].

We report on our efforts to extend this method into further directions. One direction is the application of tip-enhancement to photovoltaic devices as suggested in [2]. We obtained the first high-resolution photocurrent images of carbon nanotube devices using a metal tip to locally enhance optical-to-electrical transduction [3]. Now, we exploit this scheme to study graphene devices. This allowed for the imaging of sub-diffraction features in zero-bias photocurrent and the simultaneously recorded Raman scattering signal. An enhanced Raman scattering signal is also observed at the edges of the gold contacts.

Another possibility is the combination of tip enhanced spectroscopy with ultrafast spectroscopy techniques, such as pulse-shaping. Gated multiphoton intrapulse interference phase scan is a technique used to characterize and compensate the spectral phase of ultrashort laser pulses [4]. This technique can also be applied to characterize the near field of plasmonic particle or nano-antennas. First results show that gold nano-tips, used away from their plasmon resonance, do not significantly perturb the local field and therefore are suitable for ultrafast antenna-enhanced nanoscopy.

N. Coca-López (✉) • N. Mauser • T. Mancabelli • A. Comin • A. Hartschuh
Department Chemie and CeNS, LMU München, Butenandtstrasse 5-13 E, Munich, Germany
e-mail: nicolas.cocalopez@cup.uni-muenchen.de; achim.hartschuh@cup.uni-muenchen.de

© Springer Science+Business Media Dordrecht 2017

B. Di Bartolo et al. (eds.), *Nano-Optics: Principles Enabling Basic Research and Applications*, NATO Science for Peace and Security Series B: Physics and Biophysics, DOI 10.1007/978-94-024-0850-8_47

References

1. Mauser, N., & Hartschuh, A. (2014). Tip-enhanced near-field optical microscopy. *Chemical Society Reviews, 43*, 1248.
2. Bharadwaj, P., Deutsch, B., & Novotny, L. (2009). Optical antennas. *Advances in Optics and Photonics, 1*(3), 438–483.
3. Rauhut, N., Engel, M., Steiner, M., Krupke, R., Avouris, P., & Hartschuh, A. (2012). Antenna-enhanced photocurrent microscopy on single-walled carbon nanotubes at 30 nm resolution. *ACS Nano, 6*, 6416.
4. Comin, A., Ciesielski, R., Piredda, G., Donkers, K., & Hartschuh, A. (2014). Compression of ultrashort laser pulses via gated multiphoton intrapulse interference phase scans. *Journal of the Optical Society of America B, 5*, 1118–1125.

Chapter 48
Phase Singularities in Random Waves: Exploring Optical Statistics at the Nanoscale

L. De Angelis, R.E.C. van der Wel, B. le Feber, and L. Kuipers

Abstract Phase singularities in complex fields are deep-subwavelength optical entities of wide interest. Their distributions in two-dimensional random waves are for example reminiscent of those of particles in ionic liquids. By mapping the in-plane complex optical field above a chaotic resonator, we experimentally determine the statistics of phase singularities in random waves. All the measured quantities are in excellent agreement with theoretical predictions.

Phase singularities are locations in which the phase of a complex field is undefined [1]. In two dimensional fields these are points in the plane. The phase of the field circulates around the singularity, assuming all its possible values. The integral of this quantity on a closed path around the singularity yields an integer multiple of 2π, known as its topological charge.

With our phase- and polarization-sensitive near-field microscopewe are able to measure the in-plane complex optical field above a two dimensional chaotic resonator [2], with subwavelength resolution. As sketched in Fig. 48.1, we experimentally establish position and topological charge of phase singularities in the interference pattern of electromagnetic waves randomly oriented in the plane.

Particular attention is given to the calculation of the charge correlation function $g_Q(r)$ and the pair correlation function $g(r)$. These describe how the density of singularities varies as a function of distance from a reference singularity, respectively weighting them (g_Q) or not (g) with their topological charge. In both cases a quantitative agreement with the theory presented in [1] is found (Fig. 48.2).

L. De Angelis (✉) • R.E.C. van der Wel • L. Kuipers
FOM Institute AMOLF, Amsterdam, The Netherlands
e-mail: deangelis@amolf.nl; r.v.d.wel@amolf.nl; kuipers@amolf.nl

B. le Feber
ETH, Zurich, Switzerland
e-mail: blefeber@ethz.ch

© Springer Science+Business Media Dordrecht 2017

B. Di Bartolo et al. (eds.), *Nano-Optics: Principles Enabling Basic Research and Applications*, NATO Science for Peace and Security Series B: Physics and Biophysics, DOI 10.1007/978-94-024-0850-8_48

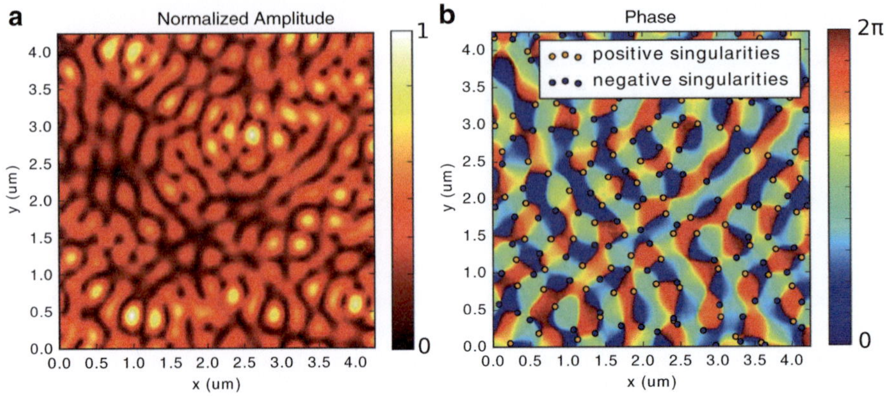

Fig. 48.1 Example of amplitude **(a)** and phase **(b)** of one projection of the optical field measured above the chaotic resonator excited with $\lambda_0 = 1550\,\text{nm}$. The field's phase circulates around phase singularities, that are pinpointed in the plot **(b)**. Their topological charge turns out to be $+1$ (positive singularities) or -1 (negative singularities)

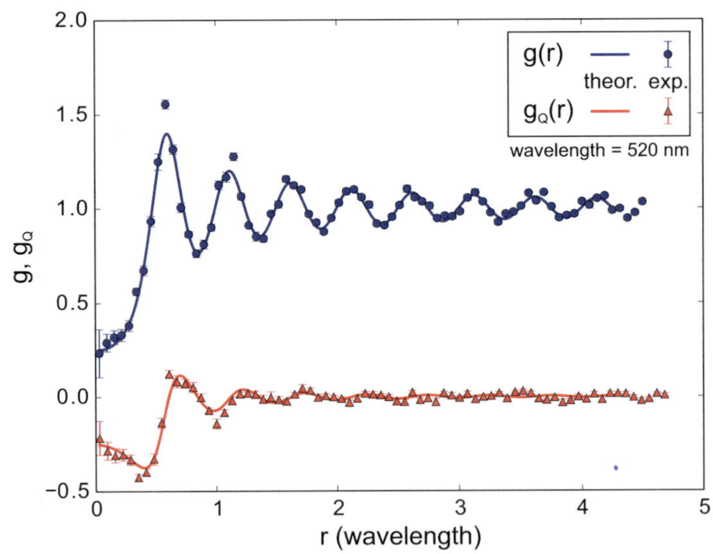

Fig. 48.2 Experimental pair correlation function $g(r)$ *(circles)* and charge correlation function $g_Q(r)$ *(triangles)* of phase singularities in random waves. Results are in excellent agreement with theoretical predictions from [1] *(lines)*

References

1. Berry, M. V., & Dennis, M. R. (2000). Phase singularities in isotropic random waves. *Proceedings of the Royal Society of London A, 456*, 2059–2079.

2. Liu, C., et al. (2015). Triggering extreme events at the nanoscale in photonic seas. *Nature Physics, 11*(4), 358–363. http://doi.org/10.1038/nphys3263.
3. De Angelis, L., Alpeggiani, F., Di Falco, A., & Kuipers, L. (2016). Spatial distribution of phase singularities in optical random vector waves. *Physical Review Letters, 117*(9), 93901. http://doi.org/10.1103/PhysRevLett.117.093901.

Chapter 49
Discrimination of Grapevine Genomic DNA Using Surface-Enhanced Raman Spectroscopy and PCA

Nicoleta E. Dina Mircescu, Cristina M. Muntean, and Nicolae Leopold

Abstract The classical methods of genomic identification rely on PCR and DNA sequencing, techniques that are destructive and time consuming. Instead, the new advances in the spectroscopic field bring in a more rapid approach for identification at strain level. Raman spectroscopy and more recently, surface-enhanced Raman spectroscopy (SERS) are used to create "molecular fingerprints" for accurate, non-intrusive genomic identification, being also used in DNA studies of plant species (Muntean CM et al., J Raman Spectrosc 44(6):817–822, 2013; Muntean CM, et al., J Raman Spectrosc 42(11):1925–1931, 2011). The essential information from complex spectroscopic patterns is usually extracted by using chemometrics, such as PCA (Heraud P, et al., J Chemom 20(5):193–197, 2006). However, the discrimination using PCA on SERS spectra collected from plant extracted DNA samples is scarcely reported in the literature so far (Qiu L et al., Vib Spectrosc 72:134–141, 2014).

SERS spectra were collected from DNA samples extracted from eight grapevine varieties of Romanian provenance by using the 532 nm laser line. Each spectrum was obtained by co-adding 5 scans of 10 s exposure each. Spectra pre-treatment is an important factor, which influences dramatically the PCA results. Different SERS spectra pre-treatment methods were proposed in the literature, although there is no general formula that could be applied for every type of sample and measurement approach (Heraud P, et al., J Chemom 20(5):193–197, 2006).

In this study we have used the Unscrambler® X Version 10.1 (Camo Software, Norway) for applying different pre-treatment methods on the spectral data, in order to achieve the best results in terms of varieties grouping. The most accurate grouping of the eight sets of spectral data was achieved, when the original spectra were processed using the following algorithms: baseline correction, multiplicative scattering

N.E. Dina Mircescu (✉) • C.M. Muntean
National Institute of R&D of Isotopic and Molecular Technology, Donat 67-103, Cluj-Napoca, Romania
e-mail: nicoleta.dina@itim-cj.ro

N. Leopold
Faculty of Physics, Babeș-Bolyai University, Kogalniceanu 1, 400084 Cluj-Napoca, Romania

© Springer Science+Business Media Dordrecht 2017
B. Di Bartolo et al. (eds.), *Nano-Optics: Principles Enabling Basic Research and Applications*, NATO Science for Peace and Security Series B: Physics and Biophysics, DOI 10.1007/978-94-024-0850-8_49

correction (MSC) and area normalization. The PCA scores plot indicated a fair discrimination between the eight grapevine varieties with different provenance, based mainly on the contribution of the spectral region between 300 and 1800 cm^{-1}, considered as the DNA spectral fingerprint region.

Acknowledgments The authors wish to thank to Drs. Ana Coste and Adela Halmagyi from the Institute of Biological Research, Cluj-Napoca, Romania, for DNA samples preparation. This work was supported by a grant of the Ministry of National Education, National Authority for Scientific Research CNCS – UEFISCDI, Romania, project number PN-II-ID-PCE-2012-4-0115.

References

1. Muntean, C. M., et al. (2013). Surface-enhanced Raman scattering assessment of DNA from leaf tissues adsorbed on silver colloidal nanoparticles. *Journal of Raman Spectroscopy, 44*(6), 817–822.
2. Muntean, C. M., et al. (2011). Surface-enhanced Raman spectroscopy of genomic DNA from in vitro grown plant species. *Journal of Raman Spectroscopy, 42*(11), 1925–1931.
3. Heraud, P., et al. (2006). Effects of pre-processing of Raman spectra on in vivo classification of nutrient status of microalgal cells. *Journal of Chemometrics, 20*(5), 193–197.
4. Qiu, L., et al. (2014). Analysis of plant genomic DNAs and the genetic relationship among plants by using surface-enhanced Raman spectroscopy. *Vibrational Spectroscopy, 72*, 134–141.

Chapter 50
High-Order Multipole Resonances in Cuboidal Surface Phonon Polariton Nanoresonators

Chase T. Ellis, J. G. Tischler, O. J. Glembocki, D. N. Chigrin, F. J. Bezares, R. Kasica, L. Shirey, J. C. Owrutsky, and J. D. Caldwell

Abstract It has been demonstrated that nanoresonators fabricated on the surface of polar dielectric materials, such as silicon carbide, are able to sustain plasmonic-like effects in the mid- to long- wave infrared spectral range with impressive figures of merit (Hillenbrand R, Taubner T, Keilmann F, Nature 418:159–162, 2002; Caldwell JD, Glembocki OJ, et al., Nano Lett 13:3690–3697, 2013; Wang T, Li P, et al., Nano Lett 13:5051–5055, 2013). Such phenomena is achieved by exploiting the TO and LO phonons to resonantly excite collective oscillations of bound lattice (Caldwell JD, Lindsay L, et al., Nanophotonics 4:2192–8614, 2015). The fact that these excitations are mediated by bound charges, rather than free charges - such as the case with plasmonic metals, results in extremely low optical losses and enhanced resonant phenomena. As such, polar dielectric nanoresonators may play a role in improving infrared nanophotonic technologies, such as waveguides, sources, near-field optics, solar cells, chemical sensors, biosensors, and photonic circuitry. However, fully realizing this potential, hinges on the ability to precisely control the near-field behavior of polar dielectric nanoresonators. In this work, we use a combination of optical measurements and finite element method simulations to investigate the far- and near-field resonant behavior of structurally-asymmetric, cuboidally-shaped, 4H-SiC nanoresonators with fixed height (h = 950 nm), fixed length (l = 400 nm), and varying width ($w = 400 - 6400$ nm). Overall, we observe over 12 polarization-sensitive resonances that can be tuned across the Reststrahlen band of 4H-SiC ($796 - 964$ cm^{-1}) (Ellis, C.T. et al. Scientific Reports 6:32959,

C.T. Ellis (✉) • J.G. Tischler • O.J. Glembocki • L. Shirey • J.C. Owrutsky • J.D. Caldwell
U.S. Naval Research Laboratory, Washington, DC, USA
e-mail: chase.ellis.ctr@nrl.navy.mil

D.N. Chigrin
Institute of Physics (IA), RWTH Aachen University, Aachen, Germany

F.J. Bezares
ICFO – The Institute of Photonic Sciences, Barcelona, Spain

R. Kasica
Center for Nanoscale Science and Technology, National Institutes of Standards and Technology, Gaithersburg, MD, USA

© Springer Science+Business Media Dordrecht 2017

B. Di Bartolo et al. (eds.), *Nano-Optics: Principles Enabling Basic Research and Applications*, NATO Science for Peace and Security Series B: Physics and Biophysics, DOI 10.1007/978-94-024-0850-8_50

2016) by changing the nanopillar aspect ratio ($AR = w/l = 1-16$). Futhermore, we find that these resonances exhibit a wide range of near-field radiation patterns that vary from a simple transverse dipole mode that is preserved for all *ARs* to complex, high-order multipoles with modal profiles that evolve with aspect ratio.

References

1. Hillenbrand, R., Taubner, T., & Keilmann, F. (2002). *Nature, 418*, 159–162.
2. Caldwell, J. D., Glembocki, O. J., et al. (2013). *Nano Letters, 13*, 3690–3697.
3. Wang, T., Li, P., et al. (2013). *Nano Letters, 13*, 5051–5055.
4. Caldwell, J. D., Lindsay, L., et al. (2015). *Nanophotonics, 4*, 2192–8614.
5. Ellis, C. T., et al. (2016). *Scientific Reports, 6*, 32959.

Chapter 51
Polarization Properties of the SERS Radiation Scattered by Linear Nanoantennas with Two Distinct Localized Plasmon Resonances

Antonino Foti, C. D'Andrea, A. Toma, B. Fazio, E. Messina, O. M. Maragò, Enzo Di Fabrizio, M. Lamy de La Chepelle, and P. G. Gucciardi

Abstract Playing with the field polarization permits to gain important insight on the near-field interaction between molecules and nanoantennas, as well as to develop new concepts of broadband plasmonic sensors (Fazio B. et al. ACS Nano 5(7):5945–59562, 2011; D'Andrea C. et al. J Phys Chem C 118(16):8571–8580, 2014; D'Andrea C. et al. ACS Nano 7(4):3522–3531, 2013; Foti A. et al. Nanospectroscopy 1:26–32, 2015). By studying the polarization state of the Surface Enhanced Raman Scattering it is, in particular, possible to better understand the re-radiation effects in linear nanoantennas featuring spectrally distinct localized plasmon resonances (LSPR). Here we show that, through the selective excitation of the short- or the long- axis LSPR, it is possible to control and totally rotate (90°) the polarization of the SERS photons of randomly oriented molecules adsorbed on linear nanoantennas. By a proper design of the nanoantenna, this can be accomplished switching the excitation wavelength from the visible to the

A. Foti (✉)
IPCF-CNR, Istituto per i Processi Chimico-Fisici, Messina, Italy

Scuola di Dottorato di Ricerca in Fisica, University of Messina, Messina, Italy
e-mail: foti@ipcf.cnr.it

C. D'Andrea
IPCF-CNR, Istituto per i Processi Chimico-Fisici, Messina, Italy

IFAC-CNR, Istituto di Fisica Applicata "Nello Carrara", Florence, Italy

B. Fazio • E. Messina • O.M. Maragò • P.G. Gucciardi
IPCF-CNR, Istituto per i Processi Chimico-Fisici, Messina, Italy

A. Toma
Istituto Italiano di Tecnologia, Genoa, Italy

E. Di Fabrizio
Physical Science and Engineering Division (PSE), King Abdullah University of Science and Technology (KAUST), Thuwal 23955-6900, Saudi Arabia

M.L. de La Chepelle
Laboratoire CSPBAT, CNRS (UMR 7244), Université Paris 13, Bobigny, France

© Springer Science+Business Media Dordrecht 2017

B. Di Bartolo et al. (eds.), *Nano-Optics: Principles Enabling Basic Research and Applications*, NATO Science for Peace and Security Series B: Physics and Biophysics, DOI 10.1007/978-94-024-0850-8_51

NIR, while using orthogonal fields polarizations, so to match the two distinct LSPRs. Modeling the nanoantenna enhancement with wavelength-dependent field enhancement tensors we recover the relations ruling the parallel-polarized and unpolarized- SERS intensity dependence found experimentally when selectively driving the two LSPRs. Our results support a physical picture in which the strongly coupled molecule-nanoantenna system behaves as an hybrid structure whose energy levels are determined by the molecule while the far-field radiation properties are determined by the antenna modes.

These results give us an important point of view for a better comprehension of the scattering process after the strong coupling between molecules and nanoantennas, which could improve the field of molecular detection and label-free biodetection by means of plasmonic nanostructures.

References

1. Fazio, B., et al. (2011). *ACS Nano, 5*(7), 5945–59562.
2. D'Andrea, C., et al. (2014). *Journal of Physical Chemistry C, 118*(16), 8571–8580.
3. D'Andrea, C., et al. (2013). *ACS Nano, 7*(4), 3522–3531.
4. Foti, A., et al. (2015). *Nanospectroscopy, 1*, 26–32.

Chapter 52
Fundamental Study and Analytical Applications of Nanoparticle-Enhanced Laser-Induced Breakdown Spectroscopy (NELIBS) of Metals, Semiconductors and Insulators

Rosalba Gaudiuso, Can Koral, Marcella Dell'Aglio, Olga De Pascale, and Alessandro De Giacomo

Abstract Nanoparticle-Enhanced Laser-Induced Breakdown Spectroscopy (NELIBS) is a recently proposed method to efficiently increase the LIBS emission signal of metals up to 2 orders of magnitude, by depositing metal nanoparticles (NPs) on the sample surface (De Giacomo A, Gaudiuso R, Koral C, Dell'Aglio M, De Pascale O Anal Chem 85). This considerable emission enhancement has been ascribed to two effects: (1) an improvement in the ablation effect, and (2) a more efficient production of seed electrons by field emission, in turn due to the enhancement of the laser electromagnetic field induced by the NPs themselves (De Giacomo A, Gaudiuso R, Koral C, Dell'Aglio M, De Pascale O Acta Part B, 98).

Here, we report our investigations about the effect played by several experimental parameters, i.e., laser energy; laser spot diameter; concentration, dimension, and kind of nanoparticles (NPs) in the case of NELIBS of metals. We also discuss NELIBS of non-metallic samples, whose emission enhancement is lower (up to 2–3 times) and follows a different mechanism. A special case of particular interest is that of transparent media, which can be efficiently made more absorptive by NP

R. Gaudiuso (✉)
Delaware State University, 1200 North Dupont Highway, Dover, DE 19901, USA
e-mail: rosalba.gaudiuso@nanotec.cnr.it

M. Dell'Aglio • O. De Pascale
Nanotec-CNR (Institute of Nanotechnology – National Research Council), via Amendola 122/D, 70126 Bari, Italy

C. Koral
Department of Chemistry, University of Bari, via Orabona 4, 70125 Bari, Italy

A. De Giacomo
Nanotec-CNR (Institute of Nanotechnology – National Research Council), via Amendola 122/D, 70126 Bari, Italy

Department of Chemistry, University of Bari, via Orabona 4, 70125 Bari, Italy

© Springer Science+Business Media Dordrecht 2017

B. Di Bartolo et al. (eds.), *Nano-Optics: Principles Enabling Basic Research and Applications*, NATO Science for Peace and Security Series B: Physics and Biophysics, DOI 10.1007/978-94-024-0850-8_52

deposition. Moreover, we demonstrated the suitability of NELIBS to quantitative analysis of various samples (metals, alloys and non-metallic samples), both with the classical approach f calibration lines, and with calibration-free methods.

References

1. De Giacomo, A., Gaudiuso, R., Koral, C., Dell'Aglio, M., De Pascale, O., Nanoparticle-Enhanced laser induced breakdown spectroscopy of metallic samples. *Analytical Chemistry*, 85.
2. De Giacomo, A., Gaudiuso, R., Koral, C., Dell'Aglio, M., De Pascale, O., Nanoparticle Enhanced Laser Induced Breakdown Spectroscopy (NELIBS): effect of nanoparticles deposited on sample surface on laser ablation and plasma emission, *Spectrochim. Acta Part B*, 98.

Chapter 53
New Antibacterial Photoactive Nanocomposite Additives for Endodontic Cements and Fillings

Y. Gerasymchuk, A. Łukowiak, A. Kędziora, A. Wedzyńska, G. Bugla-Płoskońska, D. Piątek, T. Bachanek, V. Chernii, L. Tomachynski, and W. Stręk

Abstract Dentists practice shows, that in the process of treatment of root canal treatment, quite often the secondary infection of canal takes place. This problem, on the one hand, can be caused by inaccurate purification (disinfection) of canals, untight fulfillment of canal, or incidence of bacterial strains (especially anaerobic) in the canals, which are resistant to conventional disinfectants and antiseptics therapy (Sundqvist G, Figdor D, Persson S, Sjögren U Oral Surg Oral Med Oral Pathol Oral Radiol Endod 85:86–93, 1998). In order to solve this problem, we realize the following tasks: (i) from the clinical point – collecting of biopsy bacterial strains from root canals of patients, (ii) from the microbiological point – identification of bacterial strains, measurement of their microbiological characteristics (which shows, among other, the resistance of bacteria to antibiotics), and carrying out for this strains and the corresponding commercial strains a comprehensive study of the antimicrobial activity of the studied composite materials, (iii) from the physical point – identification and explanation the effects of light on antibacterial activity of the composite material, (iv) from the material science point – development of

Y. Gerasymchuk (✉) • A. Łukowiak • W. Stręk
Department of Spectroscopy of Excited States, Institute of Low Temperature and Structure Research Polish Academy of Sciences, 2 Okolna St., 50-422 Wroclaw, Poland
e-mail: j.gerasymchuk@int.pan.wroc.pl

A. Wedzyńska
Institute of Low Temperature and Structure Research Polish Academy of Sciences, University of Wroclaw, 2 Okolna St., 50-422 Wroclaw, Poland

A. Kędziora • G. Bugla-Płoskońska
Institute of Genetics and Microbiology, Department of Microbiology, Faculty of Biological Sciences, University of Wroclaw, 63/77 Przybyszewskiego St, 51-148 Wroclaw, Poland

D. Piątek • T. Bachanek
Department of Conservative Dentistry and Endodontics, F. Skubiszewski Medical University, Al. Racławickie 1, 20-059 Lublin, Poland

V. Chernii • L. Tomachynski
V.I. Vernadskii Institute of General and Inorganic Chemistry, 32/34 Palladin Ave, Kiev, Ukraine

© Springer Science+Business Media Dordrecht 2017
B. Di Bartolo et al. (eds.), *Nano-Optics: Principles Enabling Basic Research and Applications*, NATO Science for Peace and Security Series B: Physics and Biophysics, DOI 10.1007/978-94-024-0850-8_53

canal-filling materials doped with composites, including the measurement of their antimicrobial activity. The final step of investigation should be experimental clinical studies of the developed cements (doped with composites) on biopsy materials.

Our proposition is to create a filler material for endodontic root canal treatment, which would guarantee antimicrobial protection at least to time of closing of the open apical constriction of roots, and preventing the growth of bacteria, and the secondary periodontal inflammation.

In the proposed materials we incorporated the idea of stable (not decomposable in tissues) antimicrobial agent, which have a sustained and non-invasive activated (under the light irradiation) antiseptic activity.

The new photosensitive composite material was based on highly oxidized graphite oxide, which is modified with silver nanoparticles and photosensitizer – bis(lysinato)-zirconium(IV)-phthalocyanine complex, as a dopant for dental cements (Kędziora A, Gerasymchuk Y, Sroka E, Bugla-Płoskońska G, Doroszkiewicz W, Rybak Z, Hreniak D, Stręk W Polim Med 43:129–134, 2013; Tomachynski L, Chernii V, Gorbenko H, Filonenko V, Volkov S Ukrainica Chimica Zhurnal, 1:862–867, 2004; Kovalska V, Losytskyy M, Chernii V, Volkova K, Tretyakova I, Cherepanov V, Yarmoluk S, Volkov S Bioorg Med Chem 20:330–334, 2012). The structural and morphological properties of new composite were examined by different methods (electron microscopy, powder X-ray diffraction, molecular and atomic spectroscopy measurements).

Finally, the antibacterial activity was determined (on the first stage – for three aerobic stains *Staphylococcus aureus*, *Pseudomonas aeruginosa*, and *Escherichia coli*) (Kędziora A, Gerasymchuk Y, Sroka E, Bugla-Płoskońska G, Doroszkiewicz W, Rybak Z, Hreniak D, Stręk W Polim Med 43:129–134, 2013). The activity was enhanced under the influence of the UV and IR light irradiation, because of the activation of system through generation of reactive oxygen species. The realization of this idea is associated with several microbiological studies for the separate components and whole composite material, because each component, when it was included to composite system, should fulfill its function without disturbing the action of other ingredients, and as a whole material, it should have an enhanced effect than previously used analogues or traditional antiseptic agents (eq. antibiotics).

Proposed solutions and materials will be investigated from point of usage of them in conservative dentistry and endodoncy, but in the future they may also be used not only in other fields of medicine (eq. development of dressing materials for the treatment of hard to heal wounds and decubitus ulcer, for production and/or covering of implants with biocompatible and/or bioactive film, production of filtration materials, including filters, used for protecting of the sterility of the air in the medical institutions or pharmaceutical factories, development of photocatalytic materials).

References

1. Sundqvist, G., Figdor, D., Persson, S., & Sjögren, U. (1998). *Oral Surgery, Oral Medicine, Oral Pathology, Oral Radiology, and Endodontology, 85*, 86–93.
2. Kędziora, A., Gerasymchuk, Y., Sroka, E., Bugla-Płoskońska, G., Doroszkie-wicz, W., Rybak, Z., Hreniak, D., & Stręk, W. (2013). *Polymers in Medicine, 43*, 129–134.
3. Tomachynski, L. A., Chernii, V. Y., Gorbenko, H. N., Filonenko, V. V., & Volkov, S. V. (2004). *Chemistry Biodiversity, 1*, 862–867.
4. Kovalska, V., Cherepanov, V., Losytskyy, M., Chernii, S., Senenko, A., Chernii, V., Tretyakova, I., Yarmoluk, S., & Volkov, S. (2012). *Bioorganic & Medicinal Chemistry, 20*, 330–334.

Chapter 54
Low-Loss Phonon Polaritons in Nanostructured Dielectrics

Alexander J. Giles, S. Dai, Chase T. Ellis, J. C. Owrutsky, J. G. Tischler, O. J. Glembocki, F. J. Bezares, R. Kasica, L. Shirey, M. M. Fogler, Dimitri Basov, and J. D. Caldwell

Abstract Plasmonics provides great promise for nanophotonic applications. However, the high optical losses inherent in metal-based plasmonic systems have limited progress. Thus, it is critical to identify alternative low-loss materials. One alternative is polar dielectrics that support surface phonon polariton (SPhP) modes, where the confinement of infrared light is aided by optical phonons. SiC nanopillar arrays support such modes, exhibiting a dipolar resonance transverse to the nanopillar axis and a monopolar resonance associated with the longitudinal axis dependent upon the SiC substrate. Both exhibit exceptionally narrow linewidths (7–24 cm^{-1}), with quality factors of 40–135, which exceed the theoretical limit of plasmonic systems, with extreme subwavelength confinement of (λres3/Veff)1/3 = 50–200. These observations promise to reinvigorate research in SPhP phenomena and their use for nanophotonic applications. Another approach is the use of hyperbolic materials, which have been a focus of the nanophotonics community for their potential to realize sub-diffractional imaging and focusing of light, and novel optical properties, such as a negative index of refraction. The recent observation that hexagonal boron nitride (hBN) is a natural, high efficiency hyperbolic material has led to a surge in research within this field. Due to the low-loss nature, van der Waals bonding and extreme crystal anisotropy, the hyperbolic polaritons within hBN are not only promising for novel applications within the

A.J. Giles (✉) • C.T. Ellis • J.C. Owrutsky • J.G. Tischler • O.J. Glembocki • L. Shirey • J.D. Caldwell
U.S. Naval Research Laboratory, Washington, DC, USA
e-mail: alexander.giles.ctr@nrl.navy.mil

S. Dai • M.M. Fogler • D. Basov
University of California at San Diego, La Jolla, CA, USA

F.J. Bezares
The Institute of Photonic Sciences, Barcelona, Spain

R. Kasica
Center for Nanoscale Technology, National Institutes of Standards and Technology, Gaithersburg, MD, USA

© Springer Science+Business Media Dordrecht 2017
B. Di Bartolo et al. (eds.), *Nano-Optics: Principles Enabling Basic Research and Applications*, NATO Science for Peace and Security Series B: Physics and Biophysics, DOI 10.1007/978-94-024-0850-8_54

mid-infrared, but is also extremely well suited for fundamental investigations into their resonant behaviors. We have used scattering near-field optical microscopy (s-SNOM) to directly probe the local surface electromagnetic fields of three-dimensionally confined nanostructures of hBN, reporting the first experimental observation of frequency dependent internal angular reflection within a hyperbolic nanostructure, a phenomenon previously theoretically predicted.

Acknowledgments A.J.G. acknowledges support from the NRC NRL Postdoctoral Fellowship Program, while project funding was provided via the Office of Naval Research and distributed by the Nanoscience Institute at the Naval Research Society.

References

1. Caldwell, J. D., et al. (2014). *Nature Communications, 5*, 5221.
2. Dai, S., et al. (2014). *Science, 343*, 1125–1129.
3. Caldwell, J. D. et al. (2014). *Nanophotonics.*
4. Caldwell, J. D., et al. (2013). *Nano Letters, 13*, 3690.

Chapter 55
Fabrication of SERS Substrates by Roll-to-Roll Hot Embossing

Anne Habermehl, Xin Liu, Carsten Eschenbaum, and Uli Lemmer

Abstract Surface-enhanced Raman spectroscopy (SERS) combines the high specificity of Raman scattering with a high sensitivity due to an enhancement of the electromagnetic field by metallic nanostructures. Multiple micro- and nanostructuring methods have been used to fabricate SERS-structures [1–3]. We present an approach to create enhancing arrays of nanopillars [4] by roll-to-roll (R2R) hot-embossing and to functionalize them with a gold layer. The R2R hot embossing process is shown in principle in Fig. 55.1 on the left. The setup mainly consists in two rotating cylinders (one micro- or nanostrcutured) that can be pressed against each other and additionally heated up to exceed the glass transition temperature of the polymer foil in between. The embossing parameters mainly determine the results [5]. We have investigated the influence of the embossing temperature and pressure. With the identified parameter setting we can show the uniform transfer of the nanopillars like shown in Fig. 55.1 on the right. The setup and the result of a Raman measurement of the analyte rhodamine 6G (Rh6G) on a SERS substrate and an unstructured reference substrate is shown in Fig. 55.2. In a following step this SERS substrate can be integrated into a microfluidic chip by manufacturing one master for the microstructure and the nanostructures. Our work show a way to a low cost fabrication of tailor-made sensor chips for SERS analyse.

A. Habermehl (✉) • X. Liu
Light Technology Institute, Engesserstr. 13, 76131 Karlsruhe, Germany
e-mail: anne.habermehl@kit.edu; xin.liu@kit.edu

C. Eschenbaum • U. Lemmer
Light Technology Institute, Engesserstr. 13, 76131 Karlsruhe, Germany
Institute of Microstructure Technology, Hermann-von-Helmholtz-Platz 1, 76344 Eggenstein-Leopoldshafen, Germany
e-mail: carsten.eschenbaum@kit.edu; uli.lemmer@kit.edu

© Springer Science+Business Media Dordrecht 2017

B. Di Bartolo et al. (eds.), *Nano-Optics: Principles Enabling Basic Research and Applications*, NATO Science for Peace and Security Series B: Physics and Biophysics, DOI 10.1007/978-94-024-0850-8_55

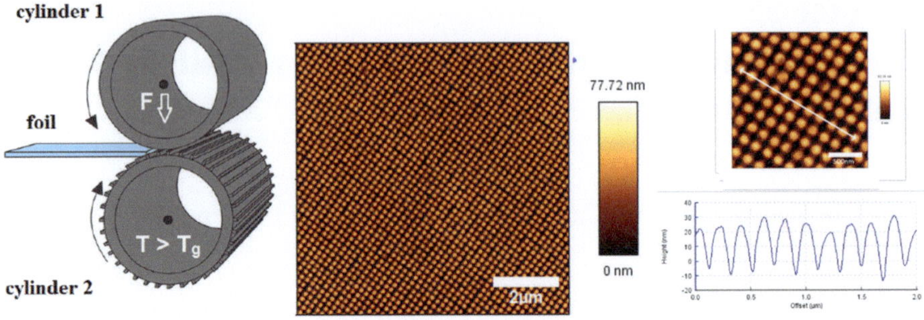

Fig. 55.1 Principle of R2R hot embossing *(left)* and atomic force microscopy of R2R hot embossed nanopillars in large *(middle)* and in detail with cross section of nanopillars *(right)*

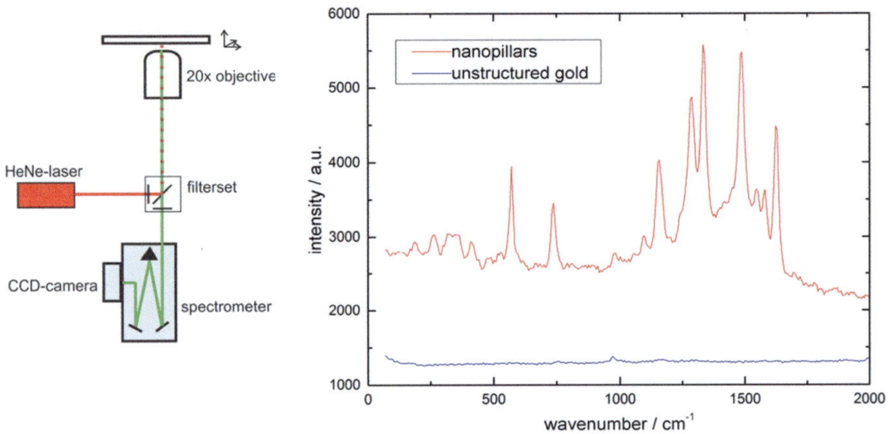

Fig. 55.2 Raman characterization setup and Raman spectrum of 30 µl of 100 µM Rh6G solution dried on the nanopillar substrate (*red*) and an unstructured reference substrate (*blue*), integration time 10 s, excitation 1, 3 mW at 632, 8 nm

Acknowledgements We acknowledge support by the Karlsruhe School of Optics and Photonics (KSOP).

References

1. Haynes, C. L., & Van Duyne, R. P. (2001). Nanosphere lithography: a versatile nanofabrication tool for studies of size-dependent nanoparticle Optics. *Journal of Physical Chemistry B, 105*, 5599–5611.
2. Wu, W., Hu, M., Ou, F. S., Li, Z., & Williams, R. S. (2010). Cones fabricated by 3D nanoimprint lithography for highly sensitive surface enhanced Raman spectroscopy. *Nanotechnology, 21*, 255502.
3. Kahl, M., Vogesa, E., Kostrewab, S., Vietsb, C., & Hillb, W. (1998). Periodically structured metallic substrates for SERS. *Sensors and Actuators B, 51*(1–3), 285–291.

4. Liu, X., et al. (2015). Tailored surface-enhanced Raman nanopillar arrays fabricated by laser-assisted replication for biomolecular detection using organic semiconductor lasers. *ACS Nano, 9*(1), 260.
5. Worgull, M. (2009). *Theory and technology of microreplication* Oxford/Burlington: William Andrew.

Chapter 56
Coupling Semiconducting Nanowires to Plasmonic Nanoantennas

Mathieu Jeannin, Pamela Rueda-Fonseca, Rudeesun Songmuang, Edith Bellet-Amalric, Kuntheak Kheng, and Gilles Nogues

Abstract We propose to develop semiconductor-metallic hybrid devices by placing metallic nanoparticles in the vicinity of nanowires, as pictured in Fig. 56.1. Acting as plasmonic nanoantennas they strongly couple to the local electromagnetic field and allow us to control their emission properties, e.g. enhancement of radiative decay rate and tunability/modification of the radiation diagram [3, 4]. We present here results and on-going experiments in the parallel studies of such nanowires and nanoantennas.

In the past few years, the field of plasmonics has experienced an on-going expansion due to the possibility to investigate light-matter coupling phenomena at the nanoscale, and within the quantum regime [1]. Among the wide variety of quantum emitters, semiconducting nanowires offer many promising properties due to the possibility of growing several kinds of high quality heterostructures, allowing for single- photon emission and large tunability of emission properties (Purcell enhancement, radiation guiding etc.) up to room temperature conditions [2]. Efficient coupling of these two technologies could improve our knowledge of fundamental light-matter interactions, as well as pave the way to new optoelectronic devices.

We propose to develop semiconductor-metallic hybrid devices by placing metallic nanoparticles in the vicinity of nanowires, as pictured in Fig. 56.1. Acting as plasmonic nanoantennas they strongly couple to the local electromagnetic field and allow us to control their emission properties, e.g. enhancement of radiative decay rate and tunability/modification of the radiation diagram [3, 4].

M. Jeannin (✉) • G. Nogues
CNRS, Institut Néel, 25 Avenue des Martyrs, 38042 Grenoble, France
e-mail: mathieu.jeannin@neel.cnrs.fr

P. Rueda-Fonseca • R. Songmuang • E. Bellet-Amalric • K. Kheng
CEA, INAC-SP2M, 17 Avenue des Martyrs, 38000 Grenoble, France

© Springer Science+Business Media Dordrecht 2017

B. Di Bartolo et al. (eds.), *Nano-Optics: Principles Enabling Basic Research and Applications*, NATO Science for Peace and Security Series B: Physics and Biophysics, DOI 10.1007/978-94-024-0850-8_56

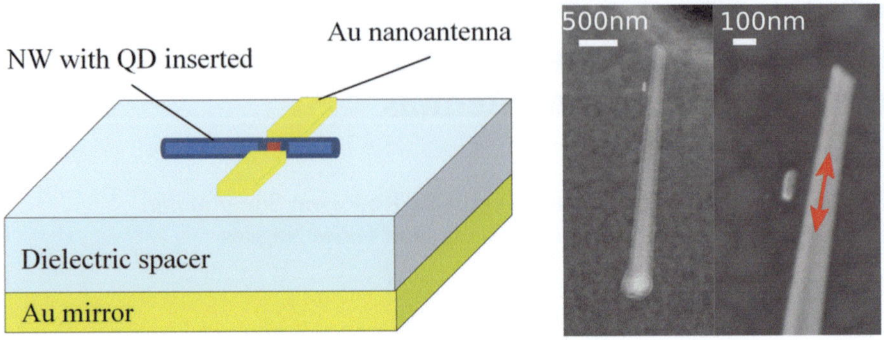

Fig. 56.1 Schematic representation of a coupled Nanowire-Antenna system, and SEM image of one realization of such system. The *red arrow* indicated the position of the quantum dot and the polarization direction of the emitted light

We present here preliminary results and on-going experiments in the parallel studies of such nanowires and nanoantennas. Polarization and k-resolved spectroscopy and imaging on single nanowires aim at characterizing the exciton properties in the nanowires (dipole orientation and polarization of the emitted field). The plasmonic nanostructures properties are investigated using cathodoluminescence in order to study the localized surface plasmon modes. Numerical calculations are performed to define design parameters and comfort experimental results. Light-antenna coupling has been theoretically studied in simple systems. Possibilities of modifying the decay rate of the structures as well as obtaining selective quenching or enhancement for different exciton orientation are found. Finally, we detail nanofabrication methods such as in-situ electron beam lithography of the structure and nanomanipulation of single nanowires, allowing us to precisely locate and arrange the semiconducting and metallic structures.

References

1. Tame, M. S., McEnery, K. R., Özdemir, S. K., Lee, J., Maier, S .A., & Kim, M. S. (2013). Quantum plasmonics. *Nature Physics, 9*, 329–340.
2. Bounouar, S., Elouneg-Jamroz, M., den Hertog, M., Morchutt, C., Bellet-Amalric, E., André, R., Bougerol, C., Genuist, Y., Poizat, J.-Ph., Tatarenko, S., & Kheng, K. (2012). Ultrafast room temperature single-photon source from nanowire-quantum dots. *Nano Letters, 12*, 2977–2981.
3. Nogues, G., Merotto, Q., Bachelier, G., Lee, E. H., & Dong Song, J. (2013). Fabrication and tuning of plasmonic optical nanoantennas around droplet epitaxy quantum dots by cathodoluminescence. *Applied Physics Letters, 102*, 231112–231116.
4. Belacel, C., Habert, B., Bigourdan, F., Marquier, F., Hugonin, J.-P., Michaelis de Vasconcellos, S., Lafosse, X., Coolen, L., Schwob, C., Javaux, C., Dubertret, B., Greffet, J.-J., Senellart, P., & Maitre, A. (2013). Controlling spontaneous emission with plasmonic optical patch antennas. *Nano Letters, 13*, 1516–1521.

Chapter 57
Optical Emission Spectroscopy of Combined Laser Ablation-Hollow Cathode Glow Discharge Plasma Source

Stefan Karatodorov, Valentin Mihailov, and Margarita Grozeva

Abstract The combination of laser ablation and glow discharge (LA-GD) as a spectroscopic source for elemental analysis is an increasing field of study in recent years. This is due to the fact that the laser ablation-glow discharge provides a more stable and reproducible spectra of the sample in comparison with the transient laser ablation. An important advantage of the laser ablation-glow discharge technique is the possibility for independent control of the processes of sample introduction and sample excitation. The combination of laser ablation and glow discharge is also capable of sampling non-conducting or low-sputtering-yield materials which for stand-alone glow discharge usually requires preliminary sample preparation procedures or use of pulsed power supply. Exploiting the characteristics of this technique has resulted in increase in sensitivity and limit of detection (Tereszchuk K, Vadillo J, Lazerna J Appl Surf Sci 255). Here results on one of the realization of the laser ablation-glow discharge technique – combination of laser ablation and hollow cathode discharge are reported (Karatodorov S, Mihailov V, Grozeva M, Open Chem 13). Spatial separation of the laser ablation introduction from the hollow cathode discharge excitation is achieved. The sample analytical line emission intensity dependence on background gas pressure, sample cathode distance and discharge current is studied. This allowed optimization of the technique for increasing its analytical characteristics to be done.

Acknowledgments The work is done with financial support of project INERA (REGPOT-2012-2013-1).

S. Karatodorov (✉) • V. Mihailov • M. Grozeva
Acad. G. Nadjakov Institute of Solid State Physics, Bulgarian Academy of Sciences, 72 Tsarigradsko Chaussee, 1784 Sofia, Bulgaria
e-mail: skarat@issp.bas.bg

© Springer Science+Business Media Dordrecht 2017
B. Di Bartolo et al. (eds.), *Nano-Optics: Principles Enabling Basic Research and Applications*, NATO Science for Peace and Security Series B: Physics and Biophysics, DOI 10.1007/978-94-024-0850-8_57

References

1. Tereszchuk, K., Vadillo, J., Lazerna, J. *Applied Surface Science,* 255.
2. Karatodorov, S., Mihailov, V., Grozeva, M. *Open Chemical,* 13.

Chapter 58
Femtosecond Transient Absorption Spectroscopy of Photochromic Thiol-Functionalized Terphenylthiazole-Based Diarylethene Molecules

Alina Khodko, Vadim Khomenko, Oleksandr Mamuta, Sergii Snegir, Pei Yu, Emmanuelle Lacaze, Alexandr Marchenko, and Nataliya Kachalova

Abstract The ultrafast cyclization dynamics of photochromic thiol-functionalized terphenylthiazole-based diarylethene molecules was investigated by femtosecond pump-probe spectroscopy. The third harmonic generation (266 nm) of femtosecond Ti:Sapphire laser (Mira-900F) was used as a pump radiation. The supercontinuum generation (450–700 nm) was used as a probe radiation. The time of cyclization reaction is close to 0.9 ps for diarylethene derivative without alkyl-thiol chain, and is near 1.4 ps for derivative with alkyl-thiol chain. These resuts proves that terphenylthiazole-based diarylethenes are the perspective optoelectronic elements with the ultrafast optical response.

The enhancement of the electro-optical response of the photochromic molecules opens vast perspectives for development of molecular resistors with optically-modulated resistance, ultra-compact and sensitive heavy metals sensors with

---

A. Khodko (✉) • V. Khomenko • O. Mamuta • A. Marchenko • N. Kachalova
Institute of Physics, National Academy of Sciences of Ukraine, Prospekt Nauky, 46, Kyiv, Ukraine
e-mail: khodkoalina@gmail.com; khomenko.vadim@gmail.com; mamuta_aleksandr@mail.ru; marchenko_alexandr@yahoo.com; kachalova.nataliya@gmail.com

S. Snegir
Chuiko Institute of Surface Chemistry, National Academy of Sciences of Ukraine, General Naumov str., 17, Kyiv, Ukraine
e-mail: ssnegir@gmail.com

P. Yu
Institut de Chimie Moléculaire et des Matériaux d'Orsay, UMR 8182, Université Paris Sud 11, 91405 Orsay, France

E. Lacaze
CNRS, UMR 7588, Institut des Nanosciences de Paris, 4 place Jussieu, F-75005 Paris, France
e-mail: emmanuelle.lacaze@insp.jussieu.fr

© Springer Science+Business Media Dordrecht 2017

B. Di Bartolo et al. (eds.), *Nano-Optics: Principles Enabling Basic Research and Applications*, NATO Science for Peace and Security Series B: Physics and Biophysics, DOI 10.1007/978-94-024-0850-8_58

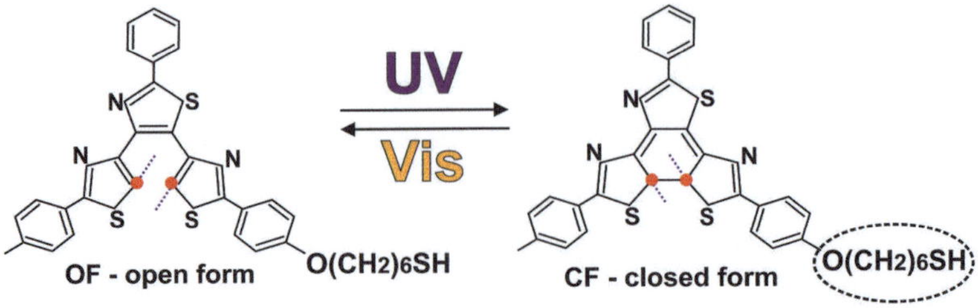

Fig. 58.1 The changes of chemical structure of terphenylthiazole-based diarylethenes under ultraviolet (UV) or visible (Vis) radiation. The alkyl-thiol chain ($O(CH_2)_6SH$) is represented in a dashed ellipse.

electro-optical control, molecular-based data storage devices with extremely high data density, etc [1]. Various organic molecules (rotaxanes, naphthalocyanines, azobenzenes, etc.) [2] undergo reversible light-induced properties. Among them, diarylethenes are particularly appropriate because of their outstanding thermal stability, good fatigue resistance, high light response, and small structural changes between the open- and closed-ring isomers. The complementary investigations of the ultrafast cyclization dynamics that depends on structural characteristics, temperature, and chemical composition of the environment are required for their further applications in optoelectronics and molecular engineering [3].

The purpose of study was to investigate the ultrafast photoinduced ring-closing reaction of two similar modifications of thiol-functionalized terphenylthiazole-based diarylethene (TBD) molecules with and without alkyl-thiol chain [4] (Fig. 58.1) dissolved in ethanol solution. The photoinduced dynamics were explored by femtosecond transient absorption spectroscopy at the Center for collective use "Femtosecond Laser Complex" at the Institute of Physics (Kyiv, Ukraine). The amplified femtosecond pulses ($E = 2.5$ mJ, $\tau = 140$ fs, $\lambda = 800$ nm, $\nu = 250$ Hz) from Ti:Sapphire laser (Coherent, USA) were split into two channels: pump initiated the cyclization reaction, while probe tracked the ultrafast changes in optical density, which have occurred in the sample. Since the cyclization reaction occurs under UV illumination, we used the third harmonic generation ($P = 4$ mW, $\tau = 250$fs, $\lambda = 266$ nm), obtained via the set of BBO crystals (i.e., beta barium borate), as a pump radiation. As a probe radiation, the supercontinuum (SC, 450–700 nm), generated in Al_2O_3 crystal by femtosecond pulses at 800 nm, was used. We probed the photoinduced optical density changes near the maximum of CF absorption band. The probe beam was split into signal and idler beams before crossing the sample, then both beams were recorded by the multichannel spectrograph (Imaging Spectrograph SP-2558, Acton, USA). To avoid accumulation of CF molecules, the solution was pumped through a quartz flow cell with 2 mm optical path and a buffer volume using a peristaltic pump with 2 ml/s flow rate. All studies were performed at room temperature $T = 22 \pm 0.2$ °C.

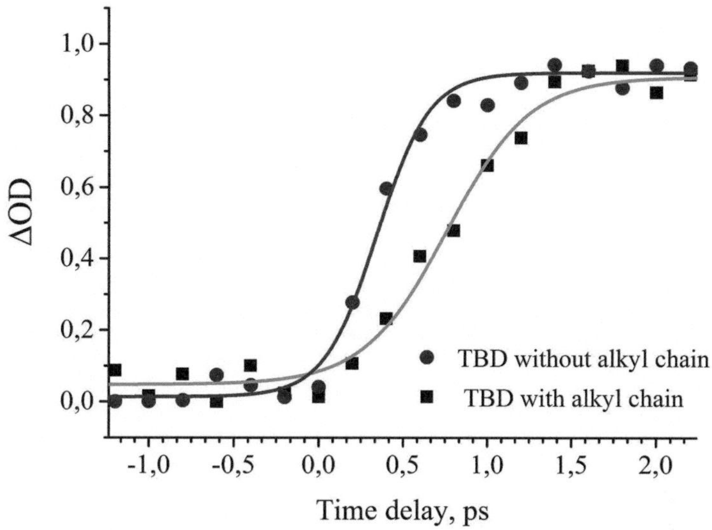

Fig. 58.2 The time profiles displaying the optical density changes while photoinduced cyclization reaction of terphenylthiazole-based diarylethenes (TBD) without and with alkyl-thiol chain. The ring-closing dynamics was fitted by the nonlinear Boltzmann function (solid line)

Time profiles of transient absorption spectrum for photochromic diarylethene molecules presented in Fig. 58.2. An analysis of experimental data shows that the time constant of cyclization reaction for photochromic TBD molecules with alkyl-thiol chain is close to 1.4 ps. However, the time constant of cyclization reaction for the TBD molecule without alkyl-thiol chain is close to 0.9 ps.

It was experimentally observed that presence of alkyl-thiol chain increase the time constant of the cyclization reaction by 0.5 ps. It can be explained due to the difference in relaxation channels of TBD molecules with and without alkyl chain after UV-induced illumination. An additional theoretical calculations are required for understanding of this difference. However, already now the obtained results can be considered as a basis for design of flexible organic electric nanodevices with specified characteristics.

Previous results based on difurylethene-base diarylethene derivatives were presented on the International National Conference "Chemistry, Physics and Technology of Surface" [5].

Acknowledgements All measurements were performed at the Center for collective use of equipment "Femtosecond Laser Complex" of National Academy of Sciences of Ukraine. The work has been funded by the NATO Project NUKR.SFPP 984702 – "Metal Nanocrystals for Highly Sensitive Detection of Biochemical Agents"

References

1. Irie, M., (2000). Diarylethenes for memories and switches. *Chemical Reviews, 100*, 1685–1716.
2. Kim, D., Jeong, H., Lee, H., Hwang, W.-T., Wolf, J., Scheer, E., Huhn, T., Jeong, H., & Lee, T. (2014). Flexible molecular-scale electronic devices composed of diarylethene photoswitching molecules. *Advanced Materials, 26*, 3968–3973.
3. Feringa, L.B. (2007). The art of building small: From molecular switches to molecular motors. *Journal of Organic Chemistry, 72*, 6635–6652.
4. Snegir, S.V., Yu, P., Maurel, F., Kapitanchuk, O. L., Marchenko, A.A., & Lacaze, E. (2014). Switching at the nanoscale: Light- and STM-tip-induced switch of a thiolated diarylethene self-assembly on Au(111). *Langmuir, 30*(45), 13556–13563.
5. Khodko, A. A., Khomenko, V. V., Voitsekhovich, V. S., Mamuta, O. D., Snegir, S. V., Marchenko, O. A., & Kachalova, N. M. (2015). *Investigation of the dynamics of ultafast processes in photochromic diarylethene derivative molecules.* In *National conference with international participation, dedicated to the 85th anniversary of academician of Ukraine O.O. Chuiko "Chemistry, Physics and Technology of Surface"*, Kyiv (pp. 205), 13–15 May 2015.

Chapter 59
Nanostructural Inhomogeneities in Chalcogenide Glasses Probed by Positron Annihilation Methods

Halyna Klym

Abstract Positron annihilation (PA) is known to be high informative tool to study atomistic imperfections like free-volume defects in different solids whatever their structural organization (Krause-Rehberg R, Leipner HS, Positron annihilation in semiconductors. Defect studies, Springer-Verlag, Berlin-Heidelberg-New York, 1999). In application to semiconductors, this method realized in the variant of PA lifetime (PAL) spectroscopy, allows useful identification of intrinsic free volumes. The PA technique in the measuring mode of Doppler broadening of annihilation radiation (DBAR) allows additional identification of dominant positron trapping sites in the tested objects. In this work, we shall analyze possibilities of combined PAL-DBAR measurements to study different inhomogeneities in GeS_2-Ga_2S_3 chalcogenide glasses (ChG) with different amount of GeS_2 and Ga_2S_3 ($80GeS_2$-$20Ga_2S_3$, $82GeS_2$-$18Ga_2S_3$, $84GeS_2$-$16Ga_2S_3$); nanovoid transformation in $80GeS_2$-$20Ga_2S_3$ glasses affected by CsCl additions and crystallization tendencies in $80GeSe_2$–$20Ga_2Se_3$ ChG caused by different duration of annealing time. These ChG possesses an excellent optical transparency along with good glass-forming ability, thermomechanical and chemical stability, which makes it one of the most unprecedented media for different IR optics applications.

It is shown that in the case of GeS_2-Ga_2S_3 base ChG Ga-related void sub-system plays a decisive role in positron trapping process, while the overall density variation is defined mainly by Ge-related sub-system. *S-W* correlation plot is grouped along nearly the same straight line tending from $80GeS_2$-$20Ga_2S_3$, $82GeS_2$-$18Ga_2S_3$ and $84GeS_2$-$16Ga_2S_3$ glasses in the direction of reduced *S* and enhanced *W* parameters. Such behavior corresponds to a so-called abnormal tendency in the k_d-ρ correlation, when all defect-related components exhibit obvious deviations, which can be ascribed to some changes in the defect environment (Klym H, Ingram A, Shpotyuk O, Calvez L, Petracovschi E, Kulyk B, Serkiz R, Szatanik R, Nanoscale Res Lett 10:49, 2015).

H. Klym (✉)
Lviv Polytechnic National University, 12 Bandera str, Lviv 79013, Ukraine
e-mail: halyna.i.klym@lpnu.ua; klymha@yahoo.com

© Springer Science+Business Media Dordrecht 2017

B. Di Bartolo et al. (eds.), *Nano-Optics: Principles Enabling Basic Research and Applications*, NATO Science for Peace and Security Series B: Physics and Biophysics, DOI 10.1007/978-94-024-0850-8_59

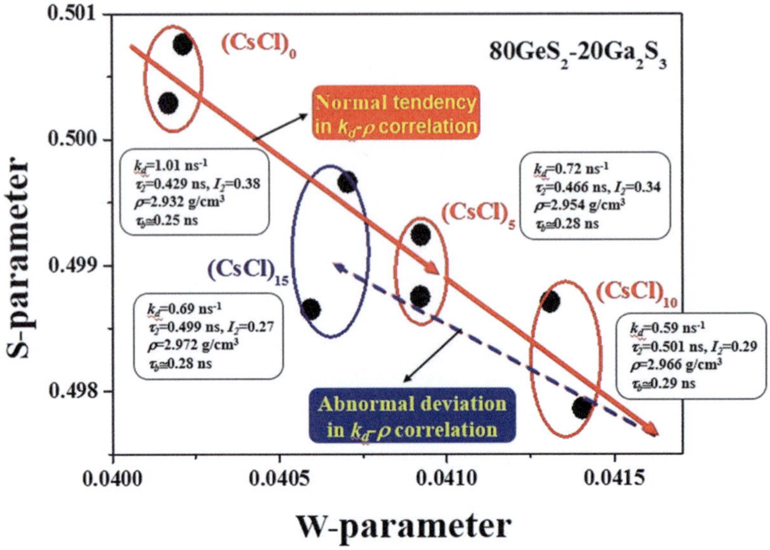

Fig. 59.1 Compositional deviations in $(80GeS_2\text{-}20Ga_2S_3)_{100\text{-}x}(CsCl)_x$ glasses on *S*-*W* plot

The structural changes caused by CsCl additives in $80GeS_2$-$20Ga_2S_3$ ChG can be adequately described by positron trapping modes determined within two-state model (Klym H, Ingram A, Shpotyuk O, Szatanik R, Petracovschi E, Calvez L, Lin C Solid State Phenom 230: 221–227, 2015). The results testify in a favor of rather unchanged nature of corresponding free-volume voids responsible for positron trapping in the studied glasses, when mainly concentration of these traps is a subject to most significant changes with composition. There are normal tendency in *S*-*W* evolution for GeS_2-Ga_2S_3-CsCl ChG (Fig. 59.1), when overall free volume of positron trapping voids is responsible for atomic density of the glasses.

The annealing of $80GeSe_2$-$20Ga_2Se_3$ glasses at 380 °C for 50, 80 and 100 h are associated with specific fragmentation of larger free-volume entities (acting as positron trapping voids) into a greater number of smaller ones. This process can be presented as abnormal tendency in the correlation between positron trapping rate k_d in defects and atomic density of the crystallized material ρ. The initiating stage of the crystallization occurring under 10 and 25 h annealing has an opposite nature, demonstrating a preferential crystallites nucleation and agglomeration of related voids with normal tendency in k_d-ρ correlation of *S*-*W* parameters (Fig. 59.2).

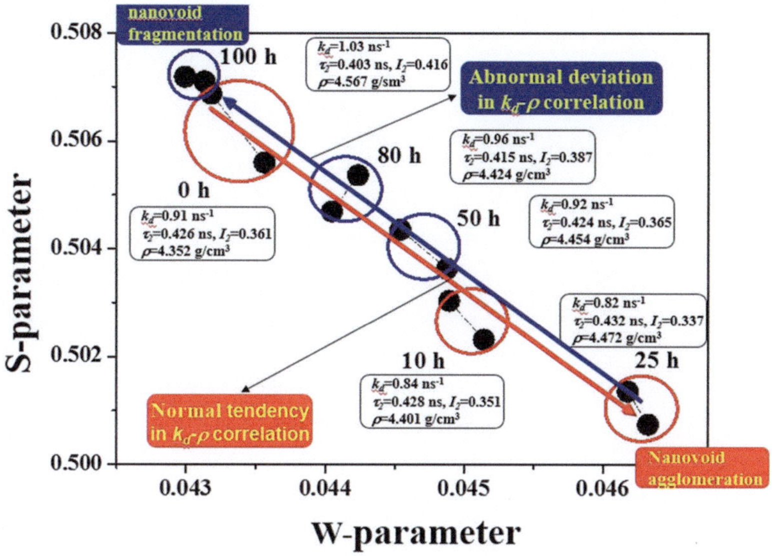

Fig. 59.2 *S-W* correlation plot for 80$GeSe_2$-20Ga_2Se_3 base glass (0 h) and annealed (Klym H, Ingram A, Shpotyuk O, Szatanik R, Petracovschi E, Calvez L, Lin C, Solid State Phenomena 230:221–227, 2015)

References

1. Krause-Rehberg, R., & Leipner, H. S. (1999). *Positron annihilation in semiconductors. Defect studies*. Berlin/Heidelberg/New York: Springer.
2. Klym, H., Ingram, A., Shpotyuk, O., Calvez, L., Petracovschi, E., Kulyk, B., Serkiz, R., & Szatanik, R. (2015). *Nanoscale Research Letters, 10*, 49.
3. Klym, H., Ingram, A., Shpotyuk, O., Szatanik, R., Petracovschi, E., Calvez, L., & Lin, C. (2015). *Solid State Phenomena, 230*, 221–227.

Chapter 60
Chiral Plasmonic Core-Shell Nanohelices

N.D. Kosters, A.K. de Hoogh, N. Rotenberg, H. Acar, H. Zeijlemaker, and L. Kuipers

Abstract Chirality is a geometrical property of objects with a mirror image that cannot be superimposed on itself. Chiral three-dimensional structures have recently received interest due to possible applications in nanophotonic technologies. For instance, chiral plasmonic planar metamaterials, have shown to significantly improve biosensing [1]. For possible applications in the visible wavelength regime, it is necessary that chiral plasmonic structures have nanoscale geometries. Recently, researchers showed a successful technique to fabricate nanometer sized metallic helices using electron beam induced deposition [2]. In this work, we fabricate core-shell (fused silica-gold) nanohelices using this technique and study the influence of geometrical parameters on optical activity. Here, optical activity is defined as $(T_{LCP} - T_{RCP})/(T_{LCP} + T_{RCP})$ with $T_{LCP}(T_{RCP})$ representing the transmission of left- (right-) handed circularly polarized light. A trend is observed when performing transmission measurements: short helices show a less strong optical response compared to tall helices. Preliminary simulations (FDTD) are used to describe the effect of gold shell thickness.

N.D. Kosters (✉) • A.K. de Hoogh • H. Acar • H. Zeijlemaker • L. Kuipers
FOM Institute AMOLF, Science Park 104, Amsterdam, The Netherlands
e-mail: kosters@amolf.nl; hoogh@amolf.nl; kuipers@amolf.nl

N. Rotenberg
Max Planck Institute for the Science of Light, 91058 Erlangen, Germany
e-mail: nir.rotenberg@mpl.mpg.de

© Springer Science+Business Media Dordrecht 2017

B. Di Bartolo et al. (eds.), *Nano-Optics: Principles Enabling Basic Research and Applications*, NATO Science for Peace and Security Series B: Physics and Biophysics, DOI 10.1007/978-94-024-0850-8_60

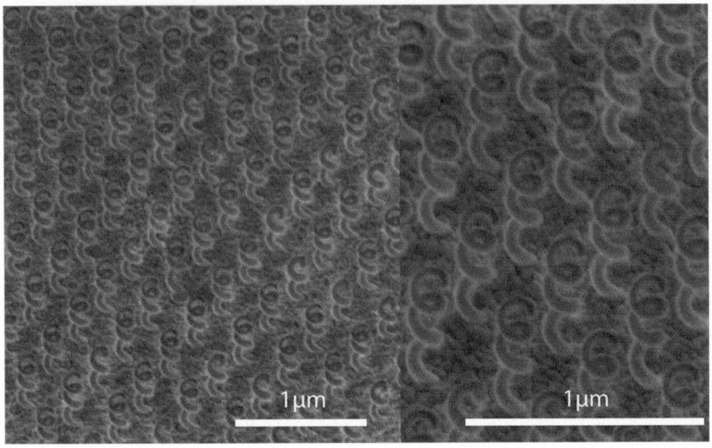

Fig. 60.1 Array of silica nanohelices fabricated on sample of glass. Pure gold is sputtered on the structure to form a homogeneous gold shell around the silica nanohelices

References

1. Hendry, E., Carpy, T., Johnston, J., Popland, M., Mikhaylovskiy, R., Lapthorn, A., Kelly, S., Barron, L., Gadegaard, N., & Kadodwala, M. (2010). Ultrasensitive detection and characterization of biomolecules using superchiral fields. *Nature Nanotechnology, 11*, 783–787.
2. Esposito, M., Tasco, V., Cuscuna, M., Todisco, F., Benedetti, A., Tarantini, I., De Giorgi, M., Sanvitto, D., & Passaseo, A. (2014). Nanoscale 3D chiral plasmonic helices with circular dichroism at visible frequencies. *ACS Photonics, 2*, 105–114.

Chapter 61
Two-Dimensional Dye Self-Assemblies on Graphene: Optical Signature

Sylvain Le Liepvre, F. Charra, A.-J. Attias, C. Fiorini, and L. Douillard

Abstract The field of molecular self-assembly has gone through tremendous advances past years. Starting from simple two-dimensional geometries, the third dimension seems now achievable using molecules like the Janus Tecton(Du P, Bleger D, Charra F, Bouchiat V, Kreher D, Mathevet F, et al. Beilstein J Nanotechnol 6:632–639, 2015; Du P, Jaouen M, Bocheux A, Bourgogne C, Han Z, Bouchiat V, et al. Angew Chem Int Ed 53(38):10060–10066, 2014). Furthermore, recent graphene fabrication process improvements brought a new convenient substrate for self-assemblies (MacLeod M, Rosei F Small 10(6):1038–1049, 2014). Thanks to substrate transfer processes (Li XS, Cai WW, An JH, Kim S, Nah J, Yang DX, et al. Science 324(5932):1312–1314, 2009; Li X, Zhu Y, Cai W, Borysiak M, Han B, Chen D, et al. Nano Lett 9(12):4359–4363, 2009), it is now possible to integrate two-dimensional self-assemblies in various systems, such as nanophotonics devices taking advantage of the graphene transparency. We study the absorption and fluorescence optical properties of six different self-assembled systems on graphene and show that multiple effects have to be considered to explain the optical behavior of these systems. An important red shift and a broadening of the absorption spectrum peaks is observed. We obtain the first fluorescence graphene functionalization taking advantage of the dye quenching dependence with the distance between the dye and the graphene. An estimation of the gain that could be obtained in a plasmonic waveguide shows that two-dimensional self-assemblies may open a new route towards plasmon amplification.

S. Le Liepvre (✉) • F. Charra • C. Fiorini • L. Douillard
DSM/IRAMIS/SPEC/LEPO, CNRS UMR 3680, CEA Saclay, Gif-sur-Yvette, France
e-mail: sylvain.le-liepvre@cea.fr

A.-J. Attias
Laboratoire de chimie des polymères, CNRS UMR 7610, Université Pierre et Marie Curie, Gif-sur-Yvette, France

© Springer Science+Business Media Dordrecht 2017
B. Di Bartolo et al. (eds.), *Nano-Optics: Principles Enabling Basic Research and Applications*, NATO Science for Peace and Security Series B: Physics and Biophysics, DOI 10.1007/978-94-024-0850-8_61

References

1. Du, P., Bleger, D., Charra, F., Bouchiat, V., Kreher, D., Mathevet, F., & Attias, A.-J. (2015). *Beilstein Journal of Nanotechnology, 6*, 632–639.
2. Du, P., Jaouen, M., Bocheux, A., Bourgogne, C., Han, Z., Bouchiat, V., Kreher, D., Mathevet, F., Fiorini-Debuisschert, C., Charra, F., & Attias, A.-J. (2014). *Angewandte Chemie International Edition, 53*(38), 10060–10066.
3. MacLeod, M., & Rosei, F. (2014). *Small, 10*(6), 1038–1049.
4. Li, X. S., Cai, W. W., An, J. H., Kim, S., Nah, J., Yang, D. X., Piner, R., Velamakanni, A., Jung, I., Tutuc, E., Banerjee, S. K., Colombo, L., & Ruoff, R. S. (2009). *Science, 324*(5932), 1312–1314.
5. Li, X., Zhu, Y., Cai, W., Borysiak, M., Han, B., Chen, D., Piner, R. D., Colombo, L., & Ruoff, R. S. (2009). *Nano Letters, 9*(12), 4359–4363.

Chapter 62
Q-Switching of Ytterbium Lasers by A Graphene Saturable Absorber

Pavel Loiko, Josep Maria Serres, Xavier Mateos, Haohai Yu, Huaijin Zhang, Junhai Liu, Konstantin Yumashev, Uwe Griebner, Valentin Petrov, Magdalena Aguiló, and Francesc Díaz

Abstract Yb:YGG, Yb:LuGG and Yb:CNGG lasers are Q-switched by a graphene-based saturable absorber. 190 ns/1.9 μJ pulses are generated at 1045.1 nm using Yb: CNGG and the maximum output power is 440 mW with 24 % slope efficiency.

Graphene, a two-dimensional material composed of a single-layer of carbon atoms arranged in a honeycomb lattice, is nowadays actively studied as a broadband saturable absorber (SA) for near-IR passively Q-switched (PQS) lasers (Bao QL, Zhang H, Wang Y, Ni Z, Yan Y, Shen ZX, Loh KP, Tang DY, Adv Funct Mater 19:3077–3083, 2009). In the present work, we report on graphene-SA PQS diode-pumped Yb bulk lasers based on cubic gallium garnets, Yb:YGG, Yb:LuGG and Yb:CNGG, as gain media.

P. Loiko (✉)
Center for Optical Materials and Technologies, Belarusian National Technical University, 65/17 Nezavisimosti Ave., 220013 Minsk, Belarus

Física i Cristal·lografia de Materials i Nanomaterials (FiCMA-FiCNA), Universitat Rovira i Virgili (URV), Campus Sescelades, c/Marcel·lí Domingo, s/n., E-43007 Tarragona, Spain
e-mail: kinetic@tut.by

J.M. Serres • X. Mateos • M. Aguiló • F. Díaz
Física i Cristal·lografia de Materials i Nanomaterials (FiCMA-FiCNA), Universitat Rovira i Virgili (URV), Campus Sescelades, c/Marcel·lí Domingo, s/n., E-43007 Tarragona, Spain

H. Yu • H. Zhang
State Key Laboratory of Crystal Materials, Shandong University, 250100 Jinan, China

J. Liu
College of Physics, Qingdao University, 308 Ning-Xia Road, 266071 Qingdao, China

K. Yumashev
Center for Optical Materials and Technologies, Belarusian National Technical University, 65/17 Nezavisimosti Ave., 220013 Minsk, Belarus

U. Griebner • V. Petrov
Max Born Institute for Nonlinear Optics and Short Pulse Spectroscopy, Max-Born-Str. 2a, D-12489 Berlin, Germany

© Springer Science+Business Media Dordrecht 2017

B. Di Bartolo et al. (eds.), *Nano-Optics: Principles Enabling Basic Research and Applications*, NATO Science for Peace and Security Series B: Physics and Biophysics, DOI 10.1007/978-94-024-0850-8_62

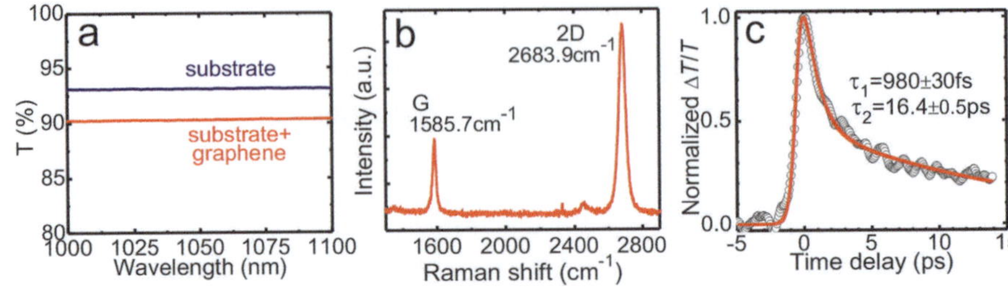

Fig. 62.1 Characterization of the graphene SA: (**a**) transmission spectrum, T(substrate + graphene)/T(substrate) = 97 %. (**b**) Raman spectrum, (**c**) pump-probe trace (*circles*) and bi-exponential fit to the data (*line*), $\tau_{1,2}$ – characteristic recovery times of internal absorption

Yb:YGG and Yb:LuGG crystals were grown by the optical floating zone method and Yb:CNGG crystal was grown by the conventional Czochralski method. The active elements were mounted in a water-cooled copper holder (kept at 12 °C). A commercial CVD-grown single-layer graphene deposited on 1.05 mm thick fused silica was used as SA. Its initial transmission at ~1.04 μm was 97.0 %, Fig. 62.1a. In accordance with Raman spectroscopy, the ratio of integrated areas of G and 2D bands, I_G/I_{2D}, was ~0.2, Fig. 62.1b, confirming the presence of a single carbon layer. The characteristic recovery times of the initial absorption for graphene was measured with the pump-probe technique at 1 μm (Griebner U, Serres JM, Mateos X, Petrov V, Aguiló M, Díaz F Adv Solid State Lasers (ASSL), p. ATh2A.16, 2014) to be 980 ± 30 fs and 16.4 ± 0.5 ps, Fig. 62.1c. The laser cavity consisted of a flat pump mirror and a concave ($R = 50$ mm) output coupler OC with $T_{OC} = 10\,\%$ at the laser wavelength. The laser crystals were pumped by a fiber-coupled InGaAs laser diode at ~977 nm. The pump spot radius in the crystal was $w_p = 100$ μm.

Stable PQS operation was achieved with all laser crystals studied. The corresponding input-output dependences are shown in Fig. 62.2a. For Yb:YGG, the maximum average output power was 462 mW at 1039.8 nm with a slope efficiency η of 23 %. This crystal provided a maximum Q-switched conversion efficiency η_{conv} of 55 %. For Yb:CNGG, 440 mW of pulsed output at 1045.1 nm was extracted with a maximum $\eta = 24\,\%$.

The duration of the Q-switched pulses decreased with the pump level, Fig. 62.2b. The shortest pulse duration t_p for Yb:YGG was ~440 ns. For Yb:CNGG, pulses as short as 190 ns were achieved. The maximum pulse energy was achieved with Yb:CNGG, $E_{out} = 1.9$ μJ. This corresponded to a peak power $P_{peak} = 10.0$ W. Figure 62.2c shows the oscilloscope traces of the shortest Q-switched pulses achieved.

To conclude, we report on the first realization of passively Q-switched Yb:YGG, Yb:LuGG and Yb:CNGG lasers at ~1.04 μm using a graphene-based saturable absorber. Our study reveals the potential of graphene-based SAs for passive Q-switching of Yb bulk lasers.

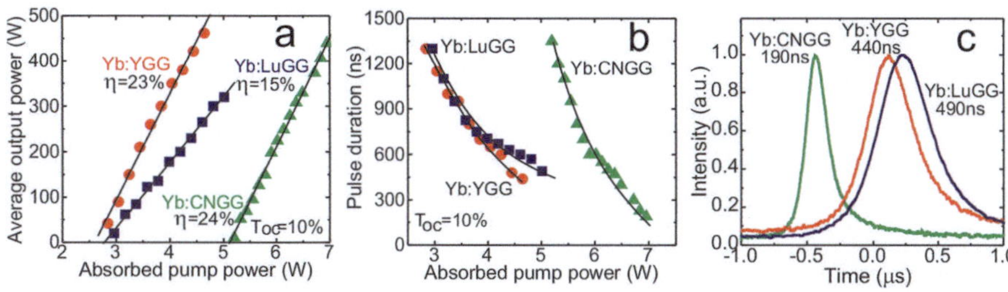

Fig. 62.2 Graphene Q-switched Yb:YGG, Yb:LuGG and Yb:CNGG diode-pumped lasers: (**a**) input–output dependences; (**b**) comparison of pulse durations; (**c**) oscilloscope traces of the shortest Q-switched pulses

References

1. Bao, Q. L., Zhang, H., Wang, Y., Ni, Z., Yan, Y., Shen, Z. X., Loh, K. P., & Tang, D. Y. (2009). *Advanced Functional Materials, 19*, 3077–3083.
2. Griebner, U. Serres, J. M. Mateos, X. Petrov, V. Aguiló, M. Díaz, F. (2014). Graphene saturable absorber Q-switched Tm:KLu(WO_4)$_2$ laser emitting at 2 μm, *Advanced Solid-State Lasers (ASSL)*, p. ATh2A.16

Chapter 63
Surface-Enhanced Fluorescence from Polypropylene Substrates

Alina Muravitskaya, Svetlana Vaschenko, Olga Kulakovich, Dmitry Guzatov, and Sergey V. Gaponenko

Abstract Surface-enhanced fluorescence of molecular probes is a highly developing field providing a number of new methods in chemical and biomedical analysis. Fluorescence enhancement effect depends on size and shape of the metallic nanoparticles, its optical absorption properties, distance between them and fluorophore (Gaponenko SV, Introduction to nanophotonics. Cambridge University Press, Cambridge, pp 456–467, 2010; Guzatov DV, Vaschenko SV, Stankevich VV, Lunevich AY, Glukhov YF, Gaponenko SV, J Phys Chem C. doi:10.1021/jp301598w). The properties of the substrate also influence on the optical response as the distribution of the nanoparticles depends on the surface properties of the substrate. The surface-enhanced fluorescence was investigated on glass (Guzatov DV, Vaschenko SV, Stankevich VV, Lunevich AY, Glukhov YF, Gaponenko SV, J Phys Chem C, doi: 10.1021/jp301598w), polystyrene 96-well microplates (Nooney RI, Stranik O, McDonagh C, MacCraith BD Langmuir, 10.1021/la801631w), quartz (Lakowicz JR, Malicka J, D'Auria S, Gryczynski I Anal Biochem. 10.1016/S0003-2697(03)00351-8) and modified polycarbonate films (Aslan K, Badugu R, Lakowicz JR, Geddes CD J Fluoresc, 10.1007/s10895-005-2515-5).

Polypropylene has become one of the most important materials due to its fine mechanical stability, good processability, high solvent resistance, and low cost. However, the weak polarity nature of its chemical structure makes the polypropylene membrane highly hydrophobic (Dai Z-W, Ling J, Huang X-J, Wan L-S, Xu Z-K J Phys Chem C, 10.1021/jp201040g). This prevented the application and investigations of this material as a substrate for plasmon-enhanced fluorescence (Aslan K, Badugu R, Lakowicz JR, Geddes CD J Fluoresc, 10.1007/s10895-005-2515-5).

In previous works surface-enhanced fluorescence of model biological conjugates BSA-FITC near silver nanoparticles deposited on glass substrates was investigated both theoretically and experimentally (Guzatov DV, Vaschenko SV,

A. Muravitskaya (✉) • S. Vaschenko • O. Kulakovich • D. Guzatov • S.V. Gaponenko
B. I. Stepanov Institute of Physics, National Academy of Sciences of Belarus, 68 Nezavisimosti av., Minsk, 220072 Minsk, Belarus
e-mail: alica.mur@mail.ru

© Springer Science+Business Media Dordrecht 2017

B. Di Bartolo et al. (eds.), *Nano-Optics: Principles Enabling Basic Research and Applications*, NATO Science for Peace and Security Series B: Physics and Biophysics, DOI 10.1007/978-94-024-0850-8_63

Stankevich VV, Lunevich AY, Glukhov YF, Gaponenko SV, J Phys Chem C, doi: 10.1021/jp301598w). The goal of this work was to study surface-enhanced fluorescence of labeled antibody on the polypropylene films modified by different amount of the polyelectrolyte layers. For this purpose multilayer nanostructures "plastic – polyelectrolytes – silver nanoparticles – polyelectrolytes – antibody" were prepared by successive deposition on plastic substrates. The fluorescence signal was measured for two sets of samples – with and without silver layer. Also was prepared reference glass sample by same technology. Silver sol was synthesized by the $AgNO_3$ citrate reduction technique. Fluorescence spectra were registered with a grating spectrograph S3801 (Solar TII, Belarus) combined with a liquid nitrogen cooled silicon CCD camera (Princeton Instruments, USA). Fluorescence excitation was performed with a cheap and affordable commercial light-emitting diode with the emission spectrum peaking at 460 nm. All optical measurements were made at room temperature (~300 K).

The Fig. 63.1 shows that the signal from polypropylene samples depends on the number of the polyelectrolyte layers. The highest enhancement in six times was obtained for three polyelectrolyte layers between silver nanoparticles and plastic surface. The decreasing of the signal with increasing of the modification can be connected with aggregation of nanoparticles and then changing of the optical properties of silver layer. The reference signal from glass sample was higher for 18 % than signal from plastic sample with three polyelectrolyte layers. Modification of the plastic surface with successive deposition of polyelectrolyte layers result in

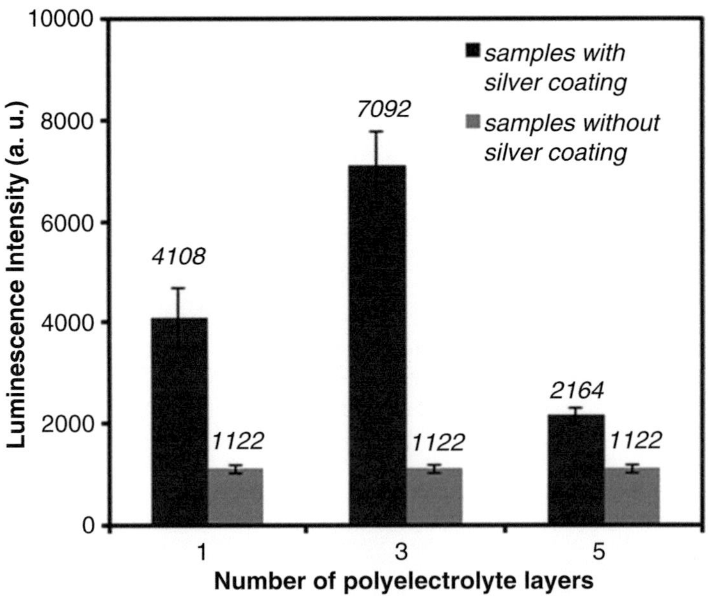

Fig. 63.1 Fluorescence dependence on the number of the polyelectrolyte layers

possibility of silver nanoparticles electrostatic deposition and then surface-enhanced fluorescence of labeled antibody on top of a silver film.

References

1. Gaponenko, S. V. (2010). *Introduction to nanophotonics* (pp. 456–467). Cambridge: Cambridge University Press.
2. Guzatov, D. V., Vaschenko, S. V., Stankevich, V. V., Lunevich, A. Y., Glukhov, Y. F., & Gaponenko, S. V. *Journal of Physical Chemistry C*, doi:10.1021/jp301598w.
3. Nooney, R. I., Stranik, O., McDonagh, C., MacCraith, B. D. *Langmuir*, 10.1021/la801631w.
4. Lakowicz, J. R., Malicka, J., D'Auria, S., & Gryczynski, I. *Analytical Biochemistry*. 10.1016/S0003-2697(03)00351-8.
5. Aslan, K., Badugu, R., Lakowicz, J. R., & Geddes, C. D. *Journal of Fluorescence*, 10.1007/s10895-005-2515-5.
6. Dai, Z.-W., Ling, J., Huang, X.-J., Wan, L.-S., & Xu, Z.-K. *Journal of Physical Chemistry C*. 10.1021/jp201040g.

Chapter 64
Holographic Laser Scanning Microscopy

Stefano L. Oscurato, Pasqualino Maddalena, Emanuele Orabona, and Antonio Ambrosio

Abstract We realize a new optical microscopy technique (Holographic Laser Scanning Microscopy, H-LSM) based on holographic laser scanning illumination of the sample. In each scanning step, a multispot illumination pattern is generated by phase modulating a laser beam via Computer Generated Holography (CGH). A CCD acquires an image of the light signal backscattered from the sample in each scanning step, and the elaboration software (Henriques R, et al. Nat Methods 7:339–340, 2010) reconstructs the image of the sample by localizing the centroid of each recorded spot (Bobroff N, Rev Sci Instrum 57:1152–1157, 1986). The elaboration software operates a spatial filtering on the recorded spots, accepting and processing only those satisfying the minimum SNR and the maximum FWHM controls. The centroid of the valid signals is displayed in the elaborated image (H-LSM image) as a single bright pixel. We implement two completely different methods in illumination schemes. A first illumination condition is realized by focusing an array of light spots onto the sample plane. This intensity pattern is obtained in objective focal plane imposing the calculated phase map (kinoform) on the laser wavefront via a Spatial Light Modulator (SLM). The array of light spots can be made scanning the sample by imposing a grating-like function to the original kinoform (Leach J, et al. Appl Opt 45:897–903, 2006). The second illumination scheme is realized by dynamically changing the random Speckle pattern originating from the random mutual interference of the laser beam wavefronts. Such Speckle pattern is easily generated by imposing a random phase profile on the laser beam using the CGH apparatus, while the mapping of the sample is realized by changing the phase profiles in time. We tested the H-LSM method on some samples presenting structures whose lateral dimensions are below the resolution limit of the optical setup. The H-LSM image of a region of a sample illuminated with an array of light spots, realizing the ordered illumination scheme, is shown in Fig. 64.1a. For comparison, in Fig. 64.1c the image obtained summing up the CCD acquired stack of the scanning procedure, is also shown. Fig. 64.1b,c,d display respectively the

---

S.L. Oscurato (✉) • P. Maddalena • E. Orabona • A. Ambrosio
CNR-SPIN and Dipartimento di Fisica "E. Pancini", Università degli Studi di Napoli Federico II, 80126 Naples, Italy
e-mail: oscurato@fisica.unina.it

© Springer Science+Business Media Dordrecht 2017

B. Di Bartolo et al. (eds.), *Nano-Optics: Principles Enabling Basic Research and Applications*, NATO Science for Peace and Security Series B: Physics and Biophysics, DOI 10.1007/978-94-024-0850-8_64

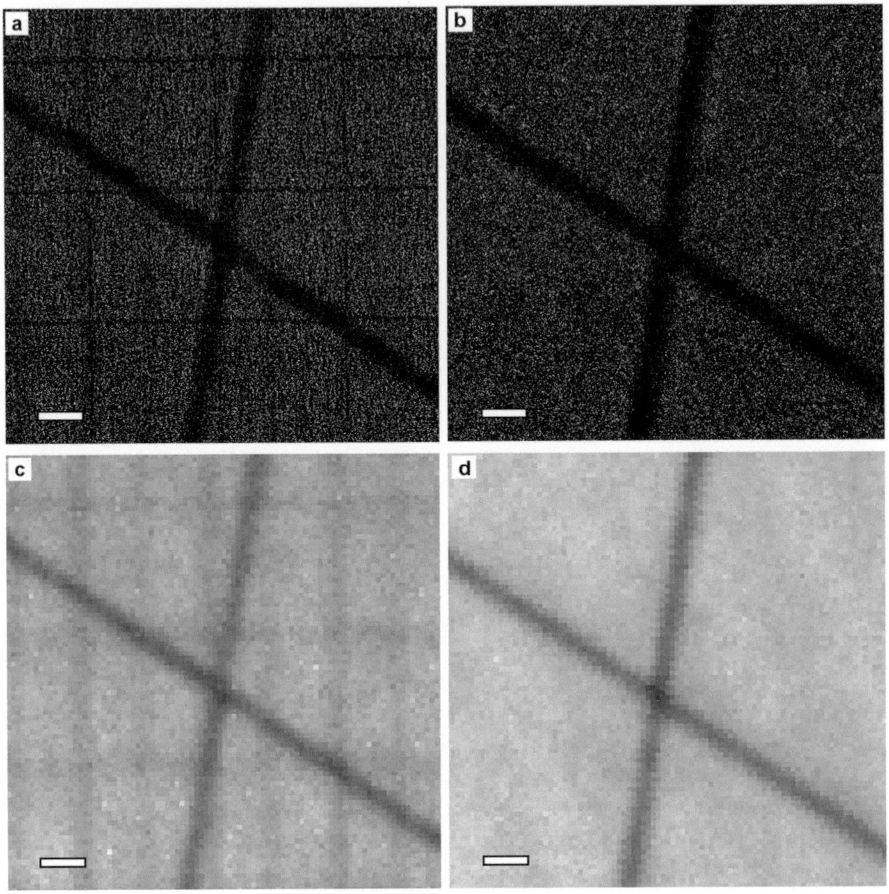

Fig. 64.1 H-LSM and Summed images of lines engraved in a Silver film. (**a**) H-LSM and (**c**) Summed images obtained in ordered illumination scheme by scanning an array of focused spots over the sample surface. (**b**) H-LSM and (**d**) Summed images reconstructed with random illumination scheme. Scale bar 500 nm

H-LSM and the summed image of the sample obtained with the random illumination scheme.

Despite a standard resolution limit, the H-LSM images show improved contrast and sharpness respect to the summed images. Moreover, the filtering on the contributing light signals operated by the elaboration software makes the method comparable to software confocal technique, with very small depth of field for the ordered illumination scheme and large depth of field in random illumination scheme compared to brightfield microscopy. The absence of moving mechanical parts provided by the holographic nature of illumination makes the proposed method useful for vibration-free measurements, with a possible miniaturized setup and 3D imaging.

References

1. Henriques, R., et al. (2010). *Nature Methods, 7*, 339–340.
2. Bobroff, N. (1986). *Review of Scientific Instruments, 57*, 1152–1157.
3. Leach, J., et al. (2006). *Applied Optics, 45*, 897–903.

Chapter 65
Physical and Chemical Features of Biochar: A Reservoir of Materials in Advanced Nanotechnologies

E. Pusceddu, A. Montanaro, S.F. Santilli, L. del Rosso, M. Aresti, P.U. Foscolo, and F. Miglietta

Abstract Biochar (BC) is a pyrogenic carbon obtained by pyrolysis and pyrogasification of biomass at different temperatures. Feedstock used to produce BC, its size range, the temperature and heating rate of pyrolysis process affect the chemical and physical characteristics, in addition to the amount of char obtained. Biochar is considered an effective and sustainable strategy for the mitigation of climate change, due to the fact that carbon is not easily returned into the atmosphere [1]. This carbon-based material contains substantial amounts (90 %) of pyrogenic C in several molecular components. Few studies take into account BC as a material containing useful carbon-based molecules in advanced nanotechnology sector. This work is focused on the study of the physical and chemical features of biochar, in particular addressed the *turbostratic-carbon* phase. X-Ray Diffraction (XRD) measurements were performed to assess different crystalline phases of biochar. In particular, we considered the turbostratic-carbon crystallites, studying their behavior as a function of production pyrolysis temperature of BC. Increasing the pyrolysis temperature, the cellulose present into starting biomass loses its orderly arrangement

E. Pusceddu (✉) • F. Miglietta
Institute of Biometeorology National Research Council – IBIMET CNR,Via Giovanni Caproni, 8 – 50145 Firenze, Italy
e-mail: emanuela.pusceddu@gmail.com

A. Montanaro • S.F. Santilli • P.U. Foscolo
Department of Industrial Engineering, University of L'Aquila, Via Campo di Pile, 67100 L'Aquila, Italy

L. del Rosso
Institute for Complex Systems National Research Council – ISC CNR, Via Madonna del piano 10, I-50019 Sesto Fiorentino, Italy

M. Aresti
Physics Department, University of Cagliari, Cittadella Universitaria, S.P. Monserrato-Sestu km 0.7, Monserrato, Italy

F. Miglietta
FoxLab, Fondazione E. Mach, via E. Mach 1, 38010 S. Michele all'Adige (TN), Italy

© Springer Science+Business Media Dordrecht 2017
B. Di Bartolo et al. (eds.), *Nano-Optics: Principles Enabling Basic Research and Applications*, NATO Science for Peace and Security Series B: Physics and Biophysics, DOI 10.1007/978-94-024-0850-8_65

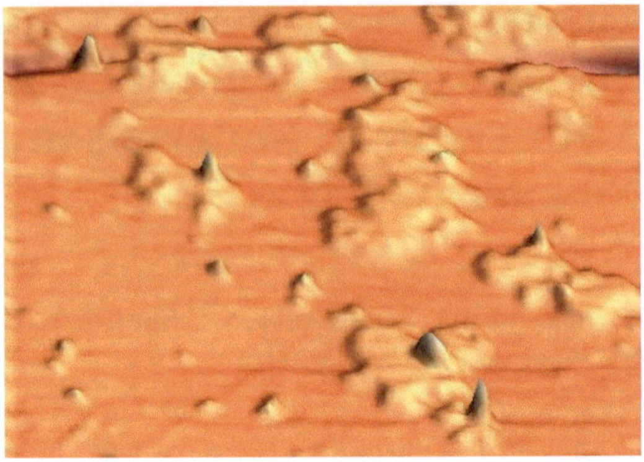

Fig. 65.1 AFM Image of biochar nanoparticles, after ultrasonic bath (90 min). BC was produced at about 600° C. The image reveals some "islands" of graphene sheets with thickness of about 1 nm

indicating an effective structural change and favours the development of carbon molecules. Two peaks were assigned, {002} and {100} *hkl* planes, corresponding to the turbostratic-carbon crystallites [2]. Rising pyrolysis temperature, X-Ray diffraction data revealed progressive stacking of graphene sheets within turbostratic-carbon crystallites. At the same time, a study of the different molecules into the turbostratic-carbon was accomplished using Raman Spectroscopy. We identified, in particular, graphite and graphene-sheets phases, in agreement with the literature [3, 4]. Subsequently, Biochar was undergoing ultrasonic bath treatment for different exposure times. This approach allows us to obtain nanoparticles of carbon molecules in demineralized water suspension. This methodology is taken into account for the graphene sheets extraction from BC nanoparticles. Atomic Force Microscopy (AFM) was used to characterize the BC nanoparticles and the topographic height sheets on mica substrate. The images reveal the presence of graphene sheets with thickness about 1 nm. In Fig. 65.1 is shown 3D AFM image of BC nanoparticles after 90 min of ultrasound treatment.

Different "islands" of graphene sheets were observed and measured; using the AFM line scan, the height profile of the sheets was measured finding 1 nm of thickness, consistent with the typical height reported for graphene sheets on mica substrate [5, 6].

Biochar can be considered as a reservoir of materials for advanced nanotechnology, such as graphene sheets.

References

1. Woolf, D., Amonette, J. E., Street-Perrott, F. A., Lehmann, J., & Joseph, S. (2010). Sustainable biochar to mitigate global climate change. *Nature Communications, 1*, 56.
2. Kim, P., Johnson, A., Edmunds, C. W., Radosevich, M., Vogt, F., Rials, T. G., & Labbe, N. (2011). Surface functionality and carbon structures in lignocellulosic-derived biochars produced by fast Pyrolysis. *Energy Fuels, 25*, 4693–4703.
3. Paris, O., Zollfrank, C., & Zickler, G. A. (2005). Decomposition and carbonisation of wood biopolymers – a microstructural study of softwood pyrolysis. *Carbon, 43*, 53–66.

4. Li, X., Hayashi, J., & Li, C-Z. (2006). FT-Raman spectroscopic study of the evolution of char structure during the pyrolysis of a Victorian brown coal. *Fuel, 85*, 1700–1707.
5. Si, Y., & Samulski, E. T. (2008). Synthesis of water soluble graphene. *Nano Letters, 8*(6), 1679–1682.
6. Feng, H., Cheng, R., Zhao, X., Duan, X., & Li, J. A low-temperature method to produce highly reduced graphene oxide. *Nature Communications, 4*, Article number:1539. doi:10.1038/ncomms2555.

Chapter 66
Modal Behaviour and Switching Properties of a Tailored Parity-Time (PT) Symmetric Grating

N. X. A. Rivolta and B. Maes

Abstract We analyse the diffraction behaviour of a parity-time (PT) symmetric transmission grating. This kind of grating consists of a periodic structure with balanced real and imaginary parts of the refractive index. The bimodal interferometric operation (Mach-Zehnder like) is modified by the introduction of balanced gain and loss, leading to efficient switching around an exceptional point.

Usually we try to avoid loss in optics, because it limits the efficiencies of many applications. However, the concept of PT symmetry (originating from quantum mechanics [1, 2]) allows for a proper use of loss, when it is judiciously balanced with gain. The main property of PT symmetric devices is that they switch from what is called a unbroken phase to a broken phase through an exceptional point. This effect is really suited for switching applications. PT symmetry already facilitated various interesting and counter-intuitive phenomena, such as optical Bloch oscillations and loss induced lasing.

We numerically analyse a diffraction grating (Fig. 66.1a) that was extensively discussed without loss/gain in [3]. This grating is tailored to achieve very high diffraction efficiencies. By introducing gain and loss (quantified by γ) in the grating we can now find out its rich transmission characteristics.

We identify two different regimes [4], which are adjusted by the not-often studied longitudinal feedback (see Fig. 66.1b) for the direct transmission T_0). The properties are explained by the two Bloch modes in the grating. Below an exceptional point ($\gamma = 0.23$) the grating acts like a Mach-Zehnder interferometer, with a gain-controlled coupling length. Beyond this point there is one mode with gain (due to symmetry breaking), which provides us with lasing resonances.

N. X. A. Rivolta (✉) • B. Maes
Micro- and Nanophotonic Materials Group, Faculty of Science, University of Mons, Mons, Belgium
e-mail: nicolas.rivolta@umons.ac.be

© Springer Science+Business Media Dordrecht 2017

B. Di Bartolo et al. (eds.), *Nano-Optics: Principles Enabling Basic Research and Applications*, NATO Science for Peace and Security Series B: Physics and Biophysics, DOI 10.1007/978-94-024-0850-8_66

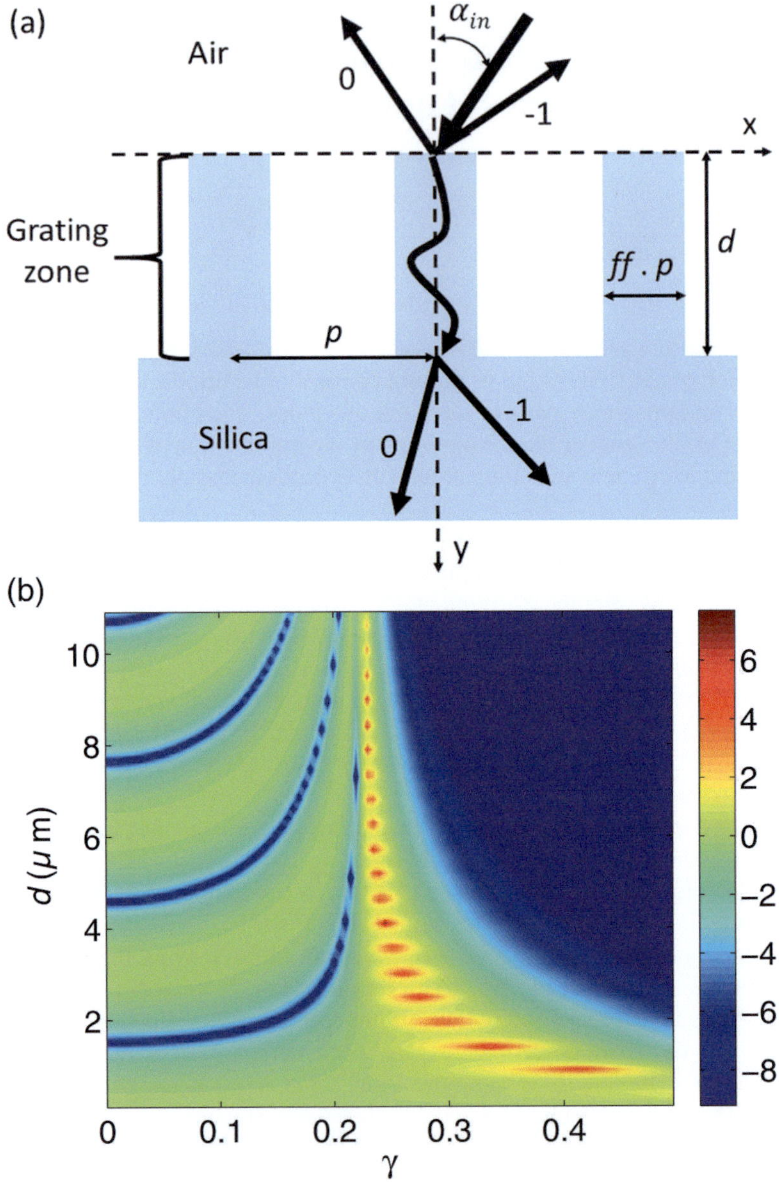

Fig. 66.1 **(a)** Diffraction grating with the reflection and transmission orders. **(b)** Transmitted light (log scale) to T_0 as a function of groove depth d and gain/loss factor γ

References

1. Bender, C. M., & Boettcher, S. (1998). Real spectra in non-Hermitian Hamiltonians having PT symmetry. *Physical Review Letters, 80*, 5243.
2. Bender, C. M. (2007). Making sense of non-Hermitian Hamiltonians. *Reports on Progress in Physics, 70*, 947.
3. Clausnitzer, T., Kämpfe, T., Kley, E.-B., Tünnermann, A., Peschel, U., Tishchenko, A. V., et al. (2005). An intelligible explanation of highly-efficient diffracion in deep dielectric rectangular transmission gratings. *Optics Express, 13*, 10448.
4. Rivolta, N. X. A., & Maes, B. (2015). Diffractive switching by interference in a tailored PT-symmetric grating. *Journal of the Optical Society of America B, 32*, 1330.

Chapter 67
Pulsed Laser-Activated Plasmonic Pyramids for Intracellular Delivery

Nabiha Saklayen, Marinus Huber, Daryl I. Vulis, Marinna Madrid, Valeria Nuzzo, and Eric Mazur

Abstract We use pulsed laser-activated plasmonic micropyramids to deliver molecules to living cells with high efficiency, viability, and throughput. Cellular therapy holds great promise for applications in gene therapy and fundamental biomedical research, and it is essential to develop a universal delivery platform that can safely deliver biomolecules to different cell types effectively. Such a platform would be an important stepping stone towards treatment of hematologic diseases such as leukemia and primary immunodeficiency disorder treatments. An idea molecular delivery platform would exhibit advantages such as high delivery efficiency, low toxicity, minimal immune reaction, and reusability. None of the currently available commercial methods, such as viral-based or electroporation, offer all desirable characteristics at once. We present a new optical method for molecular delivery that uses laser-activated microstructures. Our micropyramids produce a strong plasmonic effect under laser illumination by focusing energy in a small volume at the tip of each pyramid. This leads to the formation of microbubbles which temporarily porate the cell membrane and allow dye molecules and siRNA to diffuse into the cytoplasm. We fabricate large-area micropyramid arrays using photolithograpy, anisotropic etching of silicon, metal deposition, and template stripping. The silicon pyramid templates can be used repeatedly to fabricate gold pyramids. We optimize our laser parameters for high efficiency delivery of small dye molecules like calcein (>80 %) at high cell viability (>90 %). Alongside small dyes, we also deliver different-sized fluorescently labeled dextrans (70 kDa–2000 kDa) and fluorescent microspheres. Our method delivers molecules with high efficiency and high cell viability in different cell types, and our substrates can be reused for repeated high efficiency poration. Our scalable technique offers an innovative approach to delivering molecules to living cells for important applications in regenerative medicine.

N. Saklayen (✉) • M. Huber • D.I. Vulis • M. Madrid • E. Mazur
Harvard University, Cambridge, MA, USA
e-mail: nsaklayen@gmail.com; dvulis@seas.harvard.edu; marinnamadrid@gmail.com; mazur@physics.harvard.edu

V. Nuzzo
ECE Paris Ecole d'Ingenieurs, 37 Quai de Grenelle, 75015 Paris, France

© Springer Science+Business Media Dordrecht 2017
B. Di Bartolo et al. (eds.), *Nano-Optics: Principles Enabling Basic Research and Applications*, NATO Science for Peace and Security Series B: Physics and Biophysics, DOI 10.1007/978-94-024-0850-8_67

Chapter 68
Multi-resonant Metamaterials for Visible and Near-IR Frequencies

Nicholas Sharac, O. J. Glembocki, A. Giles, J. D. Caldwell, and S. M. Prokes

Abstract At low frequencies, metals behave as a perfect electric conducting (PEC) material and no longer exhibit plasmonic properties. By "spoofing" a metal, through the use of grooves or perforations, with feature sizes much smaller than the incident wavelength, metals can have an effective dielectric that allows plasmonic responses even in ranges outside the optical range. Additionally, while on a smooth metal film, surface plasmons will propagate along the surface, a patterned film can have localized plasmon resonance, as in the case of PEALD Ag, which has small gaps. Thus, spoof hybrid plasmonics can occur (a) in large conformal coatings (b) can allow for multi mode transmission, and (c) allow for both traveling and localized modes. Here, we explore the effects of randomness in ALD Ag, and also modify the optical response in metamaterial structures by changing the unit cell. This is explored using reflectance measurements and calculations, including band structure. This research has applications in energy harvesting, communications, and sensing.

References

1. Benson, (2011). *Nature, 480*, 193.
2. Sigma Aldrich, sigmaaldrich.com. Accessed July 2015.
3. Surface Plasmon Polaritons. http://large.stanford.edu/courses/2007/ap272/white1/. Justin White. Accessed July 2015.
4. Hammond et al. (2014). *Biosensors, 4*, 172–188.
5. Pendry J. B. et al. (2004). *Science, 305*, 847.
6. Mousavi et al. S. H. (2010). *PRL, 105*, 176803.

N. Sharac (✉) • O.J. Glembocki • A. Giles • J.D. Caldwell • S.M. Prokes
Electronics Science and Technology Division, U.S. Naval Research Laboratory, Washington, DC, USA
e-mail: nicholas.sharac.ctr@nrl.navy.mil

© Springer Science+Business Media Dordrecht 2017

B. Di Bartolo et al. (eds.), *Nano-Optics: Principles Enabling Basic Research and Applications*, NATO Science for Peace and Security Series B: Physics and Biophysics, DOI 10.1007/978-94-024-0850-8_68

Chapter 69
Nonlinear Properties of Novel Glass-Ceramics with Co^{2+}:Ga_2O_3 Nanocrystals

Nikolai Skoptsov, Pavel Loiko, Olga Dymshits, Vladimir Vitkin, Artem Kharitonov, Alexandr Zhilin, Irina Alekseeva, Svetlana Zapalova, Xavier Mateos, Alexandr Malyarevich, and Konstantin Yumashev

Abstract Transparent glass-ceramics containing γ-Ga_2O_3:Co^{2+} nanocrystals are synthesized. Their structure, absorption, absorption saturation and bleaching relaxation of Co^{2+} ions are studied. Passive Q-switching of diode-pumped Er,Yb: glass laser is realized with the developed glass-ceramics.

Crystals with Co^{2+} ions in tetrahedral sites are of practical interest due to their saturable absorption at around ~1.5 μm. Co^{2+}-doped materials are used for passive Q-switching of erbium (Er) eye-safe lasers emitting in the vicinity of this wavelength. In the present paper we report on the optical and nonlinear properties of glass-ceramics with γ-Ga_2O_3:Co^{2+} nanocrystals with the possible use as saturable absorbers of 1.5–1.7 μm Er lasers.

Glass-ceramics were prepared from the Li_2O–Ga_2O_3–SiO_2 precursor glass doped with 0.1 mol% CoO (Loiko PA, Dymshits OS, Vitkin VV, Skoptsov NA, Kharitonov AA, Zhilin AA, Alekseeva IP, Zapalova SS, Malyarevich AM, Glazunov IV, Yumashev KV Laser Phys Lett 12:035803-1-5, 2015). The precursor glass was heat-treated at the temperatures of 680–800 °C for 6 h resulting in the precipitation of nanosized cubic crystals of γ-Ga_2O_3. The crystal size of γ-Ga_2O_3 was changed from about 7–20 nm as the heat treatment temperature increased from 680 to 800 °C. A Transmission Electron Microscopy (TEM) image of the powdered glass-ceramics sample heat-treated at 730 °C is shown in Fig. 69.1a.

N. Skoptsov (✉) • P. Loiko • A. Malyarevich • K. Yumashev
Center for Optical Materials and Technologies, Belarusian National Technical University, 65/17 Nezavisimosti Ave., 220013 Minsk, Belarus
e-mail: nikolai.skoptsov@tut.by

O. Dymshits • A. Zhilin • I. Alekseeva • S. Zapalova
NITIOM S.I. Vavilov State Optical Institute, 36 Babushkina St, 192171 St. Petersburg, Russia

V. Vitkin • A. Kharitonov
University ITMO, 49 Kronverkskiy pr, 197101 St. Petersburg, Russia

X. Mateos
Física i Cristal·lografia de Materials i Nanomaterials (FiCMA-FiCNA), Universitat Rovira i Virgili (URV), Campus Sescelades, c/Marcel·lí Domingo, s/n., E-43007 Tarragona, Spain

© Springer Science+Business Media Dordrecht 2017

B. Di Bartolo et al. (eds.), *Nano-Optics: Principles Enabling Basic Research and Applications*, NATO Science for Peace and Security Series B: Physics and Biophysics, DOI 10.1007/978-94-024-0850-8_69

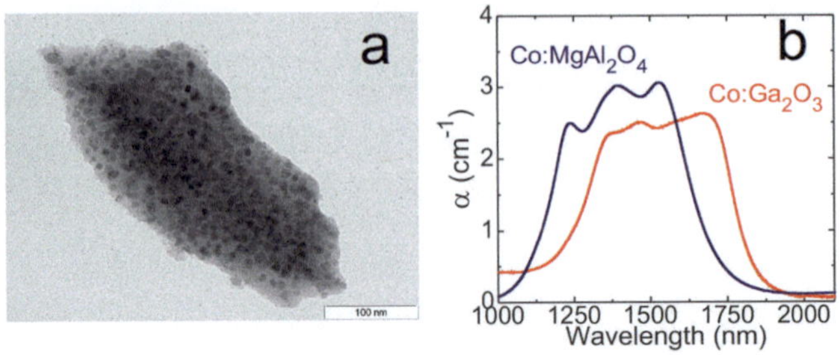

Fig. 69.1 (**a**) TEM image of powdered sample of glass-ceramics with Co:γ^- nanocrystals; (**b**) absorption band related to the $^4A_2(^4F)\rightarrow{}^4T_1(^4F)$ transition of Co^{2+} ions located in tetrahedral sites for glass-ceramics with Co:γ^- and Co:$MgAl_2O_4$ nanocrystals

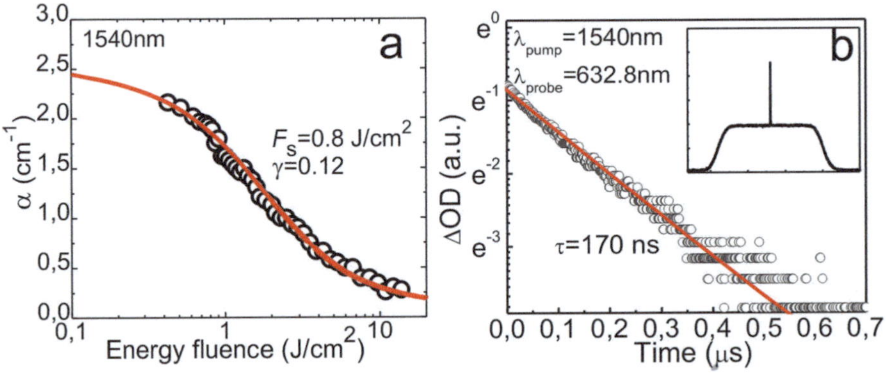

Fig. 69.2 (**a**) Saturation of nonlinear absorption for the glass-ceramics with Co:γ^- nanocrystals at 1540 nm; *points* are the experimental data, *curve* represents slow three-level absorber model; (**b**) recovery of ground state absorption of Co^{2+} ions in the γ^- nanocrystals

The absorption spectrum of the glass-ceramics heat-treated at 730 °C for 6 h is shown in Fig. 69.1b. A wide band related to the $^4A_2(^4F)\rightarrow{}^4T_1(^4F)$ transition of Co^{2+} ions located in tetrahedral sites spans up to ~1.75 μm that is much longer than for conventional glass-ceramics with Co:$MgAl_2O_4$ nanocrystals.

For the studied glass-ceramics, the absorption saturation curve was measured at 1540 nm, Fig. 69.2a. The bleaching process was modeled considering the slow three-level saturable absorber model; the best-fitting parameters are the following: saturation fluence $F_S = 0.8$ J/cm^2, saturation contrast $\gamma = 0.12$, ground-state absorption cross-section $\sigma_{GSA} = 1.7 \times 10^{-19}$ cm^2. The recovery time of the initial absorption for the Co^{2+} ions (τ = 170 ns) was determined with a pump-probe method, Fig. 69.2b.

The studied glass-ceramics were tested for Q-switching of an erbium laser. The initial transmission of the saturable absorber was $T_0 = 85\,\%$ (thickness: 0.5 mm). Laser cavity was plano-concave (total length: 65 mm, $R_{OC} = 1.5$ m, $T_{OC} = 13\,\%$). For pump pulse energy of 300 mJ, we achieved single-pulse Q-switching with the pulse energy of 1.75 mJ and pulse duration of 25 ns.

The absorption and nonlinear properties indicate that the glass-ceramics are suitable for the Q-switching of 1.5–1.7 μm eye-safe erbium lasers.

Reference

1. Loiko, P. A., Dymshits, O. S., Vitkin, V. V., Skoptsov, N. A., Kharitonov, A. A., Zhilin, A. A., Alekseeva, I. P., Zapalova, S. S., Malyarevich, A. M., Glazunov, I. V., & Yumashev, K. V. (2015). *Laser Physics Letters, 12*, 035803-1-5.

Chapter 70
Resonant Optical Trapping in Microfluidic-Integrated Hollow Photonic Crystal Cavities

Rita Therisod, Mario Tonin, and Romuald Houdré

Abstract Optical trapping of nanoparticles has many interesting applications in a variety of physical and biological studies. The ability of photonic crystal cavities to strongly confine light permits to overcome the diffraction limit that prevents the trapping of submicron sized particles with low powers in classical optical tweezers. The device studied in this project is integrated in an optofluidic chip. Distinction between 250 and 500 nm particles is shown.

R. Therisod (✉) • M. Tonin • R. Houdré
Laboratory of Advanced Semiconductors for Photonics and Electronics, École Polytechnique Fédérale de Lausanne, Lausanne, Switzerland
e-mail: rita.therisod@epfl.ch

© Springer Science+Business Media Dordrecht 2017

B. Di Bartolo et al. (eds.), *Nano-Optics: Principles Enabling Basic Research and Applications*, NATO Science for Peace and Security Series B: Physics and Biophysics, DOI 10.1007/978-94-024-0850-8_70

Chapter 71
1-D Photonic Crystals Fabricated by RF Sputtering Towards Photonic Applications

Sreeramulu Valligatla, A. Chiasera, S. Varas, A. Łukowiak, F. Scotognella, D. Narayana Rao, R. Ramponi, G. C. Righinig, and M. Ferraria

Abstract A great technological and scientific challenge is related to the fabrication of confined structures where the light is confined in systems with characteristic dimensions scale from micro to nanometers. The nanotechnology, that allows the study of innovative functional materials and gives advances in the miniaturization, has opened the way to the manufacturing of such structures and focalize the attention

---

S. Valligatla (✉)
CNR-IFN CSMFO Lab. & FBK CMM, Via alla Cascata 56/C, Povo, 38123 Trento, Italy

Dipartimento di Fisica, Università di Trento, via Sommarive 14, Povo, 38123 Trento, Italy

School of Physics, University of Hyderabad, Hyderabad 500046, India
e-mail: srihcu08@gmail.com

A. Chiasera • S. Varas
CNR-IFN CSMFO Lab. & FBK CMM, Via alla Cascata 56/C, Povo, 38123 Trento, Italy

A. Łukowiak
Department of Spectroscopy of Excited States, Institute of Low Temperature and Structure Research Polish Academy of Sciences, 2 Okolna St., 50-422 Wroclaw, Poland

F. Scotognella
Dipartimento di Fisica and Istituto di Fotonica e Nanotecnolgie CNR, Politecnico di Milano, Piazza Leonardoda Vinci 32, 20133 Milano, Italy

Center for Nano Science and Technology@PoliMi, Istituto Italiano di Tecnologia, Via Giovanni Pascoli, 70/3, 20133 Milan, Italy

D.N. Rao
School of Physics, University of Hyderabad, Hyderabad 500046, India

R. Ramponi
Dipartimento di Fisica and Istituto di Fotonica e Nanotecnolgie CNR, Politecnico di Milano, Piazza Leonardoda Vinci 32, 20133 Milano, Italy

G.C. Righinig
IFAC – CNR, MiPLab, 50019 Sesto Fiorentino, Italy

M. Ferraria
Centro di Studi e Ricerche "Enrico Fermi", Piazza del Viminale 1, 00184 Roma, Italy

© Springer Science+Business Media Dordrecht 2017

B. Di Bartolo et al. (eds.), *Nano-Optics: Principles Enabling Basic Research and Applications*, NATO Science for Peace and Security Series B: Physics and Biophysics, DOI 10.1007/978-94-024-0850-8_71

of new features of light-matter interaction (Ristic D et. al, Proc ICTON Tu. B 5.2:1–5, 2013). An example concerns planar microcavities, or one-dimensional (1-D) photonic crystals, which are the simplest photonic band – gap (PBG) devices exploitable to manage the spectroscopic properties of luminescent species such as rare earth ions (Valligatla S, Chiasera A, Varas S, Bazzanella N, Narayana Rao D, Righini GC, Ferrari M, Opt Express 20:21214, 2012) and quantum dots (Jasieniak J, Sada C, Chiasera A, Ferrari M, Martucci A, Mulvaney P, Adv Funct Mater 18:3772, 2008). The fundamental optical principle for the photonic crystals is, "localization of light" (John S, Phys Rev Lett 58:2486, 1987) so that combination of photonic crystals and nonlinear optics leads us towards new nonlinear optical devices (Ma GH, Shen J, Rajiv K, Tamg SH, Zhang ZJ, Hua ZY, Appl Phys B 80:359, 2005; Valligatla S, Chiasera A, Krishna MBM, Varas S, Narayana Rao D, Ferrari M, Righini GC, Int Conf Fiber Opt Photon 2012:W1C.2).

Here, as a first one, we report on a reproducible fabrication protocol based on rf-sputtering technique and on optical, spectroscopic and morphological characterization of the realized high quality Er^{3+}-activated dielectric microcavity consisting of alternating silica and titania films. The second example concerns a 1-D microcavity fabricated as in the previous example but with a ZnO defect layer. Further, we have observed the large enhancement in the optical nonlinearity in ns regime, for the microcavity as compared with that of reference film.

Acknowledgements This research work is performed in the framework of ITPAR Phase III (2013–2016) research project and the CNR-PAS (2014–2016) joint project "Nanostructured systems in opal configuration for the development of photonic devices".

References

1. Ristic, D. et al. (2013). *Proceedings ICTON,* Tu. B *5.2*, 1–5.
2. Valligatla, S., Chiasera, A., Varas, S., Bazzanella, N., Narayana Rao, D., Righini, G. C., & Ferrari, M. (2012). *Optics Express, 20*, 21214.
3. Jasieniak, J., Sada, C., Chiasera, A., Ferrari, M., Martucci, A., & Mulvaney, P. (2008). *Advanced Function Materials, 18*, 3772.
4. John, S. (1987). *Physical Review Letters, 58*, 2486.
5. Ma, G. H., Shen, J., Rajiv, K., Tamg, S. H., Zhang Z. J., & Hua, Z. Y. (2005). *Applied Physics B, 80*, 359.
6. Valligatla, S., Chiasera, A., Krishna, M. B. M., Varas, S., Narayana Rao, D., Ferrari, M., & Righini, G. C. *International Conference on Fiber Optics and Photonics – 2012*, W1C.2.

Chapter 72
Design of Thin Film Nanocomposite Grating Based Sensors

I. Yaremchuk, V. Fitio, Š. Meškinis, S. Tamulevičius, and Ya. Bobitski

Abstract The study DLC:Ag films consisting of nanosized metal particles embedded in a diamond carbon matrix has been presented. Optical response including retarded electrodynamics' multipole interactions of neighboring particles and the size, shape, and interparticle distance distributions in the sample were modeled using extended and renormalized Maxwell-Garnett theories. The sub-wavelength grating structures on base DLC-Ag nanocomposite have been investigated by rigorous coupled-wave analysis to evaluate resonance characteristics in a visible range.

DLC films containing different metal nanoparticles show interesting characteristics since their properties depend on the size, shape and concentration of the metallic nanoparticles. Modification of the electronic and optical properties of metallic nanoparticles opens new venues to explore the new possibilities using them for different optical elements. This work aims to provide an investigation of optical properties of DLC-Ag based nanocomposite and using it to design periodic structures for sensing applications. Actual periodic structures on base pure DLC film show resonance characteristics due to waveguide resonance (Yaremchuk I, Tamulevičius T, Fitio V, Gražulevičiūte I, Bobitski Y, Tamulevičius S, Opt Commun 301–302:1–6, 2013) at the same time DLC-Ag nanocomposite films are characterized by plasmon resonance (Yaremchuk I, Tamulevičienė A, Tamulevičius T, Šlapikas K, Balevičius Z, Tamulevičius S, Phys Status Solidi A 211:329, 2014) therefore, the incorporation of metal-dielectric composites in periodic structures enables combining the surface plasmon resonance and the resonance phenomena that take place in it.

I. Yaremchuk (✉) • V. Fitio
Department of Photonics, Lviv Polytechnic National University, S. Bandera Str. 12, 79013 Lviv, Ukraine
e-mail: iryna.y.yaremchuk@lpnu.ua

Š. Meškinis • S. Tamulevičius
Institute of Materials Science, Kaunas University of Technology, K. Baršausko St. 59, LT-51423 Kaunas, Lithuania

Ya. Bobitski
Department of Photonics, Lviv Polytechnic National University, S. Bandera Str. 12, 79013 Lviv, Ukraine

Institute of Technology, University of Rzeszow, Rejtana 16a, 35-310 Rzeszow, Poland

© Springer Science+Business Media Dordrecht 2017

B. Di Bartolo et al. (eds.), *Nano-Optics: Principles Enabling Basic Research and Applications*, NATO Science for Peace and Security Series B: Physics and Biophysics, DOI 10.1007/978-94-024-0850-8_72

The effective refractive indices of DLC:Ag films were calculated by classical and renormalized Maxwell-Garnett theories (Vasilevskiy M, Phys Status Solidi B 219:197–204, 2000) as well as compared with the experimental results (Yaremchuk I, Meškinis Š, Fitio V, Bobitski Y, Šlapikas K, Čiegis A, Balevičius Z, Selskis A, Tamulevičius S, Nanoscale Res Lett 10:157, 2015). The effective dielectric permittivity was calculated taking into account expressions for depolarization coefficients to study influence of the nonsphericity on the optical properties of nanocomposite film. It can be concluded that such theory is suitable to describe the absorption of silver nanoparticles with different deformations and radii, incorporated in diamond like carbon films and effective properties of the nanocomposite can be used for predicting their optical response. The results indicate that for bigger eccentricity there are shift, deformation, and broadening of absorption spectra. It is caused by the break of the spherical symmetry of the nanoparticles and interaction between them.

Optical characteristics of the periodic structure on base DLC-Ag film were modeled using a rigorous electromagnetic home-made code based on the S-matrix algorithm for RCWA applications to the relief diffraction (Yaremchuk I, Tamulevičius T, Fitio V, Gražulevičiūte I, Bobitski Y, Tamulevičius S, J Mod Opt 60:1781–1788, 2013). The results indicate that the sensitivity improvement of grating structure on base of DLC film can be achieved by embedding silver nanoparticles in the film (Yaremchuk I, Tamulevičienė A, Tamulevičius T, Tamulevičius S, Appl Mech Mater 490–491:53–57, 2014). The enhancement in sensitivity can be achieved increasing concentration of silver. However reflectance peaks of such structure will be more red-shifted therefore structural parameters grating must be optimized to operate in a visible range. Thus, to use the unique potential of periodic DLC-Ag nanocomposite-based structures systematic study is necessary as well as optimization of their structural and optical properties to design and fabricate the diffractive optical elements for applications in various optical systems.

This research was funded by the Ministry of Education and Science of Ukraine (DB/Tekton). I. Yaremchuk acknowledges for financial support grant within Ukrainian-Lithuanian collaboration.

References

1. Yaremchuk, I., Tamulevičius, T., Fitio, V., Gražulevičiūte, I., Bobitski, Y., & Tamulevičius, S. (2013). *Optics Communications, 301–302*, 1–6.
2. Yaremchuk, I., Tamulevičienė, A., Tamulevičius, T., Šlapikas, K., Balevičius, Z., & Tamulevičius, S. (2014). *Physica Status Solidi A, 211*, 329.
3. Vasilevskiy, M. (2000). *Physica Status Solidi B, 219*, 197–204.
4. Yaremchuk, I., Meškinis, Š., Fitio, V., Bobitski, Y., Šlapikas, K., Čiegis, A., Balevičius, Z., Selskis, A., & Tamulevičius, S. (2015). *Nanoscale Research Letters, 10*, 157.
5. Yaremchuk, I., Tamulevičius, T., Fitio, V., Gražulevičiūte, I., Bobitski, Y., & Tamulevičius, S. (2013). *Journal of Modern Optics, 60*, 1781–1788.
6. Yaremchuk, I., Tamulevičienė, A., Tamulevičius, T., & Tamulevičius, S. (2014). *Applied Mechanics and Materials, 490–491*, 53–57.

Chapter 73
Structure and Luminescence Properties of Nanofluorapatite Activated with Eu^{3+} Ions Synthesized by Hydrothermal Method

Katarzyna Zawisza, Lukasz Marciniak, Robert Pazik, and Rafael J. Wiglusz

Abstract Nanotechnology is the most intensively developing a multidisciplinary field of research combining various disciplines of science achievement. Nanomaterials show unexpected and interesting chemical and physical properties different from those of the original in the micro-sized scale. A success of nanotechnology in field of physical, chemical, and medical sciences, it has now started revolutionizing the regenerative medicine as well as theranostic.

The apatites are inorganic compounds with a general formula $M_{10}(XO_4)_6Y_2$, where M represents divalent cations (e.g. Ca^{2+}, Sr^{2+}, etc.), $XO_4 = PO_4{}^{3-}$, $VO_4{}^{3-}$, etc., and Y represents anions: F^-, OH^-, Cl^-, Br^-, etc., respectively. Calcium hydroxyapatite (HAP) is a material commonly used in bone tissue engineering because of its similarity to natural bone (Nabiyouni M, Zhou H, Luchini TJF, Bhaduri SB Mater Sci Eng C 37: 2014; Wiglusz RJ, Pozniak B, Zawisza K, Pazik R RSC Adv 5, 2015). Fluorapatite (FAP) could be an alternative biomaterial owing to its low solubility in comparison to HA, and good biocompatibility among this group. Furthermore, human tooth enamel contains some fluoride ions in the form of fluorapatite or fluor-hydroxyapatite that improve mechanical properties of teeth (Sun Y, Yang H, Tao D Ceram Int 38: 2012).

The nanofluorapatite with europium(III) impurity were obtained by microwave stimulated hydrotehermal technique and then, annealed in temperature range 500–1000 °C. The structure and morphology of the synthetic materials have been studied by X-ray diffraction (XRD), transmission electron microscopy (TEM), IR and Raman spectroscopy. The average grain size was estimated using theoretical calculations (Rietveld method) as well as experimental techniques like TEM. The photoluminescent properties of nanopowders were investigated by excitation, emission spectroscopy and emission lifetimes (Figs. 73.1 and 73.2).

K. Zawisza (✉) • L. Marciniak • R. Pazik • R.J. Wiglusz
Division of Optical Spectroscopy, Institute of Low Temperature and Structure Research, Polish Academy of Sciences, P.O. Box 1410, 50-442 Wroclaw, Poland
e-mail: K.Zawisza@int.pan.wroc.pl

© Springer Science+Business Media Dordrecht 2017

B. Di Bartolo et al. (eds.), *Nano-Optics: Principles Enabling Basic Research and Applications*, NATO Science for Peace and Security Series B: Physics and Biophysics, DOI 10.1007/978-94-024-0850-8_73

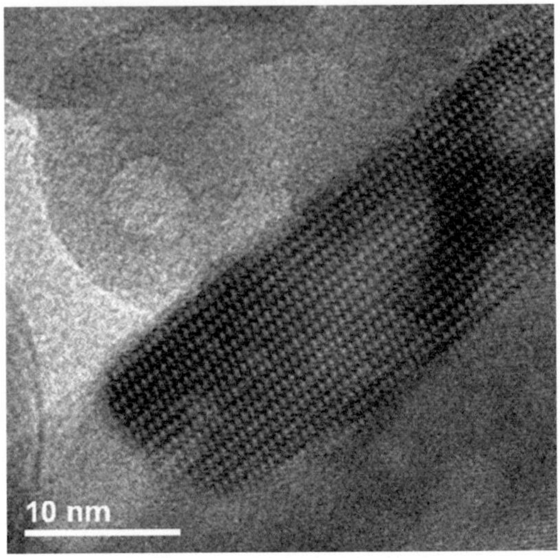

Fig. 73.1 TEM image of 0.5 % Eu-FAP at 500 °C/3 h

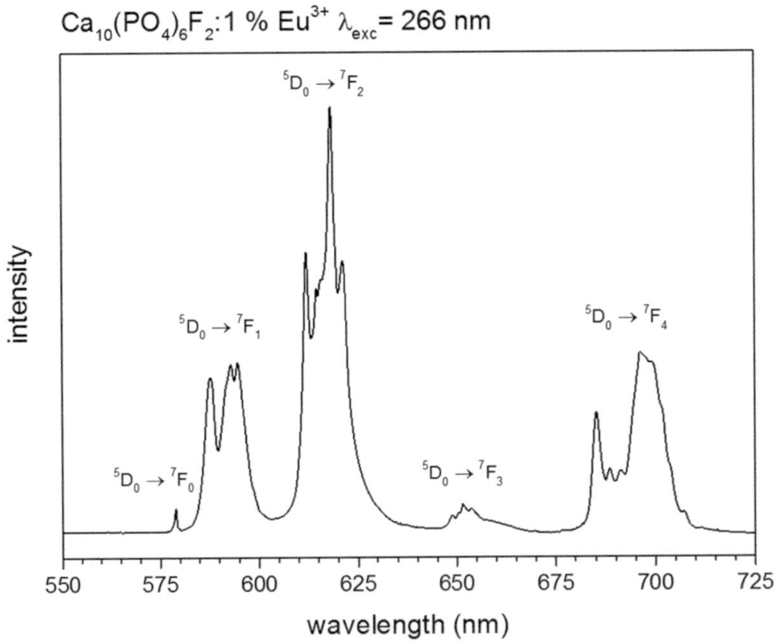

Fig. 73.2 Representative emission spectra of as-prepared nanofluorapatite activated with 1 mol% Eu^{3+} ions

References

1. Nabiyouni, M., Zhou, H., Luchini, T. J. F., & Bhaduri, S. B. (2014). *Materials Science and Engineering C, 37*, 363–368.
2. Wiglusz, R. J., Pozniak, B., Zawisza, K., & Pazik, R. (2015). *RSC Advances, 5*, 27610–27622.
3. Sun, Y., Yang, H., & Tao, D. (2012). *Ceramics International, 38*, 6937–6941.

List of Participants

1. Javier Aizpurua
   CSIC/DIPC
   San Sebastian, Spain
   aizpurua@ehu.es
2. Hatice Altug
   EPFL
   Lausanne, Switzerland
   hatice.altug@epfl.ch
3. Salma Alrasheed
   King Abdullah University of Science and Technology
   Thuwal, Saudi Arabia
   salma.alrasheed@kaust.edu.sa
4. Muhammad Sabieh Anwar
   Department of Physics
   Syed Babar Ali School of Science and Engineering
   Lahore University of Management Sciences (LUMS)
   Lahore, Pakistan
   sabieh@lums.edu.pk
5. Mauro Aresti
   Department of Physics
   University of Cagliari
   Monserrato, Italy
   mauro.aresti@dsf.unica.it
6. Juan Castro Arias
   CNRS-Laboratoire de Photonique et de Nanostructures
   Marcoussis, France
   juan_ manuel.castro_ arias@lpn.cnrs.fr

© Springer Science+Business Media Dordrecht 2017

B. Di Bartolo et al. (eds.), *Nano-Optics: Principles Enabling Basic Research and Applications*, NATO Science for Peace and Security Series B: Physics and Biophysics, DOI 10.1007/978-94-024-0850-8

7. Justice Archer
Institute of Physics
Polish Academy of Sciences
Warsaw, Poland
archer@ifpan.edu.pl
8. Steve Arnold
Polytechnic Institute of NYU
Brooklyn, NY, USA
arnold@photon.poly.edu
sarnold935@aol.com
9. Ankan Bag
Max Planck Institute for the Science of Light
Erlangen, Germany
ankan.bag@mpl.mpg.de
10. Rolindes Balda
Departamento de Fisica Applicada I-Escuela Tecnica Superior de
Ingenieros Industriales Y de Telecomunicacion-Universidad del Pais Vasco
Alda, Urquio, Bilbao, Spain
rolindes.balda@ehu.es
11. Francesco Baldacchini
Via G. Quattrucci 246
Grottaferrata, Roma, Italy
frankbald@gmail.com
12. Francesco Basso Basset
Department of Material Science and Nanotechnology
Università degli Studi di Milano-Bicocca
Milano, Italy
f.bassobasset@campus.unimib.it
13. Andrea Bertoncini
Biological and Environmental Sciences and Engineering Division
King Abdullah University of Science and Technology (KAUST)
Thuwal, Saudi Arabia
andrea.bertoncini@kaust.edu.sa
14. Georges Boulon
Laboratoire de Physico-Chimie des Materiaux Luminescents
Université Claude Bernard- Lyon I
Villeurbanne Cedex, France
georges.boulon@univ-lyon1.fr
15. John W. Bowen
The University of Reading
Reading, UK
cybjb@cyber.reading.ac.uk

16. Antonino Cala' Lesina
Department of Physics
University of Ottawa
Ottawa, Canada
Antonino.Calalesina@uottawa.ca
17. Irene Carrasco Ruiz
UNIVR – Università di Verona
Verona, Italy
irene.carrascoruiz@univr.it
18. Maura Cesaria
Department of Physics
University of Salento
Lecce, Italy
maura.cesaria@le.infn.it
19. Yang Chen
Solid State Physics
Lund University
Lund, Sweden
yang.chen@ftf.lth.se
20. Hatun Çinkaya
Department of Physics Engineering
Istanbul Technical University
Istanbul, Turkey
cinkayahatun@itu.edu.tr
21. Nellu Ciobanu
Institute of Applied Physics
Academy of Sciences of Moldova
Chişinău, Moldova
cnellu@gmail.com
22. John Collins
Department of Physics
Wheaton College
Norton, MA, USA
jcollins@wheatonma.edu
23. Elisabetta Dattola
Department of Health Sciences
Magna Grecia University of Catanzaro
Catanzaro, Italy
elisabettadattola@gmail.com
24. Lorenzo L. De Angelis
FOM Institute AMOLF
Amsterdam, The Netherlands
deangelis@amolf.nl

25. Gurram Devaraju
Department of Physics and Astronomy "Galileo Galilei"
University of Padova
Padova, Italy
g.devaraju@gmail.com
26. Baldassare Di Bartolo (Co-Director of the Institute)
Department of Physics
Boston College
Chestnut Hill, MA, USA
dibartob@bc.edu
27. Enzo Di Fabrizio
KAUST, King Abdullah University of Science and Technology
Thuwal, Saudi Arabia
Enzo.DiFabrizio@kaust.edu.sa
28. Chase T. Ellis
U.S. Naval Research Laboratory
Electronics Science and Technology Division
Washington, DC, USA
chase.ellis.ctr@nrl.navy.mil
29. Nader Engheta
Department of Electrical and Systems Engineering
School of Engineering and Applied Science
University of Pennsylvania
Philadelphia, PA, USA
engheta@ee.upenn.edu
30. Muhammad Faryad
Department of Physics
Lahore University of Management Sciences
Lahore, Pakistan
muhammad.faryad@lums.edu.pk
31. Joaquin Fernandez
Departamento de Fisica Applicada I- Escuela Tecnica Superior de
Ingenieros Industriales Y de Telecomunicacion-Universidad del Pais Vasco
Alda Urquio, Bilbao, Spain
joaquin.fernandez@ehu.es
32. Ranveig Flatabø
Department of Physics and Technology
University of Bergen
Bergen, Norway
Ranveig.Flatabo@uib.no
33. Antonino Foti
CNR – Istituto per i Processi Chimico-Fisici
Messina, Italy
foti@ipcf.cnr.it

34. Rosalba Gaudiuso
Institute of Nanotechnology
CNR
Bari, Italy
rosalba.gaudiuso@nanotec.cnr.it
35. Sergei V. Gaponenko
National Academy of Sciences
Institute of Physics
Minsk, Belarus
nanoscience@tut.by
36. Yuriy Gerasymchuk
Institute of Low Temperature and Structure Research
Polish Academy of Sciences
Wrocław, Poland
j.gerasymchuk@int.pan.wroc.pl
37. Alexander Giles
U.S. Naval Research Laboratory
Washington, D.C., USA
alexander.giles.ctr@nrl.navy.mil
38. Mikel Greben
Department of Chemical Physics and Optics
Chales University
Prague, Czech Republic
leibnits@gmail.com
39. Alexandre Grégoire
Département de Chimie et Centre d'optique photonique et laser
Université Laval, Québec, Canada
alexandre.gregoire.1@ulaval.ca
40. Sarah Griesse-Nascimento
School of Engineering and Applied Sciences
Harvard University
Cambridge, MA, USA
sarahegn@gmail.com
41. Ke Guo
FOM institute AMOLF
Amsterdam, The Netherlands
C.Guo@amolf.nl
42. Małgorzata Guzik
University of Wrocław
Wrocław, Poland
goguzik@poczta.fm

43. Anne Habermehl
Karlsruher Institut für Technologie (KIT)
Lichttechnisches Institut (LTI)
Karlsruhe, Germany
anne.habermehl@kit.edu
44. Ahmed Hassaan
University of Chemical Technology and Metallurgy (UCTM)
Sofia, Bulgaria
ahmed.hassaan@ymail.com
45. Jacobine van Hest
Debye Institute of Nanomaterial Science
Utrecht University
Utrecht, The Netherlands
J.J.H.A.vanHest@uu.nl
46. Rainer Hillenbrand
CIC nanoGUNE Consolider
San Sebastian, Spain
r.hillenbrand@nanogune.eu
47. Mathieu Jeannin
Institut Néel – CNRS
Grenoble, France
mathieu.jeannin@neel.cnrs.fr
48. Stefan Karatodorov
Institute of Solid State Physics
Bulgarian Academy of Sciences
Sofia, Bulgaria
skarat@issp.bas.bg
s.karatodorov@gmail.com
49. Ivan Karbovnyk
Department of Electronics
Ivan Franko National University of Lviv
Lviv, Ukraine
ivan_ karbovnyck@yahoo.com
50. Christian Kern
Karlsruher Institut für Technologie (KIT)
Institut für Angewandte Physik
Karlsruhe, Germany
christian.kern@kit.edu
51. Alina Khodko
Institute of Physics
National Academy of Sciences
Kyiv, Ukraine
osaozz@gmail.com

52. Matthias Kling
Ludwig-Maximilians-Universität (LMU)
München, Germany
matthias.kling@mpq.mpg.de
53. Halyna Klym
Institute of Computer Technology, Automatic and Metrology
Lviv Polytechnic National University
Lviv, Ukraine
halyna.i.klym@lpnu.ua
54. Lyudmyla Kokhtych
Coherent and Quantum Optics Dept.,
Institute of Physics of NAS of Ukraine
Kiev, Ukraine
kokhtych@gmail.com
55. Nicoline Dolfine Kosters
FOM Institute AMOLF
Amsterdam, The Netherlands
D.Kosters@amolf.nl
56. Jean-François Lemineur
Département de chimie
Université LAVAL
Quebec, Canada
jean-francois.lemineur.1@ulaval.ca
57. Sylvain Le-Liepvre
CEA Saclay/DSM/IRAMIS/SPEC
Gif-sur-Yvette, France
sylvain.le-liepvre@cea.fr
58. Qingcao(Max) Liu
Ludwig-Maximilians-Universität München
Fakultät für Physik
Garching, Germany
liuqingcao@gmail.com
59. Pavel Loiko
KTH Royal Institute of Technology
School of Information and Communication Technology
Materials and Nano Physics
Isafjordsgatan 22, Electrum 229
SE-164 40 Kista, Sweden
kinetic@tut.by
60. Nicolás Coca López
Ludwig Maximilians University
Munich, Germany
nicolas.cocalopez@cup.uni-muenchen.de

61. Marinna Madrid
Department of Physics
Harvard University
Cambridge, MA, USA
marinnamadrid@gmail.com
62. Alejandrina Martínez
Centro de Investigaciones en Optica (CIO)
Leon, Guanajuato, Mexico
alejandrina.martinez@gmail.com
63. Eric Mazur
School of Engineering and Applied Sciences
Harvard University
Cambridge, MA, USA
mazur@physics.harvard.edu
64. Régis Méjard
Laboratoire Interdisciplinaire Carnot de Bourgogne
Université de Bourgogne
DIJON Cedex, France
Regis.Mejard@u-bourgogne.fr
65. Maryna Meretska
Complex Photonic Systems (COPS)
MESA+ Institute for Nanotechnology
University of Twente
Enschede, The Netherlands
m.meretska@utwente.nl
66. Nicoleta Elena Dina Mircescu
National Institute of R& D of Isotopic and Molecular Technology
Cluj-Napoca, Romania
nicoleta.dina@itim-cj.ro
67. Patrick Müller
Karlsruhe Institute of Technology (KIT)
Institute of Applied Physics (APH) and Institute of Nanotechnology (INT)
Karlsruhe, Germany
patrick.mueller3@kit.edu
68. Alina Muravitskaya
National Academy of Science of Belarus
Minsk, Belarus
alica.mur@mail.ru
69. Lukas Novotny
Photonics Laboratory
ETH Zurich
Zurich, Switzerland
lnovotny@ethz.ch

70. Stefano Luigi Oscurato
Department of Physics
University of Naples
Napoli, Italy
oscurato@fisica.unina.it
71. Christian Ott
Lehrstuhl für Angewandte Physik
Universität Erlangen-Nürnberg
Erlangen, Germany
christian.ott@fau.de
72. Paolo Pagano
Institut des Sciences Moléculaires
Université Bordeaux, CNRS
Talence, France
paolo.pagano@u-bordeaux.fr
73. Nikhil Parappurath
FOM Institute-AMOLF
Amsterdam, The Netherlands
N.Parappurath@amolf.nl
74. Wolfram Pernice
Karlsruhe Institute of Technology
Karlsruhe, Germany
wolfram.pernice@googlemail.com
75. Anatoliy O. Pinchuk
Department of Physics and Energy Sciences
University of Colorado
Colorado Springs, CO, USA
apinchuk@uccs.edu
76. Markus Pollnau
Department of Materials and Nano Physics
School of Information and Communication Technology
KTH – Royal Institute of Technology
Kista, Sweden
m.pollnau@utwente.nl
77. Igor Ponomarov
Engineer Service LLC
Institute of Physics of NAS of Ukraine
Kyiv, Ukraine
1@igor.biz.ua
78. Emanuela Pusceddu
Institute of Biometeorology
CNR
Firenze, Italy
emanuela.pusceddu@gmail.com

79. Nicolas Rivolta
Department of Physics
University of Mons
Mons, Belgium
Nicolas.RIVOLTA@umons.ac.be
80. Nabiha Saklayen
Harvard University
Cambridge, MA, USA
nsaklayen@gmail.com
81. Maksim Sandulov
Institute of Solid State Physics
Bulgarian Academy of Sciences
Sofia, Bulgaria
msandulov@issp.bas.bg
82. Nicholas Sharac
U.S. Naval Research Laboratory
Electronics Science and Technology Division
Washington, DC, USA
nicholas.sharac.ctr@nrl.navy.mil
83. Elpidio Silvestri
Lungotevere Pietra Papa 21
00146 Roma, Italy
elpidio.silvestri@alice.it
84. Luciano Silvestri
Department of Physics
Boston College
Chestnut Hill, MA, USA
silveslu@bc.edu
85. Nikolai Skoptsov
Center for Optical Materials and Technologies
Belarusian National Technical University
Minsk, Belarus
nikolai.skoptsov@tut.by
86. Matteo Salomoni
CERN
Geneva, Switzerland
matteo.salomoni@cern.ch
87. Mark I. Stockman
Department of Physics and Astronomy
Georgia State University
Atlanta, GA, USA
mstockman@gsu.edu

88. Rita Therisod
EPFL SB ICMP LOEQ
Lausanne, Switzerland
rita.therisod@epfl.ch
89. Silvia Tofani
Consiglio Nazionale delle Ricerche -
Istituto per la Microelettronica e Microsistemi (CNR-IMM)
Roma, Italy
silvia.tofani@artov.imm.cnr.it
tofani.1253388@studenti.uniroma1.it
90. Amir Torki
Department of Electrical Engineering
KTH Royal Institute of Technology
Stockholm, Sweden
amir.torki@gmail.com
91. Rosana Martinez Turtos
CERN
Geneve, Switzerland
rosana.martinez.turtos@cern.ch
92. Alessandro Vaccari
Fondazione Bruno Kessler (FBK)
Trento, Italy
vaccari@fbk.eu
93. Sreeramulu Valligatla
CNR-IFN
Trento, Italy
srihcu08@gmail.com
94. Anna Vedda
Dipartimento di Scienza dei Materiali
Università Milano-Bicocca
Milano, Italy
anna.vedda@mater.unimib.it
95. Vanessa Verrina
Advanced Research Center for Nanolithography
Amsterdam, The Netherlands
V.Verrina@arcnl.nl
96. Jochen Vogt
Kirchhoff-Institut für Physik
Universität Heidelberg
Heidelberg, Germany
jochen.vogt@kip.uni-heidelberg.de
97. Daryl Vulis
School of Engineering and Applied Sciences
Harvard University
Cambridge, MA, USA
dvulis@seas.harvard.edu

98. Aleksandr P. Voitovich
National Academy of Sciences of Belarus
Minsk, Belarus
voitovich@imaph.bas-net.by,avoitovich@gmail.com
99. Martin Wegener
Karlsrhue Institute of Technology
Karlsruhe, Germany
martin.wegener@kit.edu
100. Andreas Wickberg
Institut für Angewandte Physik (APH)
KIT – Universität des Landes Baden-Württemberg und
Nationales Forschungszentrum in der Helmholtz-Gemeinschaft
Karlsruhe, Germany
andreas.wickberg@kit.edu
101. Jean-Pierre Wolf
University Geneva
Gèneve, Switzerland
jean-pierre.wolf@unige.ch
102. Mariusz Woźniak
Institute of Physics
Polish Academy of Sciences
Warsaw, Poland
mwozniak@ifpan.edu.pl
103. Iryna Yaremchuk
Department of Photonics
Lviv Polytechnic National University
Lviv, Ukraine
iryna.y.yaremchuk@lpnu.ua
104. Katarzyna Zawisza
Department of Spectroscopy of Excited States
Institute of Low Temperature and Structure Research
Polish Academy of Sciences
Wroclaw, Poland
K.Zawisza@int.pan.wroc.pl

Index

© Springer Science+Business Media Dordrecht 2017
B. Di Bartolo et al. (eds.), *Nano-Optics: Principles Enabling Basic Research and Applications*, NATO Science for Peace and Security Series B: Physics and Biophysics, DOI 10.1007/978-94-024-0850-8

MIX
Papier aus verantwortungsvollen Quellen
Paper from responsible sources
FSC® C105338

If you have any concerns about our products,
you can contact us on
ProductSafety@springernature.com

In case Publisher is established outside the EU,
the EU authorized representative is:
Springer Nature Customer Service Center GmbH
Europaplatz 3, 69115 Heidelberg, Germany

Printed by Libri Plureos GmbH
in Hamburg, Germany